D1264351

Springer Series in Statistics

Springer
New York
Berlin
Heidelberg
Barcelona
Hong Kong
London
Milan
Paris
Singapore
Tokyo

Springer Series in Statistics

Andersen/Borgan/Gill/Keiding: Statistical Models Based on Counting Processes.
Andrews/Herzberg: Data: A Collection of Problems from Many Fields for the Student and Research Worker.
Anscombe: Computing in Statistical Science through APL.
Berger: Statistical Decision Theory and Bayesian Analysis, 2nd edition.
Bolfarine/Zacks: Prediction Theory for Finite Populations.
Borg/Groenen: Modern Multidimensional Scaling: Theory and Applications
Brémaud: Point Processes and Queues: Martingale Dynamics.
Brockwell/Davis: Time Series: Theory and Methods, 2nd edition.
Daley/Vere-Jones: An Introduction to the Theory of Point Processes.
Dzhaparidze: Parameter Estimation and Hypothesis Testing in Spectral Analysis of Stationary Time Series.
Efromovich: Nonparametric Curve Estimation: Methods, Theory, and Applications.
Fahrmeir/Tutz: Multivariate Statistical Modelling Based on Generalized Linear Models.
Farebrother: Fitting Linear Relationships: A History of the Calculus of Observations 1750 - 1900.
Farrell: Multivariate Calculation.
Federer: Statistical Design and Analysis for Intercropping Experiments, Volume I: Two Crops.
Federer: Statistical Design and Analysis for Intercropping Experiments, Volume II: Three or More Crops.
Fienberg/Hoaglin/Kruskal/Tanur (Eds.): A Statistical Model: Frederick Mosteller's Contributions to Statistics, Science and Public Policy.
Fisher/Sen: The Collected Works of Wassily Hoeffding.
Good: Permutation Tests: A Practical Guide to Resampling Methods for Testing Hypotheses.
Goodman/Kruskal: Measures of Association for Cross Classifications.
Gouriéroux: ARCH Models and Financial Applications.
Grandell: Aspects of Risk Theory.
Haberman: Advanced Statistics, Volume I: Description of Populations.
Hall: The Bootstrap and Edgeworth Expansion.
Härdle: Smoothing Techniques: With Implementation in S.
Hart: Nonparametric Smoothing and Lack-of-Fit Tests.
Hartigan: Bayes Theory.
Hedayat/Sloane/Stufken: Orthogonal Arrays: Theory and Applications.
Heyde: Quasi-Likelihood and its Application: A General Approach to Optimal Parameter Estimation.
Heyer: Theory of Statistical Experiments.
Huet/Bouvier/Gruet/Jolivet: Statistical Tools for Nonlinear Regression: A Practical Guide with S-PLUS Examples.
Jolliffe: Principal Component Analysis.
Kolen/Brennan: Test Equating: Methods and Practices.
Kotz/Johnson (Eds.): Breakthroughs in Statistics Volume I.

(continued after index)

C. Radhakrishna Rao
Helge Toutenburg

Linear Models

Least Squares and Alternatives

Second Edition

With Contributions by Andreas Fieger

With 33 Illustrations

Springer

C. Radhakrishna Rao
Department of Statistics
Pennsylvania State University
University Park, PA 16802
USA
crr1@psuvm.psu.edu

Helge Toutenburg
Institut für Statistik
Universität München Akademiestrasse 1
80799 München
Germany
toutenb@stat.uni-muenchen.de

Library of Congress Cataloging-in-Publication Data
Rao, C. Radhakrishna (Calyampudi Radhakrishna), 1920–
Linear models: least squares and alternatives/C. Radhakrishna Rao, Helge Toutenburg. — [2nd ed.]
p. cm. — (Springer series in statistics)
Includes bibliographical references and index.
ISBN 0-387-98848-3 (alk. paper)
1. Linear models (Statistics) I. Toutenburg, Helge. II. Title. III. Series.
QA279.R3615 1999
519.5′36—dc21 99-14735

Printed on acid-free paper.

Production managed by Frank McGuckin; manufacturing supervised by Jeffrey Taub.
Photocomposed copy prepared from the authors' LaTeX files.
Printed and bound by R.R. Donnelley and Sons, Harrisonburg, VA.
Printed in the United States of America.

9 8 7 6 5 4 3 2 1

ISBN 0-387-98848-3 Springer-Verlag New York Berlin Heidelberg SPIN 10726080

Preface to the First Edition

The book is based on several years of experience of both authors in teaching linear models at various levels. It gives an up-to-date account of the theory and applications of linear models. The book can be used as a text for courses in statistics at the graduate level and as an accompanying text for courses in other areas. Some of the highlights in this book are as follows.

A relatively extensive chapter on matrix theory (Appendix A) provides the necessary tools for proving theorems discussed in the text and offers a selection of classical and modern algebraic results that are useful in research work in econometrics, engineering, and optimization theory. The matrix theory of the last ten years has produced a series of fundamental results about the definiteness of matrices, especially for the differences of matrices, which enable superiority comparisons of two biased estimates to be made for the first time.

We have attempted to provide a unified theory of inference from linear models with minimal assumptions. Besides the usual least-squares theory, alternative methods of estimation and testing based on convex loss functions and general estimating equations are discussed. Special emphasis is given to sensitivity analysis and model selection.

A special chapter is devoted to the analysis of categorical data based on logit, loglinear, and logistic regression models.

The material covered, theoretical discussion, and a variety of practical applications will be useful not only to students but also to researchers and consultants in statistics.

We would like to thank our colleagues Dr. G. Trenkler and Dr. V. K. Srivastava for their valuable advice during the preparation of the book. We wish to acknowledge our appreciation of the generous help received from Andrea Schöpp, Andreas Fieger, and Christian Kastner for preparing a fair copy. Finally, we would like to thank Dr. Martin Gilchrist of Springer-Verlag for his cooperation in drafting and finalizing the book.

We request that readers bring to our attention any errors they may find in the book and also give suggestions for adding new material and/or improving the presentation of the existing material.

University Park, PA C. Radhakrishna Rao
München, Germany Helge Toutenburg
July 1995

Preface to the Second Edition

The first edition of this book has found wide interest in the readership. A first reprint appeared in 1997 and a special reprint for the Peoples Republic of China appeared in 1998. Based on this, the authors followed the invitation of John Kimmel of Springer-Verlag to prepare a second edition, which includes additional material such as simultaneous confidence intervals for linear functions, neural networks, restricted regression and selection problems (Chapter 3); mixed effect models, regression-like equations in econometrics, simultaneous prediction of actual and average values, simultaneous estimation of parameters in different linear models by empirical Bayes solutions (Chapter 4); the method of the Kalman Filter (Chapter 6); and regression diagnostics for removing an observation with animating graphics (Chapter 7).

Chapter 8, "Analysis of Incomplete Data Sets", is completely rewritten, including recent terminology and updated results such as regression diagnostics to identify Non-MCAR processes.

Chapter 10, "Models for Categorical Response Variables", also is completely rewritten to present the theory in a more unified way including GEE-methods for correlated response.

At the end of the chapters we have given complements and exercises. We have added a separate chapter (Appendix C) that is devoted to the software available for the models covered in this book.

We would like to thank our colleagues Dr. V. K. Srivastava (Lucknow, India) and Dr. Ch. Heumann (München, Germany) for their valuable advice during the preparation of the second edition. We thank Nina Lieske for her help in preparing a fair copy. We would like to thank John Kimmel of

Springer-Verlag for his effective cooperation. Finally, we wish to appreciate the immense work done by Andreas Fieger (München, Germany) with respect to the numerical solutions of the examples included, to the technical management of the copy, and especially to the reorganization and updating of Chapter 8 (including some of his own research results). Appendix C on software was written by him, also.

We request that readers bring to our attention any suggestions that would help to improve the presentation.

University Park, PA
München, Germany
May 1999

C. Radhakrishna Rao
Helge Toutenburg

Contents

1
Introduction

Linear models play a central part in modern statistical methods. On the one hand, these models are able to approximate a large amount of metric data structures in their entire range of definition or at least piecewise. On the other hand, approaches such as the analysis of variance, which model effects such as linear deviations from a total mean, have proved their flexibility. The theory of generalized models enables us, through appropriate link functions, to apprehend error structures that deviate from the normal distribution, hence ensuring that a linear model is maintained in principle. Numerous iterative procedures for solving the normal equations were developed especially for those cases where no explicit solution is possible. For the derivation of explicit solutions in rank-deficient linear models, classical procedures are available, for example, ridge or principal component regression, partial least squares, as well as the methodology of the generalized inverse. The problem of missing data in the variables can be dealt with by appropriate imputation procedures.

Chapter 2 describes the hierarchy of the linear models, ranging from the classical regression model to the structural model of econometrics.

Chapter 3 contains the standard procedures for estimating and testing in regression models with full or reduced rank of the design matrix, algebraic and geometric properties of the OLS estimate, as well as an introduction to minimax estimation when auxiliary information is available in the form of inequality restrictions. The concepts of partial and total least squares, projection pursuit regression, and censored regression are introduced. The method of Scheffé's simultaneous confidence intervals for linear functions as well as the construction of confidence intervals for the ratio of two parametric functions are discussed. Neural networks as a nonparametric regression method and restricted regression in connection with selection problems are introduced.

Chapter 4 describes the theory of best linear estimates in the generalized regression model, effects of misspecified covariance matrices, as well as special covariance structures of heteroscedasticity, first-order autoregression, mixed effect models, regression-like equations in econometrics, and simultaneous estimates in different linear models by empirical Bayes solutions.

Chapter 5 is devoted to estimation under exact or stochastic linear restrictions. The comparison of two biased estimations according to the MDE criterion is based on recent theorems of matrix theory. The results are the outcome of intensive international research over the last ten years and appear here for the first time in a coherent form. This concerns the concept of the weak r-unbiasedness as well.

Chapter 6 contains the theory of the optimal linear prediction and gives, in addition to known results, an insight into recent studies about the MDE matrix comparison of optimal and classical predictions according to alternative superiority criteria. A separate section is devoted to Kalman filtering viewed as a restricted regression method.

Chapter 7 presents ideas and procedures for studying the effect of single data points on the estimation of β. Here, different measures for revealing outliers or influential points, including graphical methods, are incorporated. Some examples illustrate this.

Chapter 8 deals with missing data in the design matrix X. After an introduction to the general problem and the definition of the various missing data mechanisms according to Rubin, we describe various ways of handling missing data in regression models. The chapter closes with the discussion of methods for the detection of non-MCAR mechanisms.

Chapter 9 contains recent contributions to robust statistical inference based on M-estimation.

Chapter 10 describes the model extensions for categorical response and explanatory variables. Here, the binary response and the loglinear model are of special interest. The model choice is demonstrated by means of examples. Categorical regression is integrated into the theory of generalized linear models. In particular, GEE-methods for correlated response variables are discussed.

An independent chapter (Appendix A) about matrix algebra summarizes standard theorems (including proofs) that are used in the book itself, but also for linear statistics in general. Of special interest are the theorems about decomposition of matrices (A.30–A.34), definite matrices (A.35–A.59), the generalized inverse, and particularily about the definiteness of differences between matrices (Theorem A.71; cf. A.74–A.78).

Tables for the χ^2- and F-distributions are found in Appendix B.

Appendix C describes available software for regression models.

The book offers an up-to-date and comprehensive account of the theory and applications of linear models, with a number of new results presented for the first time in any book.

2
Linear Models

2.1 Regression Models in Econometrics

The methodology of regression analysis, one of the classical techniques of mathematical statistics, is an essential part of the modern econometric theory.

Econometrics combines elements of economics, mathematical economics, and mathematical statistics. The statistical methods used in econometrics are oriented toward specific econometric problems and hence are highly specialized. In economic laws, stochastic variables play a distinctive role. Hence econometric models, adapted to the economic reality, have to be built on appropriate hypotheses about distribution properties of the random variables. The specification of such hypotheses is one of the main tasks of econometric modeling. For the modeling of an economic (or a scientific) relation, we assume that this relation has a relative constancy over a sufficiently long period of time (that is, over a sufficient length of observation period), because otherwise its general validity would not be ascertainable. We distinguish between two characteristics of a structural relationship, the *variables* and the *parameters*. The variables, which we will classify later on, are those characteristics whose values in the observation period can vary. Those characteristics that do not vary can be regarded as the structure of the relation. The *structure* consists of the functional form of the relation, including the relation between the main variables, the type of probability distribution of the random variables, and the parameters of the model equations.

The *econometric model* is the epitome of all a priori hypotheses related to the economic phenomenon being studied. Accordingly, the model constitutes a catalogue of model assumptions (a priori hypotheses, a priori specifications). These assumptions express the information available a priori about the economic and stochastic characteristics of the phenomenon.

For a distinct definition of the structure, an appropriate classification of the model variables is needed. The econometric model is used to predict certain variables y called *endogenous*, given the realizations (or assigned values) of certain other variables x called *exogenous*, which ideally requires the specification of the conditional distribution of y given x. This is usually done by specifiying an *economic structure*, or a stochastic relationship between y and x through another set of unobservable random variables called *error*.

Usually, the variables y and x are subject to a time development, and the model for predicting y_t, the value of y at time point t, may involve the whole set of observations

$$y_{t-1}, y_{t-2}, \cdots , \tag{2.1}$$

$$x_t, x_{t-1}, \cdots . \tag{2.2}$$

In such models, usually referred to as dynamic models, the lagged endogenous variables (2.1) and the exogenous variables (2.2) are treated as regressors for predicting the endogenous variable y_t considered as a regressand.

If the model equations are resolved into the jointly dependent variables (as is normally assumed in the linear regression) and expressed as a function of the predetermined variables and their errors, we then have the econometric model in its *reduced form*. Otherwise, we have the *structural form* of the equations.

A model is called *linear* if all equations are linear. A model is called *univariate* if it contains only one single endogenous variable. A model with more than one endogenous variable is called *multivariate*.

A model equation of the reduced form with more than one predetermined variable is called *multivariate* or a *multiple* equation. We will get to know these terms better in the following sections by means of specific models.

Because of the great mathematical and especially statistical difficulties in dealing with econometric and regression models in the form of inequalities or even more general mathematical relations, it is customary to almost exclusively work with models in the form of equalities.

Here again, linear models play a special part, because their handling keeps the complexity of the necessary mathematical techniques within reasonable limits. Furthermore, the linearity guarantees favorable statistical properties of the sample functions, especially if the errors are normally distributed. The (linear) econometric model represents the hypothetical stochastic relationship between endogenous and exogenous variables of a complex economic law. In practice any assumed model has to be examined for its validity through appropriate tests and past evidence.

This part of model building, which is probably the most complicated task of the statistician, will not be dealt with any further in this text.

Example 2.1: As an illustration of the definitions and terms of econometrics, we want to consider the following typical example. We define the following variables:

A: deployment of manpower,

B: deployment of capital, and

Y: volume of production.

Let e be the base of the natural logarithm and c be a constant (which ensures in a certain way the transformation of the unit of measurement of A, B into that of Y). The classical Cobb-Douglas production function for an industrial sector, for example, is then of the following form:

$$Y = cA^{\beta_1} B^{\beta_2} e^{\epsilon} .$$

This function is nonlinear in the parameters β_1, β_2 and the variables A, B, and ϵ. By taking the logarithm, we obtain

$$\ln Y = \ln c + \beta_1 \ln A + \beta_2 \ln B + \epsilon .$$

Here we have

$\ln Y$	the regressand or the endogenous variable,
$\ln A$, $\ln B$	the regressors or the exogenous variables,
β_1, β_2	the regression coefficients,
$\ln c$	a scalar constant,
ϵ	the random error.

β_1 and β_2 are called *production elasticities.* They measure the power and direction of the effect of the deployment of labor and capital on the volume of production. After taking the logarithm, the function is linear in the parameters β_1 and β_2 and the regressors $\ln A$ and $\ln B$.

Hence the model assumptions are as follows: In accordance with the multiplicative function from above, the volume of production Y is dependent on only the three variables A, B, and ϵ (random error). Three parameters appear: the production elasticities β_1, β_2 and the scalar constant c. The model is multiple and is in the reduced form.

Furthermore, a possible assumption is that the errors ϵ_t are independent and identically distributed with expectation 0 and variance σ^2 and distributed independently of A and B.

2.2 Econometric Models

We first develop the model in its economically relevant form, as a system of M simultaneous linear stochastic equations in M jointly dependent variables $Y_1, \ldots, Y_M$ and K predetermined variables $X_1, \ldots, X_K$, as well as the error variables $U_1, \ldots, U_M$. The realizations of each of these variable are denoted by the corresponding small letters y_{mt}, x_{kt}, and u_{mt}, with $t = 1, \ldots, T$, the times at which the observations are taken. The system of structural equations for index t $(t = 1, \ldots, T)$ is

$$\left.\begin{array}{rcl} y_{1t}\gamma_{11} + \cdots + y_{Mt}\gamma_{M1} + x_{1t}\delta_{11} + \cdots + x_{Kt}\delta_{K1} + u_{1t} & = & 0 \\ y_{1t}\gamma_{12} + \cdots + y_{Mt}\gamma_{M2} + x_{1t}\delta_{12} + \cdots + x_{Kt}\delta_{K2} + u_{2t} & = & 0 \\ \vdots & \vdots & \vdots \\ y_{1t}\gamma_{1M} + \cdots + y_{Mt}\gamma_{MM} + x_{1t}\delta_{1M} + \cdots + x_{Kt}\delta_{KM} + u_{Mt} & = & 0 \end{array}\right\} \tag{2.3}$$

Thus, the mth structural equation is of the form $(m = 1, \ldots, M)$

$$y_{1t}\gamma_{1m} + \cdots + y_{Mt}\gamma_{Mm} + x_{1t}\delta_{1m} + \cdots + x_{Kt}\delta_{Km} + u_{mt} = 0\,.$$

Convention

A matrix A with m rows and n columns is called an $m \times n$-matrix A, and we use the symbol $\underset{m\times n}{A}$. We now define the following vectors and matrices:

$$\underset{T\times M}{Y} = \begin{pmatrix} y_{11} & \cdots & y_{M1} \\ \vdots & & \vdots \\ y_{1t} & \cdots & y_{Mt} \\ \vdots & & \vdots \\ y_{1T} & \cdots & y_{MT} \end{pmatrix} = \begin{pmatrix} y'(1) \\ \vdots \\ y'(t) \\ \vdots \\ y'(T) \end{pmatrix} = \left(\underset{T\times 1}{y_1}, \cdots, \underset{T\times 1}{y_M} \right),$$

$$\underset{T\times K}{X} = \begin{pmatrix} x_{11} & \cdots & x_{K1} \\ \vdots & & \vdots \\ x_{1t} & \cdots & x_{Kt} \\ \vdots & & \vdots \\ x_{1T} & \cdots & x_{KT} \end{pmatrix} = \begin{pmatrix} x'(1) \\ \vdots \\ x'(t) \\ \vdots \\ x'(T) \end{pmatrix} = \left(\underset{T\times 1}{x_1}, \cdots, \underset{T\times 1}{x_K} \right),$$

$$\underset{T\times M}{U} = \begin{pmatrix} u_{11} & \cdots & u_{M1} \\ \vdots & & \vdots \\ u_{1t} & \cdots & u_{Mt} \\ \vdots & & \vdots \\ u_{1T} & \cdots & u_{MT} \end{pmatrix} = \begin{pmatrix} u'(1) \\ \vdots \\ u'(t) \\ \vdots \\ u'(T) \end{pmatrix} = \left(\underset{T\times 1}{u_1}, \cdots, \underset{T\times 1}{u_M} \right),$$

$$\underset{M\times M}{\Gamma} = \begin{pmatrix} \gamma_{11} & \cdots & \gamma_{1M} \\ \vdots & & \vdots \\ \gamma_{M1} & \cdots & \gamma_{MM} \end{pmatrix} = \left(\underset{M\times 1}{\gamma_1}, \cdots, \underset{M\times 1}{\gamma_M} \right),$$

$$\underset{K\times M}{D} = \begin{pmatrix} \delta_{11} & \cdots & \delta_{1M} \\ \vdots & & \vdots \\ \delta_{K1} & \cdots & \delta_{KM} \end{pmatrix} = \left(\underset{K\times 1}{\delta_1}, \cdots, \underset{K\times 1}{\delta_M} \right).$$

We now have the matrix representation of system (2.3) for index t:

$$y'(t)\Gamma + x'(t)D + u'(t) = 0 \qquad (t = 1, \dots, T) \tag{2.4}$$

or for all T observation periods,

$$Y\Gamma + XD + U = 0\,. \tag{2.5}$$

Hence the mth structural equation for index t is

$$y'(t)\gamma_m + x'(t)\delta_m + u_{mt} = 0 \qquad (m = 1, \dots, M) \tag{2.6}$$

where γ_m and δ_m are the structural parameters of the mth equation. $y'(t)$ is a $1 \times M$-vector, and $x'(t)$ is a $1 \times K$-vector.

Conditions and Assumptions for the Model

Assumption (A)

(A.1) The parameter matrix Γ is regular.

(A.2) Linear a priori restrictions enable the identification of the parameter values of Γ, and D.

(A.3) The parameter values in Γ are standardized, so that $\gamma_{mm} = -1 \quad (m = 1, \dots, M)$.

Definition 2.1 *Let $t = \dots - 2, -1, 0, 1, 2, \dots$ be a series of time indices.*

(a) A ***univariate stochastic process*** *$\{x_t\}$ is an ordered set of random variables such that a joint probability distribution for the variables $x_{t_1}, \dots, x_{t_n}$ is always defined, with $t_1, \dots, t_n$ being any finite set of time indices.*

(b) A ***multivariate*** *(n-dimensional)* ***stochastic process*** *is an ordered set of $n \times 1$ random vectors $\{x_t\}$ with $x_t = (x_{t_1}, \dots, x_{t_n})$ such that for every choice $t_1, \dots, t_n$ of time indices a joint probability distribution is defined for the random vectors $x_{t_1}, \dots, x_{t_n}$.*

A stochastic process is called *stationary* if the joint probability distributions are invariant under translations along the time axis. Thus any finite set $x_{t_1}, \dots, x_{t_n}$ has the same joint probability distribution as the set $x_{t_1+r}, \dots, x_{t_n+r}$ for $r = \dots, -2, -1, 0, 1, 2, \dots$.

As a typical example of a univariate stochastic process, we want to mention the time series. Under the assumption that all values of the time series are functions of the time t, t is the only independent (exogenous) variable:

$$x_t = f(t). \tag{2.7}$$

The following special cases are of importance in practice:

$$\begin{aligned} x_t &= \alpha && \text{(constancy over time)}, \\ x_t &= \alpha + \beta t && \text{(linear trend)}, \\ x_t &= \alpha e^{\beta t} && \text{(exponential trend)}. \end{aligned}$$

For the prediction of time series, we refer, for example, to Nelson (1973) or Mills (1991).

Assumption (B)

The structural error variables are generated by an M-dimensional stationary stochastic process $\{u(t)\}$ (cf. Goldberger, 1964, p. 153).

(B.1) $\mathrm{E}\,u(t) = 0$ and thus $\mathrm{E}(U) = 0$.

(B.2) $\mathrm{E}\,u(t)u'(t) = \underset{M\times M}{\Sigma} = (\sigma_{mm'})$ with Σ positive definite and hence regular.

(B.3) $\mathrm{E}\,u(t)u'(t') = 0$ for $t \neq t'$.

(B.4) All $u(t)$ are identically distributed.

(B.5) For the empirical moment matrix of the random errors, let

$$p\lim T^{-1}\sum_{t=1}^{T} u(t)u'(t) = p\lim T^{-1}U'U = \Sigma. \tag{2.8}$$

Consider a series $\{z^{(t)}\} = z^{(1)}, z^{(2)}, \ldots$ of random variables. Each random variable has a specific distribution, variance, and expectation. For example, $z^{(t)}$ could be the sample mean of a sample of size t of a given population. The series $\{z^{(t)}\}$ would then be the series of sample means of a successively increasing sample. Assume that $z^* < \infty$ exists, such that

$$\lim_{t\to\infty} P\{|z^{(t)} - z^*| \geq \delta\} = 0 \quad \text{for every} \delta > 0.$$

Then z^* is called the *probability limit* of $\{z^{(t)}\}$, and we write $p\lim z^{(t)} = z^*$ or $p\lim z = z^*$ (cf. Definition A.101 and Goldberger, 1964, p. 115).

(B.6) The error variables $u(t)$ have an M-dimensional normal distribution.

Under general conditions for the process $\{u(t)\}$ (cf.Goldberger, 1964), (B.5) is a consequence of (B.1)–(B.3). Assumption (B.3) reduces the number of unknown parameters in the model to be estimated and thus enables the estimation of the parameters in Γ, D, Σ from the T observations (T sufficiently large).

The favorable statistical properties of the least-squares estimate in the regression model and in the econometric models are mainly independent of the probability distribution of $u(t)$. Assumption (B.6) is additionally needed for test procedures and for the derivation of interval estimates and predictions.

Assumption (C)

The predetermined variables are generated by a K-dimensional stationary stochastic process $\{x(t)\}$.

(C.1) $\mathrm{E}\,x(t)x'(t) = \Sigma_{xx}$, a $K \times K$-matrix, exists for all t. Σ_{xx} is positive definite and thus regular.

(C.2) For the empirical moment matrix (*sample moments*)

$$S_{xx} = T^{-1} \sum_{t=1}^{T} x(t)x'(t) = T^{-1} X'X\,, \tag{2.9}$$

the following limit exists, and every dependence in the process $\{x(t)\}$ is sufficiently small, so that

$$p\lim S_{xx} = \lim_{T\to\infty} S_{xx} = \Sigma_{xx}\,.$$

Assumption (C.2) is fulfilled, for example, for an ergodic stationary process. A stationary process $\{x(t)\}$ is called *ergodic* if the time mean of every realization (with probability 1) is the same and coincides with the expectation of the entire time series. Thus, according to (C.2), $\{x(t)\}$ is called ergodic if

$$\lim_{T\to\infty} S_{xx} = \Sigma_{xx}\,.$$

In practice, ergodicity can often be assumed for stationary processes. Ergodicity means that every realization (sample vector) has asymptotically the same statistical properties and is hence representative for the process.

(C.3) The processes $\{x(t)\}$ and $\{u(t)\}$ are *contemporaneously uncorrelated*; that is, for every t we have $\mathrm{E}\,(u(t)|x(t)) = \mathrm{E}\,(u(t)) = 0$. For the empirical moments we have

$$p\lim T^{-1} \sum_{t=1}^{T} x(t)u'(t) = p\lim T^{-1} X'U = 0. \tag{2.10}$$

Assumption (C.3) is based on the idea that the values of the predetermined variables are not determined by the state of the system at the actual time index t. Hence these values may not have to be dependent on the errors $u(t)$.

Assume that $\lim T^{-1}X'X$ exists. In many cases, especially when the predetermined variables consist only of exogenous variables, the alternative assumption can be made that the predetermined variables remain fixed for repeated samples. In this case, $\{x(t)\}$ is a nonstochastic series.

Using selected assumptions and according to our definition made in Section 2.2, the linear econometric model has the following form:

$$\left.\begin{array}{l} Y\Gamma + XD + U = 0, \\ \mathrm{E}(U) = 0, \mathrm{E}\,u(t)u'(t) = \Sigma, \\ \mathrm{E}\,u(t)u(t') = 0 \quad (t \neq t'), \\ \Gamma \text{ nonsingular}, \\ \Sigma \text{ positive definite}, \\ p\lim T^{-1}U'U = \Sigma, p\lim T^{-1}X'U = 0, \\ p\lim T^{-1}X'X = \Sigma_{xx} \text{ (positive definite)}. \end{array}\right\} \tag{2.11}$$

The general aim of our studies is to deal with problems of estimation, prediction, and model building for special types of models. For more general questions about econometric models, we refer to the extensive literature about estimation and identifiability problems of econometric model systems, for example Amemiya (1985), Goldberger (1964), and Dhrymes (1974; 1978), and to the extensive special literature, for example, in the journals *Econometrica*, *Essays in Economics and Econometrics*, and *Journal of Econometrics and Econometric Theory.*

2.3 The Reduced Form

The approach to the models of linear regression from the viewpoint of the general econometric model yields the so-called reduced form of the econometric model equation. The previously defined model has as many equations as endogenous variables. In addition to (A.1), we assume that the system of equations uniquely determines the endogenous variables, for every set of values of the predetermined and random variables. The model is then called *complete*. Because of the assumed regularity of Γ, we can express the endogenous variable as a linear vector function of the predetermined and random variables by multiplying from the right with Γ^{-1}:

$$Y = -XD\Gamma^{-1} - U\Gamma^{-1} = X\Pi + V\,, \tag{2.12}$$

where

$$\underset{K \times M}{\Pi} = -D\Gamma^{-1} = (\pi_1, \dots, \pi_M)\,. \tag{2.13}$$

This is the coefficient matrix of the reduced form (with π_m being K-vectors of the regression coefficients of the mth reduced-form equation), and

$$\underset{T\times M}{V} = -U\Gamma^{-1} = \begin{pmatrix} v'(1) \\ \vdots \\ v'(t) \\ \vdots \\ v'(T) \end{pmatrix} = (v_1, \ldots, v_M) \tag{2.14}$$

is the matrix of the random errors. The mth equation of the reduced form is of the following form:

$$y_m = X\pi_m + v_m. \tag{2.15}$$

The model assumptions formulated in (2.11) are transformed as follows:

$$\left.\begin{array}{l} \mathrm{E}(V) = -\,\mathrm{E}(U)\Gamma^{-1} = 0, \\ \mathrm{E}[v(t)v'(t)] = \Gamma'^{-1}\,\mathrm{E}[u(t)u'(t)]\Gamma^{-1} = \Gamma'^{-1}\Sigma\Gamma^{-1} = \Sigma_{vv}, \\ \Sigma_{vv} \text{ is positive definite (since } \Gamma^{-1} \text{ is nonsingular} \\ \qquad \text{and } \Sigma \text{ is positive definite),} \\ \mathrm{E}[v(t)v'(t')] = 0 \quad (t \neq t'), \\ p\lim T^{-1}V'V = \Gamma^{-1}(p\lim T^{-1}U'U)\Gamma^{-1} = \Sigma_{vv}, \\ p\lim T^{-1}X'V = 0, \quad p\lim T^{-1}X'X = \Sigma_{xx} \text{ (positive definite).} \end{array}\right\} \tag{2.16}$$

The reduced form of (2.11) is now

$$Y = X\Pi + V \quad \text{with assumptions (2.16).} \tag{2.17}$$

By specialization or restriction of the model assumptions, the reduced form of the econometric model yields the essential models of linear regression.

Example 2.2 (Keynes's model): Let C be the consumption, Y the income, and I the savings (or investment). The hypothesis of Keynes then is

(a) $C = \alpha + \beta Y$,

(b) $Y = C + I$.

Relation (a) expresses the consumer behavior of an income group, for example, while (b) expresses a condition of balance: The difference $Y - C$ is invested (or saved). The statistical formulation of Keynes's model is

$$\left.\begin{array}{l} C_t = \alpha + \beta Y_t + \epsilon_t \\ Y_t = C_t + I_t \end{array}\right\} \quad (t = 1, \ldots, T), \tag{2.18}$$

where ϵ_t is a random variable (error) with

$$\mathrm{E}\,\epsilon_t = 0, \quad \mathrm{E}\,\epsilon_t^2 = \sigma^2, \quad \mathrm{E}\,\epsilon_s\epsilon_t = 0 \quad \text{for} \quad t \neq s\,. \tag{2.19}$$

Additionally, autonomy of the investments is assumed:

$$\mathrm{E}\,I_t\epsilon_t = 0 \quad \forall t\,. \tag{2.20}$$

We now express the above model in the form (2.4) as

$$(C_t\, Y_t)\begin{pmatrix} -1 & 1 \\ \beta & -1 \end{pmatrix} + (1\,,\, I_t)\begin{pmatrix} \alpha & 0 \\ 0 & 1 \end{pmatrix} + (\epsilon_t\,,\, 0) = (0\,,\, 0). \tag{2.21}$$

Hence $K = M = 2$.

We calculate the reduced form:

$$\begin{aligned} \Pi &= -D\Gamma^{-1} = -\begin{pmatrix} \alpha & 0 \\ 0 & 1 \end{pmatrix}\begin{pmatrix} -1 & 1 \\ \beta & -1 \end{pmatrix}^{-1} \\ &= -\begin{pmatrix} \alpha & 0 \\ 0 & 1 \end{pmatrix}\begin{pmatrix} \frac{-1}{(1-\beta)} & \frac{-1}{(1-\beta)} \\ \frac{-\beta}{(1-\beta)} & \frac{-1}{(1-\beta)} \end{pmatrix} \\ &= \frac{1}{1-\beta}\begin{pmatrix} \alpha & \alpha \\ \beta & 1 \end{pmatrix}. \end{aligned} \tag{2.22}$$

Thus, the reduced form is (cf. (2.12))

$$(C_t\,,\, Y_t) = (1\,,\, I_t)\begin{pmatrix} \frac{\alpha}{(1-\beta)} & \frac{\alpha}{(1-\beta)} \\ \frac{\beta}{(1-\beta)} & \frac{1}{(1-\beta)} \end{pmatrix} + (v_{1t}\; v_{2t}) \tag{2.23}$$

with $v_{1t} = v_{2t} = \epsilon_t/(1-\beta)$. Here we have

C_t, Y_t	jointly dependent,
I_t	predetermined.

2.4 The Multivariate Regression Model

We now neglect the connection between the structural form (2.11) of the econometric model and the reduced form (2.17) and regard $Y = X\Pi + V$ as an M-dimensional system of M single regressions $Y_1, \ldots, Y_M$ onto the K regressors $X_1, \ldots, X_K$. In the statistical handling of such systems, the following representation holds. The coefficients (regression parameters) are usually denoted by $\tilde{\beta}$ and the error variables by $\tilde{\epsilon}$. We thus have $\Pi = (\tilde{\beta}_{km})$ and $V = (\tilde{\epsilon}_{mt})$.

Then $Y = X\Pi + V$, which in the expanded form is

$$\begin{pmatrix} y_{11} & \cdots & y_{M1} \\ \vdots & & \vdots \\ y_{1T} & \cdots & y_{MT} \end{pmatrix} = \begin{pmatrix} x_{11} & \cdots & x_{K1} \\ \vdots & & \vdots \\ x_{1T} & \cdots & x_{KT} \end{pmatrix}\begin{pmatrix} \tilde{\beta}_{11} & \cdots & \tilde{\beta}_{1M} \\ \vdots & & \vdots \\ \tilde{\beta}_{K1} & \cdots & \tilde{\beta}_{KM} \end{pmatrix} + \begin{pmatrix} \tilde{\epsilon}_{11} & \cdots & \tilde{\epsilon}_{M1} \\ \vdots & & \vdots \\ \tilde{\epsilon}_{1T} & \cdots & \tilde{\epsilon}_{MT} \end{pmatrix}$$

or (after summarizing the column vectors)

$$(y_1, \ldots, y_M) = X(\tilde{\beta}_1, \ldots, \tilde{\beta}_M) + (\tilde{\epsilon}_1, \ldots, \tilde{\epsilon}_M). \tag{2.24}$$

We write the components ($T \times 1$-vectors) rowwise as

$$\begin{pmatrix} y_1 \\ y_2 \\ \vdots \\ y_M \end{pmatrix} = \begin{pmatrix} X & 0 & \cdots & 0 \\ 0 & X & \cdots & 0 \\ \vdots & \vdots & & \vdots \\ 0 & 0 & \cdots & X \end{pmatrix} \begin{pmatrix} \tilde{\beta}_1 \\ \tilde{\beta}_2 \\ \vdots \\ \tilde{\beta}_M \end{pmatrix} + \begin{pmatrix} \tilde{\epsilon}_1 \\ \tilde{\epsilon}_2 \\ \vdots \\ \tilde{\epsilon}_M \end{pmatrix}. \tag{2.25}$$

The mth equation of this system is of the following form:

$$y_m = X\tilde{\beta}_m + \tilde{\epsilon}_m \qquad (m = 1, \dots, M). \tag{2.26}$$

In this way, the statistical dependence of each of the M regressands Y_m on the K regressors $X_1, \dots, X_K$ is explicitly described.

In practice, not every single regressor in X will appear in each of the M equations of the system. This information, which is essential in econometric models for identifying the parameters and which is included in Assumption (A.2), is used by setting those coefficients $\tilde{\beta}_{mk}$ that belong to the variable X_k, which is not included in the mth equation, equal to zero. This leads to a gain in efficiency for the estimate and prediction, in accordance with the exact auxiliary information in the form of knowledge of the coefficients. The matrix of the regressors of the mth equation generated by deletion is denoted by X_m, the coefficient vector belonging to X_m is denoted by β_m. Similarly, the error $\tilde{\epsilon}$ changes to ϵ. Thus, after realization of the identification, the mth equation has the following form:

$$y_m = X_m \beta_m + \epsilon_m \qquad (m = 1, \dots, M). \tag{2.27}$$

Here

- y_m is the T-vector of the observations of the mth regressand,
- X_m is the $T \times K_m$-matrix of the regressors, which remain in the mth equation,
- β_m is the K_m-vector of the regression coefficients of the mth equation,
- ϵ_m is the T-vector of the random errors of the mth equation.

Given (2.27) and $\tilde{K} = \sum_{m=1}^{M} K_m$, the system (2.25) of M single regressions changes to

$$\begin{pmatrix} y_1 \\ y_2 \\ \vdots \\ y_M \end{pmatrix} = \begin{pmatrix} X_1 & 0 & \cdots & 0 \\ 0 & X_2 & \cdots & 0 \\ \vdots & \vdots & & \vdots \\ 0 & 0 & \cdots & X_M \end{pmatrix} \begin{pmatrix} \beta_1 \\ \beta_2 \\ \vdots \\ \beta_M \end{pmatrix} + \begin{pmatrix} \epsilon_1 \\ \epsilon_2 \\ \vdots \\ \epsilon_M \end{pmatrix}, \tag{2.28}$$

or in matrix form,

$$\underset{MT \times 1}{y} = \underset{MT \times \tilde{K}}{Z} \underset{\tilde{K} \times 1}{\beta} + \underset{MT \times 1}{\epsilon}. \tag{2.29}$$

Example 2.3 (Dynamic Keynes's model): The consumption C_t in Example 2.2 was dependent on the income Y_t of the same time index t. We now want to state a modified hypothesis. According to this hypothesis, the income of the preceding period $t-1$ determines the consumption for index t:

(a) $C_t = \alpha + \beta Y_{t-1} + \epsilon_t$,

(b) $Y_t = C_t + I_t$.

Assume the investment is autonomous, as in Example 2.2. Then we have the following classification of variables:

jointly dependent variables:	C_t, Y_t
predetermined variables:	Y_{t-1}, I_t
endogenous variables:	Y_{t-1}, C_t, Y_t
lagged endogenous variable:	Y_{t-1}
exogenous variable:	I_t

Assumption (D)

The variables X_k include no lagged endogenous variables. The values x_{kt} of the nonstochastic (exogenous) regressors X_k are such that

$$\left.\begin{array}{l} \operatorname{rank}(X_m) = K_m \quad (m = 1, \ldots, M) \quad \text{and thus} \\ \operatorname{rank}(Z) = \tilde{K} \quad \text{with} \quad \tilde{K} = \sum_{m=1}^{M} K_m \,. \end{array}\right\} \tag{2.30}$$

Assumption (E)

The random errors ϵ_{mt} are generated by an MT-dimensional regular stochastic process. Let

$$\begin{array}{l} \mathrm{E}(\epsilon_{mt}) = 0, \quad \mathrm{E}(\epsilon_{mt}\, \epsilon_{m't'}) = \sigma^2 w_{mm'}(t, t') \\ (m, m' = 1, \ldots, M; \quad t, t' = 1, \ldots, T), \end{array} \tag{2.31}$$

and therefore

$$\mathrm{E}(\epsilon_m) = 0, \quad \mathrm{E}(\epsilon) = 0\,, \tag{2.32}$$

$$\mathrm{E}(\epsilon_m \epsilon'_{m'}) = \sigma^2 \underset{T\times T}{W_{mm'}} = \sigma^2 \begin{pmatrix} w_{mm'}(1,1) & \cdots & w_{mm'}(1,T) \\ \vdots & & \vdots \\ w_{mm'}(T,1) & \cdots & w_{mm'}(T,T) \end{pmatrix} \tag{2.33}$$

$$\mathrm{E}(\epsilon\epsilon') = \sigma^2 \underset{MT\times MT}{\Phi} = \sigma^2 \begin{pmatrix} W_{11} & \cdots & W_{1M} \\ \vdots & & \vdots \\ W_{M1} & \cdots & W_{MM} \end{pmatrix}. \tag{2.34}$$

Assumption (E.1)

The covariance matrices $\sigma^2 W_{mm}$ of the errors ϵ_m of the mth equation and the covariance matrix $\sigma^2\Phi$ of the error ϵ of the system are positive definite and hence regular.

Assumption (F)

The error variable ϵ has an MT-dimensional normal distribution $N(0, \sigma^2\Phi)$.

Given assumptions (D) and (E), the so-called multivariate (M-dimensional) multiple linear regression model is of the following form:

$$\left.\begin{array}{l} y = Z\beta + \epsilon, \\ \mathrm{E}(\epsilon) = 0, \mathrm{E}(\epsilon\epsilon') = \sigma^2\Phi, \\ Z \quad \text{nonstochastic, } \operatorname{rank}(Z) = \tilde{K}\,. \end{array}\right\} \tag{2.35}$$

The model is called *regular* if it satisfies (E.1) in addition to (2.28). If (F) is fulfilled, we then have a *multivariate normal regression.*

2.5 The Classical Multivariate Linear Regression Model

An error process uncorrelated in time $\{\epsilon\}$ is an important special case of model (2.35). For this process Assumption (E) is of the following form.

Assumption ($\tilde{\mathrm{E}}$)

The random errors ϵ_{mt} are generated by an MT-dimensional regular stochastic process. Let

$$\begin{aligned} \mathrm{E}(\epsilon_{mt}) &= 0, \quad \mathrm{E}(\epsilon_{mt}\epsilon_{m't}) = \sigma^2 w_{mm'}\,, \\ \mathrm{E}(\epsilon_{mt}\epsilon_{m't'}) &= 0 \quad (t \neq t')\,, \\ \mathrm{E}(\epsilon_m) &= 0, \quad \mathrm{E}(\epsilon) = 0\,, \\ \mathrm{E}(\epsilon_m\epsilon'_{m'}) &= \sigma^2 w_{mm'} I \\ \mathrm{E}(\epsilon\epsilon') = \sigma^2\Phi &= \sigma^2 \begin{pmatrix} w_{11}I & \cdots & w_{1M}I \\ \vdots & & \vdots \\ w_{M1}I & \cdots & w_{MM}I \end{pmatrix} \\ &= \sigma^2 \begin{pmatrix} w_{11} & \cdots & w_{1M} \\ \vdots & & \vdots \\ w_{M1} & \cdots & w_{MM} \end{pmatrix} \otimes I \\ &= \sigma^2 W_0 \otimes I \end{aligned} \tag{2.36}$$

where I is the $T \times T$ identity matrix and $\otimes$ denotes the Kronecker product (cf. Theorem A.99).

Assumption ($\tilde{\mathrm{E}}$.1)

The covariance matrix $\sigma^2\Phi$ is positive definite and hence regular.

Model (2.35) with Φ according to ($\tilde{\mathrm{E}}$) is called the classical multivariate linear regression model.

Independent Single Regressions

W_0 expresses the relationships between the M equations of the system. If the errors ϵ_m are uncorrelated not only in time, but equationwise as well, that is, if

$$\mathrm{E}(\epsilon_{mt}\epsilon_{m't'}) = \sigma^2 w_{mm'} = 0 \quad \text{for} \quad m \neq m', \tag{2.37}$$

we then have

$$W_0 = \begin{pmatrix} w_{11} & \cdots & 0 \\ \vdots & & \vdots \\ 0 & \cdots & w_{MM} \end{pmatrix}. \tag{2.38}$$

(Thus ($\tilde{\text{E}}$.1) is fulfilled for $w_{mm} \neq 0 \quad (m = 1, \ldots M)$.)

The M equations (2.27) of the system are then to be handled independently. They do not form a real system. Their combination in an M-dimensional system of single regressions has no influence upon the goodness of fit of the estimates and predictions.

2.6 The Generalized Linear Regression Model

Starting with the multivariate regression model (2.35), when $M = 1$ we obtain the generalized linear regression model. In the reverse case, every equation (2.27) of the multivariate model is for $M > 1$ a univariate linear regression model that represents the statistical dependence of a regressand Y on K regressors $X_1, \ldots, X_K$ and a random error ϵ:

$$Y = X_1\beta_1 + \ldots + X_K\beta_K + \epsilon. \tag{2.39}$$

The random error ϵ describes the influence of chance as well as that of quantities that cannot be measured, or can be described indirectly by other variables X_k, such that their effect can be ascribed to chance as well.

This model implies that the X_k represent the main effects on Y and that the effects of systematic components on Y contained in ϵ, in addition to real chance, are sufficiently small. In particular, this model postulates that the dependence of X_k and ϵ is sufficiently small so that

$$\mathrm{E}(\epsilon|X) = \mathrm{E}(\epsilon) = 0. \tag{2.40}$$

We assume that we have T observations of all variables, which can be represented in a linear model

$$y_t = \sum_{k=1}^{K} x_{tk}\beta_k + \epsilon_t = \underset{1\times K}{x_t'}\ \beta + \epsilon_t \quad (t = 1, \ldots T), \tag{2.41}$$

or in matrix representation as

$$y = X\beta + \epsilon. \tag{2.42}$$

The assumptions corresponding to (D), (E), and (F) are (G), (H), and (K), respectively.

Assumption (G)

The regressors X_k are nonstochastic. Their values x_{kt} are chosen such that $\operatorname{rank}(X) = K$.

Assumption (H)

The random errors ϵ_t are generated by a T-dimensional regular stochastic process. Let

$$\mathrm{E}(\epsilon_t) = 0, \quad \mathrm{E}(\epsilon_t \epsilon_{t'}) = \sigma^2 w_{tt'} , \tag{2.43}$$

and hence

$$\mathrm{E}(\epsilon\epsilon') = \sigma^2 \underset{T\times T}{W} = \sigma^2 (w_{tt'}) , \tag{2.44}$$

where W is positive definite.

Assumption (K)

The vector ϵ of the random errors ϵ_t has a T-dimensional normal distribution $N(0,\, \sigma^2 W)$.

Given (G), (H), and (2.42), the generalized linear regression model is of the following form:

$$\left.\begin{array}{l} y = X\beta + \epsilon, \\ \mathrm{E}(\epsilon) = 0, \quad \mathrm{E}(\epsilon\epsilon') = \sigma^2 W, \\ X \text{ nonstochastic, } \operatorname{rank}(X) = K. \end{array}\right\} \tag{2.45}$$

The model (2.45) is called *regular* if additionally Assumption (H) is fulfilled. If (K) is fulfilled, we then have a *generalized!normal regression*. If $w_{tt'} = 0$ for $t \neq t'$ and $w_{tt} = 1$ for all t in (H), we have the *classical linear regression model*

$$\left.\begin{array}{l} y = X\beta + \epsilon, \\ \mathrm{E}(\epsilon) = 0, \quad \mathrm{E}(\epsilon\epsilon') = \sigma^2 I, \\ X \text{ nonstochastic, } \operatorname{rank}(X) = K. \end{array}\right\} \tag{2.46}$$

If (H) holds and W is known, the generalized model can be reduced to the classical model: Because of (H), W has a positive definite inverse W^{-1}. According to well-known theorems (cf. Theorem A.41), product representations exist for W and W^{-1}:

$$W = MM, \quad W^{-1} = NN \quad (M, N \text{ quadratic and regular}).$$

Thus $(NN) = (MM)^{-1}$, including $NMMN = NWN = I$. If the generalized model $y = X\beta + \epsilon$ is transformed by multiplication from the left with N, the transformed model $Ny = NX\beta + N\epsilon$ fulfills the assumptions of the classical model:

$$\begin{aligned} \mathrm{E}(N\epsilon\epsilon' N) &= \sigma^2 NWN = \sigma^2 I; \quad \mathrm{E}(N\epsilon) = N\,\mathrm{E}(\epsilon) = 0, \\ \mathrm{rank}(NX) &= K \quad (\text{since } \mathrm{rank}(X) = K \text{ and } N \text{ regular}). \end{aligned}$$

For the above models, statistics deals, among other things, with problems of testing models, the derivation of point and interval estimates of the unknown parameters, and the prediction of the regressands (endogenous variables). Of special importance in practice is the modification of models in terms of stochastic specifications (stochastic regressors, correlated random errors with different types of covariance matrices), rank deficiency of the regressor matrix, and model restrictions related to the parameter space.

The emphasis of the following chapters is on the derivation of best estimates of the parameters and optimal predictions of the regressands in regular multiple regression models. Along the way, different approaches for estimation (prediction), different auxiliary information about the model parameters, as well as alternative criteria of superiority are taken into consideration.

2.7 Exercises

Exercise 1. The CES (constant elasticity of substitution) production function relating the production Y to labor X_1 and capital X_2 is given by

$$Y = [\alpha X_1^{-\beta} + (1-\alpha) X_2^{-\beta}]^{-\frac{1}{\beta}}.$$

Can it be transformed to a linear model?

Exercise 2. Write the model and name it in each of the following sets of causal relationships:

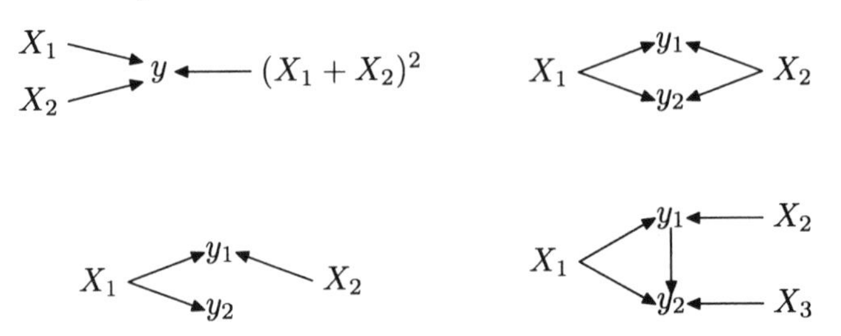

Exercise 3. If the matrix Γ in the model (2.4) is triangular, comment on the nature of the reduced form.

Exercise 4. For a system of simultaneous linear stochastic equations, the reduced form of the model is available. Can we recover the structural form from it in a logical manner? Explain your answer with a suitable illustration.

3

The Linear Regression Model

3.1 The Linear Model

The main topic of this chapter is the linear regression model and its basic principle of estimation through *least squares.* We present the algebraic, geometric, and statistical aspects of the problem, each of which has an intuitive appeal.

Let Y denote the dependent variable that is related to K independent variables $X_1, \ldots X_K$ by a function f. When the relationship is not exact, we write

$$Y = f(X_1, \ldots, X_K) + e\,. \tag{3.1}$$

When f is not linear, equation (3.1) is referred to as the nonlinear regression model. When f is linear, equation (3.1) takes the form

$$Y = X_1\beta_1 + \cdots + X_K\beta_K + e\,, \tag{3.2}$$

which is called the linear regression model.

We have T sets of observations on Y and $(X_1, \ldots, X_K)$, which we represent as follows:

$$(y, X) = \begin{pmatrix} y_1 & x_{11} & \cdots & x_{K1} \\ \vdots & \vdots & & \vdots \\ y_T & x_{1T} & \cdots & x_{KT} \end{pmatrix} = (y,\, x_{(1)}, \ldots, x_{(K)}) = \begin{pmatrix} y_1,\, x'_1 \\ \vdots \\ y_T,\, x'_T \end{pmatrix} \tag{3.3}$$

where $y = (y_1, \ldots, y_T)'$ is a T-vector and $x_i = (x_{1i}, \ldots, x_{Ki})'$ is a K-vector and $x_{(j)} = (x_{j1}, \ldots, x_{jT})'$ is a T-vector. (Note that in (3.3), the first, third and fourth matrices are partitioned matrices.)

In such a case, there are T observational equations of the form (3.2):

$$y_t = x_t'\beta + e_t\,, \quad t = 1, \ldots, T\,, \tag{3.4}$$

where $\beta' = (\beta_1, \ldots, \beta_K)$, which can be written using the matrix notation,

$$y = X\beta + e\,, \tag{3.5}$$

where $e' = (e_1, \ldots, e_T)$. We consider the problems of estimation and testing of hypotheses on β under some assumptions. A general procedure for the estimation of β is to minimize

$$\sum_{t=1}^{T} M(e_t) = \sum_{t=1}^{T} M(y_t - x_t'\beta) \tag{3.6}$$

for a suitably chosen function M, some examples of which are $M(x) = |x|$ and $M(x) = x^2$. More generally, one could minimize a global function of e such as $\max_t |e_t|$ over t. First we consider the case $M(x) = x^2$, which leads to the least-squares theory, and later introduce other functions that may be more appropriate in some situations.

3.2 The Principle of Ordinary Least Squares (OLS)

Let B be the set of all possible vectors β. If there is no further information, we have $B = \mathbb{R}^K$ (K-dimensional real Euclidean space). The object is to find a vector $b' = (b_1, \ldots, b_K)$ from B that minimizes the sum of squared residuals

$$S(\beta) = \sum_{t=1}^{T} e_t^2 = e'e = (y - X\beta)'(y - X\beta) \tag{3.7}$$

given y and X. A minimum will always exist, since $S(\beta)$ is a real-valued, convex, differentiable function. If we rewrite $S(\beta)$ as

$$S(\beta) = y'y + \beta'X'X\beta - 2\beta'X'y \tag{3.8}$$

and differentiate by β (with the help of Theorems A.91–A.95), we obtain

$$\frac{\partial S(\beta)}{\partial \beta} = 2X'X\beta - 2X'y\,, \tag{3.9}$$

$$\frac{\partial^2 S(\beta)}{\partial \beta^2} = 2X'X \quad \text{(nonnegative definite)}. \tag{3.10}$$

Equating the first derivative to zero yields what are called the *normal equations*

$$X'Xb = X'y. \tag{3.11}$$

If X is of full rank K, then $X'X$ is nonsingular and the unique solution of (3.11) is

$$b = (X'X)^{-1}X'y\,. \tag{3.12}$$

If X is not of full rank, equation (3.11) has a set of solutions

$$b = (X'X)^{-}X'y + (I - (X'X)^{-}X'X)w\,, \tag{3.13}$$

where $(X'X)^{-}$ is a g-inverse (generalized inverse) of $X'X$ and w is an arbitrary vector. [We note that a g-inverse $(X'X)^{-}$ of $X'X$ satisfies the properties $X'X(X'X)^{-}X'X = X'X$, $X(X'X)^{-}X'X = X$, $X'X(X'X)^{-}X' = X'$, and refer the reader to Section A.12 in Appendix A for the algebra of g-inverses and methods for solving linear equations, or to the books by Rao and Mitra (1971), and Rao and Rao (1998).] We prove the following theorem.

Theorem 3.1

(i) $\hat{y} = Xb$, *the empirical predictor of* y, *has the same value for all solutions* b *of* $X'Xb = X'y$.

(ii) $S(\beta)$, *the sum of squares defined in (3.7), attains the minimum for any solution of* $X'Xb = X'y$.

Proof: To prove (i), choose any b in the set (3.13) and note that

$$\begin{aligned} Xb &= X(X'X)^{-}X'y + X(I - (X'X)^{-}X'X)w \\ &= X(X'X)^{-}X'y \quad \text{(which is independent of } w\text{)}. \end{aligned}$$

Note that we used the result $X(X'X)^{-}X'X = X$ given in Theorem A.81.

To prove (ii), observe that for any β,

$$\begin{aligned} S(\beta) &= (y - Xb + X(b-\beta))'(y - Xb + X(b-\beta)) \\ &= (y-Xb)'(y-Xb) + (b-\beta)'X'X(b-\beta) + 2(b-\beta)'X'(y-Xb) \\ &= (y-Xb)'(y-Xb) + (b-\beta)'X'X(b-\beta)\,, \quad \text{using (3.11)} \\ &\geq (y-Xb)'(y-Xb) = S(b) \\ &= y'y - 2y'Xb + b'X'Xb = y'y - b'X'Xb = y'y - \hat{y}'\hat{y}\,. \end{aligned} \tag{3.14}$$

3.3 Geometric Properties of OLS

For the $T \times K$-matrix X, we define the *column space*

$$\mathcal{R}(X) = \{\theta : \theta = X\beta,\ \beta \in \mathbb{R}^K\}\,,$$

which is a subspace of $\mathbb{R}^T$. If we choose the norm $\|x\| = (x'x)^{1/2}$ for $x \in \mathbb{R}^T$, then the principle of least squares is the same as that of minimizing $\| y - \theta \|$ for $\theta \in \mathcal{R}(X)$. Geometrically, we have the situation as shown in Figure 3.1.

We then have the following theorem:

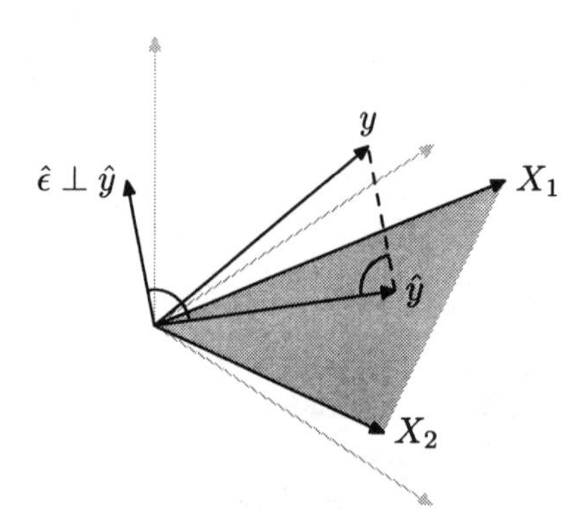

FIGURE 3.1. Geometric properties of OLS, $\theta \in \mathcal{R}(X)$ (for $T = 3$ and $K = 2$)

Theorem 3.2 *The minimum of $\| y - \theta \|$ for $\theta \in \mathcal{R}(X)$ is attained at $\hat{\theta}$ such that $(y - \hat{\theta}) \perp \mathcal{R}(X)$, that is, when $y - \hat{\theta}$ is orthogonal to all vectors in $\mathcal{R}(X)$, which is when $\hat{\theta}$ is the orthogonal projection of y on $\mathcal{R}(X)$. Such a $\hat{\theta}$ exists and is unique, and has the explicit expression*

$$\hat{\theta} = Py = X(X'X)^{-}X'y, \tag{3.15}$$

where $P = X(X'X)^{-}X'$ is the orthogonal projection operator on $\mathcal{R}(X)$.

Proof: Let $\hat{\theta} \in \mathcal{R}(X)$ be such that $(y - \hat{\theta}) \perp \mathcal{R}(X)$, that is, $X'(y - \hat{\theta}) = 0$. Then

$$\begin{aligned} \| y - \theta \|^2 &= (y - \hat{\theta} + \hat{\theta} - \theta)'(y - \hat{\theta} + \hat{\theta} - \theta) \\ &= (y - \hat{\theta})'(y - \hat{\theta}) + (\hat{\theta} - \theta)'(\hat{\theta} - \theta) \geq \| y - \hat{\theta} \|^2 \end{aligned}$$

since the term $(y - \hat{\theta})'(\hat{\theta} - \theta)$ vanishes using the orthogonality condition. The minimum is attained when $\theta = \hat{\theta}$. Writing $\hat{\theta} = X\hat{\beta}$, the orthogonality condition implies $X'(y - X\hat{\beta}) = 0$, that is, $X'X\hat{\beta} = X'y$. The equation $X'X\beta = X'y$ admits a solution, and $X\beta$ is unique for all solutions of β as shown in Theorem A.79. This shows that $\hat{\theta}$ exists.

Let $(X'X)^{-}$ be any g-inverse of $X'X$. Then $\hat{\beta} = (X'X)^{-}X'y$ is a solution of $X'X\beta = X'y$, and

$$X\hat{\beta} = X(X'X)^{-}X'y = Py,$$

which proves (3.15) of Theorem 3.2.

Note 1: If $\text{rank}(X) = s < K$, it is possible to find a matrix U of order $(K - s) \times K$ and rank $K - s$ such that $\mathcal{R}(U') \cap \mathcal{R}(X') = \{0\}$, where 0 is the null vector. In such a case, $X'X + U'U$ is of full rank K, $(X'X + U'U)^{-1}$ is a g-inverse of $X'X$, and a solution of the normal equation $X'X\beta = X'y$ can be written as

$$\hat{\beta} = (X'X + U'U)^{-1}(X'y + U'u), \tag{3.16}$$

where u is arbitrary. Also the projection operator P defined in (3.15) can be written as $P = X(X'X+U'U)^{-1}X'$. In some situations it is easy to find a matrix U satisfying the above conditions so that the g-inverse of $X'X$ can be computed as a regular inverse of a nonsingular matrix.

Note 2: The solution (3.16) can also be obtained as a conditional least-squares estimator when β is subject to the restriction $U\beta = u$ for a given arbitrary u. To prove this, we need only verify that $\hat{\beta}$ as in (3.16) satisfies the equation. Now

$$\begin{aligned} U\hat{\beta} &= U(X'X+U'U)^{-1}(X'y+U'u) \\ &= U(X'X+U'U)^{-1}U'u = u\,, \end{aligned}$$

which is true in view of result (iv) of Theorem A.81.

Note 3: It may be of some interest to establish the solution (3.16) using the calculus approach by differentiating

$$(y - X\beta)'(y - X\beta) + \lambda'(U\beta - u)$$

with respect to λ and β, where λ is a Lagrangian multiplier, which gives the equations

$$\begin{aligned} X'X\beta &= X'y + U'\lambda\,, \\ U\beta &= u\,, \end{aligned}$$

yielding the solution for β as in (3.16).

3.4 Best Linear Unbiased Estimation

3.4.1 Basic Theorems

In Sections 3.1 through 3.3, we viewed the problem of the linear model $y = X\beta + e$ as one of fitting the function $X\beta$ to y without making any assumptions on e. Now we consider e as a random variable denoted by ϵ, make some assumptions on its distribution, and discuss the estimation of β considered as an unknown vector parameter.

The usual assumptions made are

$$\mathrm{E}(\epsilon) = 0\,, \quad \mathrm{E}(\epsilon\epsilon') = \sigma^2 I\,, \tag{3.17}$$

and X is a fixed or nonstochastic matrix of order $T \times K$, with full rank K. We prove two lemmas that are of independent interest in estimation theory and use them in the special case of estimating β by linear functions of Y.

Lemma 3.3 (Rao, 1973a, p. 317) *Let T be a statistic such that* $\mathrm{E}(T) = \theta$, $\mathrm{V}(T) < \infty$, $\mathrm{V}(.)$ *denotes the variance, and where θ is a scalar parameter. Then a necessary and sufficient condition that T is MVUE (minimum* *variance unbiased estimator) of the parameter* θ *is*

$$\operatorname{cov}(T,t)=0 \quad \forall t \quad \textit{such that} \quad \mathrm{E}(t)=0 \quad \textit{and} \quad \mathrm{V}(t)<\infty\,. \tag{3.18}$$

Proof of necessity: Let T be MVUE and t be such that $E(t) = 0$ and $\mathrm{V}(t) < \infty$. Then $T + \lambda t$ is unbiased for θ for every $\lambda \in \mathbb{R}$, and

$$\begin{aligned}
\mathrm{V}(T+\lambda t) &= \mathrm{V}(T)+\lambda^2\,\mathrm{V}(t)+2\lambda\operatorname{cov}(T,t)\geq \mathrm{V}(T)\\
&\Rightarrow \lambda^2\,\mathrm{V}(t)+2\lambda\operatorname{cov}(T,t)\geq 0 \quad \forall\lambda\\
&\Rightarrow \operatorname{cov}(T,t)=0\,.
\end{aligned}$$

Proof of sufficiency: Let $\tilde{T}$ be any unbiased estimator with finite variance. Then $\tilde{T}-T$ is such that $\mathrm{E}(\tilde{T}-T)=0$, $\mathrm{V}(\tilde{T}-T)<\infty$, and

$$\begin{aligned}
\mathrm{V}(\tilde{T})=\mathrm{V}(T+\tilde{T}-T) &= \mathrm{V}(T)+\mathrm{V}(\tilde{T}-T)+2\operatorname{cov}(T,\tilde{T}-T)\\
&= \mathrm{V}(T)+\mathrm{V}(\tilde{T}-T)\geq \mathrm{V}(T)
\end{aligned}$$

if (3.18) holds.

Let $T'=(T_1,\ldots,T_k)$ be an unbiased estimate of the vector parameter $\theta'=(\theta_1,\ldots,\theta_k)$. Then the $k\times k$-matrix

$$D(T)=\mathrm{E}(T-\theta)(T-\theta)'=\begin{pmatrix} \mathrm{V}(T_1) & \operatorname{cov}(T_1,T_2) & \cdots & \operatorname{cov}(T_1,T_k)\\ \vdots & \vdots & \vdots & \vdots\\ \operatorname{cov}(T_k,T_1) & \operatorname{cov}(T_k,T_2) & \cdots & \mathrm{V}(T_k)\end{pmatrix} \tag{3.19}$$

is called the dispersion matrix of T. We say T_0 is MDUE (minimum dispersion unbiased estimator) of θ if $D(T)-D(T_0)$ is nonnegative definite, or in our notation

$$D(T)-D(T_0)\geq 0 \tag{3.20}$$

for any T such that $\mathrm{E}(T)=\theta$.

Lemma 3.4 *If* T_{i0} *is MVUE of* θ_i, $i=1,\ldots,k$, *then* $T_0'=(T_{10},\ldots,T_{k0})$ *is MDUE of* θ *and vice versa.*

Proof: Consider $a'T_0$, which is unbiased for $a'\theta$. Since $\operatorname{cov}(T_{i0},t)=0$ for any t such that $\mathrm{E}(t)=0$, it follows that $\operatorname{cov}(a'T_0,t)=0$, which shows that

$$\mathrm{V}(a'T_0)=a'D(T_0)a\leq a'D(T)a\,, \tag{3.21}$$

where T is an alternative estimator to T_0. Then (3.21) implies

$$D(T_0)\leq D(T)\,. \tag{3.22}$$

The converse is true, since (3.22) implies that the ith diagonal element of $D(T_0)$, which is $\mathrm{V}(T_{i0})$, is not greater than the ith diagonal element of $D(T)$, which is $\mathrm{V}(T_i)$.

The lemmas remain true if the estimators are restricted to a particular class that is closed under addition, such as all linear functions of observations.

Combining Lemmas 3.3 and 3.4, we obtain the fundamental equation characterizing an MDUE t of θ at a particular value θ_0:

$$\operatorname{cov}(t, z|\theta_0) = 0 \quad \forall z \quad \text{such that} \quad \mathrm{E}(z|\theta) = 0 \quad \forall \theta\,, \tag{3.23}$$

which we exploit in estimating the parameters in the linear model. If there is a t for which (3.23) holds for all θ_0, then we have a globally optimum estimator. The basic theory of equation (3.23) and its applications is first given in Rao (1989).

We revert back to the linear model

$$y = X\beta + \epsilon \tag{3.24}$$

with $\mathrm{E}(\epsilon) = 0$, $D(\epsilon) = \mathrm{E}(\epsilon\epsilon') = \sigma^2 I$, and discuss the estimation of β. Let $a + b'y$ be a linear function with zero expectation, then

$$\begin{aligned} \mathrm{E}(a + b'y) &= a + b'X\beta = 0 \quad \forall\beta \\ &\Rightarrow a = 0\,, \quad b'X = 0 \quad \text{or} \quad b \in \mathcal{R}(Z)\,, \end{aligned}$$

where Z is the matrix whose columns span the space orthogonal to $\mathcal{R}(X)$ with $\operatorname{rank}(Z) = T - \operatorname{rank}(X)$. Thus, the class of all linear functions of y with zero expectation is

$$(Zc)'y = c'Z'y\,, \tag{3.25}$$

where c is an arbitrary vector.

Case 1: Rank$(X) = K$. Rank$(Z) = T - K$ and $(X'X)$ is nonsingular, admitting the inverse $(X'X)^{-1}$. The following theorem provides the estimate of β.

Theorem 3.5 *The MDLUE (minimum dispersion linear unbiased estimator) of β is*

$$\hat{\beta} = (X'X)^{-1}X'y\,, \tag{3.26}$$

which is the same as the least squares estimator of β, and the minimum dispersion matrix is

$$\sigma^2(X'X)^{-1}\,. \tag{3.27}$$

Proof: Let $a + By$ be an unbiased estimater of β. Then

$$\mathrm{E}(a + By) = a + BX\beta = \beta \quad \forall\beta \quad \Rightarrow \quad a = 0\,, BX = I\,. \tag{3.28}$$

If By is MDLUE, using equation (3.23), it is sufficient that

$$\begin{aligned} 0 &= \operatorname{cov}(By, c'Z'y) \quad \forall c \\ &= \sigma^2 BZc \quad \forall c \\ &\Rightarrow BZ = 0 \quad \Rightarrow \quad B = AX' \quad \text{for some } A\,. \end{aligned} \tag{3.29}$$

Thus we have two equations for B from (3.28) and (3.29):

$$BX = I\,, \quad B = AX'\,.$$

Substituting AX' for B in $BX = I$:

$$A(X'X) = I \quad \Leftrightarrow \quad A = (X'X)^{-1}, \quad B = (X'X)^{-1}X', \tag{3.30}$$

giving the MDLUE

$$\hat{\beta} = By = (X'X)^{-1}X'y$$

with the dispersion matrix

$$\begin{aligned} D(\hat{\beta}) &= D((X'X)^{-1}X'y) \\ &= (X'X)^{-1}X'D(y)X(X'X)^{-1} \\ &= \sigma^2(X'X)^{-1}X'X(X'X)^{-1} = \sigma^2(X'X)^{-1}, \end{aligned}$$

which proves Theorem 3.5.

Case 2: $\text{Rank}(X) = r < K$ (deficiency in rank) and $\text{rank}(Z) = T - r$, in which case $X'X$ is singular. We denote any g-inverse of $X'X$ by $(X'X)^-$. The consequences of deficiency in the rank of X, which arises in many practical applications, are as follows.

(i) The linear model, $y = X\beta + \epsilon$, is not identifiable in the sense that there may be several values of β for which $X\beta$ has the same value, so that no particular value can be associated with the model.

(ii) The condition of unbiasedness for estimating β is $BX = I$, as derived in (3.28). If X is deficient in rank, we cannot find a B such that $BX = I$, and thus β cannot be unbiasedly estimated.

(iii) Let $l'\beta$ be a given linear parametric function and let $a + b'y$ be an estimator. Then

$$\text{E}(a + b'y) = a + b'X\beta = l'\beta \quad \Rightarrow \quad a = 0, \quad X'b = l. \tag{3.31}$$

The equation $X'b = l$ has a solution for b if $l \in \mathcal{R}(X')$. Thus, although the whole parameter is not unbiasedly estimable, it is possible to estimate all linear functions of the type $l'\beta$, $l \in \mathcal{R}(X')$. The following theorem provides the MDLUE of a given number s such linear functions

$$(l_1'\beta, \ldots, l_s'\beta) = (L'\beta)' \quad \text{with} \quad L = (l_1, \ldots, l_s). \tag{3.32}$$

A linear function $m'\beta$ is said to be nonestimable if $m \notin \mathcal{R}(X')$.

Theorem 3.6 *Let $L'\beta$ be s linear functions of β such that $\mathcal{R}(L) \subset \mathcal{R}(X')$, implying $L = X'A$ for some A. Then the MDLUE of $L'\beta$ is $L'\hat{\beta}$, where $\hat{\beta} = (X'X)^-X'y$, and the dispersion matrix of $L'\hat{\beta}$ is $\sigma^2L'(X'X)^-L$, where $(X'X)^-$ is any g-inverse of $X'X$.*

Proof: Let Cy be an unbiased estimator of $L'\beta$. Then

$$\text{E}(Cy) = CX\beta = L'\beta \quad \Rightarrow \quad CX = L'.$$

Now

$$\operatorname{cov}(Cy, Z'y) = \sigma^2 CZ = 0 \quad \Rightarrow \quad C = BX' \quad \text{for some } B\,.$$

Then $CX = L' = BX'X = L'$, giving $B = L'(X'X)^-$ as one solution, and $C = BX' = L'(X'X)^- X'$. The MDLUE of $L'\beta$ is

$$Cy = L'(X'X)^- X'y = L'\hat{\beta}\,.$$

An easy computation gives $D(L'\hat{\beta}) = \sigma^2 L'(X'X)^- L$.

Note that $\hat{\beta}$ *is not an estimate of* β. However, it can be used to compute the best estimates of estimable parametric functions of β.

Case 3: Rank$(X) = r < K$, in which case not all linear parametric functions are estimable. However there may be additional information in the form of linear relationships

$$u = U\beta + \delta \tag{3.33}$$

where U is an $s \times K$-matrix, with $\mathrm{E}(\delta) = 0$ and $D(\delta) = \sigma_0^2 I$. Note that (3.33) reduces to a nonstochastic relationship when $\sigma_0 = 0$, so that the following treatment covers both the stochastic and nonstochastic cases. Let us consider the estimation of the linear function $p'\beta$ by a linear function of the form $a'y + b'u$. The unbiasedness condition yields

$$\mathrm{E}(a'y + b'u) = a'X\beta + b'U\beta = p'\beta \quad \Rightarrow \quad X'a + U'b = p\,. \tag{3.34}$$

Then

$$\mathrm{V}(a'y + b'u) = a'a\sigma^2 + b'b\sigma_0^2 = \sigma^2(a'a + \rho b'b)\,, \tag{3.35}$$

where $\rho = \sigma_0^2/\sigma^2$, and the problem is one of minimizing $(a'a + \rho b'b)$ subject to the condition (3.34) on a and b. Unfortunately, the expression to be minimized involves an unknown quantity, except when $\sigma_0 = 0$. However, we shall present a formal solution depending on ρ. Considering the expression with a Lagrangian multiplier

$$a'a + \rho b'b + 2\lambda'(X'a + U'b - p)\,,$$

the minimizing equations are

$$a = X\lambda\,, \quad \rho b = U\lambda\,, \quad X'a + U'b = p\,.$$

If $\rho \neq 0$, substituting for a and b in the last equation gives another set of equations:

$$(X'X + \rho^{-1}U'U)\lambda = p\,, \quad a = X\lambda\,, \quad b = U\lambda \tag{3.36}$$

which is easy to solve. If $\rho = 0$, we have the equations

$$a = X\lambda\,, \quad b = U\lambda\,, \quad X'a + U'b = p\,.$$

Eliminating a, we have

$$X'X\lambda + U'b = p\,, \quad U\lambda = 0\,. \tag{3.37}$$

We solve equations (3.37) for b and λ and obtain the solution for a by using the equation $a = X\lambda$. For practical applications, it is necessary to have some estimate of ρ when $\sigma_0 \neq 0$. This may be obtained partly from the available data and partly from previous information.

3.4.2 Linear Estimators

The statistician's task is now to estimate the true but unknown vector β of regression parameters in the model (3.24) on the basis of observations (y, X) and assumptions already stated. This will be done by choosing a suitable estimator $\hat{\beta}$, which then will be used to calculate the conditional expectation $\mathrm{E}(y|X) = X\beta$ and an estimate for the error variance σ^2. It is common to choose an estimator $\hat{\beta}$ that is linear in y, that is,

$$\hat{\beta} = Cy + d\,, \tag{3.38}$$

where $C : K \times T$ and $d : K \times 1$ are nonstochastic matrices to be determined by minimizing a suitably chosen risk function.

First we have to introduce some definitions.

Definition 3.7 *$\hat{\beta}$ is called a homogeneous estimator of β if $d = 0$; otherwise $\hat{\beta}$ is called inhomogeneous.*

In Section 3.2, we have measured the model's goodness of fit by the sum of squared errors $S(\beta)$. Analogously we define, for the random variable $\hat{\beta}$, the quadratic loss function

$$L(\hat{\beta}, \beta, A) = (\hat{\beta} - \beta)' A (\hat{\beta} - \beta)\,, \tag{3.39}$$

where A is a symmetric and ≥ 0 (i.e., at least nonnegative definite) $K \times K$-matrix. (See Theorems A.36–A.38 where the definitions of $A > 0$ for positive definiteness and $A \geq 0$ for nonnegative definiteness are given.)

Obviously the loss (3.39) depends on the sample. Thus we have to consider the average or expected loss over all possible samples, which we call the risk.

Definition 3.8 *The quadratic risk of an estimator $\hat{\beta}$ of β is defined as*

$$R(\hat{\beta}, \beta, A) = \mathrm{E}(\hat{\beta} - \beta)' A (\hat{\beta} - \beta). \tag{3.40}$$

The next step now consists of finding an estimator $\hat{\beta}$ that minimizes the quadratic risk function over a class of appropriate functions. Therefore we have to define a criterion to compare estimators:

Definition 3.9 *($R(A)$ superiority) An estimator $\hat{\beta}_2$ of β is called $R(A)$ superior or an $R(A)$-improvement over another estimator $\hat{\beta}_1$ of β if*

$$R(\hat{\beta}_1, \beta, A) - R(\hat{\beta}_2, \beta, A) \geq 0\,. \tag{3.41}$$

3.4.3 Mean Dispersion Error

The quadratic risk is closely related to the matrix-valued criterion of the mean dispersion error (MDE) of an estimator. The MDE is defined as the matrix

$$\mathrm{M}(\hat{\beta}, \beta) = \mathrm{E}(\hat{\beta} - \beta)(\hat{\beta} - \beta)'. \tag{3.42}$$

We again denote the covariance matrix of an estimator $\hat{\beta}$ by $\mathrm{V}(\hat{\beta})$:

$$\mathrm{V}(\hat{\beta}) = \mathrm{E}(\hat{\beta} - \mathrm{E}(\hat{\beta}))(\hat{\beta} - \mathrm{E}(\hat{\beta}))'. \tag{3.43}$$

If $\mathrm{E}(\hat{\beta}) = \beta$, then $\hat{\beta}$ will be called unbiased (for β). If $\mathrm{E}(\hat{\beta}) \neq \beta$, then $\hat{\beta}$ is called biased. The difference between $\mathrm{E}(\hat{\beta})$ and β is

$$\mathrm{Bias}(\hat{\beta}, \beta) = \mathrm{E}(\hat{\beta}) - \beta. \tag{3.44}$$

If $\hat{\beta}$ is unbiased, then obviously $\mathrm{Bias}(\hat{\beta}, \beta) = 0$.

The following decomposition of the mean dispersion error often proves to be useful:

$$\begin{aligned} \mathrm{M}(\hat{\beta}, \beta) &= \mathrm{E}[(\hat{\beta} - \mathrm{E}(\hat{\beta})) + (\mathrm{E}(\hat{\beta}) - \beta)][(\hat{\beta} - \mathrm{E}(\hat{\beta})) + (\mathrm{E}(\hat{\beta}) - \beta)]' \\ &= \mathrm{V}(\hat{\beta}) + (\mathrm{Bias}(\hat{\beta}, \beta))(\mathrm{Bias}(\hat{\beta}, \beta))', \end{aligned} \tag{3.45}$$

that is, the MDE of an estimator is the sum of the covariance matrix and the squared bias (in its matrix version, i.e., $(\mathrm{Bias}(\hat{\beta}, \beta))(\mathrm{Bias}(\hat{\beta}, \beta))'$).

MDE Superiority

As the MDE contains all relevant information about the quality of an estimator, comparisons between different estimators may be made on the basis of their MDE matrices.

Definition 3.10 (MDE I criterion) *Let $\hat{\beta}_1$ and $\hat{\beta}_2$ be two estimators of β. Then $\hat{\beta}_2$ is called MDE-superior to $\hat{\beta}_1$ (or $\hat{\beta}_2$ is called an MDE-improvement to $\hat{\beta}_1$) if the difference of their MDE matrices is nonnegative definite, that is, if*

$$\Delta(\hat{\beta}_1, \hat{\beta}_2) = \mathrm{M}(\hat{\beta}_1, \beta) - \mathrm{M}(\hat{\beta}_2, \beta) \geq 0\,. \tag{3.46}$$

MDE superiority is a local property in the sense that (besides its dependency on σ^2) it depends on the particular value of β.

The quadratic risk function (3.40) is just a scalar-valued version of the MDE:

$$R(\hat{\beta}, \beta, A) = \mathrm{tr}\{A\, \mathrm{M}(\hat{\beta}, \beta)\}\,. \tag{3.47}$$

One important connection between $R(A)$ superiority and MDE superiority has been given by Theobald (1974) and Trenkler (1981):

Theorem 3.11 *Consider two estimators $\hat{\beta}_1$ and $\hat{\beta}_2$ of β. The following two statements are equivalent:*

$$\Delta(\hat\beta_1, \hat\beta_2) \geq 0, \quad (3.48)$$

$$R(\hat\beta_1, \beta, A) - R(\hat\beta_2, \beta, A) = \operatorname{tr}\{A\Delta(\hat\beta_1, \hat\beta_2)\} \geq 0 \quad (3.49)$$

for all matrices of the type $A = aa'$.

Proof: Using (3.46) and (3.47) we get

$$R(\hat\beta_1, \beta, A) - R(\hat\beta_2, \beta, A) = \operatorname{tr}\{A\Delta(\hat\beta_1, \hat\beta_2)\}. \quad (3.50)$$

From Theorem A.43 it follows that $\operatorname{tr}\{A\Delta(\hat\beta_1, \hat\beta_2)\} \geq 0$ for all matrices $A = aa' \geq 0$ if and only if $\Delta(\hat\beta_1, \hat\beta_2) \geq 0$.

3.5 Estimation (Prediction) of the Error Term ϵ and σ^2

The linear model (3.24) may be viewed as the decomposition of the observation y into a nonstochastic part $X\beta$, also called the signal, and a stochastic part ϵ, also called the noise (or error), as discussed in Rao (1989). Since we have estimated $X\beta$ by $X\hat\beta$, we may consider the residual

$$\hat\epsilon = y - X\hat\beta = (I - P_X)y, \quad (3.51)$$

where $P_X = X(X'X)^- X'$ is the projection operator on $\mathcal{R}(X)$, as an estimator (or predictor) of ϵ, with the mean prediction error

$$\begin{aligned} \mathrm{D}(\hat\epsilon) &= \mathrm{D}(y - X\hat\beta) = \mathrm{D}(I - P_X)y \\ &= \sigma^2(I - P_X)(I - P_X) = \sigma^2(I - P_X). \quad (3.52) \end{aligned}$$

However, the following theorem provides a systematic approach to the problem.

Theorem 3.12 *The MDLU predictor of ϵ is $\hat\epsilon$ as defined in (3.51).*

Proof: Let $C'y$ be an unbiased predictor of ϵ. Then

$$\mathrm{E}(C'y) = C'X\beta = 0 \quad \forall\beta \quad \Rightarrow \quad C'X = 0. \quad (3.53)$$

The dispersion of error is

$$\mathrm{D}(\epsilon - C'y) = \mathrm{D}(\epsilon - C'\epsilon) = \sigma^2(I - C')(I - C).$$

Putting $I - C' = M$, the problem is that of finding

$$\min MM' \quad \text{subject to} \quad MX = X. \quad (3.54)$$

Since P_X and Z span the whole $\mathbb{R}^T$, we can write

$$M' = P_X A + ZB \quad \text{for some } A \text{ and } B,$$

giving

$$\begin{aligned} X' = X'M' &= X'A, \\ MM' &= A'P_X A + B'Z'ZB \\ &= A'X(X'X)^- X'A + B'Z'ZB \\ &= X(X'X)^- X' + B'Z'ZB \geq P_X \end{aligned}$$

with equality when $B = 0$. Then

$$M' = P_X A = X(X'X)^- X'A = X(X'X)^- X',$$

and the best predictor of ϵ is

$$\hat{\epsilon} = C'Y = (I - M)y = (I - P_X)y.$$

Using the estimate $\hat{\epsilon}$ of ϵ we can obtain an unbiased estimator of σ^2 as

$$\frac{1}{T-r}\hat{\epsilon}'(I - P_X)\hat{\epsilon} = \frac{1}{T-r}y'(I - P_X)y \tag{3.55}$$

since (with rank $(X) = r$)

$$\begin{aligned} s^2 = \frac{1}{T-r}\,\mathrm{E}\,y'(I - P_X)y &= \frac{1}{T-r}\mathrm{tr}(I - P_X)\,\mathrm{D}(y) \\ &= \frac{\sigma^2}{T-r}\mathrm{tr}(I - P_X) = \sigma^2\frac{T-r}{T-r} = \sigma^2 . \end{aligned}$$

3.6 Classical Regression under Normal Errors

All results obtained so far are valid irrespective of the actual distribution of the random disturbances ϵ, provided that $\mathrm{E}(\epsilon) = 0$ and $\mathrm{E}(\epsilon\epsilon') = \sigma^2 I$. Now, we assume that the vector ϵ of random disturbances ϵ_t is distributed according to a T-dimensional normal distribution $N(0, \sigma^2 I)$, with the probability density

$$\begin{aligned} f(\epsilon; 0, \sigma^2 I) &= \prod_{t=1}^{T}(2\pi\sigma^2)^{-\frac{1}{2}}\exp\left(-\frac{1}{2\sigma^2}\epsilon_t^2\right) \\ &= (2\pi\sigma^2)^{-\frac{T}{2}}\exp\left\{-\frac{1}{2\sigma^2}\sum_{t=1}^{T}\epsilon_t^2\right\}. \end{aligned} \tag{3.56}$$

Note that the components ϵ_t $(t = 1, \ldots, T)$ are independent and identically distributed as $N(0, \sigma^2)$. This is a special case of a general T-dimensional normal distribution $N(\mu, \Sigma)$ with density

$$f(\xi; \mu, \Sigma) = \{(2\pi)^T|\Sigma|\}^{-\frac{1}{2}}\exp\left\{-\frac{1}{2}(\xi - \mu)'\Sigma^{-1}(\xi - \mu)\right\}. \tag{3.57}$$

The classical linear regression model under normal errors is given by

$$\left.\begin{array}{l} y = X\beta + \epsilon, \\ \epsilon \sim N(0, \sigma^2 I), \\ X \text{ nonstochastic, } \operatorname{rank}(X) = K. \end{array}\right\} \quad (3.58)$$

3.6.1 The Maximum-Likelihood (ML) Principle

Definition 3.13 *Let $\xi = (\xi_1, \ldots, \xi_n)'$ be a random variable with density function $f(\xi; \Theta)$, where the parameter vector $\Theta = (\Theta_1, \ldots, \Theta_m)'$ is an element of the parameter space Ω comprising all values that are a priori admissible.*

The basic idea of the maximum-likelihood principle is to consider the density $f(\xi; \Theta)$ for a specific realization of the sample ξ_0 of ξ as a function of Θ:

$$L(\Theta) = L(\Theta_1, \ldots, \Theta_m) = f(\xi_0; \Theta).$$

$L(\Theta)$ will be referred to as the likelihood function of Θ given ξ_0.

The ML principle postulates the choice of a value $\hat{\Theta} \in \Omega$ that maximizes the likelihood function, that is,

$$L(\hat{\Theta}) \geq L(\Theta) \quad \text{for all } \Theta \in \Omega.$$

Note that $\hat{\Theta}$ may not be unique. If we consider all possible samples, then $\hat{\Theta}$ is a function of ξ and thus a random variable itself. We will call it the maximum-likelihood estimator of Θ.

3.6.2 ML Estimation in Classical Normal Regression

Following Theorem A.82, we have for y from (3.58)

$$y = X\beta + \epsilon \sim N(X\beta, \sigma^2 I), \quad (3.59)$$

so that the likelihood function of y is given by

$$L(\beta, \sigma^2) = (2\pi\sigma^2)^{-\frac{T}{2}} \exp\left\{-\frac{1}{2\sigma^2}(y - X\beta)'(y - X\beta)\right\}. \quad (3.60)$$

Since the logarithmic transformation is monotonic, it is appropriate to maximize $\ln L(\beta, \sigma^2)$ instead of $L(\beta, \sigma^2)$, as the maximizing argument remains unchanged:

$$\ln L(\beta, \sigma^2) = -\frac{T}{2}\ln(2\pi\sigma^2) - \frac{1}{2\sigma^2}(y - X\beta)'(y - X\beta). \quad (3.61)$$

If there are no a priori restrictions on the parameters, then the parameter space is given by $\Omega = \{\beta; \sigma^2 : \beta \in \mathbb{R}^K; \sigma^2 > 0\}$. We derive the ML estimators of β and σ^2 by equating the first derivatives to zero (Theorems

A.91–A.95):

$$\text{(I)}\quad \frac{\partial \ln L}{\partial \beta} = \frac{1}{2\sigma^2}2X'(y - X\beta) = 0\,, \tag{3.62}$$

$$\text{(II)}\quad \frac{\partial \ln L}{\partial \sigma^2} = -\frac{T}{2\sigma^2} + \frac{1}{2(\sigma^2)^2}(y - X\beta)'(y - X\beta) = 0\,. \tag{3.63}$$

The *likelihood equations* are given by

$$\left.\begin{array}{ll} \text{(I)} & X'X\hat{\beta} = X'y\,, \\ \text{(II)} & \hat{\sigma}^2 = \frac{1}{T}(y - X\hat{\beta})'(y - X\hat{\beta})\,. \end{array}\right\} \tag{3.64}$$

Equation (I) of (3.64) is identical to the well-known normal equation (3.11). Its solution is unique, as $\operatorname{rank}(X) = K$ and we get the unique ML estimator

$$\hat{\beta} = b = (X'X)^{-1}X'y\,. \tag{3.65}$$

If we compare (II) with the unbiased estimator s^2 (3.55) for σ^2, we see immediately that

$$\hat{\sigma}^2 = \frac{T - K}{T}s^2, \tag{3.66}$$

so that $\hat{\sigma}^2$ is a biased estimator. The asymptotic expectation is given by (cf. Theorem A.102 (i))

$$\lim_{T\to\infty} \mathrm{E}(\hat{\sigma}^2) = \bar{\mathrm{E}}(\hat{\sigma}^2) = \mathrm{E}(s^2) = \sigma^2. \tag{3.67}$$

Thus we can state the following result.

Theorem 3.14 *The maximum-likelihood estimator and OLS estimator of β are identical in the model (3.59) of classical normal regression. The ML estimator $\hat{\sigma}^2$ of σ^2 is asymptotically unbiased.*

Note: The Cramér-Rao bound defines a lower bound (in the sense of definiteness of matrices) for the covariance matrix of unbiased estimators. In the model of normal regression, the Cramér-Rao bound is given by

$$\mathrm{V}(\tilde{\beta}) \geq \sigma^2(X'X)^{-1},$$

where $\tilde{\beta}$ is an arbitrary estimator. The covariance matrix of the ML estimator is just identical to this lower bound, so that b is the minimum dispersion unbiased estimator in the linear regression model under normal errors.

3.7 Testing Linear Hypotheses

In this section we consider the problem of testing a general linear hypothesis

$$H_0\colon R\beta = r \tag{3.68}$$

with R a $K \times s$-matrix and $\text{rank}(R) = K - s$, against the alternative

$$H_1\colon R\beta \neq r \tag{3.69}$$

where it will be assumed that R and r are nonstochastic and known.

The hypothesis H_0 expresses the fact that the parameter vector β obeys $(K - s)$ exact linear restrictions, which are linearly independent, as it is required that $\text{rank}(R) = K - s$. The general linear hypothesis (3.68) contains two main special cases:

Case 1: $s = 0$. The $K \times K$-matrix R is regular by the assumption $\text{rank}(X) = K$, and we may express H_0 and H_1 in the following form:

$$H_0\colon \beta = R^{-1}r = \beta^*, \tag{3.70}$$

$$H_1\colon \beta \neq \beta^*. \tag{3.71}$$

Case 2: $s > 0$. We choose an $s \times K$-matrix G complementary to R such that the $K \times K$-matrix $\begin{pmatrix} G \\ R \end{pmatrix}$ is regular of rank K. Let

$$X \begin{pmatrix} G \\ R \end{pmatrix}^{-1} = \underset{T\times K}{\tilde{X}} = \begin{pmatrix} \underset{T\times s}{\tilde{X}_1}, & \underset{T\times(K-s)}{\tilde{X}_2} \end{pmatrix},$$

$$\underset{s\times 1}{\tilde{\beta}_1} = G\beta\,, \qquad \underset{(K-s)\times 1}{\tilde{\beta}_2} = R\beta\,.$$

Then we may write

$$\begin{aligned} y = X\beta + \epsilon &= X\begin{pmatrix} G \\ R \end{pmatrix}^{-1}\begin{pmatrix} G \\ R \end{pmatrix}\beta + \epsilon \\ &= \tilde{X}\begin{pmatrix} \tilde{\beta}_1 \\ \tilde{\beta}_2 \end{pmatrix} + \epsilon \\ &= \tilde{X}_1\tilde{\beta}_1 + \tilde{X}_2\tilde{\beta}_2 + \epsilon\,. \end{aligned}$$

The latter model obeys all assumptions (3.59). The hypotheses H_0 and H_1 are thus equivalent to

$$H_0\colon \tilde{\beta}_2 = r\,; \quad \tilde{\beta}_1 \text{ and } \sigma^2 > 0 \text{ arbitrary,} \tag{3.72}$$

$$H_1\colon \tilde{\beta}_2 \neq r\,; \quad \tilde{\beta}_1 \text{ and } \sigma^2 > 0 \text{ arbitrary.} \tag{3.73}$$

Ω stands for the whole parameter space (either H_0 or H_1 is valid) and $\omega \subset \Omega$ stands for the subspace in which only H_0 is true; thus

$$\left.\begin{aligned} \Omega &= \{\beta; \sigma^2 : \beta \in \mathbb{R}^K, \sigma^2 > 0\}\,, \\ \omega &= \{\beta; \sigma^2 : \beta \in \mathbb{R}^K \text{ and } R\beta = r;\ \sigma^2 > 0\}\,. \end{aligned}\right\} \tag{3.74}$$

As a test statistic we will use the likelihood ratio

$$\lambda(y) = \frac{\max_\omega L(\Theta)}{\max_\Omega L(\Theta)}, \tag{3.75}$$

which may be derived in the following way.

Let $\Theta = (\beta, \sigma^2)$, then

$$\begin{aligned}\max_{\beta,\sigma^2} L(\beta,\sigma^2) &= L(\hat{\beta},\hat{\sigma}^2) \\ &= (2\pi\hat{\sigma}^2)^{-\frac{T}{2}} \exp\left\{-\frac{1}{2\hat{\sigma}^2}(y - X\hat{\beta})'(y - X\hat{\beta})\right\} \\ &= (2\pi\hat{\sigma}^2)^{-\frac{T}{2}} \exp\left\{-\frac{T}{2}\right\} \end{aligned} \tag{3.76}$$

and therefore

$$\lambda(y) = \left(\frac{\hat{\sigma}^2_\omega}{\hat{\sigma}^2_\Omega}\right)^{-\frac{T}{2}}, \tag{3.77}$$

where $\hat{\sigma}^2_\omega$ and $\hat{\sigma}^2_\Omega$ are ML estimators of σ^2 under H_0 and in Ω.

The random variable $\lambda(y)$ can take values between 0 and 1, which is obvious from (3.75). If H_0 is true, the numerator of $\lambda(y)$ should be greater than the denominator, so that $\lambda(y)$ should be close to 1 in repeated samples. On the other hand, $\lambda(y)$ should be close to 0 if H_1 is true.

Consider the linear transform of $\lambda(y)$:

$$\begin{aligned} F &= \left\{(\lambda(y))^{-\frac{2}{T}} - 1\right\}(T-K)(K-s)^{-1} \\ &= \frac{\hat{\sigma}^2_\omega - \hat{\sigma}^2_\Omega}{\hat{\sigma}^2_\Omega} \cdot \frac{T-K}{K-s}. \end{aligned} \tag{3.78}$$

If $\lambda \to 0$, then $F \to \infty$, and if $\lambda \to 1$, we have $F \to 0$, so that F *is close to 0* if H_0 is true and F *is sufficiently large* if H_1 is true.

Now we will determine F and its distribution for the two special cases of the general linear hypothesis.

Case 1: $s = 0$

The ML estimators under H_0 (3.70) are given by

$$\hat{\beta} = \beta^* \quad \text{and} \quad \hat{\sigma}^2_\omega = \frac{1}{T}(y - X\beta^*)'(y - X\beta^*). \tag{3.79}$$

The ML estimators over Ω are available from Theorem 3.14:

$$\hat{\beta} = b \quad \text{and} \quad \hat{\sigma}^2_\Omega = \frac{1}{T}(y - Xb)'(y - Xb). \tag{3.80}$$

Some rearrangements then yield

$$\left.\begin{aligned} b-\beta^* &= (X'X)^{-1}X'(y-X\beta^*)\,, \\ (b-\beta^*)'X'X &= (y-X\beta^*)'X\,, \\ y-Xb &= (y-X\beta^*)-X(b-\beta^*)\,, \\ (y-Xb)'(y-Xb) &= (y-X\beta^*)'(y-X\beta^*) \\ &\quad +(b-\beta^*)'X'X(b-\beta^*) \\ &\quad -2(y-X\beta^*)'X(b-\beta^*) \\ &= (y-X\beta^*)'(y-X\beta^*) \\ &\quad -(b-\beta^*)'X'X(b-\beta^*)\,. \end{aligned}\right\} \tag{3.81}$$

It follows that

$$T(\hat{\sigma}^2_\omega-\hat{\sigma}^2_\Omega)=(b-\beta^*)'X'X(b-\beta^*)\,, \tag{3.82}$$

leading to the test statistic

$$F=\frac{(b-\beta^*)'X'X(b-\beta^*)}{(y-Xb)'(y-Xb)}\cdot\frac{T-K}{K}\,. \tag{3.83}$$

Distribution of F

Numerator: The following statements are in order:

$$\begin{aligned} & b-\beta^*=(X'X)^{-1}X'[\epsilon+X(\beta-\beta^*)] && \text{[by (3.81)]}, \\ & \tilde{\epsilon}=\epsilon+X(\beta-\beta^*)\sim N(X(\beta-\beta^*),\sigma^2 I) && \text{[Theorem A.82]}, \\ & X(X'X)^{-1}X' \text{ idempotent and of rank } K, && \\ & (b-\beta^*)'X'X(b-\beta^*)=\tilde{\epsilon}'X(X'X)^{-1}X'\tilde{\epsilon} && \\ & \quad\sim\sigma^2\chi^2_K(\sigma^{-2}(\beta-\beta^*)'X'X(\beta-\beta^*)) && \text{[Theorem A.84]} \\ & \quad\text{and } \sim\sigma^2\chi^2_K \text{ under } H_0. && \end{aligned}$$

Denominator:

$$\left.\begin{aligned} & (y-Xb)'(y-Xb)=(T-K)s^2=\epsilon'(I-P_X)\epsilon && \text{[cf. (3.55)]}, \\ & \epsilon'(I-P_X)\epsilon\sim\sigma^2\chi^2_{T-K} && \text{[Theorem A.87]}. \end{aligned}\right\} \tag{3.84}$$

as $I-P_X=I-X(X'X)^{-1}X'$ is idempotent of rank $T-K$ (cf. Theorem A.61 (vi)).

We have

$$(I-P_X)X(X'X)^{-1}X'=0 \quad \text{[Theorem A.61 (vi)]}, \tag{3.85}$$

such that numerator and denominator are independently distributed (Theorem A.89).

Thus, the ratio F has the following properties (Theorem A.86):

- F is distributed as $F_{K,T-K}(\sigma^{-2}(\beta-\beta^*)'X'X(\beta-\beta^*))$ under H_1, and
- F is distributed as central $F_{K,T-K}$ under H_0: $\beta=\beta^*$.

If we denote by $F_{m,n,1-q}$ the $(1-q)$-quantile of $F_{m,n}$ (i.e., $P(F\leq F_{m,n,1-q})=1-q$), then we may derive a uniformly most powerful test, given a fixed level of significance α (cf. Lehmann, 1986, p. 372):

$$\left.\begin{array}{ll}\text{Region of acceptance of } H_0\text{:} & 0 \leq F \leq F_{K,T-K,1-\alpha}\,, \\ \text{Critical region:} & F > F_{K,T-K,1-\alpha}\,. \end{array}\right\} \tag{3.86}$$

A selection of F-quantiles is provided in Appendix B.

Case 2: $s > 0$

Next we consider a decomposition of the model in order to determine the ML estimators under H_0 (3.72) and compare them with the corresponding ML estimator over Ω. Let

$$\beta' = (\underset{1\times s}{\beta_1'}\,,\ \underset{1\times(K-s)}{\beta_2'}) \tag{3.87}$$

and, respectively,

$$y = X\beta + \epsilon = X_1\beta_1 + X_2\beta_2 + \epsilon\,. \tag{3.88}$$

We set

$$\tilde{y} = y - X_2 r\,. \tag{3.89}$$

Because $\text{rank}(X) = K$, we have

$$\text{rank}\underset{T\times s}{(X_1)} = s\,, \quad \text{rank}\underset{T\times(K-s)}{(X_2)} = K - s\,, \tag{3.90}$$

such that the inverse matrices $(X_1'X_1)^{-1}$ and $(X_2'X_2)^{-1}$ do exist.

The ML estimators under H_0 are then given by

$$\hat{\beta}_2 = r, \quad \hat{\beta}_1 = (X_1'X_1)^{-1}X_1'\tilde{y} \tag{3.91}$$

and

$$\hat{\sigma}_\omega^2 = \frac{1}{T}(\tilde{y} - X_1\hat{\beta}_1)'(\tilde{y} - X_1\hat{\beta}_1). \tag{3.92}$$

Separation of b

At first, it is easily seen that

$$\begin{aligned} b &= (X'X)^{-1}X'y \\ &= \begin{pmatrix} X_1'X_1 & X_1'X_2 \\ X_2'X_1 & X_2'X_2 \end{pmatrix}^{-1} \begin{pmatrix} X_1'y \\ X_2'y \end{pmatrix}. \end{aligned} \tag{3.93}$$

Making use of the formulas for the inverse of a partitioned matrix yields (Theorem A.19)

$$\begin{pmatrix} (X_1'X_1)^{-1}[I + X_1'X_2D^{-1}X_2'X_1(X_1'X_1)^{-1}] & -(X_1'X_1)^{-1}X_1'X_2D^{-1} \\ -D^{-1}X_2'X_1(X_1'X_1)^{-1} & D^{-1} \end{pmatrix}, \tag{3.94}$$

where

$$D = X_2' M_1 X_2 \tag{3.95}$$

and

$$M_1 = I - X_1(X_1'X_1)^{-1}X_1' = I - P_{X_1}\,. \tag{3.96}$$

M_1 is (analogously to $(I - P_X)$) idempotent and of rank $T - s$; further we have $M_1 X_1 = 0$. The $(K-s) \times (K-s)$-matrix

$$D = X_2'X_2 - X_2'X_1(X_1'X_1)^{-1}X_1'X_2 \tag{3.97}$$

is symmetric and regular, as the normal equations are uniquely solvable. The estimators b_1 and b_2 of b are then given by

$$b = \begin{pmatrix} b_1 \\ b_2 \end{pmatrix} = \begin{pmatrix} (X_1'X_1)^{-1}X_1'y - (X_1'X_1)^{-1}X_1'X_2D^{-1}X_2'M_1y \\ D^{-1}X_2'M_1y \end{pmatrix}. \tag{3.98}$$

Various relations immediately become apparent from (3.98):

$$\left.\begin{aligned} b_2 &= D^{-1}X_2'M_1y, \\ b_1 &= (X_1'X_1)^{-1}X_1'(y - X_2b_2), \\ b_2 - r &= D^{-1}X_2'M_1(y - X_2r) \\ &= D^{-1}X_2'M_1\tilde{y} \\ &= D^{-1}X_2'M_1(\epsilon + X_2(\beta_2 - r))\,, \end{aligned}\right\} \tag{3.99}$$

$$\left.\begin{aligned} b_1 - \hat{\beta}_1 &= (X_1'X_1)^{-1}X_1'(y - X_2b_2 - \tilde{y}) \\ &= -(X_1'X_1)^{-1}X_1'X_2(b_2 - r) \\ &= -(X_1'X_1)^{-1}X_1'X_2D^{-1}X_2'M_1\tilde{y}\,. \end{aligned}\right\} \tag{3.100}$$

Decomposition of $\hat{\sigma}^2_\Omega$

We write (using symbols u and v)

$$\begin{aligned} (y - Xb) &= (y - X_2r - X_1\hat{\beta}_1) &-& \Big(X_1(b_1 - \hat{\beta}_1) + X_2(b_2 - r)\Big) \\ &= \qquad u &-& \qquad v\,. \end{aligned} \tag{3.101}$$

Thus we may decompose the ML estimator $T\hat{\sigma}^2_\Omega = (y - Xb)'(y - Xb)$ as

$$(y - Xb)'(y - Xb) = u'u + v'v - 2u'v\,. \tag{3.102}$$

We have

$$u = y - X_2r - X_1\hat{\beta}_1 = \tilde{y} - X_1(X_1'X_1)^{-1}X_1'\tilde{y} = M_1\tilde{y}\,, \tag{3.103}$$

$$u'u = \tilde{y}'M_1\tilde{y}\,, \tag{3.104}$$

$$\begin{aligned} v &= X_1(b_1 - \hat{\beta}_1) + X_2(b_2 - r) \\ &= -X_1(X_1'X_1)^{-1}X_1'X_2D^{-1}X_2'M_1\tilde{y} \quad \text{[by (3.99)]} \\ &\quad + X_2D^{-1}X_2'M_1\tilde{y} \quad \text{[by (3.100)]} \\ &= M_1X_2D^{-1}X_2'M_1\tilde{y}\,, \end{aligned} \tag{3.105}$$

$$\begin{aligned} v'v &= \tilde{y}' M_1 X_2 D^{-1} X_2' M_1 \tilde{y} \\ &= (b_2 - r)' D (b_2 - r) \,, & (3.106) \\ u'v &= v'v \,. & (3.107) \end{aligned}$$

Summarizing, we may state

$$\begin{aligned} (y - Xb)'(y - Xb) &= u'u - v'v & (3.108) \\ &= (\tilde{y} - X_1\hat{\beta}_1)'(\tilde{y} - X_1\hat{\beta}_1) - (b_2 - r)'D(b_2 - r) \end{aligned}$$

or,

$$T(\hat{\sigma}_\omega^2 - \hat{\sigma}_\Omega^2) = (b_2 - r)'D(b_2 - r) \,. \tag{3.109}$$

We therefore get in case 2: $s > 0$:

$$F = \frac{(b_2 - r)'D(b_2 - r)}{(y - Xb)'(y - Xb)} \frac{T - K}{K - s} \,. \tag{3.110}$$

Distribution of F

Numerator: We use the following relations:

$$\begin{aligned} A &= M_1 X_2 D^{-1} X_2' M_1 \quad \text{is idempotent,} \\ \operatorname{rank}(A) &= \operatorname{tr}(A) = \operatorname{tr}\{(M_1 X_2 D^{-1})(X_2' M_1)\} \\ &= \operatorname{tr}\{(X_2' M_1)(M_1 X_2 D^{-1})\} \quad \text{[Theorem A.13 (iv)]} \\ &= \operatorname{tr}(I_{K-s}) = K - s \,, \\ b_2 - r &= D^{-1} X_2' M_1 \tilde{\epsilon} \quad \text{[by (3.99)]}, \\ \tilde{\epsilon} &= \epsilon + X_2(\beta_2 - r) \\ &\sim N(X_2(\beta_2 - r), \sigma^2 I) \quad \text{[Theorem A.82]}, \\ (b_2 - r)'D(b_2 - r) &= \tilde{\epsilon}' A \tilde{\epsilon} \\ &\sim \sigma^2 \chi^2_{K-s}(\sigma^{-2}(\beta_2 - r)'D(\beta_2 - r)) & (3.111) \\ &\sim \sigma^2 \chi^2_{K-s} \quad \text{under } H_0. & (3.112) \end{aligned}$$

Denominator: The denominator is equal in both cases; that is

$$(y - Xb)'(y - Xb) = \epsilon'(I - P_X)\epsilon \quad \sim \quad \sigma^2 \chi^2_{T-K} \,. \tag{3.113}$$

Because

$$(I - P_X)X = (I - P_X)(X_1, X_2) = ((I - P_X)X_1, (I - P_X)X_2) = (0, 0) \,, \tag{3.114}$$

we find

$$(I - P_X)M_1 = (I - P_X) \tag{3.115}$$

and

$$(I - P_X)A = (I - P_X)M_1 X_2 D^{-1} X_2' M_1 = 0 \,, \tag{3.116}$$

so that the numerator and denominator of F (3.110) are independently distributed [Theorem A.89]. Thus [see also Theorem A.86] the test statistic F is distributed under H_1 as $F_{K-s,T-K}(\sigma^{-2}(\beta_2 - r)'D(\beta_2 - r))$ and as central $F_{K-s,T-K}$ under H_0.

The region of acceptance of H_0 at a level of significance α is then given by

$$0 \leq F \leq F_{K-s,T-K,1-\alpha}\,. \tag{3.117}$$

Accordingly, the critical area of H_0 is given by

$$F > F_{K-s,T-K,1-\alpha}\,. \tag{3.118}$$

3.8 Analysis of Variance and Goodness of Fit

3.8.1 Bivariate Regression

To illustrate the basic ideas, we shall consider the model with a dummy variable **1** and a regressor x:

$$y_t = \beta_0 + \beta_1 x_t + \epsilon_t \quad (t = 1, \ldots, T). \tag{3.119}$$

Ordinary least-squares estimators of β_0 and β_1 are given by

$$b_1 = \frac{\sum(x_t - \bar{x})(y_t - \bar{y})}{\sum(x_t - \bar{x})^2}\,, \tag{3.120}$$

$$b_0 = \bar{y} - b_1\bar{x}\,. \tag{3.121}$$

The best predictor of y on the basis of a given x is

$$\hat{y} = b_0 + b_1 x\,. \tag{3.122}$$

Especially, we have for $x = x_t$

$$\begin{aligned} \hat{y}_t &= b_0 + b_1 x_t \\ &= \bar{y} + b_1(x_t - \bar{x}) \end{aligned} \tag{3.123}$$

(cf. (3.121)). On the basis of the identity

$$y_t - \hat{y}_t = (y_t - \bar{y}) - (\hat{y}_t - \bar{y}), \tag{3.124}$$

we may express the sum of squared residuals (cf. (3.14)) as

$$\begin{aligned} S(b) = \sum(y_t - \hat{y}_t)^2 &= \sum(y_t - \bar{y})^2 + \sum(\hat{y}_t - \bar{y})^2 \\ &\quad - 2\sum(y_t - \bar{y})(\hat{y}_t - \bar{y}). \end{aligned}$$

Further manipulation yields

$$\begin{aligned}
\sum (y_t - \bar{y})(\hat{y}_t - \bar{y}) &= \sum (y_t - \bar{y}) b_1 (x_t - \bar{x}) && \text{[cf. (3.123)]} \\
&= b_1^2 \sum (x_t - \bar{x})^2 && \text{[cf. (3.120)]} \\
&= \sum (\hat{y}_t - \bar{y})^2 && \text{[cf. (3.124)]}
\end{aligned}$$

Thus we have

$$\sum (y_t - \bar{y})^2 = \sum (y_t - \hat{y}_t)^2 + \sum (\hat{y}_t - \bar{y})^2 . \tag{3.125}$$

This relation has already been established in (3.14). The left-hand side of (3.125) is called the *sum of squares about the mean* or *corrected sum of squares of* Y (i.e., SS corrected) or SYY.

The first term on the right-hand side describes the deviation: observation minus predicted value, namely, the residual sum of squares:

$$SS\,\text{Residual:} \qquad RSS = \sum (y_t - \hat{y}_t)^2 , \tag{3.126}$$

whereas the second term describes the proportion of variability explained by regression.

$$SS\,\text{Regression:} \qquad SS_{\text{Reg}} = \sum (\hat{y}_t - \bar{y})^2 . \tag{3.127}$$

If all observations y_t are located on a straight line, we have obviously $\sum (y_t - \hat{y}_t)^2 = 0$ and thus SS corrected $= SS_{\text{Reg}}$.

Accordingly, the goodness of fit of a regression is measured by the ratio

$$R^2 = \frac{SS_{\text{Reg}}}{SS\,\text{corrected}} . \tag{3.128}$$

We will discuss R^2 in some detail. The degrees of freedom (df) of the sum of squares are

$$\sum_{t=1}^{T} (y_t - \bar{y})^2 : \quad df = T - 1$$

and

$$\sum_{t=1}^{T} (\hat{y}_t - \bar{y})^2 = b_1^2 \sum (x_t - \bar{x})^2 : \quad df = 1 ,$$

as *one* function in y_t—namely b_1—is sufficient to calculate SS_{Reg}. In view of (3.125), the degree of freedom for the sum of squares $\sum (y_t - \hat{y}_t)^2$ is just the difference of the other two df's, that is, $df = T - 2$.

All sums of squares are mutually independently distributed as χ^2_{df} if the errors ϵ_t are normally distributed. This enables us to establish the following analysis of variance (ANOVA) table:

Source of variation	Sum of squares	*df*	Mean square
Regression	SS regression	1	MS_{Reg}
Residual	RSS	$T-2$	$RSS/(T-2)=s^2$
Total	SS corrected $= SYY$	$T-1$	

We will use the following abbreviations:

$$SXX = \sum (x_t - \bar{x})^2, \tag{3.129}$$

$$SYY = \sum (y_t - \bar{y})^2, \tag{3.130}$$

$$SXY = \sum (x_t - \bar{x})(y_t - \bar{y}). \tag{3.131}$$

The sample correlation coefficient then may be written as

$$r_{XY} = \frac{SXY}{\sqrt{SXX}\sqrt{SYY}}. \tag{3.132}$$

Moreover, we have (cf. (3.120))

$$b_1 = \frac{SXY}{SXX} = r_{XY}\sqrt{\frac{SYY}{SXX}}. \tag{3.133}$$

The estimator of σ^2 may be expressed by using (3.126) as:

$$s^2 = \frac{1}{T-2}\sum \hat{\epsilon}_t^2 = \frac{1}{T-2} RSS. \tag{3.134}$$

Various alternative formulations for RSS are in use as well:

$$\begin{aligned} RSS &= \sum (y_t - (b_0 + b_1 x_t))^2 \\ &= \sum [(y_t - \bar{y}) - b_1(x_t - \bar{x})]^2 \\ &= SYY + b_1^2 SXX - 2b_1 SXY \\ &= SYY - b_1^2 SXX & (3.135) \\ &= SYY - \frac{(SXY)^2}{SXX}. & (3.136) \end{aligned}$$

Further relations become immediately apparent:

$$SS \text{ corrected} = SYY \tag{3.137}$$

and

$$\begin{aligned} SS_{\text{Reg}} &= SYY - RSS \\ &= \frac{(SXY)^2}{SXX} = b_1^2\, SXX. & (3.138) \end{aligned}$$

Checking the Adequacy of Regression Analysis

If model (3.119)

$$y_t = \beta_0 + \beta_1 x_t + \epsilon_t$$

is appropriate, the coefficient b_1 should be significantly different from zero. This is equivalent to the fact that X and Y are significantly correlated.

Formally, we compare the models (cf. Weisberg, 1980, p. 17)

$$H_0\colon\ y_t\beta_0 + \epsilon_t\,,$$
$$H_1\colon\ y_t\beta_0 + \beta_1 x_t + \epsilon_t\,,$$

by comparing testing H_0: $\beta_1 = 0$ against H_1: $\beta_1 \neq 0$.

We assume normality of the errors $\epsilon \sim N(0, \sigma^2 I)$. If we recall (3.97), that is

$$\begin{aligned} D &= x'x - x'1(1'1)^{-1}1'x\,, \quad 1' = (1, \ldots, 1) \\ &= \sum x_t^2 - \frac{(\sum x_t)^2}{T} = \sum (x_t - \bar{x})^2 = SXX\,, \end{aligned} \tag{3.139}$$

then the likelihood ratio test statistic (3.110) is given by

$$\begin{aligned} F_{1,T-2} &= \frac{b_1^2 SXX}{s^2} \\ &= \frac{SS_{\text{Reg}}}{RSS} \cdot (T-2) \\ &= \frac{MS_{\text{Reg}}}{s^2}\,. \end{aligned} \tag{3.140}$$

The Coefficient of Determination

In (3.128) R^2 has been introduced as a measure of goodness of fit. Using (3.138) we get

$$R^2 = \frac{SS_{\text{Reg}}}{SYY} = 1 - \frac{RSS}{SYY}\,. \tag{3.141}$$

The ratio SS_{Reg}/SYY describes the proportion of variability that is explained by regression in relation to the total variability of y. The right-hand side of the equation is 1 minus the proportion of variability that is not covered by regression.

Definition 3.15 *R^2 (3.141) is called the coefficient of determination.*

By using (3.123) and (3.138), we get the basic relation between R^2 and the sample correlation coefficient

$$R^2 = r_{XY}^2\,. \tag{3.142}$$

Confidence Intervals for b_0 and b_1

The covariance matrix of OLS is generally of the form $V_b = \sigma^2(X'X)^{-1} = \sigma^2 S^{-1}$. In model (3.119) we get

$$S = \begin{pmatrix} 1'1 & 1'x \\ 1'x & x'x \end{pmatrix} = \begin{pmatrix} T & T\bar{x} \\ T\bar{x} & \sum x_t^2 \end{pmatrix}, \tag{3.143}$$

$$S^{-1} = \frac{1}{SXX}\begin{pmatrix} \frac{1}{T}\sum x_t^2 & -\bar{x} \\ -\bar{x} & 1 \end{pmatrix}, \tag{3.144}$$

and therefore

$$\mathrm{Var}(b_1) = \sigma^2 \frac{1}{SXX} \tag{3.145}$$

$$\mathrm{Var}(b_0) = \frac{\sigma^2}{T}\cdot\frac{\sum x_t^2}{SXX} = \frac{\sigma^2}{T}\frac{\sum x_t^2 - T\bar{x}^2 + T\bar{x}^2}{SXX}$$

$$= \sigma^2\left(\frac{1}{T} + \frac{\bar{x}^2}{SXX}\right). \tag{3.146}$$

The estimated standard deviations are

$$\mathrm{SE}(b_1) = s\sqrt{\frac{1}{SXX}} \tag{3.147}$$

and

$$\mathrm{SE}(b_0) = s\sqrt{\frac{1}{T} + \frac{\bar{x}^2}{SXX}} \tag{3.148}$$

with s from (3.134).

Under normal errors $\epsilon \sim N(0, \sigma^2 I)$ in model (3.119), we have

$$b_1 \sim N\left(\beta_1, \sigma^2 \cdot \frac{1}{SXX}\right). \tag{3.149}$$

Thus it holds that

$$\frac{b_1 - \beta_1}{s}\sqrt{SXX} \sim t_{T-2}. \tag{3.150}$$

Analogously we get

$$b_0 \sim N\left(\beta_0, \sigma^2\left(\frac{1}{T} + \frac{\bar{x}^2}{SXX}\right)\right), \tag{3.151}$$

$$\frac{b_0 - \beta_0}{s}\sqrt{\frac{1}{T} + \frac{\bar{x}^2}{SXX}} \sim t_{T-2}. \tag{3.152}$$

This enables us to calculate confidence intervals at level $1 - \alpha$

$$b_0 - t_{T-2,1-\alpha/2}\cdot \mathrm{SE}(b_0) \le \beta_0 \le b_0 + t_{T-2,1-\alpha/2}\cdot \mathrm{SE}(b_0) \tag{3.153}$$

and

$$b_1 - t_{T-2,1-\alpha/2}\cdot \mathrm{SE}(b_1) \le \beta_1 \le b_1 + t_{T-2,1-\alpha/2}\cdot \mathrm{SE}(b_1). \tag{3.154}$$

These confidence intervals correspond to the region of acceptance of a two-sided test at the same level.

(i) Testing H_0: $\beta_0 = \beta_0^*$: The test statistic is

$$t_{T-2} = \frac{b_0 - \beta_0^*}{\mathrm{SE}(b_0)}. \tag{3.155}$$

H_0 is not rejected if

$$|t_{T-2}| \leq t_{T-2,1-\alpha/2}$$

or, equivalently, if (3.153) with $\beta_0 = \beta_0^*$ holds.

(ii) Testing H_0*:* $\beta_1 = \beta_1^*$*:* The test statistic is

$$t_{T-2} = \frac{b_1 - \beta_1^*}{\mathrm{SE}(b_1)} \tag{3.156}$$

or, equivalently,

$$t_{T-2}^2 = F_{1,T-2} = \frac{(b_1 - \beta_1^*)^2}{(\mathrm{SE}(b_1))^2} \,. \tag{3.157}$$

This is identical to (3.140) if H_0: $\beta_1 = 0$ is being tested.

H_0 will not be rejected if

$$|t_{T-2}| \leq t_{T-2,1-\alpha/2}$$

or, equivalently, if (3.154) with $\beta_1 = \beta_1^*$ holds.

3.8.2 Multiple Regression

If we consider more than two regressors, still under the assumption of normality of the errors, we find the methods of analysis of variance to be most convenient in distinguishing between the two models $y = \mathbf{1}\beta_0 + X\beta_* + \epsilon = \tilde{X}\beta + \epsilon$ and $y = \mathbf{1}\beta_0 + \epsilon$. In the latter model we have $\hat{\beta}_0 = \bar{y}$, and the related residual sum of squares is

$$\sum (y_t - \hat{y}_t)^2 = \sum (y_t - \bar{y})^2 = SYY \,. \tag{3.158}$$

In the former model, $\beta = (\beta_0, \beta_*)'$ will be estimated by $b = (\tilde{X}'\tilde{X})^{-1}\tilde{X}'y$.

The two components of the parameter vector β in the full model may be estimated by

$$b = \begin{pmatrix} \hat{\beta}_0 \\ \hat{\beta}_* \end{pmatrix}, \; \hat{\beta}_* = (X'X)^{-1}X'y, \; \hat{\beta}_0 = \bar{y} - \hat{\beta}_*'\bar{x} \,. \tag{3.159}$$

Thus we have

$$\begin{aligned} RSS &= (y - \tilde{X}b)'(y - \tilde{X}b) \\ &= y'y - b'\tilde{X}'\tilde{X}b \\ &= (y - \mathbf{1}\bar{y})'(y - \mathbf{1}\bar{y}) - \hat{\beta}_*'(X'X)\hat{\beta}_* + T\bar{y}^2. \end{aligned} \tag{3.160}$$

The proportion of variability explained by regression is (cf. (3.138))

$$SS_{\mathrm{Reg}} = SYY - RSS \tag{3.161}$$

with RSS from (3.160) and SYY from (3.158). Then the ANOVA table is of the form

Source of variation	Sum of squares	*df*	Mean square
Regression on $X_1, \ldots, X_K$	SS_{Reg}	K	SS_{Reg}/K
Residual	RSS	$T-K-1$	$RSS/(T-K-1)$
Total	SYY	$T-1$	

The multiple coefficient of determination

$$R^2 = \frac{SS_{\text{Reg}}}{SYY} \tag{3.162}$$

again is a measure of the proportion of variability explained by regression of y on $X_1, \ldots, X_K$ in relation to the total variability SYY.

The F-test for

$$H_0\colon \beta_* = 0$$

versus

$$H_1\colon \beta_* \neq 0$$

(i.e., $H_0\colon y = 1\beta_0 + \epsilon$ versus $H_1\colon y = 1\beta_0 + X\beta_* + \epsilon$) is based on the test statistic

$$F_{K,T-K-1} = \frac{SS_{\text{Reg}}/K}{s^2}. \tag{3.163}$$

Often, it is of interest to test for significance of single components of β. This type of a problem arises, for example, in stepwise model selection, with respect to the coefficient of determination.

Criteria for Model Choice

Draper and Smith (1966) and Weisberg (1980) have established a variety of criteria to find the right model. We will follow the strategy, proposed by Weisberg.

Ad hoc criteria

Denote by $X_1, \ldots, X_K$ all available regressors, and let $\{X_{i1}, \ldots, X_{ip}\}$ be a subset of $p \leq K$ regressors. We denote the respective residual sum of squares by RSS_K and RSS_p. The parameter vectors are

$$\begin{aligned}
&\beta \text{ for } X_1, \cdots, X_K\,,\\
&\beta_1 \text{ for } X_{i1}, \cdots, X_{ip}\,,\\
&\beta_2 \text{ for } (X_1, \cdots, X_K)\backslash(X_{i1}, \cdots, X_{ip})\,.
\end{aligned}$$

A choice between the two models can be examined by testing $H_0\colon \beta_2 = 0$. We apply the F-test since the hypotheses are nested:

$$F_{(K-p),T-K} = \frac{(RSS_p - RSS_K)/(K-p)}{RSS_K/(T-K)}. \tag{3.164}$$

We prefer the full model against the partial model if H_0: $\beta_2 = 0$ is rejected, that is, if $F > F_{1-\alpha}$ (with degrees of freedom $K - p$ and $T - K$).

Model choice based on an adjusted coefficient of determination

The coefficient of determination (see (3.161) and (3.162))

$$R_p^2 = 1 - \frac{RSS_p}{SYY} \tag{3.165}$$

is inappropriate to compare a model with K and one with $p < K$, because R_p^2 always increases if an additional regressor is incorporated into the model, irrespective of its values. The full model always has the greatest value of R_p^2.

Theorem 3.16 *Let $y = X_1\beta_1 + X_2\beta_2 + \epsilon = X\beta + \epsilon$ be the full model and $y = X_1\beta_1 + \epsilon$ be a submodel. Then we have*

$$R_X^2 - R_{X_1}^2 \geq 0. \tag{3.166}$$

Proof: Let

$$R_X^2 - R_{X_1} = \frac{RSS_{X_1} - RSS_X}{SYY},$$

so that the assertion (3.166) is equivalent to

$$RSS_{X_1} - RSS_X \geq 0.$$

Since

$$\begin{aligned} RSS_X &= (y - Xb)'(y - Xb) \\ &= y'y + b'X'Xb - 2b'X'y \\ &= y'y - b'X'y \end{aligned} \tag{3.167}$$

and, analogously,

$$RSS_{X_1} = y'y - \hat{\beta}_1' X_1' y ,$$

where

$$b = (X'X)^{-1}X'y$$

and

$$\hat{\beta}_1 = (X_1'X_1)^{-1}X_1'y$$

are OLS estimators in the full and in the submodel, we have

$$RSS_{X_1} - RSS_X = b'X'y - \hat{\beta}_1' X_1' y . \tag{3.168}$$

Now with (3.93)–(3.99),

$$b'X'y = (b_1', b_2') \begin{pmatrix} X_1'y \\ X_2'y \end{pmatrix}$$

$$\begin{aligned} &= (y' - b_2'X_2')X_1(X_1'X_1)^{-1}X_1'y + b_2'X_2'y \\ &= \hat{\beta}_1'X_1'y + b_2'X_2'M_1y\,. \end{aligned}$$

Thus (3.168) becomes

$$\begin{aligned} RSS_{X_1} - RSS_X &= b_2'X_2'M_1y \\ &= y'M_1X_2D^{-1}X_2'M_1y \geq 0\,, \end{aligned} \tag{3.169}$$

which proves (3.166).

On the basis of Theorem 3.16 we define the statistic

$$F\text{-change} = \frac{(RSS_{X_1} - RSS_X)/(K-p)}{RSS_X/(T-K)}\,, \tag{3.170}$$

which is distributed as $F_{K-p,T-K}$ under H_0: "submodel is valid." In model choice procedures, F-change tests for significance of the change of R_p^2 by adding additional $K-p$ variables to the submodel.

In multiple regression, the appropriate adjustment of the ordinary coefficient of determination is provided by the coefficient of determination adjusted by the degrees of freedom of the multiple model:

$$\bar{R}_p^2 = 1 - \left(\frac{T-1}{T-p}\right)(1 - R_p^2)\,. \tag{3.171}$$

Note: If there is no constant β_0 present in the model, then the numerator is T instead of $T-1$, so that $\bar{R}_p^2$ may possibly take negative values. This cannot occur when using the ordinary R_p^2.

If we consider two models, the smaller of which is supposed to be fully contained in the bigger, and we find the relation

$$\bar{R}_{p+q}^2 < \bar{R}_p^2\,,$$

then the smaller model obviously shows a better goodness of fit.

Further criteria are, for example, Mallows's C_p (cf. Weisberg, 1980, p. 188) or criteria based on the residual mean dispersion error $\hat{\sigma}_p^2 = RSS_p/(T-p)$. There are close relations between these measures.

Confidence Intervals

As in bivariate regression, there is a close relation between the region of acceptance of the F-test and confidence intervals for β in the multiple regression model.

Confidence Ellipsoids for the Whole Parameter Vector β

Considering (3.83) and (3.86), we get for $\beta^* = \beta$ a confidence ellipsoid at level $1-\alpha$:

$$\frac{(b-\beta)'X'X(b-\beta)}{(y-Xb)'(y-Xb)} \cdot \frac{T-K}{K} \leq F_{K,T-K,1-\alpha}\,. \tag{3.172}$$

Confidence Ellipsoids for Subvectors of β

From (3.110) we have

$$\frac{(b_2-\beta_2)'D(b_2-\beta_2)}{(y-Xb)'(y-Xb)} \cdot \frac{T-K}{K-s} \le F_{K-s,T-K,1-\alpha} \tag{3.173}$$

as a $(1-\alpha)$-confidence ellipsoid for β_2.

Further results may be found in Judge, Griffiths, Hill, and Lee (1980); Goldberger (1964); Pollock (1979); Weisberg (1980); and Kmenta (1971).

3.8.3 *A Complex Example*

We now want to demonstrate model choice in detail by means of the introduced criteria on the basis of a data set. Consider the following model with $K = 4$ real regressors and $T = 10$ observations:

$$y = \mathbf{1}\beta_0 + X_1\beta_1 + X_2\beta_2 + X_3\beta_3 + X_4\beta_4 + \epsilon\,.$$

The data set (y, X) is

Y	X_1	X_2	X_3	X_4
18	3	7	20	−10
47	7	13	5	19
125	10	19	−10	100
40	8	17	4	17
37	5	11	3	13
20	4	7	3	10
24	3	6	10	5
35	3	7	0	22
59	9	21	−2	35
50	10	24	0	20

The sample moments are displayed in the following table.

	Mean	Std. deviation	Variance
X_1	6.200	2.936	8.622
X_2	13.200	6.647	44.178
X_3	3.300	7.846	61.567
X_4	23.100	29.471	868.544
Y	45.500	30.924	956.278

The following matrix contains the correlations, the covariances, the one-tailed p-values of the t-tests $t_{T-2} = r\sqrt{(T-2)/(1-r^2)}$ for H_0: "correlation equals zero," and the cross-products $\sum_{t=1}^{T} X_{1t}Y_t$. For example, the upper right element has:

$$\begin{aligned} \text{Correlation}(X_1, Y) &= .740 \\ \text{Covariance}(X_1, Y) &= 67.222 \end{aligned}$$

$$
\begin{aligned}
p\text{-value} &= .007\\
\text{Cross-product} &= 605.000
\end{aligned}
$$

	X_1	X_2	X_3	X_4	Y
X_1	1.000	.971	–.668	.652	.740
	8.622	18.956	–15.400	56.422	67.222
		.000	.017	.021	.007
	77.600	170.600	–138.600	507.800	605.000
X_2	.971	1.000	–.598	.527	.628
	8.956	44.178	–31.178	103.000	129.000
	.000		.034	.059	.026
	170.600	397.600	–280.600	928.800	1161.000
X_3	–.668	-.598	1.000	–.841	–.780
	–15.400	–31.178	61.567	–194.478	–189.278
	.017	.034		.001	.004
	–138.600	–280.600	554.100	–1750.300	–1703.500
X_4	.652	.527	–.841	1.000	.978
	56.422	103.200	–194.478	868.544	890.944
	.021	.059	.001		.000
	507.800	928.800	–1750.300	7816.900	8018.500
Y	.740	.628	–.780	.978	1.000
	67.222	129.000	–189.278	890.944	956.278
	.007	.026	.004	.000	
	605.000	1161.000	–1703.500	8018.500	8606.500

We especially recognize that

- X_1 and X_2 have a significant positive correlation ($r = .971$),
- X_3 and X_4 have a significant negative correlation ($r = -.841$),
- all X-variables have a significant correlation with Y.

The significance of the correlation between X_1 and X_3 or X_4, and between X_2 and X_3 or X_4 lies between .017 and .059, which is quite large as well.

We now apply a stepwise procedure for finding the best model.

Step 1 of the Procedure

The stepwise procedure first chooses the variable X_4, since X_4 shows the highest correlation with Y (the p-values are X_4: .000, X_1: .007, X_2: .026, X_3: .004). The results of this step are listed below.

Multiple R	.97760		
R^2	.95571	R^2-change	.95571
Adjusted R^2	.95017	F-change	172.61878
Standard error	6.90290	Signif. F-change	.00000

The ANOVA table is:

	df	Sum of squares	Mean square
Regression	1	8225.29932	8225.2993
Residual	8	381.20068	47.6500

with $F = 172.61878$ (Signif. F: .0000). The determination coefficient for the model $y = \mathbf{1}\hat{\beta}_0 + X_4\hat{\beta}_4 + \epsilon$ is

$$R_2^2 = \frac{SS_{\text{Reg}}}{SYY} = \frac{8225.29932}{8225.29932 + 381.20068} = .95571\,,$$

and the adjusted determination coefficient is

$$\bar{R}_2^2 = 1 - \left(\frac{10-1}{10-2}\right)(1 - .95571) = .95017\,.$$

The table of the estimates is as follows

	$\hat{\beta}$	SE($\hat{\beta}$)	95% confidence interval lower	upper
X_4	1.025790	.078075	.845748	1.205832
Constant	21.804245	2.831568	15.274644	28.333845

Step 2 of the Procedure

Now the variable X_1 is included. The adjusted determination coefficient increases to $\bar{R}_3^2 = .96674$.

Multiple R	.98698		
R^2	.97413	R^2-change	.01842
Adjusted R^2	.96674	F-change	4.98488
Standard error	5.63975	Signif. F-change	.06070

The ANOVA table is:

	df	Sum of squares	Mean square
Regression	2	8383.85240	4191.9262
Residual	7	222.64760	31.8068

with $F = 131.79340$ (Signif. F: .0000).

Step 3 of the Procedure

Now that X_3 is included, the adjusted determination coefficient increases to $\bar{R}_4^2 = .98386$.

Multiple R	.99461		
R^2	.98924	R^2-change	.01511
Adjusted R^2	.98386	F-change	8.42848
Standard error	3.92825	Signif. F-change	.02720

The ANOVA table is:

	df	Sum of squares	Mean square
Regression	3	8513.91330	2837.9711
Residual	6	92.58670	15.4311

with $F = 183.91223$ (Signif. F: .00000).

The test statistic F-change was calculated as follows:

$$\begin{aligned} F_{1,6} &= \frac{RSS_{(X_4,X_1,1)} - RSS_{(X_4,X_1,X_3,1)}}{RSS_{(X_4,X_1,X_3,1)}/6} \\ &= \frac{222.64760 - 92.58670}{15.4311} \\ &= 8.42848. \end{aligned}$$

The 95% and 99% quantiles of the $F_{1,6}$-distribution are 5.99 and 13.71, respectively. The p-value of F-change is .0272 and lies between 1% and 5%. Hence, the increase in determination is significant on the 5% level, but not on the 1% level.

The model choice procedure stops at this point, and the variable X_2 is not taken into consideration. The model chosen is $y = \mathbf{1}\beta_0 + \beta_1 X_1 + \beta_3 X_3 + \beta_4 X_4 + \epsilon$ with the statistical quantities shown below.

			95% confidence interval	
	$\hat{\beta}$	$SE(\hat{\beta})$	lower	upper
X_4	1.079	.084	.873	1.285
X_1	2.408	.615	.903	3.913
X_3	.937	.323	.147	1.726
Constant	2.554	4.801	−9.192	14.301

The Durbin-Watson test statistic is $d = 3.14$, which exceeds d_u^*. (Table 4.1 displays values for T=15, 20, 30, ...), hence H_0: $\rho = 0$ cannot be rejected.

3.8.4 Graphical Presentation

We now want to display the structure of the (y, X)-matrix by means of the bivariate scatterplots. The plots shown in Figures 3.2 to 3.5 confirm the relation between X_1, X_2 and X_3, X_4, and the X_i and Y, but they also show the strong influence of single observations for specific data constellations. This influence is examined more closely with methods of the sensitivity analysis (Chapter 7).

The F-tests assume a normal distribution of the errors or y. This assumption is checked with the Kolmogorov-Smirnov test. The test statistic has a value of 0.77 (p-value .60). Hence, normality is not rejected at the 5% level.

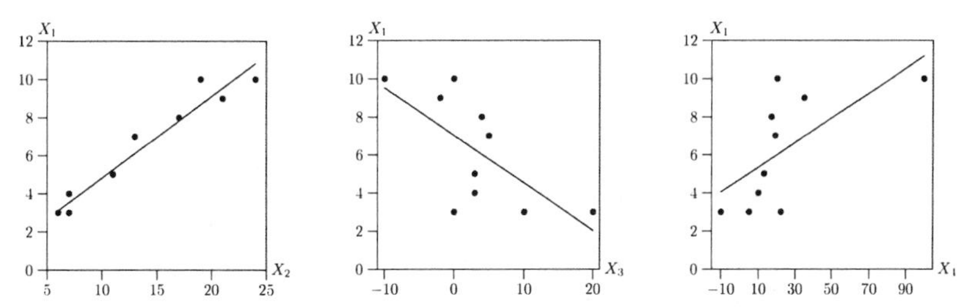

FIGURE 3.2. Scatterplots and regression for X_1 on X_2, X_3 and X_4, respectively

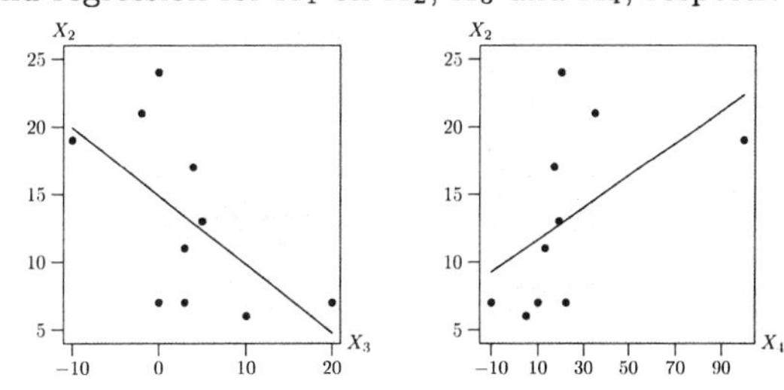

FIGURE 3.3. Scatterplots and regression for X_2 on X_3 and X_4, respectively

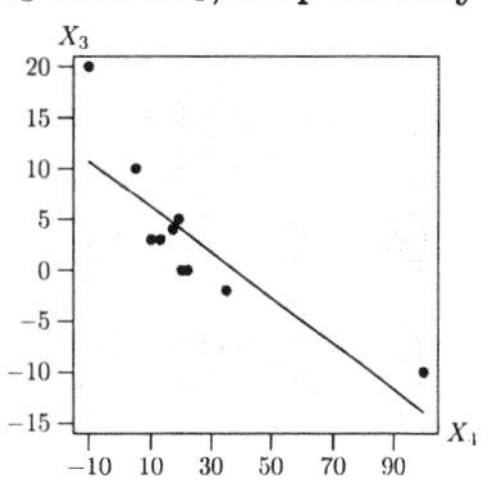

FIGURE 3.4. Scatterplot and regression for X_3 on X_4

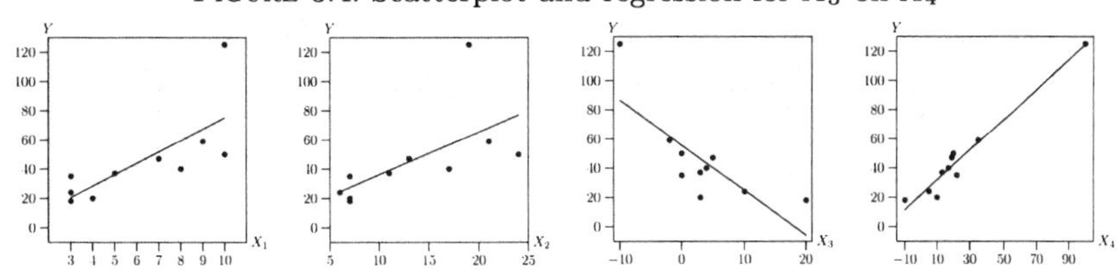

FIGURE 3.5. Scatterplot and regression for Y on X_1, X_2, X_3 and X_4, respectively

3.9 The Canonical Form

To simplify considerations about the linear model—especially when X is deficient in rank, leading to singularity of $X'X$—the so-called canonical form is frequently used (Rao, 1973a, p. 43).

The spectral decomposition (Theorem A.30) of the symmetric matrix $X'X$ is

$$X'X = P\Lambda P' \tag{3.174}$$

with $P = (p_1, \ldots, p_K)$ and $PP' = I$. Model (3.58) can then be written as

$$\begin{aligned} y &= XPP'\beta + \epsilon \\ &= \tilde{X}\tilde{\beta} + \epsilon \end{aligned} \tag{3.175}$$

with $\tilde{X} = XP$, $\tilde{\beta} = P'\beta$, and $\tilde{X}'\tilde{X} = P'X'XP = \Lambda = \text{diag}(\lambda_1, \ldots, \lambda_K)$, so that the column vectors of $\tilde{X}$ are orthogonal. The elements of $\tilde{\beta}$ are called regression parameters of the principal components.

Let $\hat{\beta} = Cy$ be a linear estimator of β with the MDE matrix $M(\hat{\beta}, \beta)$. In the transformed model we obtain for the linear estimator $P'\hat{\beta} = P'Cy$ of the parameter $\tilde{\beta} = P'\beta$

$$\begin{aligned} M(P'\hat{\beta}, \tilde{\beta}) &= \text{E}(P'\hat{\beta} - P'\beta)(P'\hat{\beta} - P'\beta)' \\ &= P'M(\hat{\beta}, \beta)P. \end{aligned} \tag{3.176}$$

Hence, relations between two estimates remain unchanged. For the scalar MDE (cf. Chapter 5) we have

$$\text{tr}\{M(P'\hat{\beta}, \tilde{\beta})\} = \text{tr}\{M(\hat{\beta}, \beta)\}, \tag{3.177}$$

so that the scalar MDE is independent of the parameterization (3.175).

For the covariance matrix of the OLS estimate b of β in the original model, we have

$$\text{V}(b) = \sigma^2 (X'X)^{-1} = \sigma^2 \sum \lambda_i^{-1} p_i p_i'. \tag{3.178}$$

The OLS estimate b^* of $\tilde{\beta}$ in the model (3.175) is

$$\begin{aligned} b^* &= (\tilde{X}'\tilde{X})^{-1}\tilde{X}'y \\ &= \Lambda^{-1}\tilde{X}'y \end{aligned} \tag{3.179}$$

with the covariance matrix

$$\text{V}(b^*) = \sigma^2 \Lambda^{-1}. \tag{3.180}$$

Hence the components of b^* are uncorrelated and have the variances $\text{var}(b_i^*) = \sigma^2 \lambda_i^{-1}$. If $\lambda_i > \lambda_j$, then $\tilde{\beta}_i$ is estimated more precisely than $\tilde{\beta}_j$:

$$\frac{\text{var}(b_i^*)}{\text{var}(b_j^*)} = \frac{\lambda_j}{\lambda_i} < 1. \tag{3.181}$$

The geometry of the reparameterized model (3.175) is examined extensively in Fomby, Hill, and Johnson (1984, pp. 289–293). Further remarks can be found in Vinod and Ullah (1981, pp. 5–8). In the case of problems concerning multicollinearity, reparameterization leads to a clear representation of dependence on the eigenvalues λ_i of $X'X$. Exact or strict multicollinearity means $|X'X| = 0$ in the original model and $|\tilde{X}'\tilde{X}| = |\Lambda| = 0$ in the reparameterized model, so that at least one eigenvalue is equal to zero. For weak multicollinearity in the sense of $|\tilde{X}'\tilde{X}| \approx 0$, the smallest eigenvalue or the so-called

$$\text{condition number} \quad k = \left(\frac{\lambda_{\max}}{\lambda_{\min}}\right)^{\frac{1}{2}} \tag{3.182}$$

is used for diagnostics (cf. Weisberg, 1985, p. 200; Chatterjee and Hadi, 1988, pp. 157–178).

Belsley, Kuh, and Welsch (1980, Chapter 3) give a detailed discussion about the usefulness of these and other measures for assessing weak multicollinearity.

3.10 Methods for Dealing with Multicollinearity

In this section we want to introduce more algebraically oriented methods: principal components regression, ridge estimation, and shrinkage estimators. Other methods using exact linear restrictions and procedures with auxiliary information are considered in Chapter 5.

3.10.1 Principal Components Regression

The starting point of this procedure is the reparameterized model (3.175)

$$y = XPP'\beta + \epsilon = \tilde{X}\tilde{\beta} + \epsilon\,.$$

Let the columns of the orthogonal matrix $P = (p_1, \ldots, p_K)$ of the eigenvectors of $X'X$ be numbered according to the magnitude of the eigenvalues $\lambda_1 \geq \lambda_2 \geq \ldots \geq \lambda_K$. Then $\tilde{x}_i = Xp_i$ is the ith principal component and we get

$$\tilde{x}_i'\tilde{x}_i = p_i'X'Xp_i = \lambda_i\,. \tag{3.183}$$

We now assume exact multicollinearity. Hence $\text{rank}(X) = K - J$ with $J \geq 1$. We get (A.31 (vii))

$$\lambda_{K-J+1} = \ldots = \lambda_K = 0\,. \tag{3.184}$$

According to the subdivision of the eigenvalues into the groups $\lambda_1 \geq \ldots \geq \lambda_{K-J} > 0$ and the group (3.184), we define the subdivision

$$P = (P_1, P_2)\,, \quad \Lambda = \begin{pmatrix} \Lambda_1 & 0 \\ 0 & 0 \end{pmatrix}, \quad \tilde{X} = (\tilde{X}_1, \tilde{X}_2) = (XP_1, XP_2)\,,$$

$$\tilde{\beta} = \begin{pmatrix} \tilde{\beta}_1 \\ \tilde{\beta}_2 \end{pmatrix} = \begin{pmatrix} P_1'\beta \\ P_2'\beta \end{pmatrix}$$

with $\tilde{X}_2 = 0$ according to (3.183). We now obtain

$$\begin{aligned} y &= \tilde{X}_1\tilde{\beta}_1 + \tilde{X}_2\tilde{\beta}_2 + \epsilon && (3.185)\\ &= \tilde{X}_1\tilde{\beta}_1 + \epsilon. && (3.186) \end{aligned}$$

The OLS estimate of the $(K-J)$-vector $\tilde{\beta}_1$ is $b_1 = (\tilde{X}_1'\tilde{X}_1)^{-1}\tilde{X}_1'y$. The OLS estimate of the full vector $\tilde{\beta}$ is

$$\begin{aligned} \begin{pmatrix} b_1 \\ 0 \end{pmatrix} &= (X'X)^{-}X'y \\ &= (P\Lambda^{-}P')X'y, \end{aligned} \tag{3.187}$$

with Theorem A.63

$$\Lambda^{-} = \begin{pmatrix} \Lambda_1^{-1} & 0 \\ 0 & 0 \end{pmatrix} \tag{3.188}$$

being a g-inverse of Λ.

Remark: The handling of exact multicollinearity by means of principal components regression corresponds to the transition from the model (3.185) to the reduced model (3.186) by putting $\tilde{X}_2 = 0$. This transition can be equivalently achieved by putting $\tilde{\beta}_2 = 0$ and hence by a linear restriction

$$0 = (0, I)\begin{pmatrix} \tilde{\beta}_1 \\ \tilde{\beta}_2 \end{pmatrix}.$$

The estimate b_1 can hence be represented as a restricted OLS estimate (cf. Section 5.2).

A cautionary note on PCR. In practice, zero eigenvalues can be destinguished only by the small magnitudes of the observed eigenvalues. Then, one may be tempted to omit all the principal components with the corresponding eigenvalues below a certain threshold value. But then, there is a possibility that a principal component with a small eigenvalue is a good predictor of the response variable and its omission may decrease the efficiency of prediction drastically.

3.10.2 Ridge Estimation

In case of $\text{rank}(X) = K$, the OLS estimate has the minimum-variance property in the class of all unbiased, linear, homogeneous estimators. Let $\lambda_1 \geq \lambda_2 \geq \ldots \geq \lambda_K$ denote the eigenvalues of S. Then we have for the scalar MDE of b

$$\text{tr}\{M(b,\beta)\} = \text{tr}\{\text{V}(b)\} = \sigma^2 \sum_{i=1}^{K} \lambda_i^{-1}. \tag{3.189}$$

In the case of weak multicollinearity, at least one eigenvalue λ_i is relatively small, so that $\text{tr}\{\text{V}(b)\}$ and the variances of all components b_j of $b = (b_1, \ldots, b_K)'$ are large:

$$\begin{aligned} b_j &= e_j' b, \\ \text{var}(b_j) &= e_j' \text{V}(b) e_j, \quad \text{and, hence,} \end{aligned}$$

$$\begin{aligned} \mathrm{var}(b_j) &= \sigma^2 \sum_{i=1}^{K} \lambda_i^{-1} e_j' p_i p_i' e_j \\ &= \sigma^2 \sum_{i=1}^{K} \lambda_i^{-1} p_{ij}^2 \end{aligned} \tag{3.190}$$

with the jth unit vector e_j and the ith eigenvector $p_i' = (p_{i1}, \ldots, p_{ij}, \ldots, p_{iK})$.

The scalar MDE

$$\mathrm{tr}\{M(b, \beta)\} = \mathrm{E}(b - \beta)'(b - \beta)$$

can be interpreted as the mean Euclidean distance between the vectors b and β, hence multicollinearity means a global unfavorable distance to the real parameter vector. Hoerl and Kennard (1970) used this interpretation as a basis for the definition of the ridge estimate

$$b(k) = (X'X + kI)^{-1} X'y, \tag{3.191}$$

with $k \geq 0$, the nonstochastic quantity, being the control parameter. Of course, $b(0) = b$ is the ordinary LS estimate.

Using the abbreviation

$$G_k = (X'X + kI)^{-1}, \tag{3.192}$$

$\mathrm{Bias}(b(k), \beta)$ and $\mathrm{V}(b(k))$ can be expressed as follows:

$$\mathrm{E}(b(k)) = G_k X'X\beta = \beta - kG_k\beta, \tag{3.193}$$

$$\mathrm{Bias}(b(k), \beta) = -kG_k\beta, \tag{3.194}$$

$$\mathrm{V}(b(k)) = \sigma^2 G_k X'X G_k. \tag{3.195}$$

Hence the MDE matrix is

$$M(b(k), \beta) = G_k(\sigma^2 X'X + k^2\beta\beta')G_k \tag{3.196}$$

and using $X'X = P\Lambda P'$, we get

$$\mathrm{tr}\{M(b(k), \beta)\} = \sum_{i=1}^{K} \frac{\sigma^2\lambda_i + k^2\beta_i^2}{(\lambda_i + k)^2} \tag{3.197}$$

(cf. Goldstein and Smith, 1974).

Proof: Let $X'X = P\Lambda P'$ be the spectral decomposition of $X'X$. We then have (Theorems A.30, A.31)

$$\begin{aligned} X'X + kI = G_k^{-1} &= P(\Lambda + kI)P', \\ G_k &= P(\Lambda + kI)^{-1}P', \end{aligned}$$

and in general

$$\mathrm{tr}\{\mathrm{diag}(l_1, \cdots, l_k)\beta\beta' \,\mathrm{diag}(l_1, \cdots, l_k)\} = \sum \beta_i^2 l_i^2 .$$

With $l_i = (\lambda_i + k)^{-1}$, we obtain relation (3.197).

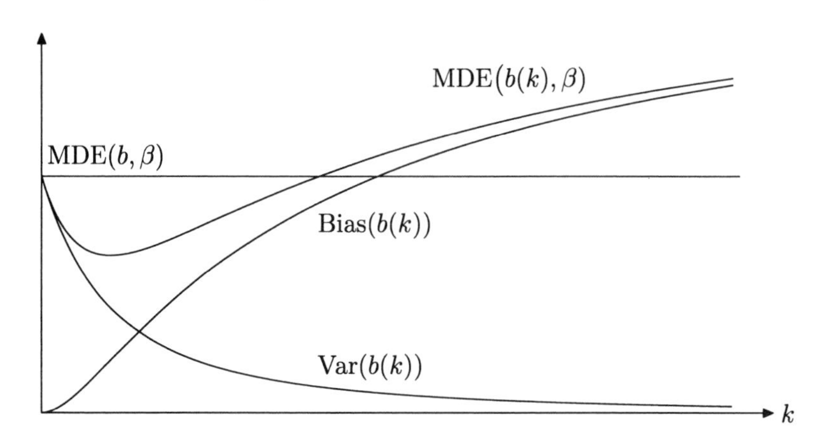

FIGURE 3.6. Scalar MDE function for $b = (X'X)^{-1}X'y$ and $b(k) = G_k X'y$ in dependence on k for $K = 1$

The scalar MDE of $b(k)$ for fixed σ^2 and a fixed vector β is a function of the ridge parameter k, which starts at $\sum \sigma^2/\lambda_i = \text{tr}\{\text{V}(b)\}$ for $k = 0$, takes its minimum for $k = k_{\text{opt}}$ and then it increases monotonically, provided that $k_{\text{opt}} < \infty$ (cf. Figure 3.6).

We now transform $M(b, \beta) = M(b) = \sigma^2(X'X)^{-1}$ as follows:

$$\begin{aligned} M(b) &= \sigma^2 G_k (G_k^{-1}(X'X)^{-1}G_k^{-1})G_k \\ &= \sigma^2 G_k (X'X + k^2(X'X)^{-1} + 2kI)G_k\,. \end{aligned} \tag{3.198}$$

From Definition 3.10 we obtain the interval $0 < k < k^*$ in which the ridge estimator is MDE-I-superior to the OLS b, according to

$$\begin{aligned} \Delta(b, b(k)) &= M(b) - M(b(k), \beta) \\ &= kG_k[\sigma^2(2I + k(X'X)^{-1}) - k\beta\beta']G_k. \end{aligned} \tag{3.199}$$

Since $G_k > 0$, we have $\Delta(b, b(k)) \geq 0$ if and only if

$$\sigma^2(2I + k(X'X)^{-1}) - k\beta\beta' \geq 0\,, \tag{3.200}$$

or if the following holds (Theorem A.57):

$$\sigma^{-2}k\beta'(2I + k(X'X)^{-1})^{-1}\beta \leq 1\,. \tag{3.201}$$

As a sufficient condition for (3.200), independent of the model matrix X, we obtain

$$2\sigma^2 I - k\beta\beta' \geq 0 \tag{3.202}$$

or—according to Theorem A.57—equivalently,

$$k \leq \frac{2\sigma^2}{\beta'\beta}\,. \tag{3.203}$$

The range of k, which ensures the MDE-I superiority of $b(k)$ compared to b, is dependent on $\sigma^{-1}\beta$ and hence unknown.

If auxiliary information about the length (norm) of β is available in the form

$$\beta'\beta \leq r^2 , \tag{3.204}$$

then

$$k \leq \frac{2\sigma^2}{r^2} \tag{3.205}$$

is sufficient for (3.203) to be valid. Hence possible values for k, in which $b(k)$ is better than b, can be found by estimation of σ^2 or by specification of a lower limit or by a combined a priori estimation $\sigma^{-2}\beta'\beta \leq \tilde{r}^2$.

Swamy, Mehta, and Rappoport (1978) and Swamy and Mehta (1977) investigated the following problem:

$$\min_{\beta}\{\sigma^{-2}(y - X\beta)'(y - X\beta)|\beta'\beta \leq r^2\} .$$

The solution of this problem

$$\hat{\beta}(\mu) = (X'X + \sigma^2\mu I)^{-1}X'y , \tag{3.206}$$

is once again a ridge estimate and $\hat{\beta}'(\mu)\hat{\beta}(\mu) = r^2$ is fulfilled. Replacing σ^2 by the estimate s^2 provides a practical solution for the estimator (3.206) but its properties can be calculated only approximately.

Hoerl and Kennard (1970) derived the ridge estimator by the following reasoning. Let $\hat{\beta}$ be any estimator and $b = (X'X)^{-1}X'y$ the OLS. Then the error sum of squares estimated with $\hat{\beta}$ can be expressed, according to the property of optimality of b, as

$$\begin{aligned} S(\hat{\beta}) &= (y - X\hat{\beta})'(y - X\hat{\beta}) \\ &= (y - Xb)'(y - Xb) + (b - \hat{\beta})'X'X(b - \hat{\beta}) \\ &= S(b) + \Phi(\hat{\beta}) , \end{aligned} \tag{3.207}$$

since the term

$$\begin{aligned} 2(y - Xb)'X(b - \hat{\beta}) &= 2y'(I - X(X'X)^{-1}X')X(b - \hat{\beta}) \\ &= 2MX(b - \hat{\beta}) = 0 \end{aligned}$$

since $MX = 0$.

Let $\Phi_0 > 0$ be a fixed given value for the error sum of squares. Then a set $\{\hat{\beta}\}$ of estimates exists that fulfill the condition $S(\hat{\beta}) = S(b) + \Phi_0$. In this set $\{\hat{\beta}\}$ we look for the estimate $\hat{\beta}$ with minimal length:

$$\min_{\hat{\beta}} \{\hat{\beta}'\hat{\beta} + \frac{1}{k}[(b - \hat{\beta})'X'X(b - \hat{\beta}) - \Phi_0]\}, \tag{3.208}$$

where $1/k$ is a Lagrangian multiplier. Differentiation of this function with respect to $\hat{\beta}$ and $1/k$ leads to the normal equations

$$\hat{\beta} + \frac{1}{k}(X'X)(\hat{\beta} - b) = 0 ,$$

and hence

$$\begin{aligned}\hat{\beta} &= (X'X + kI)^{-1}(X'X)b \\ &= G_k X'y\,, \end{aligned} \tag{3.209}$$

as well as

$$\Phi_0 = (b - \hat{\beta})'X'X(b - \hat{\beta})\,. \tag{3.210}$$

Hence, the solution of the problem (3.208) is the ridge estimator $\hat{\beta} = b(k)$ (3.209). The ridge parameter k is to be determined iteratively so that (3.210) is fulfilled.

For further representations about ridge regression see Vinod and Ullah (1981) and Trenkler and Trenkler (1983).

3.10.3 Shrinkage Estimates

Another class of biased estimators, which was very popular in research during the 1970s, is defined by the so-called shrinkage estimator (Mayer and Wilke, 1973):

$$\hat{\beta}(\rho) = (1+\rho)^{-1}b\,, \quad \rho \geq 0 \quad (\rho \text{ known})\,, \tag{3.211}$$

which "shrinks" the OLS estimate:

$$\begin{aligned}\mathrm{E}\left(\hat{\beta}(\rho)\right) &= (1+\rho)^{-1}\beta, \\ \mathrm{Bias}\left(\hat{\beta}(\rho), \beta\right) &= -\rho(1+\rho)^{-1}\beta\,, \\ \mathrm{V}\left(\hat{\beta}(\rho)\right) &= \sigma^2(1+\rho)^{-2}(X'X)^{-1}\,, \end{aligned}$$

and

$$M\left(\hat{\beta}(\rho), \beta\right) = (1+\rho)^{-2}(\mathrm{V}(b) + \rho^2\beta\beta')\,. \tag{3.212}$$

The MDE-I comparison with the OLS leads to

$$\Delta(b, \hat{\beta}(\rho)) = (1+\rho)^{-2}\rho\sigma^{-2}\left[(\rho+2)(X'X)^{-1} - \sigma^{-2}\rho\beta\beta'\right] \geq 0$$

if and only if (Theorem A.57)

$$\frac{\sigma^{-2}\rho}{(\rho+2)}\beta'X'X\beta \leq 1\,.$$

Then

$$\sigma^{-2}\beta'X'X\beta \leq 1 \tag{3.213}$$

is a sufficient condition for the MDE-I superiority of $\hat{\beta}(\rho)$ compared to b.

This form of restriction will be used as auxiliary information for the derivation of minimax-linear estimates in Section 3.13.

Note: Results about the shrinkage estimator in the canonical model can be found in Farebrother (1978).

3.10.4 Partial Least Squares

Univariate partial least squares is a particular method of analysis in models with possibly more explanatory variables than samples. In spectroscopy one aim may be to predict a chemical composition from spectra of some material. If all wavelengths are considered as explanatory variables, then traditional stepwise OLS procedure soon runs into collinearity problems caused by the number of explanatory variables and their interrelationships (cf. Helland, 1988).

The aim of partial least squares is to predict the response by a model that is based on linear transformations of the explanatory variables. Partial least squares (PLS) is a method of constructing regression models of type

$$\hat{Y} = \beta_0 + \beta_1 T_1 + \beta_2 T_2 + \cdots + \beta_p T_p \,, \tag{3.214}$$

where the T_i are linear combinations of the explanatory variables $X_1, X_2, \ldots, X_K$ such that the sample correlation for any pair T_i, T_j $(i \neq j)$ is 0. We follow the procedure given by Garthwaite (1994). First, all the data are centered. Let $\bar{y}, \bar{x_1}, \ldots, \bar{x_k}$ denote the sample means of the columns of the $T \times (K+1)$-data matrix

$$(y, X) = (y_1, x_1, \ldots, x_k)\,,$$

and define the variables

$$U_1 = Y - \bar{x_i}\,, \tag{3.215}$$

$$V_{1i} = X_i - \bar{x_i} \quad (i = 1, \ldots, K)\,. \tag{3.216}$$

Then the data values are the T-vectors

$$u_1 = y - \bar{y}1\,, \quad (\bar{u}_1 = 0)\,, \tag{3.217}$$

$$v_{i1} = x_i - \bar{x}_i 1\,, \quad (\bar{v}_{1i} = 0)\,. \tag{3.218}$$

The linear combinations T_j, called factors, latent variables, or *components*, are then determined sequentially. The procedure is as follows:

(i) U_1 is first regressed against V_{11}, then regressed against V_{12}, ..., then regressed against V_{1K}. The K univariate regression equations are

$$\hat{U}_{1i} = b_{1i} V_{1i} \quad (i = 1, \ldots, K)\,, \tag{3.219}$$

$$\text{where} \quad b_{1i} = \frac{v'_{1i} u_1}{v'_{1i} v_{1i}}\,. \tag{3.220}$$

Then each of the K equations in (3.220) provides an estimate of U_1. To have one resulting estimate, one may use a simple average $\sum_{i=1}^{K} b_{1i} V_{1i} / K$ or a weighted average such as

$$T_1 = \sum_{i=1}^{K} w_{1i} b_{1i} V_{1i} \tag{3.221}$$

with the data value

$$t_1 = \sum_{i=1}^{K} w_{1i} b_{1i} v_{1i} \,. \tag{3.222}$$

(ii) The variable T_1 should be a useful predictor of U_1 and hence of Y. The information in the variable X_i that is not in T_1 may be estimated by the residuals from a regression of X_i on T_1, which are identical to the residuals, say Y_{2i}, if V_{1i} is regressed on T_1, that is,

$$V_{2i} = V_{1i} - \frac{t_1' v_{1i}}{t_1' t_1} T_1 \,. \tag{3.223}$$

To estimate the amount of variability in Y that is not explained by the predictor T_1, one may regress U_1 on T_1 and take the residuals, say U_2.

(iii) Define now the individual predictors

$$\hat{U}_{2i} = b_{2i} V_{2i} \quad (i = 1, \ldots, K) \,, \tag{3.224}$$

where

$$b_{2i} = \frac{v_{2i}' u_2}{v_{2i}' v_{2i}} \tag{3.225}$$

and the weighted average

$$T_2 = \sum_{i=1}^{K} w_{2i} b_{2i} V_{2i} \,. \tag{3.226}$$

(iv) *General iteration step.* Having performed this algorithm k times, the remaining residual variability in Y is U_{k+1} and the residual information in X_i is $V_{(k+1)i}$, where

$$U_{k+1} = U_k - \frac{t_k' u_k}{t_k' t_k} T_k \tag{3.227}$$

and

$$V_{(k+1)i} = V_{ki} - \frac{t_k' v_{ki}}{t_k' t_k} T_k \,. \tag{3.228}$$

Regressing U_{k+1} against $V_{(k+1)i}$ for $I = 1, \ldots, K$ gives the individual predictors

$$\hat{U}_{(k+1)i} = b_{(k+1)i} V_{(k+1)i} \tag{3.229}$$

with

$$b_{(k+1)i} = \frac{v_{(k+1)i}' u_{k+1}}{v_{(k+1)i}' v_{(k+1)i}}$$

and the $(k+1)$th component

$$T_{k+1} = \sum_{i=1}^{K} w_{(k+1)i} b_{(k+1)i} V_{(k+1)i} \,. \tag{3.230}$$

(v) Suppose that this process has stopped in the pth step, resulting in the PLS regression model given in (3.214). The parameters $\beta_0, \beta_1, \ldots, \beta_p$ are estimated by univariate OLS. This can be proved as follows. In matrix notation we may define

$$V_{(k)} = (V_{k1}, \ldots, V_{kK}) \quad (k = 1, \ldots, p) \,, \tag{3.231}$$

$$\hat{U}_{(k)} = (b_{k1} V_{k1}, \ldots, b_{kK} V_{kK}) \quad (k = 1, \ldots, p) \,, \tag{3.232}$$

$$w_{(k)} = (w_{k1}, \ldots, w_{kK})' \quad (k = 1, \ldots, p) \,, \tag{3.233}$$

$$T_{(k)} = \hat{U}_{(k)} w_{(k)} \quad (k = 1, \ldots, p) \,, \tag{3.234}$$

$$V_{(k)} = V_{(k-1)} - \frac{v'_{(k-1)} t_{k-1}}{t'_{k-1} t_{k-1}} T_{k-1} \,. \tag{3.235}$$

By construction (cf. (3.228)) the sample residuals $v_{(k+1)i}$ are orthogonal to $v_{ki}, v_{(k-1)i}, \ldots, v_{1i}$, implying that $v'_{(k)} v_{(j)} = 0$ for $k \neq j$, hence, $\hat{u}'_{(k)} \hat{u}_{(j)} = 0$ for $k \neq j$, and finally,

$$t'_k t_j = 0 \quad (k \neq j) \,. \tag{3.236}$$

This is the well-known feature of the PLS (cf. Wold, Wold, Dunn, and Ruhe, 1984; Helland, 1988) that the sample components t_i are pairwise uncorrelated. The simple consequence is that parameters β_k in equation (3.214) may be estimated by simple univariate regressions of Y against T_k. Furthermore, the preceding estimates $\hat{\beta}_k$ stay unchanged if a new component is added.

Specification of the Weights

In the literature, two weighting policies are discussed. First, one may set $w_{ij} = 1/K$ to give each predictor $\hat{U}_{ki}$ $(i = 1, \ldots, K)$ the same weight in any kth step. The second policy in practice is the choice

$$w_{ki} = v'_{ki} v_{ki} \quad (\text{for all } k, i) \,. \tag{3.237}$$

As $\bar{v}_{ki} = 0$, the sample variance of V_{ki} is $\widehat{\text{var}}(V_{ki}) = v'_{ki} v_{ki} / (T - 1)$. Using w_{ki} defined in (3.237) gives $w_{ki} b_{ki} = v'_{ki} u_k$ and

$$T_k = \sum_{i=1}^{K} (v'_{ki} u_k) V_{(k)i} \,. \tag{3.238}$$

The T-vector v_{ki} is estimating the amount of information in X_i that was not included in the preceding component T_{k-1}. Therefore, its vector norm $v'_{ki} v_{ki}$ is a measure for the contribution of X_i to T_k.

Size of the Model

Deciding the number of components (p) usually is done via some *cross-validation* (Stone, 1974; Geisser, 1974). The data set is divided into groups. At each step k, the model is fitted to the data set reduced by one of the groups. Predictions are calculated for the deleted data, and the sum of squares of predicted minus observed values for the deleted data is calculated. Next, the second data group is left out, and so on, until each data point has been left out once and only once. The total sum of squares (called PRESS) of predictions minus observations is a measure of the predictive power of the kth step of the model. If for a chosen constant

$$\text{PRESS}_{(k+1)} - \text{PRESS}_{(k)} < \text{constant},$$

then the procedure stops. In simulation studies, Wold et al. (1984) and Garthwaite (1994) have compared the predictive power of PLS, stepwise OLS, principal components estimator (PCR), and other methods. They found PLS to be better than OLS and PCR and comparable to, for example, ridge regression.

Multivariate extension of PLS is discussed by Garthwaite (1994). Helland (1988) has discussed the equivalence of alternative univariate PLS algorithms.

3.11 Projection Pursuit Regression

The term *projection pursuit* (Friedman and Tukey, 1974) describes a technique for the exploratory analysis of multivariate data. This method searches for interesting linear projections of a multivariate data set onto a linear subspace, such as, for example, a plane or a line. These low-dimensional orthogonal projections are used to reveal the structure of the high-dimensional data set.

Projection pursuit regression (PPR) constructs a model for the regression surface $y = f(X)$ using projections of the data onto planes that are spanned by the variable y and a linear projection $a'X$ of the independent variables in the direction of the vector a. Then one may define a function of merit (Friedman and Stuetzle, 1981) or a projection index (Friedman and Tukey, 1974; Jones and Sibson, 1987) $I(a)$ depending on a. Projection pursuit attempts to find directions a that give (local) optima of $I(a)$. The case $a = 0$ is excluded, and a is constrained to be of unit length (i.e., any a is scaled by dividing by its length).

In linear regression the response surface is assumed to have a known functional form whose parameters have to be estimated based on a sample (y_t, x_t'). The PPR procedure models the regression surface iteratively as a sum of smooth functions of linear combinations $a'X$ of the predictors, that is, the regression surface is approximated by a sum of smooth functions

$$\phi(x) = \sum_{m=1}^{M} S_{a_m}(a_m' X) \tag{3.239}$$

(M is the counter of the runs of iteration). The algorithm is as follows (Friedman and Stuetzle, 1981):

(i) Collect a sample (y_t, x_t'), $t = 1, \ldots, T$, and assume the y_t to be centered.

(ii) Initialize residuals $r_t = y_t$, $t = 1, \ldots, T$ and set counter $M = 0$.

(iii) Choose a vector a and project the predictor variables onto one dimension $z_t = a' x_t$, $t = 1, \ldots, T$, and calculate a univariate nonparametric regression $S_a(a' x_t)$ of current residuals r_t on z_t as ordered in ascending values of z_t. These nonparametric functions are based on local averaging such as

$$S(z_t) = \text{AVE}(y_i)\,, \quad j - k \le i \le j + k\,, \tag{3.240}$$

where k defines the bandwidth of the smoother.

(iv) Define as a function of merit $I(a)$, for example, the fraction of unexplained variance

$$I(a) = 1 - \sum_{t=1}^{T} \frac{(r_t - S_a(a' x_t))^2}{\sum_{t=1}^{T} r_t^2}\,. \tag{3.241}$$

(v) Optimize $I(a)$ over the direction a.

(vi) Stop if $I(a) \le \epsilon$ (a given lower bound of smoothness). Otherwise update as follows:

$$\begin{aligned} r_t &\leftarrow r_t - S_a^M(a_M' x_t)\,, \quad t = 1, \ldots, T\,, \\ M &\leftarrow M + 1\,. \end{aligned} \tag{3.242}$$

Interpretation: The PPR algorithm may be seen to be a successive refinement of smoothing the response surface by adding the optimal smoother $S_a^M(a' X)$ to the current model.

Remark: Huber (1985) and Jones and Sibson (1987) have included projection pursuit regression in a general survey of attempts at getting *interesting* projections of high-dimensional data and nonparametric fittings such as principal components, multidimensional scaling, nonparametric regression, and density estimation.

3.12 Total Least Squares

In contrast to our treatment in the other chapters, we now change assumptions on the independent variables, that is, we allow the X_i to be measured with errors also. The method of fitting such models is known as orthogonal regression or errors-in-variables regression, also called total least squares. The idea is as follows (cf. van Huffel and Zha, 1993).

Consider an overdetermined set of $m > n$ linear equations in n unknowns x $(A : m \times n,\ x : n \times 1,\ a : m \times 1)$

$$Ax = a\,. \tag{3.243}$$

Then the ordinary least-squares problem may be written as

$$\min_{\hat{a} \in \mathbb{R}^m} \|a - \hat{a}\|_2 \quad \text{subject to} \quad \hat{a} \in \mathcal{R}(A)\,, \tag{3.244}$$

where $\|x\|_2$ is the L_2-norm or Euclidean norm of a vector x. Let $\hat{a}$ be a solution of (3.244), then any vector x satisfying $Ax = \hat{a}$ is called a LS solution (LS = least squares). The difference

$$\Delta a = a - \hat{a} \tag{3.245}$$

is called the LS correction. The assumptions are that errors occur only in the vector a and that A is exactly known.

If we also allow for perturbations in A, we are led to the following definition.

The *total least-squares (TLS) problem* for solving an overdetermined linear equation $Ax = a$ is defined by

$$\min_{(\hat{A}, \hat{a}) \in \mathbb{R}^{m \times (n+1)}} \|(A, a) - (\hat{A}, \hat{a})\|_F \tag{3.246}$$

subject to

$$\hat{a} \in R(\hat{A})\,, \tag{3.247}$$

where

$$\|M\|_F = [\text{tr}(MM')]^{\frac{1}{2}} \tag{3.248}$$

is the Frobenius norm of a matrix M.

If a minimizer $(\hat{A}, \hat{a})$ is found, then any x satisfying $\hat{A}x = \hat{a}$ is called a TLS solution, and

$$[\Delta\hat{A}, \Delta\hat{a}] = (A, a) - (\hat{A}, \hat{a}) \tag{3.249}$$

is called the TLS correction.

Indeed, the TLS problem is more general than the LS problem, for the TLS solution is obtained by approximating the columns of the matrix A by $\hat{A}$ and a by $\hat{a}$ until $\hat{a}$ is in the space $\mathcal{R}(\hat{A})$ and $\hat{A}x = \hat{a}$.

Basic Solution to TLS

We rewrite $Ax = a$ as

$$(A,a)\begin{pmatrix} x \\ -1 \end{pmatrix} = 0\,. \tag{3.250}$$

Let the singular value decomposition (SVD; cf. Theorem A.32) of the $(m, n+1)$-matrix (A, a) be

$$\begin{aligned}(A,a) &= ULV' \\ &= \sum_{i=1}^{n+1} l_i u_i v_i'\,, \end{aligned} \tag{3.251}$$

where $l_1 \geq \ldots \geq l_{n+1} \geq 0$. If $l_{n+1} \neq 0$, then (A,a) is of rank $n+1$, $\mathcal{R}\big((A,a)'\big) = \mathbb{R}^{n+1}$, and (3.250) has no solution.

Lemma 3.17 (Eckart-Young-Mirsky matrix approximation theorem) *Let A : $n \times n$ be a matrix of* $\operatorname{rank}(A) = r$, *and let $A = \sum_{i=1}^{r} l_i u_i v_i'$, $l_i > 0$, be the singular value decomposition of A. If $k < r$ and $A_k = \sum_{i=1}^{k} l_i u_i v_i'$, then*

$$\min_{\operatorname{rank}(\hat{A})=k} \|A - \hat{A}\|_2 = \|A - \hat{A}_k\|_2 = l_{k+1}$$

and

$$\min_{\operatorname{rank}(\hat{A})=k} \|A - \hat{A}\|_F = \|A - \hat{A}_k\|_F = \sqrt{\sum_{i=k+1}^{p} l_i^2}\,,$$

where $p = \min(m, n)$.

Proof: See Eckart and Young (1936), Mirsky (1960), Rao (1979; 1980).

Based on this theorem, the best rank n approximation $(\hat{A}, \hat{a})$ of (A, a) in the sense of minimal deviation in variance is given by

$$(\hat{A}, \hat{a}) = U\hat{L}V'\,, \quad \text{where} \quad \hat{L} = (l_1, \ldots, l_n, 0)\,. \tag{3.252}$$

The minimal TLS correction is then given by

$$l_{n+1} = \min_{\operatorname{rank}(\hat{A},\hat{a})=n} \|(A,a) - (\hat{A},\hat{a})\|_F\,. \tag{3.253}$$

So we have

$$(A,a) - (\hat{A},\hat{a}) = (\Delta\hat{A}, \Delta\hat{a}) = l_{n+1} u_{n+1} v_{n+1}'\,. \tag{3.254}$$

Then the approximate equation (cf. (3.250))

$$(\hat{A},\hat{a})\begin{pmatrix} x \\ -1 \end{pmatrix} = 0 \tag{3.255}$$

is compatible and has solution

$$\begin{pmatrix} \hat{x} \\ -1 \end{pmatrix} = \frac{-1}{v_{n+1,n+1}} v_{n+1}\,, \tag{3.256}$$

where $v_{n+1,n+1}$ is the $(n+1)$th component of the vector v_{n+1}. Finally, $\hat{x}$ is solution of the TLS equation $\hat{A}x = \hat{a}$.

On the other hand, if l_{n+1} is zero, then $\text{rank}(A, a) = n$, $v_{n+1} \in \mathcal{N}\{(A, a)\}$, and the vector $\hat{x}$ defined in (3.256) is the exact solution of $Ax = a$.

3.13 Minimax Estimation

3.13.1 Inequality Restrictions

Minimax estimation is based on the idea that the quadratic risk function for the estimate $\hat{\beta}$ is not minimized over the entire parameter space $\mathbb{R}^K$, but only over an area $B(\beta)$ that is restricted by a priori knowledge. For this, the supremum of the risk is minimized over $B(\beta)$ in relation to the estimate (minimax principle).

In many of the models used in practice, the knowledge of a priori restrictions for the parameter vector β may be available in a natural way. Stahlecker (1987) shows a variety of examples from the field of economics (such as input-output models), where the restrictions for the parameters are so-called workability conditions of the form $\beta_i \geq 0$ or $\beta_i \in (a_i, b_i)$ or $\text{E}(y_t|X) \leq a_t$ and more generally

$$A\beta \leq a \,. \tag{3.257}$$

Minimization of $S(\beta) = (y - X\beta)'(y - X\beta)$ under inequality restrictions can be done with the simplex algorithm. Under general conditions we obtain a numerical solution. The literature deals with this problem under the generic term *inequality restricted least squares* (cf. Judge and Takayama, 1966; Dufour, 1989; Geweke, 1986; Moors and van Houwelingen, 1987). The advantage of this procedure is that a solution $\hat{\beta}$ is found that fulfills the restrictions. The disadvantage is that the statistical properties of the estimates are not easily determined and no general conclusions about superiority can be made. If all restrictions define a convex area, this area can often be enclosed in an ellipsoid of the following form:

$$B(\beta) = \{\beta : \beta' T \beta \leq k\} \tag{3.258}$$

with the origin as center point or in

$$B(\beta, \beta_0) = \{\beta : (\beta - \beta_0)' T (\beta - \beta_0) \leq k\} \tag{3.259}$$

with the center point vector β_0.

For example, (3.257) leads to $\beta' A' A \beta \leq a^2$, and hence to the structure $B(\beta)$.

Inclusion of Inequality Restrictions in an Ellipsoid

We assume that for all components β_i of the parameter vector β, the following restrictions in the form of intervals are given a priori:

$$a_i \le \beta_i \le b_i \quad (i = 1, \dots, K)\,. \tag{3.260}$$

The empty restrictions ($a_i = -\infty$ and $b_i = \infty$) may be included. The limits of the intervals are known. The restrictions (3.260) can alternatively be written as

$$\frac{|\beta_i - (a_i + b_i)/2|}{1/2(b_i - a_i)} \le 1 \quad (i = 1, \dots, K)\,. \tag{3.261}$$

We now construct an ellipsoid $(\beta - \beta_0)'T(\beta - \beta_0) = 1$, which encloses the cuboid (3.261) and fulfills the following conditions:

(i) The ellipsoid and the cuboid have the same center point, $\beta_0 = \frac{1}{2}(a_1 + b_1, \dots, a_K + b_K)$.

(ii) The axes of the ellipsoid are parallel to the coordinate axes, that is, $T = \operatorname{diag}(t_1, \dots, t_K)$.

(iii) The corner points of the cuboid are on the surface of the ellipsoid, which means we have

$$\sum_{i=1}^{K} \left(\frac{a_i - b_i}{2}\right)^2 t_i = 1\,. \tag{3.262}$$

(iv) The ellipsoid has minimal volume:

$$V = c_K \prod_{i=1}^{K} t_i^{-\frac{1}{2}}, \tag{3.263}$$

with c_K being a constant dependent on the dimension K.

We now include the linear restriction (3.262) for the t_i by means of Lagrangian multipliers λ and solve (with $c_K^{-2} V_K^2 = \prod t_i^{-1}$)

$$\min_{\{t_i\}} \tilde{V} = \min_{\{t_i\}} \left\{ \prod_{i=1}^{K} t_i^{-1} - \lambda \left[\sum_{i=1}^{K} \left(\frac{a_i - b_i}{2}\right)^2 t_i - 1 \right] \right\}. \tag{3.264}$$

The normal equations are then

$$\frac{\partial \tilde{V}}{\partial t_j} = -t_j^{-2} \prod_{i \ne j} t_i^{-1} - \lambda \left(\frac{a_j - b_j}{2}\right)^2 = 0 \tag{3.265}$$

and

$$\frac{\partial \tilde{V}}{\partial \lambda} = \sum \left(\frac{a_i - b_i}{2}\right)^2 t_i - 1 = 0\,. \tag{3.266}$$

From (3.265) we get

$$\begin{aligned}\lambda &= -t_j^{-2}\prod_{i\neq j} t_i^{-1}\left(\frac{2}{a_j-b_j}\right)^2 \quad \text{(for all } j=1,\dots,K)\\ &= -t_j^{-1}\prod_{i=1}^{K} t_i^{-1}\left(\frac{2}{a_j-b_j}\right)^2,\end{aligned} \tag{3.267}$$

and for any two i, j we obtain

$$t_i\left(\frac{a_i-b_i}{2}\right)^2 = t_j\left(\frac{a_j-b_j}{2}\right)^2, \tag{3.268}$$

and hence—after summation—according to (3.266),

$$\sum_{i=1}^{K}\left(\frac{a_i-b_i}{2}\right)^2 t_i = Kt_j\left(\frac{a_j-b_j}{2}\right)^2 = 1. \tag{3.269}$$

This leads to the required diagonal elements of T:

$$t_j = \frac{4}{K}(a_j-b_j)^{-2} \quad (j=1,\dots,K).$$

Hence, the optimal ellipsoid $(\beta-\beta_0)'T(\beta-\beta_0)=1$, which contains the cuboid, has the center point vector

$$\beta_0' = \frac{1}{2}(a_1+b_1,\dots,a_K+b_K) \tag{3.270}$$

and the following matrix, which is positive definite for finite limits a_i, b_i $(a_i \neq b_i)$,

$$T = \text{diag}\frac{4}{K}\left((b_1-a_1)^{-2},\dots,(b_K-a_K)^{-2}\right). \tag{3.271}$$

Interpretation: The ellipsoid has a larger volume than the cuboid. Hence, the transition to an ellipsoid as a priori information represents a weakening, but comes with an easier mathematical handling.

Example 3.1: (Two real regressors) The center-point equation of the ellipsoid is (cf. Figure 3.7)

$$\frac{x^2}{a^2}+\frac{y^2}{b^2}=1,$$

or

$$(x,y)\begin{pmatrix}\frac{1}{a^2} & 0\\ 0 & \frac{1}{b^2}\end{pmatrix}\begin{pmatrix}x\\ y\end{pmatrix}=1$$

with

$$T = \text{diag}\left(\frac{1}{a^2},\frac{1}{b^2}\right) = \text{diag}(t_1,t_2)$$

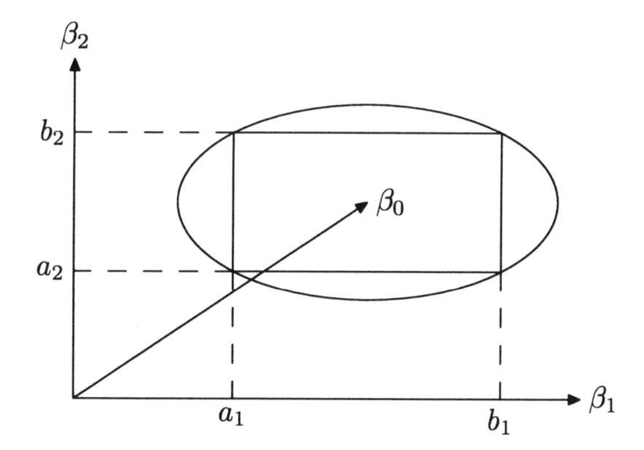

FIGURE 3.7. A priori rectangle and enclosing ellipsoid

and the area $F = \pi ab = \pi t_1^{-\frac{1}{2}} t_2^{-\frac{1}{2}}$.

3.13.2 The Minimax Principle

Consider the quadratic risk $R_1(\hat{\beta}, \beta, A) = \operatorname{tr}\{AM(\hat{\beta}, \beta)\}$ and a class $\{\hat{\beta}\}$ of estimators. Let $B(\beta) \subset \mathbb{R}^K$ be a convex region of a priori restrictions for β. The criterion of the minimax estimator leads to the following.

Definition 3.18 *An estimator $b^* \in \{\hat{\beta}\}$ is called a minimax estimator of β if*

$$\min_{\{\hat{\beta}\}} \sup_{\beta \in B} R_1(\hat{\beta}, \beta, A) = \sup_{\beta \in B} R_1(b^*, \beta, A) . \tag{3.272}$$

Linear Minimax Estimators

We now confine ourselves to the class of linear homogeneous estimators $\{\hat{\beta} = Cy\}$. For these estimates the risk can be expressed as (cf. (4.15))

$$R_1(Cy, \beta, A) = \sigma^2 \operatorname{tr}(ACC') + \beta' T^{\frac{1}{2}} \tilde{A} T^{\frac{1}{2}} \beta \tag{3.273}$$

with

$$\tilde{A} = T^{-\frac{1}{2}} (CX - I)' A (CX - I) T^{-\frac{1}{2}} , \tag{3.274}$$

and $T > 0$ is the matrix of the a priori restriction

$$B(\beta) = \{\beta : \beta' T \beta \le k\} . \tag{3.275}$$

Using Theorem A.44 we get

$$\sup_{\beta} \frac{\beta' T^{\frac{1}{2}} \tilde{A} T^{\frac{1}{2}} \beta}{\beta' T \beta} = \lambda_{\max}(\tilde{A})$$

and hence

$$\sup_{\beta' T \beta \leq k} R_1(Cy, \beta, A) = \sigma^2 \mathrm{tr}(ACC') + k\lambda_{\max}(\tilde{A}) . \qquad (3.276)$$

Since the matrix $\tilde{A}$ (3.274) is dependent on the matrix C, the maximum eigenvalue $\lambda_{\max}(\tilde{A})$ is dependent on C as well, but not in an explicit form that could be used for differentiation. This problem has received considerable attention in the literature. In addition to iterative solutions (Kuks, 1972; Kuks and Olman, 1971, 1972) the suggestion of Trenkler and Stahlecker (1987) is of great interest. They propose to use the inequality $\lambda_{\max}(\tilde{A}) \leq \mathrm{tr}(\tilde{A})$ to find an upper limit of $R_1(Cy, \beta, A)$ that is differentiable with respect to C, and hence find a substitute problem with an explicit solution. A detailed discussion can be found in Schipp (1990).

An explicit solution can be achieved right away if the weight matrices are confined to matrices of the form $A = aa'$ of rank 1, so that the $R_1(\hat{\beta}, \beta, A)$ risk equals the weaker $R_2(\hat{\beta}, \beta, a)$ risk (cf. (4.4)).

Linear Minimax Estimates for Matrices $A = aa'$ of Rank 1

In the case where $A = aa'$, we have

$$\tilde{A} = [T^{-\frac{1}{2}}(CX - I)'a][a'(CX - I)T^{-\frac{1}{2}}] = \tilde{a}\tilde{a}' , \qquad (3.277)$$

and according to the first Corollary to Theorem A.28 we obtain $\lambda_{\max}(\tilde{A}) = \tilde{a}'\tilde{a}$. Therefore, (3.276) becomes

$$\sup_{\beta' T \beta \leq k} R_2(Cy, \beta, a) = \sigma^2 a'CC'a + ka'(CX - I)T^{-1}(CX - I)'a . \qquad (3.278)$$

Differentiation with respect to C leads to (Theorems A.91, A.92)

$$\frac{1}{2}\frac{\partial}{\partial C}\{\sup_{\beta' T \beta \leq k} R_2(Cy, \beta, a)\} = (\sigma^2 I + kXT^{-1}X')C'aa' - kXT^{-1}aa' . \qquad (3.279)$$

Since a is any fixed vector, (3.279) equals zero for all matrices aa' if and only if

$$C'_* = k(\sigma^2 I + kXT^{-1}X')^{-1}XT^{-1} . \qquad (3.280)$$

After transposing (3.280) and multiplying from the left with $(\sigma^2 T + kS)$, we obtain

$$\begin{aligned} (\sigma^2 T + kS)C_* &= kX'[\sigma^2 I + kXT^{-1}X'][\sigma^2 I + kXT^{-1}X']^{-1} \\ &= kX' , \end{aligned}$$

which leads to the solution $(S = X'X)$

$$C_* = (S + k^{-1}\sigma^2 T)^{-1}X' . \qquad (3.281)$$

Using the abbreviation

$$D_* = (S + k^{-1}\sigma^2 T) , \qquad (3.282)$$

we have the following theorem.

Theorem 3.19 (Kuks, 1972) *In the model $y = X\beta + \epsilon$, $\epsilon \sim (0, \sigma^2 I)$, with the restriction $\beta' T\beta \leq k$ with $T > 0$, and the risk function $R_2(\hat{\beta}, \beta, a)$, the linear minimax estimator is of the following form:*

$$\begin{aligned} b_* &= (X'X + k^{-1}\sigma^2 T)^{-1} X'y \\ &= D_*^{-1} X'y \end{aligned} \tag{3.283}$$

with

$$\text{Bias}(b_*, \beta) = -k^{-1}\sigma^2 D_*^{-1} T\beta\,, \tag{3.284}$$

$$V(b_*) = \sigma^2 D_*^{-1} S D_*^{-1} \tag{3.285}$$

and the minimax risk

$$\sup_{\beta' T\beta \leq k} R_2(b_*, \beta, a) = \sigma^2 a' D_*^{-1} a\,. \tag{3.286}$$

Theorem 3.20 *Given the assumptions of Theorem 3.19 and the restriction $(\beta - \beta_0)' T(\beta - \beta_0) \leq k$ with center point $\beta_0 \neq 0$, the linear minimax estimator is of the following form:*

$$b_*(\beta_0) = \beta_0 + D_*^{-1} X'(y - X\beta_0) \tag{3.287}$$

with

$$\text{Bias}(b_*(\beta_0), \beta) = -k^{-1}\sigma^2 D_*^{-1} T(\beta - \beta_0), \tag{3.288}$$

$$V(b_*(\beta_0)) = V(b_*)\,, \tag{3.289}$$

and

$$\sup_{(\beta-\beta_0)' T(\beta-\beta_0) \leq k} R_2(b_*(\beta_0), \beta, a) = \sigma^2 a' D_*^{-1} a\,. \tag{3.290}$$

Proof: The proof is similar to that used in Theorem 3.19, with $\beta - \beta_0 = \tilde{\beta}$.

Interpretation: A change of the center point of the a priori ellipsoid has an influence only on the estimator itself and its bias. The minimax estimator is not operational, because of the unknown σ^2. The smaller the value of k, the stricter is the a priori restriction for fixed T. Analogously, the larger the value of k, the smaller is the influence of $\beta' T\beta \leq k$ on the minimax estimator. For the borderline case we have

$$B(\beta) = \{\beta : \beta' T\beta \leq k\} \to \mathbb{R}^K \quad \text{as} \quad k \to \infty$$

and

$$\lim_{k\to\infty} b_* \to b = (X'X)^{-1} X'y\,. \tag{3.291}$$

Comparison of $\hat{\beta}_*$ and b

(i) Minimax Risk Since the OLS estimator is unbiased, its minimax risk is

$$\sup_{\beta' T\beta \leq k} R_2(b, \cdot, a) = R_2(b, \cdot, a) = \sigma^2 a' S^{-1} a \,. \tag{3.292}$$

The linear minimax estimator b_* has a smaller minimax risk than the OLS estimator, because of its optimality, according to Theorem 3.19. Explicitly, this means (Toutenburg, 1976)

$$\begin{aligned} R_2(b, \cdot, a) &- \sup_{\beta' T\beta \leq k} R_2(b_*, \beta, a) \\ &= \sigma^2 a'(S^{-1} - (k^{-1}\sigma^2 T + S)^{-1})a \geq 0 \,, \end{aligned} \tag{3.293}$$

since $S^{-1} - (k^{-1}\sigma^2 T + S)^{-1} \geq 0$ (cf. Theorem A.40 or Theorem A.52).

(ii) MDE-I Superiority With (3.288) and (3.289) we get

$$\begin{aligned} M(b_*, \beta) &= V(b_*) + \text{Bias}(b_*, \beta)\,\text{Bias}(b_*, \beta)' \\ &= \sigma^2 D_*^{-1}(S + k^{-2}\sigma^2 T\beta\beta' T')D_*^{-1} \,. \end{aligned} \tag{3.294}$$

Hence, b_* is MDE-I-superior to b if

$$\Delta(b, b_*) = \sigma^2 D_*^{-1}[D_* S^{-1} D_* - S - k^{-2}\sigma^2 T\beta\beta' T']D_*^{-1} \geq 0 \,, \tag{3.295}$$

hence if and only if

$$\begin{aligned} B &= D_* S^{-1} D_* - S - k^{-2}\sigma^2 T\beta\beta' T' \\ &= k^{-2}\sigma^4 T[\{S^{-1} + 2k\sigma^{-2}T^{-1}\} - \sigma^{-2}\beta\beta']T \geq 0 \\ &= k^{-2}\sigma^4 T C^{\frac{1}{2}}[I - \sigma^{-2}C^{-\frac{1}{2}}\beta\beta' C^{-\frac{1}{2}}]C^{\frac{1}{2}}T \geq 0 \end{aligned} \tag{3.296}$$

with $C = S^{-1} + 2k\sigma^{-2}T^{-1}$. This is equivalent (Theorem A.57) to

$$\sigma^{-2}\beta'(S^{-1} + 2k\sigma^{-2}T^{-1})^{-1}\beta \leq 1 \,. \tag{3.297}$$

Since $(2k\sigma^{-2}T^{-1})^{-1} - (S^{-1} + 2k\sigma^{-2}T^{-1}) \geq 0$,

$$k^{-1} \leq \frac{2}{\beta'\beta} \tag{3.298}$$

is sufficient for the MDE-I superiority of the minimax estimator b_* compared to b. This condition corresponds to the condition (3.203) for the MDE-I superiority of the ridge estimator $b(k)$ compared to b.

We now have the following important interpretation: The linear minimax estimator b_* is a ridge estimate $b(k^{-1}\sigma^2)$. Hence, the restriction $\beta' T\beta \leq k$ has a stabilizing effect on the variance. The minimax estimator is operational if σ^2 can be included in the restriction $\beta' T\beta \leq \sigma^2 k = \tilde{k}$:

$$b_* = (X'X + \tilde{k}^{-1}T)^{-1}X'y \,.$$

Alternative considerations, as in Chapter 6, when σ^2 is not known in the case of mixed estimators, have to be made (cf. Toutenburg, 1975a; 1982, pp. 95–98).

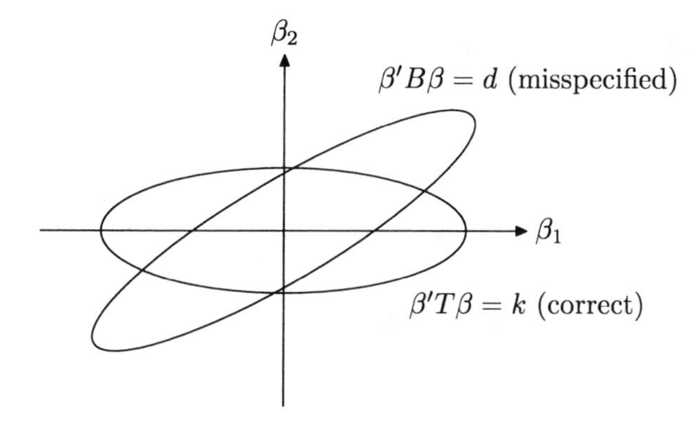

FIGURE 3.8. Misspecification by rotation and distorted length of the axes

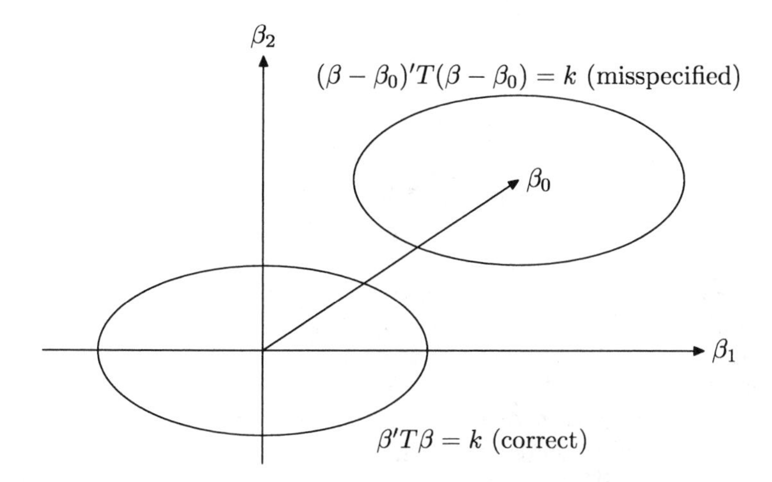

FIGURE 3.9. Misspecification by translation of the center point

From (3.297) we can derive a different sufficient condition: $kT^{-1} - \beta\beta' \geq 0$, equivalent to $\beta' T\beta \leq k$. Hence, the minimax estimator b_* is always MDE-I-superior to b, in accordance with Theorem 3.19, if the restriction is satisfied, that is, if it is chosen correctly.

The problem of robustness of the linear minimax estimator relative to misspecification of the a priori ellipsoid is dealt with in Toutenburg (1984; 1990)

Figures 3.8 and 3.9 show typical situations for misspecifications.

3.14 Censored Regression

3.14.1 Overview

Consider the regression model (cf. (3.24))

$$y_t = x_t'\beta + \epsilon_t\,, \quad t = 1, \ldots, T\,. \tag{3.299}$$

There are numerous examples in economics where the dependent variable y_t is censored, and what is observable is, for example,

$$\begin{aligned} y_t^* &= 1 \quad \text{if} \quad y_t \geq 0\,, \\ y_t^* &= 0 \quad \text{if} \quad y_t < 0\,, \end{aligned} \tag{3.300}$$

or

$$\begin{aligned} y_t^* &= y \quad \text{if} \quad y > 0\,, \\ y_t^* &= 0 \quad \text{if} \quad y \leq 0\,. \end{aligned} \tag{3.301}$$

Model (3.300) is called the binary choice model, and model (3.301), the Tobit model. The problem is to estimate β from such models, generally referred to as limited dependent variable models. For specific examples of such models in economics, the reader is referred Maddala (1983). A variety of methods have been proposed for the estimation of β under models (3.299, 3.300) and (3.299, 3.301) when the e_t's have normal and unknown distributions.

Some of the well-known methods in the case of the Tobit model (3.299, 3.301) are the maximum likelihood method under a normality assumption (as described in Maddala, 1983, pp. 151–156; Amemiya, 1985, Chapter 10; Heckman, 1976), distribution-free least-squares type estimators by Buckley and James (1979) and Horowitz (1986); quantile including the LAD (least absolute deviations) estimators by Powell (1984); and Bayesian computing methods by Polasek and Krause (1994). A survey of these methods and Monte Carlo comparisons of their efficiencies can be found in the papers by Horowitz (1988) and Moon (1989). None of these methods provides closed-form solutions. They are computationally complex and their efficiencies depend on the distribution of the error component in the model and the intensity of censoring. No clear-cut conclusions emerge from these studies on the relative merits of various methods, especially when the sample size is small. Much work remains to be done in this area.

In the present section, we consider some recent contributions to the asymptotic theory of estimation of the regression parameters and tests of linear hypotheses based on the LAD method, with minimal assumptions.

3.14.2 LAD Estimators and Asymptotic Normality

We consider the Tobit model (3.299, 3.301), which can be written in the form

$$y_t^+ = (x_t'\beta + \epsilon_t)^+ , \quad t = 1, \dots, T , \tag{3.302}$$

where $y_t^+ = y_t I$, $(y_t > 0)$, and $I(\cdot)$ denotes the indicator function of a set, and assume that

(A.1) $\epsilon_1, \epsilon_2, \dots$ are i.i.d. random variables such that the distribution function F of ϵ_1 has median zero and positive derivative $f(0)$ at zero.

(A.2) The parameter space B to which β_0, the true value of β, belongs is a bounded open set of $\mathbb{R}^K$ (with a closure $\bar{B}$).

Based on the fact $\text{med}(Y_t^+) = (x_t'\beta_0)^+$, Powell (1984) introduced and studied the asymptotic properties of the LAD estimate $\hat{\beta}_T$ of β_0, which is a Borel-measurable solution of the minimization problem

$$\sum_{t=1}^{T} |Y_t^+ - (x_t'\hat{\beta}_T)^+| = \min \left\{ \sum_{t=1}^{T} |Y_t^+ - (x_t'\beta)^+| \, : \, \beta \in \bar{B} \right\} . \tag{3.303}$$

Since $\sum_{t=1}^{T} |Y_t - (x_t'\beta)^+|$ is not convex in β, the analysis of $\hat{\beta}_T$ is quite difficult. However, by using uniform laws of large numbers, Powell established the strong consistency of $\hat{\beta}_T$ when x_t's are independent variables with $\mathrm{E}\,\|x_t\|^3$ being bounded, where $\|\cdot\|$ denotes the Euclidean norm of a vector. He also established its asymptotic normal distribution under some conditions.

With the help of the maximal inequalities he developed, Pollard (1990) improved the relevant result of Powell on asymptotic normality by relaxing Powell's assumptions and simplified the proof to some extent. Pollard permitted vectors $\{x_t\}$ to be deterministic. We investigate the asymptotic behavior of $\hat{\beta}_T$ under weaker conditions. We establish the following theorem, where we write

$$\mu_t = x_t'\beta_0 \quad \text{and} \quad S_T = \sum_{t=1}^{T} I(\mu_t > 0) x_t x_t' . \tag{3.304}$$

Theorem 3.21 *Assume that (A.1), (A.2) hold, and the following assumptions are satisfied:*

(A.3) For any $\sigma > 0$, there exists a finite $\alpha > 0$ such that

$$\sum_{t=1}^{T} \|x_t\|^2 I(\|x_t\| > \alpha) < \sigma \lambda_{\min}(S_T) \quad \textit{for } T \textit{ large},$$

where $\lambda_{\min}(S_T)$ is the smallest eigenvalue of S_T.

(A.4) For any $\sigma > 0$, there is a $\delta > 0$ such that

$$\sum_{t=1}^{T} \|x_t\|^2 I(|\mu_t| \le \delta) \le \sigma \lambda_{\min}(S_T) \quad \text{for } T \text{ large}.$$

(A.5)

$$\lambda_{\min} \frac{(S_T)}{(\log T)^2} \to \infty, \quad as \quad T \to \infty.$$

Then

$$2f(0) S_T^{\frac{1}{2}} (\hat{\beta}_T - \beta_0) \xrightarrow{L} N(0, I_K)$$

where I_K denotes the identity matrix of order K.

Note: If (A.1)–(A.4) and (A.5*): $\lambda_{\min}(S_T)/\log T \to \infty$ hold, then

$$\lim_{T \to \infty} \hat{\beta}_T = \beta_0 \quad \text{in probability}.$$

For a proof of Theorem 3.21, the reader is referred to Rao and Zhao (1993).

3.14.3 Tests of Linear Hypotheses

We consider tests of linear hypotheses such as

$$H_0\colon H'(\beta - \beta_0) = 0 \quad \text{against} \quad H_1\colon H'(\beta - \beta_0) \neq 0, \tag{3.305}$$

where H is a known $K \times q$-matrix of rank q, and β_0 is a known K-vector $(0 < q < K)$. Let

$$\beta_T^* = \arg \inf_{H'(\beta-\beta_0)=0} \sum_{t=1}^{T} |(x_t' b)^+ - y_t^+|, \tag{3.306}$$

$$\hat{\beta}_T = \arg \inf_b \sum_{t=1}^{T} |(x_t' b)^+ - y_t^+|, \tag{3.307}$$

where all the infima are taken over $b \in \bar{B}$. Define the likelihood ratio, Wald and Rao's score statistics:

$$M_T = \sum_{t=1}^{T} |(x_t' \beta_T^*)^+ - y_t^+| - \sum_{t=1}^{T} |(x_t' \hat{\beta}_T)^+ - y_t^+|, \tag{3.308}$$

$$W_T(\hat{\beta}_T - \beta_0)' H (H' S_T^{-1} H)^{-1} H' (\hat{\beta}_T - \beta_0), \tag{3.309}$$

$$R_T = \xi(\beta_T^*)' S_T^{-1} \xi(\beta_T^*), \tag{3.310}$$

where S_T is as defined in (3.304) and

$$\xi(b) = \sum_{t=1}^{T} I(x_i' b > 0) \operatorname{sgn}(x_t' b - y_t^+) x_t$$

$$= \sum_{t=1}^{T} I(x_t'b > 0)\,\text{sgn}(x_t b - y_t) x_t\,.$$

The main theorem concerning tests of significance is as follows, where we write

$$x_{tT} = S_T^{-\frac{1}{2}} x_t\,, \quad H_T = S_T^{-\frac{1}{2}} H (H' S_T^{-1} H)^{-\frac{1}{2}}\,,$$

$$\sum_{t=1}^{T} I(\mu_t > 0) x_{tT} x_{tT}' = I_K\,, \quad H_T' H_T = I_q\,.$$

Theorem 3.22 *Suppose that the assumptions (A.1)–(A.5) are satisfied. If β is the true parameter and H_0 holds, then each of $4f(0)M_T$, $4[f(0)]^2 W_T$, and R_T can be expressed as*

$$\left\| \sum_{t=1}^{T} I(\mu_t > 0)\,\text{sgn}(e_t) H_T' x_{tT} \right\|^2 + o_K(1)\,. \tag{3.311}$$

Consequently, $4f(0)M_T$, $4f(0)^2 W_T$, and R_T have the same limiting chi-square distribution with the degrees of freedom q.

In order for the results of Theorem 3.22 to be useful in testing the hypothesis H_0 against H_1, some "consistent" estimates of S_T and $f(0)$ should be obtained. We say that $\hat{S}_T$ is a "consistent" estimate of the matrix S_T if

$$S_T^{-\frac{1}{2}} \hat{S}_T S_T^{-\frac{1}{2}} \to I_K \quad \text{as} \quad T \to \infty\,. \tag{3.312}$$

It is easily seen that

$$\hat{S}_T = \sum_{t=1}^{T} I(x_t' \hat{\beta}_T > 0) x_t x_t'$$

can be taken as an estimate of S_T. To estimate $f(0)$, we take $h = h_T > 0$ such that $h_T \to 0$ and use

$$\begin{aligned} \hat{f}_T(0) &= h \sum_{t=1}^{T} I(x_t'\hat{\beta}_T > 0)^{-1} \\ &\quad \times \sum_{t=1}^{T} I(x_t'\hat{\beta}_T > 0)\, I(x_t'\beta_T < y_t^+ \le x_t'\hat{\beta}_T + h) \end{aligned} \tag{3.313}$$

as an estimate of $f(0)$, which is similar to that suggested by Powell (1984). Substituting $\hat{S}_T$ for S_T and $\hat{f}_T$ for $f(0)$ in (3.308), (3.309), and (3.310), we denote the resulting statistics by $\hat{M}_T$, $\hat{W}_T$, and $\hat{R}_T$, respectively. Due to consistency of $\hat{S}_T$ and $\hat{f}_T(0)$, all the statistics

$$4\hat{f}_T(0)\hat{M}_T\,, \quad 4[\hat{f}_T(0)]^2 \hat{W}_T\,, \quad \text{and} \quad \hat{R}_T \tag{3.314}$$

have the same asymptotic chi-square distribution on q degrees of freedom.

Note: It is interesting to observe that the nuisance parameter $f(0)$ does not appear in the definition of $\hat{R}_T$. We further note that

$$4\hat{f}_T(0)\hat{M}_T = 4[\hat{f}_T(0)]^2\hat{W}_T + o_K(1), \tag{3.315}$$

and under the null hypothesis, the statistic

$$U_T = 4\left(\frac{\hat{M}_T}{\hat{W}_T}\right)^2 \hat{W}_T = 4\frac{\hat{M}_T^2}{\hat{W}_T} \overset{L}{\Rightarrow} \chi_q^2. \tag{3.316}$$

We can use U_T, which does not involve $f(0)$, to test H_0. It would be of interest to examine the relative efficiencies of these tests by Monte Carlo simulation studies.

3.15 Simultaneous Confidence Intervals

In the regression model

$$\underset{T\times 1}{y} = \underset{T\times K}{X}\ \underset{K\times T}{\beta} + \underset{T\times 1}{\epsilon}$$

with $\mathrm{E}(\epsilon) = 0, \mathrm{E}(\epsilon\epsilon') = \sigma^2 I$, the least squares estimator of β is $\hat{\beta} = (X'X)^{-1}X'y$ and $\mathrm{V}(\hat{\beta}) = \sigma^2(X'X)^{-1} = \sigma^2 H$ (say). To test the hypothesis $\beta = \beta_0$, we have seen that the test criterion is

$$F = \frac{(\hat{\beta} - \beta_0)H^{-1}(\hat{\beta} - \beta_0)}{Ks^2} \sim F_{K,T-K} \tag{3.317}$$

where $(T-K)s^2 = y'y - \hat{\beta}'X'y$, and $F_{K,T-K}$ is the F-statistic with K and $T-K$ degrees of freedom.

We give a characterization of the above F-test, which leads to the construction of Scheffé's simultaneous confidence intervals on linear functions of β. Consider a single linear function $l'\beta$ of β. The least squares estimator of $l'\beta$ is $l'\hat{\beta}$ with covariance $\sigma^2 l'Hl$. Then the t-statistic to test a hypothesis on $l'\beta$ is

$$t = \frac{l'\hat{\beta} - l'\beta}{\sqrt{s^2 l'Hl}}. \tag{3.318}$$

Now we choose l to maximize

$$t^2 = \frac{l'(\hat{\beta} - \beta)(\hat{\beta} - \beta)'l}{s^2 l'Hl}. \tag{3.319}$$

Using the Cauchy-Schwarz inequality (see Theorem A.54), we see the maximum value of t^2 is

$$\frac{(\hat{\beta} - \beta)'H^{-1}(\hat{\beta} - \beta)}{s^2},$$

which is KF, where F is as defined in (3.317). Thus, we have

$$\frac{(\hat{\beta}-\beta)'H^{-1}(\hat{\beta}-\beta)}{Ks^2} = \frac{1}{Ks^2}\max_l \frac{l'(\hat{\beta}-\beta)(\hat{\beta}-\beta)'l}{l'Hl} \sim F_{K,T-K}.$$

If $F_{1-\alpha}$ is the $(1-\alpha)$ quantile of $F_{K,T-K}$, we have

$$P\left\{\max_l \frac{|l'(\hat{\beta}-\beta)(\hat{\beta}-\beta)'l|}{\sqrt{l'Hl}} \leq s\sqrt{KF_{1-\alpha}}\right\} = 1-\alpha$$

that is,

$$P\left\{\left|l'(\hat{\beta}-\beta)(\hat{\beta}-\beta)'l\right| \leq s\sqrt{KF_{1-\alpha}l'Hl} \text{ for all } l\right\} = 1-\alpha$$

or

$$P\left\{l'\beta \in l'\hat{\beta} \pm s\sqrt{KF_{1-\alpha}l'Hl} \text{ for all } l\right\} = 1-\alpha. \tag{3.320}$$

Equation (3.320) provides confidence intervals for all linear functions $l'\beta$. Then, as pointed out by Scheffé (1959),

$$P\left\{l'\beta \in l'\hat{\beta} \pm s\sqrt{KF_{1-\alpha}l'Hl} \text{ for any given subset of } l\right\} \geq 1-\alpha\,, \tag{3.321}$$

which ensures that the simultaneous confidence intervals for linear functions $l'\beta$ where l belongs to any set (finite or infinite) has a probability not less than $1-\alpha$.

3.16 Confidence Interval for the Ratio of Two Linear Parametric Functions

Let $\theta_1 = P_1'\beta$ and $\theta_2 = P_2'\beta$ be two linear parametric functions and we wish to find a confidence interval of $\lambda = \frac{\theta_1}{\theta_2}$.

The least squares estimators of θ_1 and θ_2 are

$$\hat{\theta}_1 = P_1'\hat{\beta} \quad \text{and} \quad \hat{\theta}_2 = P_2'\hat{\beta}$$

with the variance-covariance matrix

$$\sigma^2\begin{pmatrix} P_1'HP_1 & P_1'HP_2 \\ P_2'HP_1 & P_2'HP_2 \end{pmatrix} = \sigma^2\begin{pmatrix} a & b \\ b' & c \end{pmatrix}, \text{ say.}$$

Then

$$\mathrm{E}(\hat{\theta}_1 - \lambda\hat{\theta}_2) = 0\,, \quad \mathrm{var}(\hat{\theta}_1 - \lambda\hat{\theta}_2) = \sigma^2(a - 2\lambda b + \lambda^2 c)\,.$$

Hence

$$F = \frac{(\hat{\theta}_1 - \lambda\hat{\theta}_2)^2}{s^2(a - 2\lambda b + \lambda^2 c)} \sim F_{1,T-K}$$

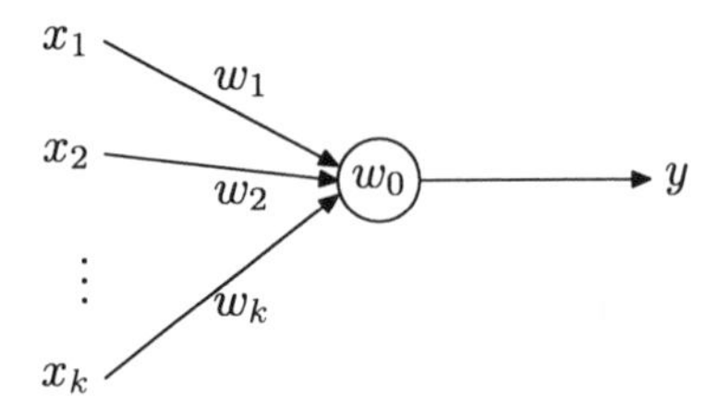

FIGURE 3.10. A single-unit perceptron

and

$$P\left\{(\hat{\theta}_1 - \lambda\hat{\theta}_2)^2 - F_{1-\alpha}s^2(a - 2\lambda b + \lambda^2 c) \leq 0\right\} = 1 - \alpha\,. \qquad (3.322)$$

The inequality within the brackets in (3.322) provides a $(1-\alpha)$ confidence region for λ. Because the expression in (3.322) is quadratic in λ, the confidence region is the interval between the roots of the quadratic equation or outside the interval, depending on the nature of the coefficients of the quadratic equation.

3.17 Neural Networks and Nonparametric Regression

The simplest feed-forward neural network is the so-called single-unit perceptron displayed in Figure 3.10. This perceptron consists of K input units $x_1, \ldots, x_K$ and one output unit y. The input values x_i are weighted with weights w_i $(i = 1, \ldots, K)$ so that the expected response y is related to the vector $x = (x_1, \ldots, x_K)$ of covariates according to

$$y = w_0 + \sum_{i=1}^{K} w_i x_i\,. \qquad (3.323)$$

In general, neural networks are mathematical models representing a system of interlinked computational units. Perceptrons have strong association with regression and discriminant analysis. Unsupervised networks are used for pattern classification and pattern recognition. An excellent overview on neural networks in statistics may be found in Cheng and Titterington (1994). In general, the input-output relationship at a neuron may be written as

$$y = f(x, w) \qquad (3.324)$$

where $f(\cdot)$ is a known function. $f(\cdot)$ is called the activation function. Assume that we have observations $(x^{(1)}, y^{(1)}), \ldots, (x^{(n)}, y^{(n)})$ of n individuals in a so-called training sample. Then the vector of weights w has to be determined such that the so-called energy or learning function

$$\mathrm{E}(w) = \sum_{j=1}^{n} \left(y^{(j)} - f(x^{(j)}, w)\right)^2 \tag{3.325}$$

is minimized with respect to w. This is just a least squares problem. To find the weight $\hat{w}$ minimizing $\mathrm{E}(w)$ we have to solve the following system of estimation equations ($k = 0, \ldots, K$)

$$\frac{\partial\,\mathrm{E}(w)}{\partial w_k} = \sum_{j=1}^{n} (y^{(j)} - f(x^{(j)}, w)) \frac{\partial f(x^j, w)}{\partial w_k} = 0. \tag{3.326}$$

In practice, numerical methods are used to minimize $\mathrm{E}(w)$. Well-known techniques that have been implemented are the generalized delta rule or error back-propagation (Rumelhart, Hinton, and Williams, 1986), gradient methods such as the method of steepest descent (Thisted, 1988), genetic algorithms, Newton-Raphson algorithms, and variants of them.

If a multilayer perceptron is considered, then it may be interpreted as a system of nonlinear regression functions that is estimated by optimizing some measure of fit. Recent developments in this field are the projection-pursuit regression (see Section 3.11) and its modifications by (Tibshirani, 1992) using so-called slide functions, and the generalized additive models (see Hastie and Tibshirani (1990)).

During the last five years a lot of publications have demonstrated the successful application of neural networks to problems of practical relevance. Among them, in the field of medicine the analysis based on a logistic regression model (see Section 10.3.1) is of special interest.

3.18 Logistic Regression and Neural Networks

Let y be a binary outcome variable and $x = (x_1, \ldots, x_K)$ a vector of covariates. As activation function $f(.)$ we choose the logistic function $l(v) = \exp(v)/(1 + \exp(v))$. Then the so-called logistic perceptron $y = l(w_0 + \sum_{i=1}^{K} w_i x_i)$ is modeling the relationship between y and x. The estimation equations (3.326) become (Schumacher, Roßner, and Vach, 1996)

$$\frac{\partial\,\mathrm{E}(w)}{\partial w_k} = \sum_{j=1}^{n} 2f(x^{(j)}, w)(1 - f(x^{(j)}, w)) x_k^{(j)} (y^{(j)} - f(x^{(j)}, w)) = 0. \tag{3.327}$$

For solving (3.327), the least-squares back-propagation method (Rumelhart et al., 1986) is used. It is defined by

$$\hat{w}^{(v+1)} = \hat{w}^{(v)} - \eta \partial\,\mathrm{E}(\hat{w}^{(v)})$$

for $v = 0, 1, \ldots$ and $\hat{w}^{(0)}$ a chosen starting value. The positive constant η is called the learning rate. The nonlinear model of the logistic perceptron is identical to the logistic regression model (10.61) so that the weights w can be interpreted like the regression coefficients β. (For further discussion, see Vach, Schumacher, and Roßner (1996).)

3.19 Restricted Regression

3.19.1 *Problem of Selection*

In plant and animal breeding we have the problem of selecting individuals for propagation on the basis of observed measurements $x_1, \ldots, x_p$ in such a way that there is improvement in a desired characteristic y_0 in the future generations. At the suggestion of R. A. Fisher that the best selection index is the regression of y_0 on $x_1, \ldots, x_p$ with individuals having a larger value preferred in selection, Fairfield Smith (1936) worked out the computational details, and Rao (1953) provided the theoretical background for the solution.

In practice, it may so happen that improvement in the main characteristic is accompanied by deterioration (side effects) in certain other desired characteristics, $y_1, \ldots, y_q$. This problem was addressed by Kempthorne and Nordskog (1959) and Tallis (1962), who modified the selection index to ensure that *no change* in $y_1, \ldots, y_q$ occurs, and subject to this condition maximum possible improvement in y_0 is achieved. Using the techniques of quadratic programming, Rao (1962; 1964) showed that a selection index can be constructed to provide maximum improvement in y_0 while ensuring that there are possible improvements in $y_1, \ldots, y_q$, but no deterioration. The theory and computations described in this section are taken from the above cited papers of Rao.

3.19.2 *Theory of Restricted Regression*

Let $x' = (x_1, \ldots, x_p)$ be the vector of predictors, Λ be the dispersion matrix of x, and c_i be the column vectors of the covariances $c_{i1}, \ldots, c_{ip}$, of y_i with $x_1, \ldots, x_p$, for $i = 0, 1, 2, \ldots, q$. Denote by C the partitioned matrix $(c_0, c_1, \ldots, c_q)$, and denote the dispersion matrix of $y' = (y_0, \ldots, y_q)$ by $\Sigma = (\sigma_{ij})$, $i, j = 0, 1, \ldots, q$. Let us assume that the rank of C is $q + 1$, Λ is nonsingular, and $p \geq q + 1$. If b is a p-vector, correlation of y_i and $b'x$ is

$$\frac{(b'c_i)}{\sqrt{\sigma_{ii} b' \Lambda b}} .$$

The problem is to choose b such that

$$\frac{(b'c_0)}{\sqrt{\sigma_{00} b' \Lambda b}} \tag{3.328}$$

is a maximum subject to the conditions

$$b'c_0 > 0, \quad b'c_i \geq 0, \quad i = 1, \ldots, q. \tag{3.329}$$

Note that maximizing (3.328) without any restriction leads to $b'x$, which is the linear regression of y_0 on $(x_1, \ldots, x_p)$ apart from the constant term. In such a case the selection index is $b'x$ and individuals with large values of $b'x$ are selected for future propagation.

If the constraints (3.329) are imposed to avoid side effects, then the problem is one of nonlinear programming for which the following theorems are useful.

Lemma 3.23 *Given a p-vector b satisfying the conditions (3.329), there exists a $(q+1)$-vector g such that*

(i) $m = \Lambda^{-1}Cg$ *satisfies conditions (3.329), and*

(ii) $\dfrac{m'c_0}{\sqrt{m'\Lambda m}} \geq \dfrac{b'c_1}{\sqrt{b'\Lambda b}}$.

Proof: Choose a matrix D such that $(\Lambda^{-1}C : \Lambda^{-1}D)$ is of full rank and $C'\Lambda^{-1}D = 0$, so that the spaces generated by $\Lambda^{-1}C$ and $\Lambda^{-1}D$ are orthogonal under the inner product $\alpha'\Lambda\beta$ for any two vectors α and β. Then there exist vectors g and h such that any vector b can be decomposed as

$$b = \Lambda^{-1}Cg + \Lambda^{-1}Dh = m + \Lambda^{-1}Dh.$$

To prove (i) observe that

$$0 \leq b'c_i = m'c_i + c_i'\Lambda^{-1}Dh = m'c_i, \quad i = 0, \ldots, q.$$

To prove (ii) we have $b'\Lambda b = m'\Lambda m + h'D'\Lambda^{-1}Dh \geq m'\Lambda m$, and since $b'c_0 = m'c_0$, we have

$$\frac{m'c_0}{\sqrt{m'\Lambda m}} \geq \frac{b'c_0}{\sqrt{b'\Lambda b}}.$$

Lemma 3.23 reduces the problem to that of determining m of the form $\Lambda^{-1}Cg$ where g is of a smaller order than m.

Lemma 3.24 *The problem of determining g such that with $m = \Lambda^{-1}Cg$, the conditions (3.329) are satisfied and $m'c_0/\sqrt{m'\Lambda m}$ is a maximum is equivalent to the problem of minimizing a nonnegative quadratic form $(u - \xi)'B(u - \xi)$ with u restricted to nonnegative vectors, where B and ξ are computed from the known quantities C and Λ.*

Proof: Let $v' = (v_0, v_1, \ldots, v_q)$ be a $(q+1)$-vector with all nonnegative elements and let g be a solution of

$$C'm = C'\Lambda^{-1}Cg = v$$

giving

$$g = Av\,,\ m = \Lambda^{-1}CAv \tag{3.330}$$

$$\frac{m'c_0}{\sqrt{m'\Lambda m}} = \frac{v_0}{\sqrt{v'Av}} \tag{3.331}$$

where $A = (C'\Lambda^{-1}C)^{-1}$. Writing $v_i/v_o = u_i, i = 1, \ldots, q$, and denoting the elements of the $(q+1) \times (q+1)$-matrix A by (a_{ij}),we can write the square of the reciprocal of (3.331) as

$$\delta + (u - \xi)'B(u - \xi) = \delta + Q(u)$$

where

$$B = (a_{ij})\,, \quad i, j = 1, \ldots, q$$

and $\xi' = (\xi_1, \ldots, \xi_q)$ is a solution of

$$-B\xi = \alpha_0\,, \quad \alpha_1' = (a_{01}, \ldots, a_{0q}) \tag{3.332}$$

and

$$\delta = a_{00} - \sum\sum\nolimits_{i,j\geq 1} a_{ij}\xi_i\xi_j\,.$$

The solution of (3.332) is

$$\xi_i = \frac{c_i'\Lambda^{-1}c_0}{c_o'\Lambda^{-1}c_0}\,, \quad i = 1, \ldots, q$$

and

$$\delta = (c_0'\Lambda^{-1}c_0)^{-1}\,,$$

which are the simple functions of c_i and Λ^{-1}. Now

$$\begin{aligned}\sup_g \frac{m'c_0}{\sqrt{m'\Lambda m}} &= \sup_{u\geq 0}\{\delta + Q(u)\}^{-\frac{1}{2}} \\ &= \{\delta + \inf_{u\geq 0} Q(u)\}^{-\frac{1}{2}}\,.\end{aligned}$$

The problem is thus reduced to that of minimizing the nonnegative quadratic form $Q(u)$ with the restriction that the elements of u are nonnegative.

If $u_0' = (u_{10}, \ldots, u_{q0})$ is the minimizing vector, then the optimum m is found from (3.331) as

$$m = \Lambda^{-1}CAv_0$$

and the selection index is

$$v_0'AC'\Lambda^{-1}x\,,\ v_0' = (1, u_{10}, \ldots, u_{q0})\,. \tag{3.333}$$

3.19.3 Efficiency of Selection

The correlation between y_0 and the best selection index (multiple regression) when there are no restrictions is

$$R_1 = \frac{1}{\sqrt{\delta \sigma_{00}}} .$$

With the restriction that the changes in mean values of other variables are to be in specified directions if possible, or otherwise zero, the correlation between y_0 and the best selection index is

$$R_2 = \frac{1}{\sqrt{\sigma_{00}\{\delta + \min_{u \geq 0} Q(u)\}}} .$$

If the restriction is such that no change in mean values of $y_1, \ldots, y_q$ is derived, then the selection index is obtained by putting $u = 0$, giving the correlation coefficient

$$R_3 = \frac{1}{\sqrt{\sigma_{00}\{\delta + \xi' B \xi\}}} .$$

It may be seen that

$$R_1 \geq R_2 \geq R_3 ,$$

which implies that selection efficiency possibly increases by generalizing the restriction of no changes to possible changes in desired directions.

The correlation coefficient between the selection index and the variables $y_i (i \neq 0)$ is

$$\frac{u_i \sqrt{\sigma_{00}}}{\sqrt{\sigma_{ii}}} R_2 , \quad i = 1, \ldots, q$$

which enables the estimation of changes in the mean value of $y_i, i = 1, \ldots, q$. When $u_i = 0$, the expected change is zero, as expected.

3.19.4 Explicit Solution in Special Cases

When $q = 1$, the solution is simple. The quadratic form $Q(u)$ reduces to

$$a_{11}\Big(u_1 - \frac{c_0' \Lambda^{-1} c_1}{c_0' \Lambda^{-1} c_0}\Big)^2 . \tag{3.334}$$

If $c_0' \Lambda^{-1} c_1 \geq 0$, then the minimum of (3.334) for nonnegative u_1 is zero, and the multiple regression of y_0 on $x_1, \ldots, x_p$ is $c_0' \Lambda^{-1} x$, apart from the constant term.

If $c_0' \Lambda^{-1} c_1 < 0$, then the minimum is attained when $u_1 = 0$, and using (3.334) the selection index is found to be

$$c_0' \Lambda^{-1} x - \frac{c_0' \Lambda^{-1} c_1}{c_1' \Lambda^{-1} c_1} c_1' \Lambda^{-1} x \tag{3.335}$$

which is a linear combination of the multiple regressions of y_0 and y_1 on $x_1, \ldots, x_p$. The square of the correlation between y_0 and (3.335) is

$$\sigma_{00}^{-1}\left[c_0'\Lambda^{-1}c_0 - \frac{(c_0'\Lambda^{-1}c_1)^2}{c_1'\Lambda^{-1}c_1}\right] \tag{3.336}$$

and that between y_1 and its regression on $x_1, \ldots, x_p$ is

$$\sigma_{00}^{-1}c_0'\Lambda c_0\,,$$

and the reduction in correlation due to restriction on y_1, when $c_0'\Lambda^{-1}c_1 < 0$ is given by the second term in (3.336).

The next practically important case is that of $q = 3$. The quadratic form to be minimized is

$$Q(u_1, u_2) = a_{11}(u_1 - \xi_1)^2 + 2a_{12}(u_1 - \xi_1)(u_2 - \xi_2) + a_{22}(u_2 - \xi_2)^2\,.$$

A number of cases arise depending on the signs of $\xi_1, \xi_2, \ldots$

Case (i) Suppose that $\xi_1 \geq 0, \xi_2 \geq 0$. The minimum of Q is zero and the multiple regression of y_1 on $x_1, \ldots, x_p$ is the selection function.

Case (ii) Suppose that $\xi_1 < 0, \xi_2 \geq 0$. The minimum of Q is attained on the boundary $u_1 = 0$. To determine the value of u_2, we solve the equation

$$\frac{1}{2}\frac{dQ(0, u_2)}{du_2} = a_{22}(u_2 - \xi_2) - a_{12}\xi_1 = 0\,,$$

obtaining

$$u_2 = \frac{a_{12}}{a_{22}}\xi_1 + \xi_2\,. \tag{3.337}$$

If $a_{12}\xi_1 + a_{22}\xi_2 \geq 0$, then the minimum value of Q is attained when $u_{10} = 0$ and u_{20} has the right-hand side value in (3.337). If $a_{12}\xi_1 + a_{22}\xi_2 < 0$, then the minimum is attained at $u_{10} = 0, u_{20} = 0$. The selection function is determined as indicated in (3.333). The case of $\xi_1 \geq 0, \xi_2 < 0$ is treated in a similar way.

Case (iii) Suppose that $\xi_1 < 0, \xi_2 < 0$. There are three possible pairs of values at which the minimum might be attained:

$$\begin{aligned}
u_{10} &= 0\,, \quad u_{20} = \frac{a_{12}}{a_{22}}\xi_1 + \xi_2\,, \\
u_{10} &= \frac{a_{12}}{a_{11}}\xi_2 + \xi_1\,, \quad u_{20} = 0\,, \\
u_{10} &= 0\,, \quad u_{20} = 0\,.
\end{aligned}$$

Out of these we need consider only the pairs where both coordinates are nonnegative and then choose that pair for which Q is a minimum.

When $q > 3$, the number of different cases to be considered is large. When each $\xi_i \geq 0$, the minimum of Q is zero. But in the other cases the algorithms developed for general quadratic programming (Charnes and

Cooper, 1961, pp. 682–687) may have to be adopted. It may, however, be observed that by replacing $u' = (u_1, \ldots, u_q)$ by $w' = (w_1^2, \ldots, w_q^2)$ in Q, the problem reduces to that of minimizing a quartic in $w_1, \ldots w_q$ without any restrictions. No great simplification seems to result by transforming the problem in this way. As mentioned earlier, the practically important cases correspond to $q = 2$ and 3 for which the solution is simple, as already indicated. The selective efficiency may go down rapidly with increase in the value of q.

For additional literature on selection problems with restrictions, the reader is referred to Rao (1964).

3.20 Complements

3.20.1 Linear Models without Moments: Exercise

In the discussion of linear models in the preceding sections of this chapter, it is assumed that the error variables have second-order moments. What properties does the OLSE, $\hat{\beta} = (X'X)^{-1}X'y$, have if the first- and second-order moments do not exist? The question is answered by Jensen (1979) when ϵ has a spherical distribution with the density

$$\mathcal{L}(y) = \sigma^{-T}\Psi_T\{(y - X\beta)'(y - X\beta)/\sigma^2\}. \tag{3.338}$$

We represent this class by $S_k(X\beta, \sigma^2 I)$, where k represents the integral order of moments that ϵ admits. If $k = 0$, no moments exist. Jensen (1979) proved among other results the following.

Theorem 3.25 (Jensen, 1979) *Consider $\hat{\beta} = (X'X)^{-1}X'y$ as an estimator β in the model $y = X\beta + \epsilon$. Then*

(i) *If $\mathcal{L}(y) \in S_0(X\beta, \sigma^2 I)$, then $\hat{\beta}$ is median unbiased for β and $\hat{\beta}$ is at least as concentrated about β as any other median unbiased estimator of β.*
[Note that an s-vector $t \in \mathbb{R}^s$ is said to be modal unbiased for $\theta \in \mathbb{R}^s$ if $a't$ is modal unbiased for $a'\theta$ for all a.]

(ii) *If $\mathcal{L}(y) \in S_1(X\beta, \sigma^2 I)$, then $\hat{\beta}$ is unbiased for β and is at least as concentrated around β as any other unbiased linear estimator.*

(iii) *If $\mathcal{L}(y) \in S_0(X\beta, \sigma^2 I)$ and in addition unimodal, then $\hat{\beta}$ is modal unbiased for β.*

3.20.2 Nonlinear Improvement of OLSE for Nonnormal Disturbances

Consider the linear regression model (3.24). The Gauss-Markov Theorem states that $b = (X'X)^{-1}X'y$ is the best linear unbiased estimator for β, that is, $\text{Var}(\tilde{b}) - \text{Var}(b)$ is nonnegative definite for any other linear unbiased estimator $\tilde{b}$. If ϵ is multinormally distributed, then b is even the best unbiased estimator.

Hence, if ϵ is not multinormally distributed, there is a potential of nonlinear unbiased estimators for β that improve upon b.

- What is the most general description of such estimators?
- What is the best estimator within this class?

Remark. This problem was proposed by G. Trenkler. Related work may be found in Kariya (1985) and Koopmann (1982).

3.20.3 A Characterization of the Least Squares Estimator

Consider the model $y = X\beta + \epsilon$ with $\text{Cov}(\epsilon) = \sigma^2 I$, $\text{rank}(X) = K$, the size of vector β, and a submodel $y_{(i)} = X_{(i)}\beta + \epsilon_{(i)}$ obtained by choosing $k \leq T$ rows of the original model. Further, let

$$\hat{\beta} = (X'X)^{-1}X'y\,, \quad \hat{\beta}_{(i)} = (X'_{(i)}X_{(i)})^{-}X'_{(i)}y_{(i)} \tag{3.339}$$

be the LSEs from the original and the submodel respectively. Subramanyam (1972) and Rao and Precht (1985) proved the following result.

Theorem 3.26 *Denoting* $d_{(i)} = |X'_{(i)}X_{(i)}|$, *we have*

$$\hat{\beta} = \frac{\sum_{i=1}^{c} d_{(i)}\hat{\beta}_{(i)}}{\sum_{i=1}^{c} d_{(i)}} \tag{3.340}$$

where c *is the number of all possible subsets of size* k *from* $\{1, \ldots, T\}$.

The result (3.340), which expresses $\hat{\beta}$ as a weighted average of $\hat{\beta}_{(i)}$, is useful in regression diagnostics. We may calculate all possible $\hat{\beta}_{(i)}$ and look for consistency among them. If some appear to be much different from others, then we may examine the data for outliers or existence of clusters and consider the possibility of combining them with a different set of weights (some may be zero) than those in (3.340). Further results of interest in this direction are contained in Wu (1986).

3.20.4 A Characterization of the Least Squares Estimator: A Lemma

Consider the model $\epsilon_i = y_i - x_i'\beta$, $i = 1, 2, \ldots$, in which $\epsilon_1, \epsilon_2, \ldots$, are independently and identically distributed with mean 0 and variance σ^2, and the x_i''s are K-vectors of constants. Let the $K \times n$-matrix $X' = (x_1, \ldots, x_n)$ of constants be of rank K. Define $h_{ii}(n) = x_i'(X'X)^{-1}x_i$ and $b = (X'X)^{-1}X'y$ where $y = (y_1, \ldots, y_n)'$. Then for any $r \times K$-matrix C of constants and of rank $r \leq K$,

$$\sigma^{-2}(Cb - C\beta)'[C(X'X)^{-1}C']^{-1}(Cb - C\beta) \to \chi_r^2$$

if (and only if) $\max_{1 \leq i \leq n} h_{ii}(n) \to \infty$.

This result and the condition on $h_{ii}(n)$ were obtained by Srivastava (1971; 1972) using a lemma of Chow (1966).

3.21 Exercises

Exercise 1. Define the principle of least squares. What is the main reason to use $e'e$ from (3.7) instead of other objective functions such as $\max_t |e_t|$ or $\sum_{t=1}^{T} |e_t|$?

Exercise 2. Discuss the statement: In well-designed experiments with quantitative x-variables it is not necessary to use procedures for reducing the number of included x-variables after the data have been obtained.

Exercise 3. Find the least squares estimators of β in $y = X\beta + \epsilon$ and $y = \alpha 1 + X\beta + \epsilon^*$, where 1 denotes a column vector with all elements unity. Compare the dispersion matrices as well as the residual sums of squares.

Exercise 4. Consider the two models $y_1 = \alpha_1 1 + X\beta + \epsilon_1$ and $y_2 = \alpha_2 1 + X\beta + \epsilon_2$ (with 1 as above). Assuming ϵ_1 and ϵ_2 to be independent with same distributional properties, find the least squares estimators of α_1, α_2, and β.

Exercise 5. In a bivariate linear model, the OLSE's are given by b_0 (3.120) and b_1 (3.121). Calculate the covariance matrix $\mathrm{V}\binom{b_0}{b_1}$. When are b_0 and b_1 uncorrelated?

Exercise 6. Show that the estimator minimizing the generalized variance (determinant of variance-covariance matrix) in the class of linear and unbiased estimators of β in the model $y = X\beta + \epsilon$ is nothing but the least squares estimator.

Exercise 7. Let $\hat{\beta}_1$ and $\hat{\beta}_2$ be the least squares estimators of β from $y_1 = X\beta + \epsilon_1$ and $y_2 = X\beta + \epsilon_2$. If β is estimated by $\hat{\beta} = w\hat{\beta}_1 + (1-w)\hat{\beta}_2$ with $0 < w < 1$, determine the value of w that minimizes the trace of the dispersion matrix of $\hat{\beta}$. Does this value change if we minimize $\mathrm{E}(\hat{\beta} - \beta)'X'X(\hat{\beta} - \beta)$?

Exercise 8. Demonstrate that the best quadratic estimator of σ^2 is $(T - K + 2)^{-1} y'(I - P)y$, where P is the projection matrix on $\mathcal{R}(X)$.

Exercise 9. Let the following model be given:

$$y = \beta_0 + \beta_1 x_1 + \beta_2 x_2 + \beta_3 x_3 + \epsilon.$$

(i) Formulate the hypothesis H_0: $\beta_2 = 0$ as a linear restriction $r = R\beta$ on β.

(ii) Write down the test statistic for testing H_0: $\beta_2 = 0$.

Excercise 10. Describe a procedure for testing the equality of first p elements of β_1 and β_2 in the model $y_1 = X_1\beta_1 + \epsilon_1$ and $y_2 = X_2\beta_2 + \epsilon_2$. Assume that $\epsilon_1 \sim N(0, \sigma^2 I_{n_1})$ and $\epsilon_2 \sim N(0, \sigma^2 I_{n_2})$ are stochastically independent.

Exercise 11. If $\hat{\theta}_i$ is a MVUE (minimum variance unbiased estimator) of $\theta_i, i = 1, \dots, k$, then $a_1\hat{\theta}_1 + \dots + a_k\hat{\theta}_k$ is a MVUE of $a_1\theta_1 + \dots + a_k\theta_k$ for any $a_1, \dots, a_k$.

4 The Generalized Linear Regression Model

4.1 Optimal Linear Estimation of β

In Chapter 2 the generalized linear regression model is introduced as a special case ($M = 1$) of the multivariate (M-dimensional) model. If assumption (H) of Section 2.6 holds, we may write the generalized linear model as

$$\left.\begin{array}{l} y = X\beta + \epsilon\,, \\ \mathrm{E}(\epsilon) = 0, \quad \mathrm{E}(\epsilon\epsilon') = \sigma^2 W\,, \\ W \text{ positive definite,} \\ X \text{ nonstochastic, } \operatorname{rank}(X) = K\,. \end{array}\right\} \tag{4.1}$$

A noticeable feature of this model is that the $T \times T$ symmetric matrix W introduces $T(T+1)/2$ additional unknown parameters in the estimation problem. As the sample size T is fixed, we cannot hope to estimate all the parameters $\beta_1, \ldots, \beta_K, \sigma^2$, and w_{ij} $(i \leq j)$ simultaneously. If possible, we may assume that W is known. If not, we have to restrict ourselves to error distributions having a specific structure so that the number of parameters is reduced, such as, for instance, in heteroscedasticity or autoregression (see the following sections). We first consider the estimation of β when W is assumed to be fixed (and known).

We again confine ourselves to estimators that are linear in the response vector y, that is, we choose the set-up (cf. (3.38))

$$\hat{\beta} = Cy + d\,. \tag{4.2}$$

The matrix C and the vector d are nonstochastic and are determined through optimization of one of the following scalar risk functions:

$$\begin{aligned} R_1(\hat{\beta}, \beta, A) &= \mathrm{E}(\hat{\beta} - \beta)' A (\hat{\beta} - \beta) \qquad (4.3)\\ & \quad (A \text{ a positive definite } K \times K\text{-matrix}),\\ R_2(\hat{\beta}, \beta, a) &= \mathrm{E}[(\hat{\beta} - \beta)' a]^2 \qquad (4.4)\\ & \quad (a \neq 0 \text{ a fixed } K\text{-vector}),\\ R_3(\hat{\beta}, \beta) &= \mathrm{E}(y - X\hat{\beta})' W^{-1} (y - X\hat{\beta}). \qquad (4.5) \end{aligned}$$

Remarks:

(i) The function $R_1(\hat{\beta}, \beta, A)$ is the quadratic risk given in (3.40) (see Definition 3.8). The matrix A may be interpreted as an additional parameter, or it may be specified by the user. In order to have unique solutions $(\hat{C}, \hat{d})$ and possibly independent of A, we restrict the set of matrices to be positive definite. Minimizing the risk $R_1(\hat{\beta}, \beta, A)$ with respect to $\hat{\beta}$ is then equivalent to optimal estimation of the parameter β itself.

(ii) Minimizing the risk $R_2(\hat{\beta}, \beta, a) = R_1(\hat{\beta}, \beta, aa')$ means essentially the optimal estimation of the linear function $a'\beta$ instead of β.

(iii) Minimizing the risk $R_3(\hat{\beta}, \beta)$ boils down to the optimal estimation of the conditional expectation $\mathrm{E}(y|X) = X\beta$, that is, to the optimal classical prediction of mean values of y. The weight matrix W^{-1} standardizes the structure of the disturbances.

Using these risk functions enables us to define the following criteria for the optimal estimation of β:

Criterion C_i $(i = 1, 2$ or $3)$: $\hat{\beta}$ is said to be the linear estimator with minimum risk $R_i(\hat{\beta})$—or $\hat{\beta}$ is said to be R_i-optimal—if

$$R_i(\hat{\beta}, \beta, \cdot) \leq R_i(\tilde{\beta}, \beta, \cdot) \qquad (4.6)$$

for X, W fixed and for all β, σ^2 where $\tilde{\beta}$ is any other linear estimator for β.

4.1.1 R_1-Optimal Estimators

Heterogeneous R_1-Optimal Estimator

From (4.2) the estimation error in $\hat{\beta}$ is clearly expressible as

$$\hat{\beta} - \beta = (CX - I)\beta + d + C\epsilon, \qquad (4.7)$$

from which we derive

$$R_1(\hat{\beta}, \beta, A) = \mathrm{E}[(CX - I)\beta + d + C\epsilon]' A [(CX - I)\beta + d + C\epsilon]$$

$$= \quad [(CX-I)\beta+d]'A[(CX-I)\beta+d] + \mathrm{E}(\epsilon'C'AC\epsilon)\,. \tag{4.8}$$

The second term in (4.8) is free from d. Therefore the optimal value of d is that which minimizes the first term. As the first term cannot be negative, it attains its minimum when

$$\hat{d} = -(\hat{C}X - I)\beta\,. \tag{4.9}$$

Now we observe that

$$\begin{aligned} \min_C \mathrm{E}(\epsilon'C'AC\epsilon) &= \min_C \mathrm{tr}\{AC(\mathrm{E}\,\epsilon\epsilon')C'\} \\ &= \min_C \sigma^2 \mathrm{tr}\{ACWC'\}\,, \end{aligned} \tag{4.10}$$

so that an application of Theorems A.93 to A.95 yields

$$\frac{\partial}{\partial C}\sigma^2 \mathrm{tr}\{ACWC'\} = 2\sigma^2 ACW\,. \tag{4.11}$$

Equating this to the null matrix, the optimal C is seen to be $\hat{C} = 0$ as A and W are positive definite and regular. Inserting $\hat{C} = 0$ in (4.9) gives $\hat{d} = \beta$, which after substitution in (4.2) yields the trivial conclusion that the R_1-optimal heterogeneous estimator of β is β itself (cf. Theil, 1971, p. 125). We call this trivial estimator $\hat{\beta}_1$:

$$\hat{\beta}_1 = \beta \tag{4.12}$$

with

$$R_1(\hat{\beta}_1, \beta, A) = 0 \quad \text{and} \quad V(\hat{\beta}_1) = 0\,. \tag{4.13}$$

$\hat{\beta}_1$ clearly has zero bias and zero risk, but zero usefulness too (Bibby and Toutenburg, 1977, p. 76). The only information given by $\hat{\beta}_1$ is that the heterogeneous structure of a linear estimator will not lead us to a feasible solution of the estimation problem. Let us next see what happens when we confine ourselves to the class of homogeneous linear estimators.

Homogeneous R_1-Optimal Estimator

Putting $d = 0$ in (4.2) gives

$$\hat{\beta} - \beta = (CX - I)\beta + C\epsilon\,, \tag{4.14}$$

$$R_1(\hat{\beta}, \beta, A) = \beta'(X'C' - I)A(CX - I)\beta + \sigma^2 \mathrm{tr}\{ACWC'\} \tag{4.15}$$

$$\frac{\partial R_1(\hat{\beta}, \beta, A)}{\partial C} = 2A[C(X\beta\beta'X' + \sigma^2 W) - \beta\beta'X'] \tag{4.16}$$

(cf. Theorems A.92, A.93). The matrix $X\beta\beta'X' + \sigma^2 W$ is positive definite (Theorem A.40) and, hence, nonsingular. Equating (4.16) to a null matrix gives the optimal C as

$$\hat{C}_2 = \beta\beta'X'(X\beta\beta'X' + \sigma^2 W)^{-1}. \tag{4.17}$$

Applying Theorem A.18 (iv), we may simplify the expression for $\hat{C}_2$ by noting that

$$(X\beta\beta'X' + \sigma^2 W)^{-1} = \sigma^{-2}W^{-1} - \frac{\sigma^{-4}W^{-1}X\beta\beta'X'W^{-1}}{1+\sigma^{-2}\beta'X'W^{-1}X\beta}. \tag{4.18}$$

Letting

$$S = X'W^{-1}X, \tag{4.19}$$

we see this matrix is positive definite since $\text{rank}(X) = K$.

Therefore, the homogeneous R_1-optimal estimator is

$$\begin{aligned}\hat{\beta}_2 &= \beta\left[\sigma^{-2}\beta'X'W^{-1}y - \frac{\sigma^{-4}\beta'S\beta\beta'X'W^{-1}y}{1+\sigma^{-2}\beta'S\beta}\right] \\ &= \beta\left[\sigma^{-2} - \frac{\sigma^{-4}\beta'S\beta}{1+\sigma^{-2}\beta'S\beta}\right]\beta'X'W^{-1}y \\ &= \beta\left[\frac{\beta'X'W^{-1}y}{\sigma^2+\beta'S\beta}\right]\end{aligned} \tag{4.20}$$

(cf. Theil, 1971; Toutenburg, 1968; Rao, 1973a, p. 305; and Schaffrin 1985; 1986; 1987).

If we use the abbreviation

$$\alpha(\beta) = \frac{\beta'S\beta}{\sigma^2+\beta'S\beta} \tag{4.21}$$

and note that $\alpha(\beta) < 1$, then

$$\mathrm{E}(\hat{\beta}_2) = \beta\alpha(\beta), \tag{4.22}$$

from which it follows that, on the average, $\hat{\beta}_2$ results in underestimation of β. The estimator $\hat{\beta}_2$ is biased, that is,

$$\begin{aligned}\text{Bias}(\hat{\beta}_2, \beta) &= \mathrm{E}(\hat{\beta}_2) - \beta \\ &= (\alpha(\beta) - 1)\beta \\ &= \frac{-\sigma^2}{\sigma^2+\beta'S\beta}\beta\end{aligned} \tag{4.23}$$

and has the covariance matrix

$$V(\hat{\beta}_2) = \sigma^2\beta\beta' \cdot \frac{\beta'S\beta}{(\sigma^2+\beta'S\beta)^2}. \tag{4.24}$$

Therefore its mean dispersion error matrix is

$$M(\hat{\beta}_2, \beta) = \frac{\sigma^2\beta\beta'}{\sigma^2+\beta'S\beta}. \tag{4.25}$$

Univariate Case $K = 1$

If β is a scalar and $X = x$ is a T-vector, then $\hat{\beta}_2$ (4.20) simplifies to

$$\hat{\beta}_2 = \frac{x'y}{x'x + \sigma^2\beta^{-2}} \tag{4.26}$$

$$= b(1 + \sigma^2\beta^{-2}(x'x)^{-1})^{-1}, \tag{4.27}$$

where b is the ordinary least-squares estimator (OLSE) $b = (x'y)/(x'x)$ for β in the model $y_t = \beta x_t + \epsilon_t$. Hence, $\hat{\beta}_2$ (4.27) is of shrinkage type (cf. Section 3.10.3).

In general, the estimator $\hat{\beta}_2$ (4.20) is a function of the unknown vector $\sigma^{-1}\beta$ (vector of signal-to-noise ratios), and therefore it is not operational. Nevertheless, this estimator provides us with

(i) information about the structure of homogeneous linear estimators that may be used to construct two-stage estimators in practice, and

(ii) the minimum of the R_1 risk within the class of homogeneous linear estimators as

$$R_1(\hat{\beta}, \beta, A) = \text{tr}\{AM(\hat{\beta}_2, \beta)\}, \tag{4.28}$$

where $M(\hat{\beta}_2, \beta)$ is given in (4.25).

To have operational estimators for β, one may replace $\sigma^{-1}\beta$ in (4.27) by estimates or a prior guess or, alternatively, one may demand for unbiasedness of the linear estimator $\hat{\beta} = Cy$.

Homogeneous, Unbiased, R_1-Optimal Estimator

A homogeneous linear estimator is unbiased (see (3.28)) if

$$CX - I = 0 \tag{4.29}$$

or, equivalently, if

$$c_i'X - e_i' = 0 \quad (i = 1, \ldots, K), \tag{4.30}$$

where e_i' and c_i' are the ith row vectors of I and C, respectively. Using (4.29) in (4.15), we find that $R_1(\hat{\beta}, \beta, A)$ becomes $\sigma^2\text{tr}(ACWC')$. Therefore, the optimal C in this case is the solution obtained from

$$\min_C \tilde{R}_1 = \min_C \left\{ \sigma^2\text{tr}\{ACWC'\} - 2\sum_{i=1}^{K} \lambda_i'(c_i'X - e_i')' \right\}, \tag{4.31}$$

where $\lambda_1, \lambda_2, \ldots, \lambda_K$ are K-vectors of Lagrangian multipliers. Writing $\underset{K\times K}{\Lambda'} = (\lambda_1, \ldots, \lambda_K)$, differentiating with respect to C and Λ, and equating to null matrices, we get

$$\frac{\partial \tilde{R}_1}{\partial C} = 2\sigma^2 ACW - 2\Lambda X' = 0, \tag{4.32}$$

$$\frac{\partial \tilde{R}_1}{\partial \Lambda} = 2(CX - I) = 0\,, \tag{4.33}$$

which yield the optimal C as

$$\hat{C}_3 = (X'W^{-1}X)^{-1}X'W^{-1} = S^{-1}X'W^{-1}. \tag{4.34}$$

The matrix $\hat{C}_3$ is consistent with the condition (4.29):

$$\hat{C}_3 X = S^{-1}X'W^{-1}X = S^{-1}S = I. \tag{4.35}$$

Therefore the homogeneous, unbiased, R_1-optimal estimator is specified by

$$\hat{\beta}_3 = b = S^{-1}X'W^{-1}y\,, \tag{4.36}$$

and it has risk and covariance matrix as follows

$$R_1(b, \beta, A) = \sigma^2 \text{tr}(AS^{-1}) = \text{tr}(AV(b))\,, \tag{4.37}$$

$$V(b) = \sigma^2 S^{-1}. \tag{4.38}$$

The following theorem summarizes our findings.

Theorem 4.1 *Assume the generalized linear regression model (4.1) and the quadratic risk function*

$$R_1(\hat{\beta}, \beta, A) = \text{E}(\hat{\beta} - \beta)'A(\hat{\beta} - \beta), \quad A > 0. \tag{4.39}$$

Then the optimal linear estimators for β are

(a) *heterogeneous:* $\hat{\beta}_1 = \beta\,,$

(b) *homogeneous:* $\hat{\beta}_2 = \beta \left[\frac{\beta' X' W^{-1} y}{\sigma^2 + \beta' S \beta} \right],$

(c) *homogeneous unbiased:* $\hat{\beta}_3 = b = S^{-1}X'W^{-1}y.$

The R_1-optimal estimators are independent of A. Further, the optimal estimators are ordered by their risks as

$$R_1(\hat{\beta}_1, \beta, A) \le R_1(\hat{\beta}_2, A) \le R_1(\hat{\beta}_3, \beta, A).$$

4.1.2 R_2-Optimal Estimators

If we allow the symmetric weight matrix A of the quadratic risk $R_1(\hat{\beta}, \beta, A)$ to be nonnegative definite, we are led to the following weaker criterion.

Criterion $\tilde{C}_1$: The linear estimator $\hat{\beta}$ is said to be $\tilde{R}_1$-optimal for β if

$$\text{E}(\hat{\beta} - \beta)'A(\hat{\beta} - \beta) \le \text{E}(\tilde{\beta} - \beta)'A(\tilde{\beta} - \beta) \tag{4.40}$$

holds for (X, W) fixed and for any (β, σ^2) and for any nonnegative definite matrix A where $\tilde{\beta}$ is any other linear estimator. Therefore, any R_1-optimal estimator is $\tilde{R}_1$-optimal, too. Moreover, the following theorem proves that the criteria $\tilde{C}_1$ and C_2 are equivalent.

Theorem 4.2 *The criteria $\tilde{C}_1$ and C_2 are equivalent.*

Proof:
1. Every R_2-optimal estimator $\hat{\beta}$ is $\tilde{R}_1$-optimal: Assume A to be any nonnegative definite matrix with eigenvalues $\lambda_i \geq 0$ and the corresponding orthonormal eigenvectors p_i. Now we can express

$$A = \sum_{i=1}^{K} \lambda_i p_i p_i' . \tag{4.41}$$

If $\hat{\beta}$ is R_2-optimal, then for any estimator $\tilde{\beta}$ and for the choice $a = p_i \quad (i = 1, \dots, K)$, we have

$$\mathrm{E}(\hat{\beta} - \beta)' p_i p_i' (\hat{\beta} - \beta) \leq \mathrm{E}(\tilde{\beta} - \beta)' p_i p_i' (\tilde{\beta} - \beta) , \tag{4.42}$$

and therefore

$$\lambda_i \, \mathrm{E}(\hat{\beta} - \beta)' p_i p_i' (\hat{\beta} - \beta) \leq \lambda_i \, \mathrm{E}(\tilde{\beta} - \beta)' p_i p_i' (\tilde{\beta} - \beta) , \tag{4.43}$$

from which it follows that

$$\mathrm{E}(\hat{\beta} - \beta)' (\sum \lambda_i p_i p_i') (\hat{\beta} - \beta) \leq \mathrm{E}(\tilde{\beta} - \beta)' (\sum \lambda_i p_i p_i') (\tilde{\beta} - \beta) . \tag{4.44}$$

Therefore $\hat{\beta}$ is also $\tilde{R}_1$-optimal.

2. Every $\tilde{R}_1$-optimal estimator $\hat{\beta}$ is R_2-optimal: Choose the nonnegative definite matrix $A = aa'$, where $a \neq 0$ is any K-vector. Then the $\tilde{R}_1$ optimality of $\hat{\beta}$ implies

$$\mathrm{E}(\hat{\beta} - \beta)' aa' (\hat{\beta} - \beta) \leq \mathrm{E}(\tilde{\beta} - \beta)' aa' (\tilde{\beta} - \beta) , \tag{4.45}$$

and hence $\hat{\beta}$ is also R_2-optimal.
This completes the proof of the equivalence of the criteria $\tilde{C}_1$ and C_2.

4.1.3 R_3-Optimal Estimators

Using the risk $R_3(\hat{\beta}, \beta)$ from (4.5) and the heterogeneous linear estimator $\hat{\beta} = Cy + d$, we obtain

$$\begin{aligned} R_3(\hat{\beta}, \beta) &= \mathrm{E}(y - X\hat{\beta})' W^{-1} (y - X\hat{\beta}) \\ &= [(I - CX)\beta - d]' S [(I - CX)\beta - d] \\ &\quad + \sigma^2 \mathrm{tr}[W^{-1}(I - XC) W (I - C'X')] \\ &= u^2 + v^2 , \end{aligned} \tag{4.46}$$

for instance. As the second term v^2 is free from d, the optimal value of d is that value that minimizes the first expression u^2. As u^2 is nonnegative, the minimum value that it can take is zero. Therefore, setting $u^2 = 0$, we get the solution as

$$\hat{d} = (I - \hat{C}X)\beta , \tag{4.47}$$

where $\hat{C}$ is the yet-to-be-determined optimal value of C. This optimal value of C is obtained by minimizing v^2. Now, using Theorem A.13 (iv), we observe that

$$v^2 = \sigma^2 \text{tr}[I + C'SCW - 2C'X'], \tag{4.48}$$

and hence (Theorems A.91 to A.95)

$$\frac{1}{2\sigma^2}\frac{\partial v^2}{\partial C} = SCW - X' = 0, \tag{4.49}$$

and therefore the solution is

$$\hat{C} = S^{-1}X'W^{-1}. \tag{4.50}$$

Inserting $\hat{C}$ in (4.47), we obtain

$$\hat{d} = (I - S^{-1}X'W^{-1}X)\beta = 0. \tag{4.51}$$

Therefore, the R_3-optimal estimator is homogeneous in y. Its expression and properties are stated below.

Theorem 4.3 *The R_3-optimal estimator for β is*

$$b = S^{-1}X'W^{-1}y \tag{4.52}$$

with

$$V(b) = \sigma^2 S^{-1} \tag{4.53}$$

and

$$R_3(b, \beta) = \sigma^2 \,\text{tr}(I - W^{-1}XS^{-1}X') = \sigma^2(T - K), \tag{4.54}$$

where $S = X'W^{-1}X$.

4.2 The Aitken Estimator

In the classical model the best linear unbiased estimator (BLUE) is given by the OLSE $b_0 = (X'X)^{-1}X'y$. In the generalized linear model (4.1) we may find the BLUE for β by using a simple algebraic connection between these two models.

Because W and W^{-1} are symmetric and positive definite, there exist matrices M and N (cf. Theorem A.31 (iii)) such that

$$W = MM \quad \text{and} \quad W^{-1} = NN, \tag{4.55}$$

where $M = W^{1/2}$ and $N = W^{-1/2}$ are regular and symmetric.

Transforming the model (4.1) by premultiplication with N:

$$Ny = NX\beta + N\epsilon \tag{4.56}$$

and letting

$$Ny = \tilde{y}, \quad NX = \tilde{X}, \quad N\epsilon = \tilde{\epsilon}, \tag{4.57}$$

we see that

$$\mathrm{E}(\tilde{\epsilon}) = \mathrm{E}(N\epsilon) = 0, \quad \mathrm{E}(\tilde{\epsilon}\tilde{\epsilon}') = \mathrm{E}(N\epsilon\epsilon' N) = \sigma^2 I. \tag{4.58}$$

Therefore, the linearly transformed model $\tilde{y} = \tilde{X}\beta + \tilde{\epsilon}$ satisfies the assumptions of the classical model. The OLSE b in this model may be written as

$$\begin{aligned} b &= (\tilde{X}'\tilde{X})^{-1}\tilde{X}'\tilde{y} \\ &= (X'NN'X)^{-1}X'NN'y \\ &= (X'W^{-1}X)^{-1}X'W^{-1}y. \end{aligned} \tag{4.59}$$

Based on Theorem 3.5 we may conclude that the estimator is unbiased:

$$\begin{aligned} \mathrm{E}(b) &= (X'W^{-1}X)^{-1}X'W^{-1}\,\mathrm{E}(y) \\ &= (X'W^{-1}X)^{-1}X'W^{-1}X\beta = \beta \end{aligned} \tag{4.60}$$

and has minimal variance. This may be proved as follows.

Let $\tilde{\beta} = \tilde{C}y$ be another linear unbiased estimator for β and let

$$\tilde{C} = \hat{C} + D \tag{4.61}$$

with the optimal matrix

$$\hat{C} = S^{-1}X'W^{-1}; \tag{4.62}$$

then the unbiasedness of $\tilde{\beta}$ is ensured by $DX = 0$ including $\hat{C}WD = 0$. Then we obtain the covariance matrix of $\tilde{\beta}$ as

$$\begin{aligned} V(\tilde{\beta}) &= \mathrm{E}(\tilde{C}\epsilon\epsilon'\tilde{C}') \\ &= \sigma^2(\hat{C} + D)W(\hat{C}' + D') \\ &= \sigma^2\hat{C}W\hat{C}' + \sigma^2 DWD' \\ &= V(b) + \sigma^2 DWD', \end{aligned} \tag{4.63}$$

implying $V(\tilde{\beta}) - V(b) = \sigma^2 D'WD$ to be nonnegative definite (cf. Theorem A.41 (v)).

Theorem 4.4 (Gauss-Markov-Aitken) *If $y = X\beta + \epsilon$ where $\epsilon \sim (0, \sigma^2 W)$, the generalized least-squares estimator (GLSE)*

$$b = (X'W^{-1}X)^{-1}X'W^{-1}y \tag{4.64}$$

is unbiased and is the best linear unbiased estimator for β. Its covariance matrix is given by

$$V(b) = \sigma^2(X'W^{-1}X)^{-1} = \sigma^2 S^{-1}. \tag{4.65}$$

The estimator b is R_3-optimal as well as the homogeneous, unbiased R_1- and R_2-optimal solution.

For the other unknown parameter σ^2 and the covariance matrix, the following estimators are available:

$$s^2 = \frac{(y - Xb)'W^{-1}(y - Xb)}{T - K} \tag{4.66}$$

and

$$\hat{V}(b) = s^2 S^{-1}. \tag{4.67}$$

These estimators are unbiased for σ^2 and $\sigma^2 S^{-1}$, respectively:

$$\mathrm{E}(s^2) = R_3(b, \beta)(T - K)^{-1} = \sigma^2 \quad \text{and} \quad \mathrm{E}(\hat{V}(b)) = \sigma^2 S^{-1}. \tag{4.68}$$

Analogous to Theorem 3.6, we obtain

Theorem 4.5 *Assume the generalized linear model (4.1). Then the best linear unbiased estimator of* $d = a'\beta$ *and its variance are given by*

$$\hat{d} = a'b\,, \tag{4.69}$$

$$\mathrm{var}(\hat{d}) = \sigma^2 a' S^{-1} a = a' V(b) a\,. \tag{4.70}$$

For a general least squares approach, see Section 4.9.

4.3 Misspecification of the Dispersion Matrix

One of the features of the ordinary least-squares estimator $b_0 = (X'X)^{-1}X'y$ is that in the classical model with uncorrelated errors, no knowledge of σ^2 is required for point estimation of β. When the residuals are correlated, it is necessary for point estimation of β to have prior knowledge or assumptions about the covariance matrix W, or at least an estimate of it.

Assuming the general linear model $y = X\beta + \epsilon$, $\epsilon \sim (0, \sigma^2 W)$ so that W is the true covariance matrix, then misspecification relates to using a covariance matrix $A \neq W$.

Reasons for this misspecification of the covariance matrix could be one of the following:

(i) The correlation structure of disturbances may have been ignored in order to use OLS estimation and hence simplify calculations. (This is done, for instance, as the first step in model building in order to obtain a rough idea of the underlying relationships.)

(ii) The true matrix W may be unknown and may have to be estimated by $\hat{W}$ (which is stochastic).

(iii) The correlation structure may be better represented by a matrix that is different from W.

In any case, the resulting estimator will have the form

$$\hat{\beta} = (X'A^{-1}X)^{-1}X'A^{-1}y, \tag{4.71}$$

where the existence of A^{-1} and $(X'A^{-1}X)^{-1}$ have to be ensured. (For instance, if $A > 0$, then the above inverse exists.) Now, the estimator $\hat{\beta}$ is unbiased for β, that is,

$$\mathrm{E}(\hat{\beta}) = \beta \tag{4.72}$$

for any misspecified matrix A as $\operatorname{rank}(X'A^{-1}X) = K$.

Further, $\hat{\beta}$ has the dispersion matrix

$$V(\hat{\beta}) = \sigma^2(X'A^{-1}X)^{-1}X'A^{-1}WA^{-1}X(X'A^{-1}X)^{-1} \tag{4.73}$$

so that using the false matrix A results in a loss in efficiency in estimating β by $\hat{\beta}$ instead of the GLSE $b = S^{-1}X'W^{-1}y$, as is evident from

$$\begin{aligned} V(\hat{\beta}) - V(b) &= \sigma^2[(X'A^{-1}X)^{-1}X'A^{-1} - S^{-1}X'W^{-1}] \\ &\quad \times W[(X'A^{-1}X)^{-1}X'A^{-1} - S^{-1}X'W^{-1}]', \end{aligned} \tag{4.74}$$

which is nonnegative definite (Theorems 4.4 and A.41 (iv)).

There is no loss in efficiency if and only if

$$(X'A^{-1}X)^{-1}X'A^{-1} = S^{-1}X'W^{-1},$$

and then $\hat{\beta} = b$.

Let us now investigate the most important case, in which the OLSE $b = (X'X)^{-1}X'y = b_0$, say, is mistakenly used instead of the true GLSE. That is, let us assume $A = I$. Letting $U = (X'X)^{-1}$, we get the increase in dispersion due to the usage of the OLSE $b_0 = UX'y$ instead of the GLSE as (see (4.74))

$$V(b_0) - V(b) = \sigma^2(UX' - S^{-1}X'W^{-1}) \times W(XU - W^{-1}XS^{-1}).$$

Therefore, it is clear that $V(b_0) = V(b)$ holds if and only if

$$UX' = S^{-1}X'W^{-1}.$$

This fact would imply that

$$UX' = S^{-1}X'W^{-1} \Leftrightarrow X'WZ = 0 \Leftrightarrow X'W^{-1}Z = 0, \tag{4.75}$$

where Z is a matrix of maximum rank such that $Z'X = 0$. Since $W > 0$, we can find a symmetric square root $W^{\frac{1}{2}}$ such that $W^{\frac{1}{2}}W^{\frac{1}{2}} = W$. Furthermore, X and Z span the whole space so that $W^{1/2}$ can be expressed as

$$\begin{aligned} W^{\frac{1}{2}} &= XA_1 + ZB_1 \\ \Rightarrow W &= XA_1A_1'X' + XA_1B_1'Z' + ZB_1A_1'X' + ZB_1B_1'Z'. \end{aligned}$$

Expressing the condition $X'WZ = 0$:

$$X'XA_1B_1'Z'Z = 0 \quad \Leftrightarrow \quad A_1B_1' = 0.$$

Similarly, $B_1 A_1' = 0$, so that

$$W = XAX' + ZBZ',$$

where A and B are nonnegative definite matrices. So we have the following theorem, which is proved under more general conditions in Rao (1967) and Rao (1968).

Theorem 4.6 *The OLSE and the GLSE are identical if and only if the following form holds:*

$$W = XAX' + ZBZ', \tag{4.76}$$

which is equivalent to the condition $X'WZ = 0$.

It is easy to see that if the regressor matrix X has one column as the unit vector, then for the choice

$$W = (1-\rho)I + \rho\mathbf{1}\mathbf{1}' \quad (0 \le \rho < 1), \tag{4.77}$$

the condition $X'WZ = 0$ holds. Thus (4.77) is one choice of W for which OLSE = GLSE (McElroy, 1967).

Note: The condition $0 \le \rho < 1$ ensures that $(1-\rho)I + \rho\mathbf{1}\mathbf{1}'$ is positive definite for all values of the sample size T. For given T, it would be replaced by $-1/(T-1) < \rho < 1$. A matrix of type (4.77) is said to be *compound symmetric.*

Clearly, an incorrect specification of W will also lead to errors in estimating σ^2 by $\hat{\sigma}^2$, which is based on $\hat{\epsilon}$. Assume that A is chosen instead of W. Then the vector of residuals is

$$\hat{\epsilon} = y - X\hat{\beta} = (I - X(X'A^{-1}X)^{-1}X'A^{-1})\epsilon,$$

and we obtain

$$\begin{aligned}
(T-K)\hat{\sigma}^2 &= \hat{\epsilon}'\hat{\epsilon} \\
&= \operatorname{tr}\{(I - X(X'A^{-1}X)^{-1}X'A^{-1}) \\
&\quad \times \epsilon\epsilon'(I - A^{-1}X(X'A^{-1}X)^{-1}X')\}, \\
\mathrm{E}(\hat{\sigma}^2)(T-K) &= \sigma^2\operatorname{tr}(W - X(X'A^{-1}X)^{-1}X'A^{-1}) \\
&\quad + \operatorname{tr}\{\sigma^2 X(X'A^{-1}X)^{-1}X'A^{-1}(I-2W) + XV(\hat{\beta})X'\}.
\end{aligned} \tag{4.78}$$

Standardizing the elements of W by $\operatorname{tr}(W) = T$, and using Theorem A.13 (i), the first expression in (4.78) equals $T-K$. For the important case $A = I$, expression (4.78) becomes

$$\begin{aligned}
\mathrm{E}(\hat{\sigma}^2) &= \sigma^2 + \frac{\sigma^2}{T-K}\operatorname{tr}[X(X'X)^{-1}X'(I-W)] \\
&= \sigma^2 + \frac{\sigma^2}{T-K}(K - \operatorname{tr}[(X'X)^{-1}X'WX]).
\end{aligned} \tag{4.79}$$

The final term represents the bias of s^2 when the OLSE is mistakenly used. This term tends to be negative if the disturbances are positively correlated, that is, there is a tendency to underestimate the true variance. Goldberger (1964, p. 239) has investigated the bias of the estimate $s^2(X'X)^{-1}$ of $V(b_0)$ in case W is the dispersion matrix of heteroscedastic or autoregressive processes. More general investigations of this problem are given in Dufour (1989).

Remark: Theorem 4.6 presents the general condition for the equality of the OLSE and the GLSE. Puntanen (1986) has presented an overview of alternative conditions. Baksalary (1988) characterizes a variety of necessary and sufficient conditions by saying that all these covariance structures may be ignored without any consequence for best linear unbiased estimation. Further interesting results concerning this problem and the relative efficiency of the OLSE are discussed in Krämer (1980) and Krämer and Donninger (1987).

4.4 Heteroscedasticity and Autoregression

Heteroscedasticity of ϵ means that the disturbances are uncorrelated but not identically distributed, that is $\{\epsilon_t\}$ is said to be heteroscedastic if

$$\mathrm{E}(\epsilon_t \epsilon_{t'}) = \begin{cases} \sigma_t^2 & \text{for } t = t', \\ 0 & \text{for } t \neq t', \end{cases} \tag{4.80}$$

or, in matrix notation,

$$\mathrm{E}(\epsilon\epsilon') = \sigma^2 W = \sigma^2 \begin{pmatrix} k_1 & 0 & \cdots & 0 \\ 0 & k_2 & \cdots & 0 \\ \vdots & \vdots & \ddots & \vdots \\ 0 & 0 & \cdots & k_T \end{pmatrix} = \sigma^2 \operatorname{diag}(k_1, \ldots, k_T), \tag{4.81}$$

where $k_t = \sigma_t^2/\sigma^2$ can vary in the interval $[0, \infty)$.

Standardizing W by $\operatorname{tr}\{W\} = T$, we obtain

$$\sum k_t = \sum \frac{\sigma_t^2}{\sigma^2} = T, \tag{4.82}$$

and hence $\sigma^2 = \sum \sigma_t^2 / T$ is the arithmetic mean of the variances. If $k_t = k$ for $t = 1, \ldots, T$, we have the classical model, also called a model with homoscedastic disturbances. Now

$$W^{-1} = \operatorname{diag}(k_1^{-1}, \ldots, k_T^{-1}), \tag{4.83}$$

and therefore the GLSE $b = S^{-1}X'W^{-1}y$, with $X' = (x_1, \ldots, x_T)$, is of the special form

$$b = \left(\sum x_t x_t' \frac{1}{k_t}\right)^{-1} \left(\sum x_t y_t \frac{1}{k_t}\right). \tag{4.84}$$

It follows that b is a weighted estimator minimizing the weighted sum of squared errors:

$$R_3(\hat{\beta}, \beta) = \hat{\epsilon}' W^{-1} \hat{\epsilon} = \sum \hat{\epsilon}_t^2 \frac{1}{k_t} \,. \tag{4.85}$$

A typical situation of heteroscedasticity is described in Goldberger (1964, p. 235). Let us assume that in the univariate model

$$y_t = \alpha + \beta x_t + \epsilon_t \quad (t = 1, \ldots, T),$$

the variance of ϵ_t is directly proportional to the square of x_t, that is,

$$\operatorname{var}(\epsilon_t) = \sigma^2 x_t^2 \,.$$

Then we have $W = \operatorname{diag}(x_1^2, \ldots, x_T^2)$, namely, $k_t = x_t^2$. Applying b as in (4.84) is then equivalent to transforming the data according to

$$\frac{y_t}{x_t} = \alpha \left(\frac{1}{x_t} \right) + \beta + \frac{\epsilon_t}{x_t}, \quad \operatorname{var}\left(\frac{\epsilon_t}{x_t} \right) = \sigma^2$$

and calculating the OLSE of (α, β). An interesting feature of this special case is that the roles of intercept term and regression coefficient in the original model are interchanged in the transformed model.

Another model of practical importance is that of aggregate data: We do not have the original samples y and X, but we do have the sample means

$$\bar{y}_t = \frac{1}{n_t} \sum_{j=1}^{n_t} y_j \,, \quad \bar{x}_{ti} = \frac{1}{n_t} \sum_{j=1}^{n_t} x_{ji}$$

so that the relationship is

$$\bar{y}_t = \sum_{i=1}^{K} \beta_i \bar{x}_{ti} + \bar{\epsilon}_t \quad (t = 1, \ldots, T),$$

where $\operatorname{var}(\bar{\epsilon}_t) = \sigma^2 / n_t$. Thus we have $W = \operatorname{diag}(1/n_1, \ldots, 1/n_T)$.

Another model of practical relevance with heteroscedastic disturbances is given by the block diagonal design. In many applications we are confronted with the specification of grouped data (see, for example, the models of analysis of variance). It may be assumed that the regression variables are observed over m periods (example: the repeated measurement model) or for m groups (example: m therapies) and in n situations. Thus the sample size of each individual is m, and the global sample size is therefore $T = mn$. Assuming that in any group the within-group variances are identical (i.e., $\mathrm{E}\, \epsilon_i \epsilon_i' = \sigma_i^2 I$ $(i = 1, \ldots, n)$) and that the between-group disturbances are uncorrelated, then we obtain the block diagonal dispersion matrix

$$\mathrm{E}(\epsilon\epsilon') = \begin{pmatrix} \sigma_1^2 I & 0 & \cdots & 0 \\ 0 & \sigma_2^2 I & \cdots & 0 \\ \vdots & \vdots & \ddots & \vdots \\ 0 & 0 & \cdots & \sigma_m^2 I \end{pmatrix} = \operatorname{diag}(\sigma_1^2 I, \ldots, \sigma_m^2 I). \tag{4.86}$$

The model may be written as

$$\begin{pmatrix} y_1 \\ y_2 \\ \vdots \\ y_m \end{pmatrix} = \begin{pmatrix} X_1 \\ X_2 \\ \vdots \\ X_m \end{pmatrix} \beta + \begin{pmatrix} \epsilon_1 \\ \epsilon_2 \\ \vdots \\ \epsilon_m \end{pmatrix}. \tag{4.87}$$

Note: This structure of a linear model occurs more generally in the m-dimensional (multivariate) regression model and in the analysis of panel data.

More generally, we may assume that the disturbances follow the so-called process of *intraclass correlation*. The assumptions on ϵ are specified as follows:

$$\epsilon_{tj} = v_j + u_{tj}\,, \quad t = 1, \ldots, m,\ j = 1, \ldots, n\,, \tag{4.88}$$

where the disturbances v_j are identical for the m realizations of each of the n individuals:

$$\begin{aligned} \mathrm{E}\, v_j = 0, \quad \mathrm{var}(v_j) &= \sigma_v^2\,, \quad j = 1, \ldots, n\,, \\ \mathrm{cov}(v_j v_j') &= 0\,, \quad j \neq j'\,. \end{aligned} \tag{4.89}$$

The disturbances u_{tj} vary over all $T = mn$ realizations and have

$$\begin{aligned} \mathrm{E}\, u_{tj} &= 0\,, \quad \mathrm{var}(u_{tj}) = \sigma_u^2\,, \\ \mathrm{cov}(u_{tj}, u_{t'j'}) &= 0\,, \quad (t,j) \neq (t',j')\,, \end{aligned} \tag{4.90}$$

and, moreover,

$$\mathrm{cov}(u_{tj}, v_{j'}) = 0 \quad \text{for all } t, j, j'\,, \tag{4.91}$$

that is, both processes $\{u\}$ and $\{v\}$ are uncorrelated.

The $T \times T$-dispersion matrix of ϵ is therefore of the form

$$\mathrm{E}\, \epsilon\epsilon' = \mathrm{diag}(\Phi, \ldots, \Phi)\,, \tag{4.92}$$

where Φ is the $m \times m$-matrix of intraclass correlation:

$$\Phi = \mathrm{E}(u_j u_j') = \sigma^2 \Psi = \sigma^2 \begin{pmatrix} 1 & \gamma & \cdots & \gamma \\ \gamma & 1 & \cdots & \gamma \\ \vdots & \vdots & & \vdots \\ \gamma & \gamma & \cdots & \gamma \end{pmatrix} \tag{4.93}$$

with

$$\sigma^2 = \sigma_v^2 + \sigma_u^2 \quad \text{and} \quad \gamma = \frac{\sigma_v^2}{\sigma^2}\,.$$

As pointed out in Schönfeld (1969), we may write

$$\Psi = (1 - \gamma)\left(I + \frac{\gamma}{1-\gamma} 11'\right) \tag{4.94}$$

so that its inverse is

$$\Psi^{-1} = \frac{1}{1-\gamma}\left(I - \frac{\gamma}{1+\gamma(m-1)}11'\right). \tag{4.95}$$

Based on this, we get the GLSE as

$$b = \left[\sum_{j=1}^{n} D(x_j, x_j)\right]^{-1} \left[\sum_{j=1}^{n} d(x_j, x_j)\right] \tag{4.96}$$

with the modified central sample moments

$$D(x_j, x_j) = \frac{1}{m}X_j'X_j - \frac{\gamma m}{1+\gamma(m-1)}\bar{x}_j\bar{x}_j{}'$$

and

$$d(x_j, x_j') = \frac{1}{m}X_j'y_j - \frac{\gamma m}{1+\gamma(m-1)}\bar{x}_j\bar{y}_j\,.$$

Remark: Testing for heteroscedasticity is possible if special rank test statistics for any of the specified models of the above are developed Huang (1970). As a general test, the F-test is available when normality of the disturbances can be assumed. On the other hand, the well-known tests for homogeneity of variances may be chosen. A common difficulty is that there is no procedure for determining the optimal grouping of the estimated disturbances $\hat{\epsilon}_t$, whereas their grouping greatly influences the test procedures.

Autoregressive Disturbances

It is a typical situation in time-series analysis that the data are interdependent, with many reasons for interdependence of the successive disturbances. Autocorrelation of first and higher orders in the disturbances can arise, for example, from observational errors in the included variables or from the estimation of missing data by either averaging or extrapolating.

Assume $\{u_t\}$ $(t = \ldots, -2, -1, 0, 1, 2, \ldots)$ to be a random process having

$$\mathrm{E}(u_t) = 0, \quad \mathrm{E}(u_t^2) = \sigma_u^2, \quad \mathrm{E}(u_t u_{t'}) = 0 \quad \text{for } t \neq t'. \tag{4.97}$$

Using $\{u_t\}$, we generate the following random process:

$$v_t - \mu = \rho(v_{t-1} - \mu) + u_t\,, \tag{4.98}$$

where $|\rho| < 1$ is the autocorrelation coefficient that has to be estimated. By repeated substitution of the model (4.98), we obtain

$$v_t - \mu = \sum_{s=0}^{\infty} \rho^s u_{t-s}\,, \tag{4.99}$$

and therefore with (4.97)

$$\mathrm{E}(v_t) \;=\; \mu + \sum_{s=0}^{\infty} \rho^s\, \mathrm{E}(u_{t-s}) = \mu\,, \tag{4.100}$$

$$\begin{aligned} \mathrm{E}(v_t - \mu)^2 &= \sum_{s=0}^{\infty}\sum_{r=0}^{\infty} \rho^{s+r}\, \mathrm{E}(u_{t-s}u_{t-r}) \\ &= \sigma_u^2 \sum_{s=0}^{\infty} \rho^{2s} = \sigma_u^2(1-\rho^2)^{-1} = \sigma^2 . \end{aligned} \tag{4.101}$$

Then the vector $v' = (v_1, \ldots, v_T)$ has the mean

$$\mathrm{E}(v') = (\mu, \ldots, \mu)$$

and dispersion matrix $\Sigma = \sigma^2 W$, where

$$W = \begin{pmatrix} 1 & \rho & \rho^2 & \cdots & \rho^{T-1} \\ \rho & 1 & \rho & \cdots & \rho^{T-2} \\ \vdots & \vdots & \vdots & & \vdots \\ \rho^{T-1} & \rho^{T-2} & \rho^{T-3} & \cdots & 1 \end{pmatrix} \tag{4.102}$$

is regular and has the inverse

$$W^{-1} = \frac{1}{1-\rho^2} \begin{pmatrix} 1 & -\rho & 0 & \cdots & 0 & 0 \\ -\rho & 1+\rho^2 & -\rho & \cdots & 0 & 0 \\ 0 & -\rho & 1+\rho^2 & \cdots & 0 & 0 \\ \vdots & \vdots & \vdots & & \vdots & \vdots \\ 0 & 0 & 0 & \cdots & 1+\rho^2 & -\rho \\ 0 & 0 & 0 & \cdots & -\rho & 1 \end{pmatrix} . \tag{4.103}$$

Letting $\epsilon_t = v_t$ and $\mu = 0$, we obtain the generalized linear regression model with autocorrelated disturbances. This model is said to be a first-order autoregression. The GLSE for β is

$$b = (X'W^{-1}X)^{-1}X'W^{-1}y , \tag{4.104}$$

where W^{-1} is given by (4.103). From (4.102) it follows that the correlation between ϵ_t and $\epsilon_{t-\tau}$ is $\sigma^2\rho^\tau$, that is, the correlation depends on the difference of time $|\tau|$ and decreases for increasing values of $|\tau|$ as $|\tau| < 1$.

Testing for Autoregression

The performance of the GLSE $b = (X'W^{-1}X)^{-1}X'W^{-1}y$ when W is misspecified was investigated in Section 4.3. Before b can be applied, however, the assumptions on W, such as (4.78), have to be checked. Since no general test is available for the hypothesis "ϵ is spherically distributed," we have to test specific hypotheses on W. If the first-order autoregressive scheme is a plausible proposition, the well-known Durbin-Watson test can be applied (see Durbin and Watson (1950, 1951)). If $\rho > 0$ is suspected, then the Durbin-Watson test for

$$H_0\colon \rho = 0 \quad \text{against} \quad H_1\colon \rho > 0$$

is based on the test statistic

$$d = \frac{\sum_{t=2}^{T}(\hat{\epsilon}_t - \hat{\epsilon}_{t-1})^2}{\sum_{t=1}^{T}\hat{\epsilon}_t^2}, \tag{4.105}$$

where $\hat{\epsilon}_t$ are the estimated residuals from the classical regression model (i.e., $W = I$). The statistic d is seen to be a function of the empirical coefficient of autocorrelation $\hat{\rho}$ of the vector of residuals $\hat{\epsilon} = y - X(X'X)^{-1}X'y$:

$$\hat{\rho} = \frac{\sum_{t=2}^{T}\hat{\epsilon}_t\hat{\epsilon}_{t-1}}{\sqrt{\sum_{t=2}^{T}\hat{\epsilon}_t^2}\sqrt{\sum_{t=2}^{T}\epsilon_{t-1}^2}}. \tag{4.106}$$

Using the approximation

$$\sum_{t=1}^{T}\hat{\epsilon}_t^2 \approx \sum_{t=2}^{T}\hat{\epsilon}_t^2 \approx \sum_{t=2}^{T}\hat{\epsilon}_{t-1}^2, \tag{4.107}$$

we obtain

$$d \approx \frac{2\sum\hat{\epsilon}_t^2 - 2\sum\hat{\epsilon}_t\hat{\epsilon}_{t-1}}{\sum\hat{\epsilon}_t^2} \approx 2(1-\hat{\rho}) \tag{4.108}$$

and therefore $0 < d < 4$. For $\hat{\rho} = 0$ (i.e., no autocorrelation) we get $d = 2$. The distribution of d obviously depends on X. Consequently, the exact critical values obtained from such a distribution will be functions of X and as such it would be difficult to prepare tables. To overcome this difficulty, we find two statistics d_l and d_u such that $d_l \leq d \leq d_u$ and their distributions do not depend on X. Let d_l^* be the critical value obtained from the distribution of d_l, and let d_u^* be the critical value found from the distribution of d_u. Some of these critical values are given in Table 4.1; see Durbin and Watson (1950, 1951) for details.

The one-sided Durbin-Watson test for H_0: $\rho = 0$ against H_1: $\rho > 0$ is as follows:

do not reject H_0	if	$d \geq d_u^*$,
reject H_0	if	$d \leq d_l^*$,
no decision	if	$d_l^* < d < d_u^*$.

If the alternative hypothesis is H_1: $\rho < 0$, the test procedure remains the same except that $\tilde{d} = (4 - d)$ is used as the test statistic in place of d.

For the two-sided alternative H_1: $\rho \neq 0$, the procedure is as follows:

do not reject H_0	if	$d\,(\text{or}\,\tilde{d}) \geq d_u^*$,
reject H_0	if	$d\,(\text{or}\,\tilde{d}) \leq d_l^*$,
no decision	if	$d_u^* < d < (4 - d_l^*)$.

Note: Some of the statistical packages include the exact critical values of the Durbin-Watson test statistic.

Estimation in Case of Autocorrelation

Two-Stage Estimation. If H_0: $\rho = 0$ is rejected, then the estimator $\hat{\rho}$ from (4.106) is inserted in W^{-1} from (4.103), resulting in the estimator $\hat{W}^{-1}$ and

$$\hat{b} = (X'\hat{W}^{-1}X)^{-1}X'\hat{W}^{-1}y. \tag{4.109}$$

If some moderate general conditions hold, this estimator is consistent, that is, we may expect that

$$\text{plim}\hat{b} = \beta. \tag{4.110}$$

It may happen that this procedure has to be repeated as an iterative process until a relative stability of the estimators $\hat{\rho}$ and $\hat{\beta}$ is achieved. The iteration starts with the OLSE $b_0 = (X'X)^{-1}X'y$. Then $\hat{\epsilon} = y - Xb_0$, $\hat{\rho}$ (4.106), and $\hat{b}$ (4.109) are calculated. Then again $\hat{\epsilon} = y - X\hat{b}$, $\hat{\rho}$ (using this last $\hat{\epsilon}$), and $\hat{b}$ are calculated. This process stops if changes in $\hat{\rho}$ and $\hat{b}$ are smaller than a given value.

Transformation of Variables. As an alternative procedure for overcoming autoregression, the following data transformation is available. The model with transformed variables has homoscedastic disturbances and may be estimated by the OLSE.

We define the following differences:

$$\Delta_\rho y_t = y_t - \rho y_{t-1}, \tag{4.111}$$

$$\Delta_\rho x_{it} = x_{it} - \rho x_{it-1}, \tag{4.112}$$

$$u_t = \epsilon_t - \rho\epsilon_{t-1}, \tag{4.113}$$

where $\text{E}(uu') = \sigma^2 I$ (see (4.97) and (4.98) with $\epsilon_t = v_t$).

Then the model

$$y = X\beta + \epsilon, \quad \epsilon \sim (0, \sigma^2 W)$$

with W from (4.102) is transformed to the classical model

$$\Delta_\rho y_t = \beta_0(1-\rho) + \beta_1\Delta_\rho x_{1t} + \ldots + \beta_K\Delta_\rho x_{Kt} + u_t. \tag{4.114}$$

Note: With the exception of β_0, all the parameters β_i are unchanged.

When ρ is known, the parameters in model (4.114) can be estimated by OLSE. If ρ is unknown, it has to be estimated by $\hat{\rho}$ (4.106). Then the parameters β_i in model (4.114) are estimated by OLSE (two-stage OLSE) when ρ is replaced by $\hat{\rho}$ (Cochrane and Orcutt, 1949). In practice, one can expect that both of the above two-stage procedures will almost coincide.

If ρ is near 1, the so-called first differences

$$\Delta y_t = y_t - y_{t-1}, \tag{4.115}$$

$$\Delta x_{it} = x_{it} - x_{it-1}, \tag{4.116}$$

$$u_t = \epsilon_t - \epsilon_{t-1} \tag{4.117}$$

TABLE 4.1. Five percent significance points for the Durbin-Watson test (Durbin and Watson, 1951).

	$K^*=1$		$K^*=2$		$K^*=3$		$K^*=4$		$K^*=5$	
T	d_l^*	d_u^*	d_l^*	d_u^*	d_l^*	d_u^*	d_l^*	d_u^*	d_l^*	d_u^*
15	1.08	1.36	0.95	1.54	0.82	1.75	0.69	1.97	0.56	2.21
20	1.20	1.41	1.10	1.54	1.00	1.68	0.90	1.83	0.79	1.99
30	1.35	1.49	1.28	1.57	1.21	1.67	1.14	1.74	1.07	1.83
40	1.44	1.54	1.39	1.60	1.34	1.66	1.29	1.72	1.23	1.79
50	1.50	1.59	1.46	1.63	1.42	1.65	1.38	1.72	1.34	1.77

Note: K^* is the number of exogeneous variables when the dummy variable is excluded.

are taken.

Remark: The transformed endogenous variables in (4.115) are almost uncorrelated. The method of first differences is therefore applied as an attempt to overcome the problem of autocorrelation.

Note: An overview of more general problems and alternative tests for special designs including power analyses may be found in Judge et al. (1980, Chapter 5).

Example 4.1: We demonstrate an application of the test procedure for autocorrelation in the following model with a dummy variable **1** and one exogeneous variable X:

$$y_t = \beta_0 + \beta_1 x_t + \epsilon_t, \quad \epsilon_t \sim N(0, \sigma_t^2), \tag{4.118}$$

or, in matrix formulation,

$$y = (\mathbf{1}, X) \begin{pmatrix} \beta_0 \\ \beta_1 \end{pmatrix} + \epsilon, \quad \epsilon \sim N(0, \sigma^2 W). \tag{4.119}$$

Let the following sample of size $T = 6$ be given:

$$y = \begin{pmatrix} 1 \\ 3 \\ 2 \\ 3 \\ 0 \\ 2 \end{pmatrix}, \quad X = \begin{pmatrix} 1 & -4 \\ 1 & 3 \\ 1 & 4 \\ 1 & 5 \\ 1 & 3 \\ 1 & 3 \end{pmatrix}.$$

We get

$$\begin{aligned} X'X &= \begin{pmatrix} 6 & 14 \\ 14 & 84 \end{pmatrix}, \quad X'y = \begin{pmatrix} 11 \\ 34 \end{pmatrix}, \\ |X'X| &= 308, \\ (X'X)^{-1} &= \frac{1}{308} \begin{pmatrix} 84 & -14 \\ -14 & 6 \end{pmatrix}, \end{aligned}$$

$$\begin{aligned}
b_0 &= (X'X)^{-1}X'y = \frac{1}{308}\begin{pmatrix} 448 \\ 50 \end{pmatrix} = \begin{pmatrix} 1.45 \\ 0.16 \end{pmatrix}, \\
\hat{y} &= Xb_0 = \begin{pmatrix} 0.81 \\ 1.93 \\ 2.09 \\ 2.25 \\ 1.93 \\ 1.93 \end{pmatrix}, \quad \hat{\epsilon} = y - Xb_0 = \begin{pmatrix} 0.19 \\ 1.07 \\ -0.09 \\ 0.75 \\ -1.93 \\ 0.07 \end{pmatrix}, \\
\hat{\rho} &= \frac{\sum_{t=2}^{6} \hat{\epsilon}_{t-1}\hat{\epsilon}_t}{\sum_{t=2}^{6} \hat{\epsilon}_{t-1}^2} = \frac{-1.54}{5.45} = -0.28, \\
d &= 2(1-\hat{\rho}) = 2.56, \\
\tilde{d} &= 4 - d = 1.44.
\end{aligned}$$

From Table 4.1 we find, for $K^* = 1$, the critical value corresponding to $T = 6$ is $d_u^* < 1.36$, and therefore H_0: $\rho = 0$ is not rejected. The autocorrelation coefficient $\hat{\rho} = -0.28$ is not significant. Therefore, $\begin{pmatrix} \beta_0 \\ \beta_1 \end{pmatrix}$ of model (4.119) is estimated by the OLSE $b_0 = \begin{pmatrix} 1.45 \\ 0.16 \end{pmatrix}$.

4.5 Mixed Effects Model: A Unified Theory of Linear Estimation

4.5.1 Mixed Effects Model

Most, if not all, linear statistical models used in practice are included in the formulation

$$y = X\beta + U\xi + \epsilon \tag{4.120}$$

where y is a T-vector of observations, X is a given $T \times K$-design matrix, β is an unknown K-vector of fixed parameters, ξ is an unknown s-vector of unknown random effects, U is a given $T \times s$-matrix, and ϵ is an unknown T-vector of random errors with the following characteristics:

$$\mathrm{E}(\epsilon) = 0, \quad \mathrm{E}(\epsilon\epsilon') = V, \quad \mathrm{E}(\xi) = 0, \quad \mathrm{E}(\xi\xi') = \Gamma, \quad \mathrm{Cov}(\epsilon, \xi) = 0. \tag{4.121}$$

The components of ξ are unobserved covariates on individuals, and the problem of interest is the estimation of linear combinations of β and ξ, as in animal breeding programs, where ξ represents intrinsic values of individuals on the basis of which some individuals are chosen for future breeding (see Henderson (1984) for applications in animal breeding programs). We assume that the matrices V and Γ are known and derive optimum estimates of fixed and random effects. In practice, V and Γ are usually unknown, and they may be estimated provided they have a special structure depending on a few unknown parameters.

4.5.2 A Basic Lemma

First we prove a basic lemma due to Rao (1989), which provides a solution to all estimation problems in linear models. We say that G_* is a minimum in a given set of n.n.d. matrices $\{G\}$ of order T, if G_* belongs to the set, and for any element $G \in \{G\}$, $G - G_*$ is a nonnegative-definite (n.n.d.) matrix, in which case we write $G_* \leq G$.

Lemma 4.7 (Rao, 1989) *Let $V : T \times T$ be n.n.d., $X : T \times K$, $F : T \times K$, and $P : K \times K$ be given matrices such that $\mathcal{R}(F) \subset \mathcal{R}(V : X)$ and $\mathcal{R}(P) \subset \mathcal{R}(X')$, and consider the $K \times K$-matrix function of $A : T \times K$*

$$f(A) = A'VA - A'F - F'A. \tag{4.122}$$

Then

$$\min_{X'A=P} f(A) = f(A_*) \tag{4.123}$$

where (A_*, B_*) is a solution of the equation

$$\left.\begin{array}{rcl} VA + XB & = & F \\ X'A & = & P\,. \end{array}\right\} \tag{4.124}$$

Furthermore,

$$f(A_*) = \min_{X'A=P} f(A) = -A_*'F - B_*'P = -F'A_* - P'B_* \tag{4.125}$$

and is unique for any solution of (4.124).

Proof: Let (A_*, B_*) be a solution of (4.124). Any A such that $X'A = P$ can be written as $A_* + ZC$ where $Z = X^{\perp}$ and C is arbitrary. Then

$$\begin{aligned} f(A) &= A'VA - A'F - F'A \\ &= (A_* + ZC)'V(A_* + ZC) - (A_* + ZC)'F - F'(A_* + ZC) \\ &= (A_*'VA_* - A_*'F - F'A_*) + C'Z'VZC \\ &\quad + (A_*'V - F')ZC + C'Z'(VA_* - F) \qquad (4.126) \\ &= (A_*'VA_* - A_*'F - F'A_*) + C'Z'VZC \\ &= f(A_*) + C'Z'VZC \qquad (4.127) \end{aligned}$$

since, using equation (4.124)

$$VA_* + XB_* = F \Rightarrow \left\{ \begin{array}{rcl} C'Z'(VA_* - F) & = & 0 \\ (A_*'V - F')ZC & = & 0 \end{array} \right. \tag{4.128}$$

so that the last two terms in (4.126) are zero. The difference $f(A) - f(A_*) = C'Z'VZC$, which is n.n.d; this proves the optimization part. Now

$$f(A_*) = A_*'VA_* - A_*'F - F'A_*$$

$$\begin{aligned} &= A_*'(VA_* - F) - F'A_* = -A_*'XB_* - F'A_* \\ &= -P'B_* - F'A_* = -A_*'F - B_*'P, \end{aligned} \tag{4.129}$$

which proves (4.125). Let

$$\begin{pmatrix} V & X \\ X' & 0 \end{pmatrix}^- = \begin{pmatrix} C_1 & C_2 \\ C_2' & -C_4 \end{pmatrix}$$

for any g-inverse. Then

$$\begin{aligned} A_* &= C_1F + C_2P, \quad B_* = C_2'F - C_4P, \\ f(A_*) &= -P'(C_2'F - C_4P) - F'(C_1F + C_2P) \\ &= P'C_4P - P'C_2'F - F'C_2P - F'C_1F \\ &= \begin{cases} P'C_4P \text{ if } F = 0, \\ -F'C_1F \text{ if } P = 0. \end{cases} \end{aligned}$$

4.5.3 Estimation of $X\beta$ (the Fixed Effect)

Let $A'y$ be an unbiased estimate of $X\beta$. Then $\mathrm{E}(A'y) = X\beta \; \forall \beta \Rightarrow A'X = X'$, and

$$\begin{aligned} \mathrm{E}[(A'y - X\beta)(A'y - X\beta)'] &= \mathrm{E}[A'(U\xi + \epsilon)(U\xi + \epsilon)'A] \\ &= A'(U\Gamma U' + V)A = A'V_*A \end{aligned}$$

where $V_* = U\Gamma U' + V$. Then the problem is that of finding

$$\min_{X'A = X'} A'V_*A.$$

The optimum A is a solution of the equation

$$\begin{aligned} V_*A + XB &= 0 \\ X'A &= X', \end{aligned}$$

which is of the same form as in the only fixed-effects case except that V_* takes the place of V. If

$$\begin{pmatrix} V_* & X \\ X' & 0 \end{pmatrix}^- = \begin{pmatrix} C_1 & C_2 \\ C_2' & -C_4 \end{pmatrix}$$

then

$$A_* = C_2X', B_* = -C_4X'$$

giving the MDE of $X\beta$ as

$$A_*'y = XC_2'y \tag{4.130}$$

with the dispersion matrix

$$XC_4X'.$$

4.5.4 Prediction of $U\xi$ (the Random Effect)

Let $A'Y$ be a predictor of $Y\xi$ such that

$$\mathrm{E}(A'y - U\xi) = A'X\beta = O \Rightarrow A'X = 0\,.$$

Then

$$\begin{aligned}
\mathrm{E}(A'Y - U\xi)(A'Y - U\xi)' &= \mathrm{E}[(A' - I)U\xi\xi'U'(A - I)] + \mathrm{E}[A'\epsilon\epsilon'A] \\
&= (A' - I)U\Gamma U'(A - I) + A'VA \\
&= A'V_*A - A'U\Gamma U' - U\Gamma U'A + U\Gamma U'.
\end{aligned}$$

The problem is to find

$$\min_{X'A=0} (A'V_*A - A'W - WA + W)\,,$$

where $W = U\Gamma U'$. Applying the basic lemma of Section 4.5.1, the minimum is attained at a solution A_* of the equation

$$\begin{aligned}
V_*A + XB &= W \\
X'A &= 0.
\end{aligned}$$

In terms of the elements of the g-inverse of Theorem A.108, a solution is

$$A_* = C_1W, B_* = C_2'W$$

giving the mean-dispersion error of prediction (MDEP) of $U\xi$ as

$$A_*'y = WC_1y \tag{4.131}$$

with the dispersion of prediction error

$$-A_*'W + W = W - WC_1W\,.$$

4.5.5 Estimation of ϵ

Let $A'y$ be an estimate of ϵ such that

$$\mathrm{E}(A'y - \epsilon) = A'X\beta = 0 \Rightarrow A'X = 0.$$

Then

$$\begin{aligned}
\mathrm{E}[(A'y - \epsilon)(A'y - \epsilon)'] &= \mathrm{E}[A'U\xi\xi'U'A] + \mathrm{E}[(A' - I)\epsilon\epsilon'(A - I)] \\
&= A'V_*A - A'V - VA + V. \qquad (4.132)
\end{aligned}$$

Proceeding as in Section 4.5.2, the optimum A is

$$A_* = C_1V$$

giving the MDEP of ϵ as

$$A_*'y = VC_1y \tag{4.133}$$

with the dispersion of prediction error

$$-A_*'V + V = V - VC_1V.$$

The expressions (4.130),(4.131), and (4.133) for the estimators (predictors) of $X\beta, U\xi$, and ϵ suggest an alternative procedure of computing them through a conditioned equation. Consider the equation

$$\begin{aligned} V_* \alpha + X\beta &= y \\ X'\alpha = 0 & \end{aligned}$$

and solve for (α, β). If $(\hat{\alpha}, \hat{\beta})$ is a solution, then the estimate of $X\beta$ is $X\hat{\beta}$, of $U\xi$ is $W\hat{\alpha}$, and that of ϵ is $V\hat{\alpha}$.

The estimators obtained in this section involve the matrices V and Γ, which may not be known in practice. They cannot in general be estimated unless they have a simple structure involving a smaller number of parameters as in the case of variance components. In such a case, we may first estimate V and Γ by some method such as those described in the case of variance components by Rao and Kleffe (1988) and use them in the second stage for the estimation of $X\beta, U\xi$, and ϵ.

We discussed separately the estimation of $X\beta$, $U\xi$, and ϵ. If we need an estimate of a joint function of the fixed and random effects and the random error such as

$$P'X\beta + Q'U\xi + R'\epsilon \tag{4.134}$$

we need only substitute the separate estimates for the unknowns and obtain the estimate

$$P'X\hat{\beta} + Q'U\hat{\xi} + R'\hat{\epsilon},$$

which is the MDE for (4.134).

4.6 Regression-Like Equations in Econometrics

4.6.1 Stochastic Regression

In the following we consider some results concerning regression-like equations in econometric models. We assume the linear relationship

$$y = X\beta + \epsilon \tag{4.135}$$

where $y : T \times 1$, $X : T \times K$, $\beta : K \times 1$, and $\epsilon : T \times 1$. Unlike in the models of Chapters 3 and 4, we now assume that X is stochastic. In econometrics, the exogenous variables are usually assumed to be correlated with the random error ϵ, that is, X is supposed to be correlated with ϵ such that

$$p\lim(T^{-1}X'\epsilon) \neq 0. \tag{4.136}$$

As in Section 2.2, we assume that

$$p\lim(T^{-1}X'X) = \Sigma_{XX} \tag{4.137}$$

exists and is nonsingular. If we apply ordinary least squares to estimate β in (4.135), we get with $b = (X'X)^{-1}X'y$

$$p\lim(b) = \beta + \Sigma_{XX}^{-1} \, p\lim(T^{-1}X'\epsilon)\,, \tag{4.138}$$

and hence the OLS estimator b of β is not consistent.

4.6.2 Instrumental Variable Estimator

The method of instrumental variables (IV) is one of the techniques to get a consistent estimator of β. The idea is as follows. We suppose that in addition to the observations in y and X we have available T observations on K "instrumental variables" collected in the $T \times K$-matrix Z that are contemporaneously uncorrelated (see (2.11)) with the random error ϵ, that is,

$$\text{plim}(T^{-1}Z'\epsilon) = 0\,, \tag{4.139}$$

but are correlated with the regressors such that $\text{plim}(T^{-1}Z'X) = \Sigma_{ZX}$ exists and is nonsingular. Then the instrumental variable estimator of β is defined by

$$b^* = (Z'X)^{-1}Z'y\,. \tag{4.140}$$

This estimator is consistent:

$$\begin{aligned} b^* &= (Z'X)^{-1}Z'(X\beta + \epsilon) \\ &= \beta + (Z'X)^{-1}Z'\epsilon \\ &= \beta + (T^{-1}Z'X)^{-1}(T^{-1}Z'\epsilon), \end{aligned}$$

and hence, with (4.139) and (4.140), $\text{plim}(b^*) = \beta + \Sigma_{ZX}^{-1} * 0 = \beta$.

Using the relationship $(b^* - \beta)(b^* - \beta)' = (Z'X)^{-1}Z'\epsilon\epsilon'Z(X'Z)^{-1}$, we see that the asymptotic covariance matrix of b^* is

$$\bar{\Sigma}_{b^*b^*} = \bar{\text{E}}(b^* - \beta)(b^* - \beta)' = T^{-1}\sigma^2\Sigma_{ZX}^{-1}\Sigma_{ZZ}\Sigma_{ZX}^{-1\prime} \tag{4.141}$$

provided that $\text{plim}T(T^{-1}Z'\epsilon)(T^{-1}\epsilon Z) = \sigma^2\Sigma_{ZZ}$. It is clear that conditionally on Z and X for every T, we have $\text{E}(b^*) = \beta$ and $\text{cov}(b^*) = \sigma^2(Z'X)^{-1}(Z'Z)(X'Z)^{-1}$.

To interpret this estimator, consider the following. The least squares estimator b is the solution to the normal equations $X'Xb = X'y$, which can be obtained by premultiplying the relation $y = X\beta + \epsilon$ of observations through by X', replacing β by b, and dropping $X'\epsilon$. Quite analogous, the instrumental variable estimator b^* is the solution to the normal equations $Z'Xb^* = Z'y$, which are obtained by premultiplying the observational model $y = X\beta + \epsilon$ through by Z', replacing β by b^*, and dropping $Z'\epsilon$.

Remark. Note that an instrument is a variable that is at least uncorrelated with the random error ϵ and is correlated with the regressor variables in X. Using the generalized variance $\text{G.var}|b^*| = |\text{cov}(b^*)|$ as efficiency measure, it is proved (Dhrymes, 1974, p. 298) that the generalized variance of b^* is minimized with respect to Z if the coefficient of vector correlation between the instruments and the regressors is maximized. Of course, $Z = X$ would yield the optimal instrumental variable estimator $b^* = b = (X'X)^{-1}X'y$. This is just the OLSE, which by (4.136) fails to be consistent. Hence one has to find instruments that are highly correlated with X but are not identical to X. For a more detailed discussion of this problem, see Goldberger (1964) and Siotani, Hayakawa, and Fujikoshi (1985).

4.6.3 Seemingly Unrelated Regressions

We now consider a set of equations

$$y_i = X_i\beta_i + \epsilon_i\,, \quad i = 1, \ldots, M \tag{4.142}$$

where $y_i : T \times 1$, $X_i : T \times K_i$, $\beta_i : K_i \times 1$ and $\epsilon_i : T \times 1$. The model is already in the reduced form (see Section 2.3). However if ϵ_i and ϵ_j are correlated for some pairs of indices i, j, $(i \neq j)$, then the equations in (4.142) are correlated to each other through the random errors, although by construction they are seemingly unrelated.

Let us write the equations given in (4.142) according to

$$\begin{pmatrix} y_1 \\ y_2 \\ \vdots \\ y_M \end{pmatrix} = \begin{pmatrix} X_1 & 0 & \ldots & 0 \\ 0 & X_2 & \ldots & 0 \\ \vdots & & \ddots & \\ 0 & 0 & \ldots & X_M \end{pmatrix} \begin{pmatrix} \beta_1 \\ \beta_2 \\ \vdots \\ \beta_M \end{pmatrix} + \begin{pmatrix} \epsilon_1 \\ \epsilon_2 \\ \vdots \\ \epsilon_M \end{pmatrix} \tag{4.143}$$

as a multivariate linear regression model (see (2.29)) or more compactly as

$$y = X\beta + \epsilon$$

where $y : MT \times 1$, $X : MT \times K$, $\beta : MK \times 1$, $\epsilon : MT \times 1$, and $K = \sum_{i=1}^{M} K_i$. The covariance matrix of ϵ is

$$\mathrm{E}(\epsilon\epsilon') = \Sigma \otimes I_T \tag{4.144}$$

where $\Sigma = (\sigma_{ij})$ and $\mathrm{E}(\epsilon\epsilon') = \sigma_{ij} I_T$. $\otimes$ denotes the Kronecker product (see Theorem A.99). If Σ is known, then β is estimated by the GLSE (see (4.64)) as

$$\hat{\beta} = (X'(\Sigma \otimes I)^{-1}X)^{-1}(X'(\Sigma \otimes I)^{-1}y)\,, \tag{4.145}$$

which is the BLUE of β in case of nonstochastic regressors X. This GLSE and the least squares estimator $(X'X)^{-1}X'y$ are identical when either Σ is diagonal or $X_1 = X_2 = \ldots = X_M$, or more generally when all X_i's span the same column space; see Dwivedi and Srivastava (1978) and Bartels and Fiebig (1991) for some interesting conditions when they are equivalent.

When Σ is unknown it is replaced by an estimator $\hat{\Sigma} = (\hat{\sigma_{ij}})$. Among others, Zellner (1962, 1963) has proposed the following two methods for estimating the unknown matrix Σ.

Restricted Zellner's Estimator (RZE)

This estimator is based on the OLSE residuals

$$\hat{\epsilon}_i = y_i - X_i b_i\,, \quad b_i = (X_i'X_i)^{-1}X_i'y_i \quad (i = 1, \ldots, M)$$

of the equations in the system (4.143). The covariance σ_{ij} is estimated by

$$\hat{\sigma}_{ij} = \hat{\epsilon}_i'\hat{\epsilon}_j \Big/ \sqrt{(T - K_i)(T - K_j)}$$

resulting in $\hat{\Sigma} = (\hat{\sigma}_{ij})$, which is substituted for Σ in (4.145), leading to

$$\hat{\beta}_{\mathrm{RZE}} = (X'(\hat{\Sigma} \otimes I)^{-1}X)^{-1}(X'(\hat{\Sigma} \otimes I)^{-1}y). \tag{4.146}$$

Unrestricted Zellner's Estimator (UZE)

Define the $T \times K$-matrix $\tilde{X} = (X_1, \ldots, X_M)$ and let $\tilde{\epsilon}_i = y_i - \tilde{X}(\tilde{X}'\tilde{X})^{-1}\tilde{X}'y_i$ be the residual in the regression of y_i on $\tilde{X}$, $(i = 1, \ldots, M)$. Then σ_{ij} is estimated by

$$\tilde{\sigma}_{ij} = \tilde{\epsilon}_i'\tilde{\epsilon}_j / (T - K)$$

resulting in $\tilde{\Sigma} = (\tilde{\sigma}_{ij})$ and leading to the estimator

$$\hat{\beta}_{\mathrm{UZE}} = (X'(\tilde{\Sigma} \otimes I)^{-1}X)^{-1}(X'(\tilde{\Sigma} \otimes I)^{-1}y)\,. \tag{4.147}$$

When the random vectores ϵ_i are symmetrically distributed around the mean vector, Kakwani (1967) has pointed out that the estimators $\hat{\beta}_{\mathrm{RZE}}$ and $\hat{\beta}_{\mathrm{UZE}}$ are unbiased for β provided that $\mathrm{E}(\hat{\beta}_{\mathrm{RZE}})$ and $\mathrm{E}(\hat{\beta}_{\mathrm{UZE}})$ exist. Srivastava and Raj (1979) have derived some conditions for the existence of these mean vectors; see also Srivastava and Giles (1987). Further, if the underlying distribution is normal, Srivastava (1970) and Srivastava and Upadhyaha (1978) have observed that both the estimators have identical variance-covariance matrices to order T^{-2}. When the distribution is neither symmetric nor normal, both the estimators are generally biased; see Srivastava and Maekawa (1995) for the effect of departure from normality on the asymptotic properties.

4.7 Simultaneous Parameter Estimation by Empirical Bayes Solutions

4.7.1 Overview

In this section, the empirical Bayes procedure is employed in simultaneous estimation of vector parameters from a number of linear models. It is shown that with respect to quadratic loss function, empirical Bayes estimators are better than least squares estimators. While estimating the parameter for a particular linear model, a suggestion shall be made for distinguishing between the loss due to decision makers and the loss due to individuals.

We consider k linear models

$$y_i = X\beta_i + \epsilon, \quad i = 1, \ldots, k, \tag{4.148}$$

where y_i is a T-vector of observations, X is a known $(T \times K)$-matrix with full rank, and β_i is a K-vector and ϵ_i is a T-vector of unobservable random variables. We assume

$$\mathrm{E}(\epsilon_i|\beta_i) = 0, \quad \mathrm{D}(\epsilon_i|\beta_i) = \sigma^2 V, \tag{4.149}$$

and assume the following prior distribution of β_i

$$\mathrm{E}(\beta_i) = \beta, \quad \mathrm{D}(\beta_i) = F, \quad \mathrm{cov}(\beta_i, \beta_j) = 0, \quad i \neq j, \tag{4.150}$$

where V is of full rank and known. The following problem of simultaneous estimation of $p'\beta_i$, $i = 1, \ldots, k$, where p is any given vector, will be considered. We note that the problem of estimating β_i is the same as that of estimating a general linear function $p'\beta_i$. If we use the MDE-I criterion in estimating $p'\beta_i$, we automatically obtain estimates of β_i with a *minimum mean dispersion error matrix* (MDE).

Such a problem of simultaneous estimation arises in the construction of a selection index for choosing individuals with a high intrinsic genetic value. For instance, β_i may represent unknown genetic parameters and x_i be observable characteristics on the ith individual, while $p'\beta_i$ for a given p is the genetic value to be estimated in terms of observed y_i.

We use the following notations and results throughout this section. Consider a linear model

$$y = X\beta + \epsilon, \tag{4.151}$$

where β is a K-vector of unknown parameters, $\mathrm{E}(\epsilon) = 0$, and $\mathrm{D}(\epsilon) = \sigma^2 V$. To avoid some complications, let us assume that V is nonsingular and the rank of X is K.

The least squares estimator of β is

$$\beta^{(l)} = (X'V^{-1}X)^{-1}X'V^{-1}y \tag{4.152}$$

and a ridge regression estimator of β is

$$\beta^{(r)} = (G + X'V^{-1}X)^{-1}X'V^{-1}y \tag{4.153}$$

for some chosen nonnegative definite matrix G. (Ridge regression estimator was introduced in Section 3.10.2 in the special case $V = I$ with the particular choice $G = k^2 I$.) It may be noted that

$$\beta^{(r)} = T\beta^{(l)} \tag{4.154}$$

where $T = (G + X'V^{-1}X)^{-1}X'V^{-1}X$ has all its eigenvalues less than unity if G is not the null matrix. The following matrix identities, which are variants of Theorem A.18. (iii), will prove useful:

$$(V+XFX')^{-1} = V^{-1}-V^{-1}X(X'V^{-1}X+F^{-1})^{-1}X'V^{-1} \quad (4.155)$$
$$(V+XF)^{-1} = V^{-1}-V^{-1}X(I+FV^{-1}X)^{-1}FV^{-1} \quad (4.156)$$
$$(V+F)^{-1} = V^{-1}-V^{-1}(V^{-1}+F^{-1})^{-1}V^{-1}. \quad (4.157)$$

4.7.2 Estimation of Parameters from Different Linear Models

Let us consider k linear models $y_i = X\beta_i + \epsilon_i$, $i = 1, \ldots, k$ as mentioned in (4.148) with assumptions (4.149) and (4.150). We shall find a_0, a_1 such that

$$E(p'\beta_i - a_0 - a_1'y_i)^2 \quad (4.158)$$

is a minimum for each i for given p. The problem as stated is easily solvable when σ^2, β, and F are known. We shall review known results and also consider the problem of estimation when σ^2, β, and F are unknown but can be estimated.

Case 1 (σ^2, β, and F are known)

Theorem 4.8 *The optimum estimator of $p'\beta_i$ in the sense of (4.158) is $p'\beta_i^{(b)}$ where $\beta_i^{(b)}$ can be written in the following alternative forms (where $U = (X'V^{-1}X)^{-1}$)*

$$\beta_i^{(b)} = \beta + FX'(XFX'+\sigma^2V)^{-1}(y_i - X\beta) \quad (4.159)$$
$$= \beta + (\sigma^2F^{-1}+U^{-1})^{-1}X'V^{-1}(y_i - X\beta) \quad (4.160)$$
$$= (\sigma^2F^{-1}+U^{-1})^{-1}\sigma^2F^{-1}\beta + \beta_i^{(r)} \quad (4.161)$$
$$= \beta + F(F+\sigma^2U)^{-1}(\beta_i^{(l)} - \beta) \quad (4.162)$$
$$= \sigma^2U(F+\sigma^2U)^{-1}\beta + F(F+\sigma^2U)^{-1}\beta_i^{(l)} \quad (4.163)$$
$$= \beta_i^{(l)} - \sigma^2U(F+\sigma^2U)^{-1}(\beta_i^{(l)} - \beta), \quad (4.164)$$

where $\beta_i^{(r)}$ is the ridge regression estimator as defined in (4.153) with $G = \sigma^2F^{-1}$. The prediction error is $p'Qp$ where

$$Q = \sigma^2(\sigma^2F^{-1}+U^{-1})^{-1} \quad (4.165)$$
$$= \sigma^2F(F+\sigma^2U)^{-1}U \quad (4.166)$$
$$= \sigma^2(U - \sigma^2U(F+\sigma^2U)^{-1}U). \quad (4.167)$$

Some of the results are proved in Rao (1974) and others can be easily deduced using the identities (4.155)–(4.157). We shall refer to $\beta_i^{(b)}$ as the Bayes estimator of β_i with parameters of its prior distribution as defined in (4.150). We make the following observations.

Note 1: It may be noted that the ridge regression estimator (4.153) originally defined with $V = I$ and $G = k^2 I$ is the Bayes estimator when the prior distribution of the regression parameter has 0 (null vector) as the mean and $\sigma^2 k^2 I$ as the dispersion matrix. More generally, we find from (4.161) that the ridge regression estimator as defined in (4.153) is the Bayes estimator when the mean and dispersion matrix of prior distribution are the null vector and $\sigma^2 G^{-1}$, respectively.

Note 2: The Bayes estimator of β_i is a weighted linear combination of its least squares estimator and the mean of its prior distribution.

Note 3: The estimator $\beta_i^{(b)}$ as defined in Theorem 4.8 is optimum in the class of linear estimators. However, it is optimum in the entire class of estimators if the regression of β_i on y_i is linear. A characterization of the prior distribution of β_i is obtained in Rao (1974) using the property that the regression of β_i on y_i is linear.

Note 4: The matrix

$$\mathrm{E}(\beta_i^{(l)} - \beta_i)(\beta_i^{(l)} - \beta_i)' - \mathrm{E}(\beta_i^{(b)} - \beta_i)(\beta_i^{(b)} - \beta_i)' \tag{4.168}$$

is nonnegative definite, where β_i^l is the least squares estimator β_i in the ith model. Of course, the Bayes estimator has the minimum MDE compared to any other linear estimator.

Thus when σ^2, β, and F are known, $p'\beta_i$ is estimated by $p'\beta_i^{(b)}$ for $i = 1, \ldots, k$, and the compound loss

$$\mathrm{E}\sum_i^k (p'\beta_i - p'\beta_i^{(b)})^2 \tag{4.169}$$

is minimum compared to any other set of linear estimators. We shall consider the modifications to be made when σ^2, β, and F are unknown.

Note 5: It may be noted that for fixed β_i, the expected value of (4.168) may not be nonnegative definite. Indeed, the optimality of the Bayes estimator over the least squares estimator is not uniform for all values of β_i. It is true only for a region of β_i such that $\| \beta_i - \beta \|$, the norm of $\beta_i - \beta$ where β is the chosen prior mean of β_i, is less than a preassigned quantity depending on σ^2, F, and U.

Case 2 (σ^2, β, and F are unknown)

When σ^2, β, and F are unknown, we shall substitute for them suitable estimates in the formulae (4.159)–(4.164) for estimating β_i. The following unbiased estimates σ_*^2, β_*, and F_* of σ^2, β, and F, respectively are well known.

$$k\beta_* = \sum_1^k \beta_i^{(l)} \tag{4.170}$$

$$k(T-K)\sigma_*^2 = \sum_1^k (y_i'V^{-1}y_i - y_i'V^{-1}X\beta_i^{(l)}) = W \quad (4.171)$$

$$(k-1)(F_* + \sigma_*^2 U) = \sum_1^k (\beta_i^{(l)} - \beta_*)(\beta_i^{(l)} - \beta_*)' = B. \quad (4.172)$$

By substituting constant multiples of these estimators for σ^2, β, and F in (4.164), we obtain the empirical Bayes estimator of $p'\beta_i$ as $p'\beta_i^{(c)}$, where $\beta_i^{(c)}$ is

$$\beta_i^{(c)} = \beta_i^{(l)} - cWUB^{-1}(\beta_i^{(l)} - \beta_*), \quad i = 1, \ldots, k, \quad (4.173)$$

with $c = (k-K-2)/(kT-kK+2)$ as determined in (4.185).

Theorem 4.9 *Let β_i and ϵ_i have multivariate normal distributions, in which case W and B are independently distributed with*

$$W \sim \sigma^2\chi^2(kT-kK) \quad (4.174)$$

$$B \sim W_K(k-1, F+\sigma^2 U) \quad (4.175)$$

that is, as chi-square on $k(T-K)$ degrees of freedom and Wishart on $(k-1)$ degrees of freedom, respectively. Then

$$\mathrm{E}\sum_{i=1}^k (\beta_i^{(c)} - \beta_i)(\beta_i^{(c)} - \beta_i)'$$

$$= k\sigma^2 U - \frac{\sigma^4 k(T-K)(k-K-2)}{k(T-K)+2} U(F+\sigma^2 U)^{-1}U \quad (4.176)$$

for the optimum choice $c = (k-K-2)/(kT-kK+2)$ in (4.173) provided $k \geq K+2$.

Proof: Consider

$$\sum_{i=1}^k (\beta_i^{(c)} - \beta_i)(\beta_i^{(c)} - \beta_i)'$$

$$= \sum_{i=1}^k (\beta_i^{(l)} - \beta_i)(\beta_i^{(l)} - \beta_i)' + c^2W^2UB^{-1}U - 2cWU$$

$$+ cW\sum_{i=1}^k \beta_i(\beta_i^{(l)} - \beta_*)'B^{-1}U$$

$$+ cWUB^{-1}\sum_{i=1}^k (\beta_i^{(l)} - \beta_*)\beta_i'. \quad (4.177)$$

Let us observe that

$$\mathrm{E}(W) = k(T-K)\sigma^2 \quad (4.178)$$

$$\begin{aligned}
\mathrm{E}(W^2) &= k(T-K)(kT-kK+2)\sigma^4 && (4.179)\\
\mathrm{E}(B^{-1}) &= (k-K-2)^{-1}(F+\sigma^2 U)^{-1} && (4.180)\\
\mathrm{E}\sum_1^k \beta_i(\beta_i^{(l)}-\beta_*)'B^{-1} &= F(F+\sigma^2 U)^{-1} && (4.181)\\
\mathrm{E}\,B^{-1}\sum_1^k (\beta_i^{(l)}-\beta_*)\beta_i' &= (F+\sigma^2 U)^{-1}F. && (4.182)
\end{aligned}$$

Then (4.177) reduces to

$$k\sigma^2 U + \sigma^4 g U(F+\sigma^2 U)^{-1}U \tag{4.183}$$

where

$$g = \frac{c^2 k(T-K)(kT-kK+2)}{k-K-2} - 2ck(T-K). \tag{4.184}$$

The optimum choice of c in (4.184) is

$$c = (k-K-2)(kT-kK+2)\,, \tag{4.185}$$

which leads to the value (4.176) given in Theorem 4.9.

Note 1: The results of Theorem 4.9 are generalizations of the results in the estimation of scalar parameters considered by Rao (1974).

Note 2: Expression (4.176) for the compound loss of empirical Bayes estimators is somewhat larger than the corresponding expression for Bayes estimators, which is k times (4.149), and the difference is the additional loss due to using estimates of σ^2, β, and F when they are unknown.

Note 3: If β_i is estimated by $\beta_i^{(l)}$, then the compound MDE is

$$\mathrm{E}\sum_1^k (\beta_i^{(l)}-\beta_i)(\beta_i^{(l)}-\beta_i)' = k\sigma^2 U \tag{4.186}$$

and the difference between (4.186) and (4.176), the MDE for the empirical Bayes estimator, is

$$\frac{\sigma^4 k(T-K)(k-K-2)}{k(T-K)+2}U(F+\sigma^2 U)^{-1}U\,, \tag{4.187}$$

which is nonnegative definite.

Thus the expected compound loss for the estimation of $p'\beta_i$, $i=1,\ldots,k$, is smaller for the empirical Bayes estimator than for the least squares estimator.

Note 4: It may be easily shown that the expectation of (4.177) for fixed values of $\beta_1,\ldots,\beta_k$ is smaller than $k\sigma^2 U$, as in the univariate case (Rao, 1974). Thus the empirical Bayes estimators (4.173) are uniformly better than the least squares estimators without *any* assumption on the a priori distribution of β_i. The actual expression for the expectation of (4.177) for fixed $\beta_1, \ldots, \beta_k$ may be written in the form

$$k\sigma^2 U - \frac{\sigma^4 (k-K-2)^2 k(T-K)}{k(T-K)+2} E(UB^{-1}U), \tag{4.188}$$

which gives an indication of the actual decrease in loss by using empirical Bayes estimators.

Note 5: In the specification of the linear models we have assumed that the dispersion matrix $\sigma^2 V$ of the error vector is known apart from a constant multiplier. If V is unknown, it cannot be completely estimated from the observations $y_1, \ldots, y_k$ alone. However, if V has a suitable structure, it may be possible to estimate it.

4.8 Supplements

The class of linear unbiased estimators between OLSE and GLSE. Consider the general linear regression model

$$y = X\beta + \epsilon, \quad \epsilon \sim (0, \sigma^2 W). \tag{4.189}$$

The covariance matrix $\sigma^2 W$ of ϵ is assumed to be a known and positive definite matrix.

Consider the ordinary least squares estimator (OLSE) $b = (X'X)^{-1}X'y$ and the generalized least squares estimator (GLSE)

$$b(W) = (X'W^{-1}X)^{-1}X'W^{-1}y.$$

There exists a number of conditions under which OLSE and GLSE coincide. However, an open question is the following: What is the explicit form of all linear unbiased estimators $\tilde{b}$ for β in model (4.189) whose efficiency lies between that of OLSE and GLSE, that is, $\tilde{b} = Cy + c$, $CX = I$, and $\text{cov}(b) \leq \text{cov}(\tilde{b}) \leq \text{cov}\left(b(W)\right)$, where "$\leq$" denotes the Loewner ordering of nonnegative definite matrices?

Remark. Some work in this direction was done by Amemiya (1983), Balestra (1983), and Groß and Trenkler (1997).

4.9 Gauss-Markov, Aitken and Rao Least Squares Estimators

Consider the linear model

$$y = X\beta + \epsilon$$

$$\mathrm{E}(\epsilon) = 0 \, , \mathrm{E}(\epsilon\epsilon') = \sigma^2 W$$

where $y : T \times 1, X : T \times K, \beta : K \times 1$ and $\epsilon : T \times 1$ matrices. We review the estimation of β and σ^2 through minimization of a quadratic function of $y - X\beta$, under various assumptions on the ranks of X and W.

4.9.1 Gauss-Markov Least Squares

$W = I$ and $\mathrm{rank}(X) = K$ (i.e., X has full rank K)

Under these conditions, it is shown in Chapter 3 that the minimum dispersion linear estimator of β is

$$\hat{\beta} = \arg\min_{\beta} (y - X\beta)'(y - X\beta)$$

an explicit form of which is

$$\hat{\beta} = (X'X)^{-1}X'y$$

with

$$\mathrm{V}(\hat{\beta}) = \sigma^2 (X'X)^{-1} \, .$$

An unbiased estimator of σ^2 is

$$\hat{\sigma}^2 = (y - X\hat{\beta})'(y - X\hat{\beta})/(T - K) \, .$$

The method is referred to as Gauss-Markov least squares.

$W = I, \mathrm{rank}(X) = s < K$ (i.e., X is deficient in rank)

Under these conditions the MDLE of $L'\beta$ where $L : K \times r$ and $\mathcal{R}(L) \subset \mathcal{R}(X')$, i. e., the linear space spanned by the columns of L is contained in the linear space spanned by the columns X', is

$$L'\hat{\beta} = L'(X'X)^{-}X'y$$

where

$$\hat{\beta} = \arg\min_{\beta} (y - X\beta)'(y - X\beta)$$

with

$$\mathrm{V}(L'\hat{\beta}) = \sigma^2 L'(X'X)^{-}L$$

where in all the equations above $(X'X)^{-}$ is any g-inverse of $X'X$. An unbiased estimator of σ^2 is

$$\hat{\sigma}^2 = (y - X\hat{\beta})'(y - X\hat{\beta})/(T - s) \, .$$

Thus with no modification, Gauss-Markov least squares theory can be extended to the case where X is deficient in rank, noting that only linear functions $p'\beta$ with $p \subset \mathcal{R}(X')$ are unbiasedly estimable.

4.9.2 Aitken Least Squares

W is p.d. and $\text{rank}(X) = K$

Under these conditions, it is shown in Chapter 4 that the MDLE of β is

$$\hat{\beta} = \arg\min_{\beta}(y - X\beta)'W^{-1}(y - X\beta)$$

an explicit solution of which is

$$\hat{\beta} = (X'W^{-1}X)^{-1}X'W^{-1}y$$

with the dispersion matrix

$$\text{V}(\hat{\beta}) = \sigma^2(X'W^{-1}X)^{-1}.$$

An unbiased estimator of σ^2 is

$$\hat{\sigma}^2 = (y - X\hat{\beta})'W^{-1}(y - X\hat{\beta})/(T - K).$$

The method is referred to as Aitken least squares.

W is p.d. and $\text{rank}(X) = s < K$

Under these conditions, the MDLE of $L'\beta$ where L satisfies the same condition as above is

$$L'\hat{\beta} = L'(X'W^{-1}X)^{-}X'W^{-1}y$$

where

$$\hat{\beta} = \arg\min_{\beta}(y - X\beta)'W^{-1}(y - X\beta)$$

and $(X'W^{-1}X)^{-}$ is any g-inverse of $X'W^{-1}X$. The dispersion matrix of $L'\hat{\beta}$ is

$$\text{V}(L'\hat{\beta}) = \sigma^2 L'(X'W^{-1}X)^{-}L.$$

An unbiased estimator of σ^2 is

$$\hat{\sigma}^2 = (y - X\hat{\beta})'W^{-1}(y - X\hat{\beta})/(T - s).$$

Thus, Aitken least squares method can be extended to the case where X is deficient in rank.

Now we raise the question whether the least squares theory can be extended to the case where both W and X may be deficient in rank through minimization of a suitable quadratic function of $y - X\beta$. This problem is investigated in Rao (1973b) and the solution is as follows.

4.9.3 Rao Least Squares

$\text{rank}(W) = t \leq T$, $\text{rank}(X) = s \leq K$

First we prove a theorem.

Theorem 4.10 *Let $R = W + XUX'$ where U is an n.n.d. matrix such that $\mathcal{R}(W) \subset \mathcal{R}(R)$ and $\mathcal{R}(X) \subset \mathcal{R}(R)$. Then*

(i) $X(X'R^-X)^-X'RX = X$

(ii) $X(X'R^-X)^-X'R^-RM = 0$ *if* $X'M = 0$

(iii) $\text{tr}(R^-R - X(X'R^-X)^-X') = \text{rank}(W : X) - \text{rank}(X)$

where $()^-$ is any choice of g-inverse of the matrices involved.

The results are easy to prove using the properties of g-inverses discussed in A.12. Note that all the expressions in (i), (ii) and (iii) are invariant for any choice of g-inverse. For proving the results, it is convenient to use the Moore-Penrose g-inverse.

Theorem 4.11 *Let $\hat{\beta}$ be*

$$\hat{\beta} = \arg\min_{\beta}(y - X\beta)'R^-(y - X\beta)$$

a solution of which is

$$\hat{\beta} = (X'R^-X)^-X'R^-y$$

for any choice of g-inverses involved. Then:

(i) *The MDLE of $L'\beta$, where $L : K \times r$ and $\mathcal{R}(L) \subset \mathcal{R}(X')$, is $L'\hat{\beta}$ with the variance-covariance matrix*

$$\text{V}(L'\hat{\beta}) = \sigma^2 L'\{(X'R^-X)^- - U\}L\,.$$

(ii) *An unbiased estimator of σ^2 is*

$$\hat{\sigma}^2 = (y - X\hat{\beta})'R^-(y - X\hat{\beta})/f$$

where $f = \text{rank}(W : X) - \text{rank}(X)$.

Proof: Let $L = X'C$ since $\mathcal{R}(L) \subset \mathcal{R}(X')$. Then

$$\begin{aligned} \text{E}(L'\hat{\beta}) &= C'X(X'R^-X)^-X'R^-\,\text{E}(y) \\ &= C'X(X'R^-X)^-X'R^-X\beta \\ &= C'X\beta = L'\beta\,, \end{aligned}$$

using (i) of Theorem 4.10, so that $L'\hat{\beta}$ is unbiased for $L'\beta$.

Let $M'y$ be such that $\text{E}(M'y) = 0$, i.e., $M'X = 0$. Consider

$$\begin{aligned} \text{Cov}(L'\hat{\beta}, M'y) &= \sigma^2 C'X(X'R^-X)^-X'R^-WM \\ &= \sigma^2 C'X(X'R^-X)^-X'R^-RM \\ &= \sigma^2 C'X(X'R^-X)^-X'M = 0\,, \end{aligned}$$

using (ii) of Theorem 4.10. This is true for all M such that $\text{E}(M'y) = 0$, so that $L'\hat{\beta}$ has minimum variance-covariance matrix as an unbiased estimator of $L'\beta$.

The expression for the variance-covariance matrix of $L'\hat{\beta}$ is

$$\begin{aligned} V(L'\hat{\beta}) &= \sigma^2 C'X(X'R^-X)^-X'R^-W(C'X(X'R^-X)^-X'R^-)' \\ &= \sigma^2 C'X[(X'R^-X) - U]X'C \\ &= \sigma^2 L'[(X'R^-X) - U]L \end{aligned}$$

Finally

$$\begin{aligned} E(y - X\hat{\beta})'R^-(y - X\hat{\beta}) &= E(y - X\beta)'R^-(y - X\beta) - E(y - X\beta)'R^-(X\beta - X\hat{\beta}) \\ &= \sigma^2 \mathrm{tr}[R^-W - R^-X(X'R^-X)^-X'R^-W] \\ &= \sigma^2 \mathrm{tr}R^-[I - X(X'R^-X)^-X'R^-]R \\ &= \sigma^2[\mathrm{tr}R^-R - \mathrm{tr}X(X'R^-X)^-X'R^-] \\ &= \sigma^2[\mathrm{rank}(W : X) - \mathrm{rank}(X)], \end{aligned}$$

using (iii) of Theorem 4.10 which yields to the unbiased estimate of σ^2 given in Theorem 4.11.

Note 1. One choice is $U = b^2 I$ where b is any constant. However, any choice of U such that $\mathcal{R}(W + XUX')$ contains both $\mathcal{R}(W)$ and $\mathcal{R}(X)$ will do.

Note 2. Theorem 4.11 holds in the general situation where W is n.n.d. or p.d. and X is deficient in rank or not. Even if W is p.d., it helps in computations to choose $R = (W + XX')$ in setting up the quadratic form defined in Theorem 4.11 for minimization, and use the results of Theorem 4.11 for estimation purposes.

Thus we have a very general theory of least squares which holds good in all situations and which in particular, includes Gauss-Markov and Aitken theories. Further details on unified least squares theory can be found in Rao (1973b).

4.10 Exercises

Exercise 1. In the model $y = \alpha\mathbf{1} + X\beta + \epsilon$, with $\mathbf{1}$ denoting a column vector having all elements unity, show that the GLSE of α is given by

$$\frac{\mathbf{1}}{\mathbf{1}'\Sigma\mathbf{1}}\mathbf{1}'[\Sigma - \Sigma X(X'\Sigma_*^{-1}X)^{-1}X'\Sigma_*^{-1}]y$$

where $E(\epsilon\epsilon') = \Sigma^{-1}$ and $\Sigma_*^{-1} = \Sigma - \frac{\mathbf{1}}{\mathbf{1}'\Sigma\mathbf{1}}\Sigma\mathbf{1}\mathbf{1}'\Sigma$.

Exercise 2. If disturbances are equicorrelated in the model of Exercise 1, is GLSE of β equal to the LSE?

Exercise 3. In the model $y = X\beta + \epsilon$ with $E(\epsilon) = 0$ and $E(\epsilon\epsilon') = \sigma^2 W$, show that $d = \beta'X'(\sigma^2 W + X\beta\beta'X')^{-1}y$ is an unbiased estimator of $(1 + \sigma^2/\beta'X'W^{-1}X\beta)^{-1}$. Find its variance.

Exercise 4. If $\hat{\beta}$ is the GLSE of β in the model $y = X\beta + \epsilon$, can we express the dispersion matrix of the difference vector $(y - X\hat{\beta})$ as the difference of the dispersion matrices of y and $X\hat{\beta}$?

Exercise 5. When σ^2 in the model $y = X\beta + \epsilon$ with $\mathrm{E}(\epsilon) = 0$ and $\mathrm{E}(\epsilon\epsilon') = \sigma^2 W$ is estimated by

$$\hat{\sigma}^2 = \left(\frac{1}{T-K}\right) y'[I - X(X'X)^{-1}X']y,$$

show that $\hat{\sigma}^2$ is not an unbiased estimator of σ^2 and

$$\left(\frac{1}{T-K}\sum_{i=1}^{T-K}\mu_i\right) \leq \mathrm{E}\left(\frac{\hat{\sigma}^2}{\sigma^2}\right) \leq \left(\frac{1}{T-K}\sum_{i=T-K+1}^{T}\mu_i\right)$$

where $\mu_1 \leq \mu_2 \leq \ldots \leq \mu_T$ are the eigenvalues of W.

Exercise 6. If the disturbances in a linear regression model are autocorrelated, are the residuals also autocorrelated? Is the converse true?

Exercise 7. Suppose that the λ_i's are the eigenvalues of the matrix PAP in which

$$A = \begin{bmatrix} 1 & -1 & 0 & \ldots & 0 & 0 \\ -1 & 2 & -1 & \ldots & 0 & 0 \\ 0 & -1 & 2 & \ldots & 0 & 0 \\ \vdots & \vdots & \vdots & & \vdots & \vdots \\ 0 & 0 & 0 & \ldots & 2 & -1 \\ 0 & 0 & 0 & \ldots & -1 & 1 \end{bmatrix} \quad \text{and } P = I - X(X'X)^{-1}X'.$$

Show that the Durbin-Watson statistic can be expressed as

$$\left(\sum_{i=1}^{T-K}\lambda_i u_i^2 \Big/ \sum_{i=1}^{T-K} u_i^2\right)$$

where the u_i's are independently and identically distributed normal random variables with zero mean and unit variance.

Exercise 8. In the model $y_t = \beta x_t + \epsilon_t$ with $\mathrm{E}(\epsilon_t) = 0$ and $\mathrm{E}(\epsilon_t^2)$ proportional to x_t^2, show that the GLSE of β is the mean of ratios (y_t/x_t). What happens to this result when a constant term is included in the model?

Exercise 9. In the case of stochastic regression of Section 4.6.1, consider the least squares estimator b and instrumental variable estimator b^*. Show that the asymptotic covariance matrix of b cannot exceed the asymptotic covariance matrix of b^*.

Exercise 10. Consider a seemingly unrelated regression equation model containing only two equations, $y_1 = X_1\beta_1 + \epsilon_1$ and $y_2 = X_2\beta_2 + \epsilon_2$. If X_2 is a submatrix of X_1, show that the GLSE of β_2 is equal to the least squares estimator. How are they related in the case of β_1?

5
Exact and Stochastic Linear Restrictions

5.1 Use of Prior Information

As a starting point, which was also the basis of the standard regression procedures described in the previous chapters, we take a T-dimensional sample of the variables y and $X_1, \ldots, X_K$. If the classical linear regression model $y = X\beta + \epsilon$ with its assumptions is assumed to be a realistic picture of the underlying relationship, then the least-squares estimator $b = (X'X)^{-1}X'y$ is optimal in the sense that it has smallest variability in the class of linear unbiased estimators for β.

In statistical research there have been many attempts to provide better estimators; for example,

(i) by experimental design that provides minimal values to the variances of certain components β_i of β or the full covariance matrix $\sigma^2(X'X)^{-1}$ through a suitable choice of X,

(ii) by the introduction of biased estimators;

(iii) by the incorporation of prior information available in the form of exact or stochastic restrictions (cf. Chipman and Rao, 1964; Toutenburg, 1973; Yancey, Judge, and Bock, 1973; 1974);

(iv) by the methods of simultaneous (multivariate) estimation, if the model of interest may be connected with a system of other linear equations (cf. Nagar and Kakwani, 1969; Goldberger, Nagar, and Odeh, 1961; Toutenburg and Wargowske, 1978).

In this chapter we confine ourselves to methods related to example (iii). Moreover, we concentrate on the classical regression model and assume that $\text{rank}(X) = K$. Only in Sections 5.8 and 5.9 do we consider the dispersion matrix of the generalized linear model, namely, $\text{E}(\epsilon\epsilon') = \sigma^2 W$.

Examples of Prior Information in the Form of Restrictions

In addition to observations on the endogenous and exogenous variables (such observations are called the sample), we now assume that we have *auxiliary information* on the vector of regression coefficients. When this takes the form of inequalities, the minimax principle (see Section 3.13) or simplex algorithms can be used to find estimators, or at least numerical solutions, that incorporate the specified restrictions on β. Let us assume that the auxiliary information is such that it can be written in the form of linear equalities

$$r = R\beta\,, \tag{5.1}$$

with r a J-vector and R a $J \times K$-matrix. We assume that r and R are known and in addition that $\text{rank}(R) = J$, so that the J linear restrictions in (5.1) are independent.

Examples of linear restrictions:

- Exact knowledge of a single component β_1 of β, such as,

$$\beta_1 = \beta_1^*\,, \quad r = (\beta_1^*)\,, \quad R = (1, 0, \ldots, 0). \tag{5.2}$$

- Formulating a hypothesis on a subvector of $\beta = (\beta_1, \beta_2)'$ as, for example, H_0: $\beta_2 = 0$ with $r = R\beta$ and

$$r = 0, \quad R = (0, I)\,. \tag{5.3}$$

- Condition of reparameterization $\sum \alpha_i = \sum \beta_j = 0$ in the analysis of variance model $y_{ij} = \mu + \alpha_i + \beta_j + \epsilon_{ij}$:

$$0 = (1, \ldots, 1)\alpha = (1, \ldots, 1)\beta\,. \tag{5.4}$$

- Knowledge of the ratios between certain coefficients, such as, $\beta_1 : \beta_2 : \beta_3 = ab : b : 1$, which may be reformulated as

$$r = \begin{pmatrix} 0 \\ 0 \end{pmatrix}, \quad R = \begin{pmatrix} 1 & -a & 0 \\ 0 & 1 & -b \end{pmatrix} \begin{pmatrix} \beta_1 \\ \beta_2 \\ \beta_3 \end{pmatrix}.$$

5.2 The Restricted Least-Squares Estimator

To use sample and auxiliary information simultaneously, we have to minimize the sum of squared errors $S(\beta)$ under the linear restriction $r = R\beta$;

that is, we have to minimize

$$S(\beta,\lambda) = (y - X\beta)'(y - X\beta) - 2\lambda'(R\beta - r) \tag{5.5}$$

with respect to β and λ. Here λ is a K-vector of Lagrangian multipliers. Using Theorems A.91–A.93 gives

$$\frac{1}{2}\frac{\partial S(\beta,\lambda)}{\partial \beta} = -X'y + X'X\beta - R'\lambda = 0\,, \tag{5.6}$$

$$\frac{1}{2}\frac{\partial S(\beta,\lambda)}{\partial \lambda} = R\beta - r = 0\,. \tag{5.7}$$

Denoting the solution to this problem by $\hat{\beta} = b(R)$, we get from (5.6)

$$b(R) = (X'X)^{-1}X'y + (X'X)^{-1}R'\lambda. \tag{5.8}$$

Including the restriction (5.7) yields

$$Rb(R) = r = Rb + R(X'X)^{-1}R'\lambda, \tag{5.9}$$

and, using $R(X'X)^{-1}R' > 0$ (cf. Theorem A.39 (vi)), the optimal λ is derived as

$$\hat{\lambda} = (R(X'X)^{-1}R')^{-1}(r - Rb)\,. \tag{5.10}$$

Inserting $\hat{\lambda}$ in (5.8) and using the abbreviation $S = X'X$, we get

$$b(R) = b + S^{-1}R'[RS^{-1}R']^{-1}(r - Rb)\,. \tag{5.11}$$

The restricted least-squares estimator (RLSE) $b(R)$ is the sum of the unrestricted LSE b and a correction term that makes sure the exact restriction $r = R\beta$ holds for the estimator of β

$$\begin{aligned} Rb(R) &= Rb + [RS^{-1}R'][RS^{-1}R']^{-1}(r - Rb) \\ &= r\,. \end{aligned} \tag{5.12}$$

Moments of $b(R)$:

If $r = R\beta$ holds, then $b(R)$ is unbiased:

$$\begin{aligned} \mathrm{E}\left(b(R)\right) &= \beta + S^{-1}R'[RS^{-1}R']^{-1}(r - R\beta) \\ &= \beta\,. \end{aligned}$$

Moreover, we have

$$\mathrm{V}\left(b(R)\right) = \sigma^2 S^{-1} - \sigma^2 S^{-1}R'[RS^{-1}R']^{-1}RS^{-1}, \tag{5.13}$$

which shows that the covariance matrix of $b(R)$ depends only on R. It is seen that the estimator $b(R)$ always has a smaller variance compared with the estimator b in the following sense:

$$\mathrm{V}(b) - \mathrm{V}\left(b(R)\right) = \sigma^2 S^{-1}R'[RS^{-1}R']^{-1}RS^{-1} \geq 0\,. \tag{5.14}$$

Therefore, the use of exact linear restrictions leads to a gain in efficiency.

Remark: It can be shown that $b(R)$ is the best linear unbiased estimator of β in the class

$$\{\hat{\beta} = Cy + Dr\} = \left\{\hat{\beta} = (C, D)\begin{pmatrix} y \\ r \end{pmatrix}\right\}$$

of linear estimators (cf. Theil, 1971, p. 536; Toutenburg, 1975b, p. 99). This class of estimators is heterogeneous in y (i.e., $\hat{\beta} = Cy + d$ with $d = Dr$) but homogeneous in $\begin{pmatrix} y \\ r \end{pmatrix}$.

Special Case: Exact Knowledge of a Subvector

The comparison of a submodel $y = X_1\beta_1 + \epsilon$ with a full model $y = X_1\beta_1 + X_2\beta_2 + \epsilon$ was fully discussed in Section 3.7.

In the submodel we have $\beta_2 = 0$, which may be written as $r = R\beta$ with

$$r = 0, \quad R = (0, I). \tag{5.15}$$

Let

$$S = \begin{pmatrix} X_1'X_1 & X_1'X_2 \\ X_2'X_1 & X_2'X_2 \end{pmatrix}, \quad S^{-1} = \begin{pmatrix} S^{11} & S^{12} \\ S^{21} & S^{22} \end{pmatrix},$$

where the S^{ij} may be taken from (3.94). Let b_1 and b_2 denote the components of b corresponding to β_1 and β_2 (see (3.98)). Then the restricted LSE $b(R)$ from (5.11) for the restriction (5.15) may be given in a partitioned form:

$$\begin{aligned} b(0, I) &= \begin{pmatrix} b_1 \\ b_2 \end{pmatrix} - \begin{pmatrix} S^{11} & S^{12} \\ S^{21} & S^{22} \end{pmatrix}\begin{pmatrix} 0 \\ I \end{pmatrix} \\ &\quad \times \left[(0, I)\begin{pmatrix} S^{11} & S^{12} \\ S^{21} & S^{22} \end{pmatrix}\begin{pmatrix} 0 \\ I \end{pmatrix}\right]^{-1}(0, I)\begin{pmatrix} b_1 \\ b_2 \end{pmatrix} \\ &= \begin{pmatrix} b_1 - S^{12}(S^{22})^{-1}b_2 \\ b_2 - S^{22}(S^{22})^{-1}b_2 \end{pmatrix} \\ &= \begin{pmatrix} (X_1'X_1)^{-1}X_1'y \\ 0 \end{pmatrix}. \end{aligned}$$

We have used $(S^{22})^{-1} = (D^{-1})^{-1} = D$ and formula (3.99).

As a component of the restricted LSE under the restriction $(0, I)\begin{pmatrix} \beta_1 \\ \beta_2 \end{pmatrix} = 0$, the subvector β_1 is estimated by the OLSE of β_1 in the submodel

$$\hat{\beta}_1 = (X_1'X_1)^{-1}X_1'y, \tag{5.16}$$

as can be expected.

If $\beta_2 = \beta_2^* \neq 0$ is given as exact prior information, then the restricted estimator has the form

$$b(0, I) = \begin{pmatrix} \hat{\beta}_1 \\ \beta_2^* \end{pmatrix}. \tag{5.17}$$

5.3 Stepwise Inclusion of Exact Linear Restrictions

The set $r = R\beta$ of linear restrictions has $J < K$ linearly independent restrictions

$$r_j = R_j'\beta\,, \quad j = 1, \ldots, J. \tag{5.18}$$

Here we shall investigate the relationships between the restricted least-squares estimators for either two nested (i.e., linearly dependent) or two disjoint (i.e., independent) sets of restrictions.

Assume $r_1 = R_1\beta$ and $r_2 = R_2\beta$ to be disjoint sets of J_1 and J_2 exact linear restrictions, respectively, where $J_1 + J_2 = J$. We denote by

$$r = \begin{pmatrix} r_1 \\ r_2 \end{pmatrix} = \begin{pmatrix} R_1 \\ R_2 \end{pmatrix} \beta = R\beta \tag{5.19}$$

the full set of restrictions. Let us assume full column ranks, that is, $\text{rank}(R_1) = J_1$, $\text{rank}(R_2) = J_2$, and $\text{rank}(R) = J$. If $b(R_1)$, $b(R_2)$, and $b(R)$ are the restricted LSEs corresponding to the restriction matrices R_1, R_2, and R, respectively, we obtain

$$\text{V}\left(b(R)\right) \leq \text{V}\left(b(R_i)\right) \leq \text{V}(b)\,, \quad i = 1, 2 \tag{5.20}$$

(in the sense that the difference of two dispersion matrices is nonnegative definite).

The relationships $\text{V}(b) - \text{V}\left(b(R_i)\right) \geq 0$ and $\text{V}(b) - \text{V}\left(b(R)\right) \geq 0$ are a consequence of (5.14). Hence, we have to check that

$$\text{V}\left(b(R_1)\right) - \text{V}\left(b(R)\right) \geq 0 \tag{5.21}$$

holds true, which implies that adding further restrictions to a set of restrictions generally leads to a gain in efficiency.

Using the structure of (5.19), we may rewrite the restricted LSE for the complete set $r = R\beta$ as follows:

$$b(R) = b + S^{-1}(R_1', R_2') \begin{pmatrix} R_1 S^{-1} R_1' & R_1 S^{-1} R_2' \\ R_2 S^{-1} R_1' & R_2 S^{-1} R_2' \end{pmatrix}^{-1} \begin{pmatrix} r_1 - R_1 b \\ r_2 - R_2 b \end{pmatrix}. \tag{5.22}$$

With the abbreviations

$$A = RS^{-1}R' = \begin{pmatrix} E & F \\ F' & G \end{pmatrix} \tag{5.23}$$

$$\begin{aligned} R_1 S^{-1} R_1' = E, \quad & R_1 S^{-1} R_2' = F, \\ R_2 S^{-1} R_2' = G, \quad & H = G - F'E^{-1}F \end{aligned} \tag{5.24}$$

(E is nonsingular since $\text{rank}(R_1) = J_1$), and using Theorem A.19, we get the following partitioned form of the dispersion matrix (5.13) of $b(R)$:

$$\begin{aligned} \sigma^{-2}\text{V}\left(b(R)\right) &= S^{-1} - S^{-1}(R_1', R_2') \\ &\quad \times \begin{pmatrix} E^{-1} + E^{-1}FH^{-1}F'E^{-1} & -E^{-1}FH^{-1} \\ -H^{-1}F'E^{-1} & H^{-1} \end{pmatrix} \end{aligned}$$

$$\times \begin{pmatrix} R_1 \\ R_2 \end{pmatrix} S^{-1}. \tag{5.25}$$

Now, the covariance of $b(R_1)$ and $b(R)$ is

$$\mathrm{E}(b(R_1) - \beta)(b(R) - \beta)' = \mathrm{cov}\left(b(R_1), b(R)\right). \tag{5.26}$$

Using

$$b(R_1) - \beta = S^{-1}(I - R_1' E^{-1} R_1 S^{-1}) X' \epsilon, \tag{5.27}$$

$$b(R) - \beta = S^{-1}(I - (R_1', R_2') A^{-1} \begin{pmatrix} R_1 \\ R_2 \end{pmatrix} S^{-1}) X' \epsilon \tag{5.28}$$

along with

$$(I, E^{-1}F)A^{-1} = (E^{-1}, 0) \tag{5.29}$$

$$R_1'(I, E^{-1}F)A^{-1} \begin{pmatrix} R_1 \\ R_2 \end{pmatrix} = R_1' E^{-1} R_1, \tag{5.30}$$

we arrive at the following result:

$$\mathrm{cov}\left(b(R_1), b(R)\right) = \mathrm{V}\left(b(R)\right). \tag{5.31}$$

By Theorem A.41 (v), we know that

$$\left(b(R_1) - b(R)\right)\left(b(R_1) - b(R)\right)' \geq 0$$

holds for any sample and, hence, for the expectation also.

Now, using (5.31), we get the relationship (5.21):

$$\begin{aligned} &\mathrm{E}[b(R_1) - \beta - (b(R) - \beta)][b(R_1) - \beta - (b(R) - \beta)]' \\ &= \mathrm{V}\left(b(R_1)\right) + \mathrm{V}\left(b(R)\right) - 2\,\mathrm{cov}\left(b(R_1), b(R)\right) \\ &= \mathrm{V}\left(b(R_1)\right) - \mathrm{V}\left(b(R)\right) \geq 0. \end{aligned} \tag{5.32}$$

Thus we find the following result:

Theorem 5.1 *Let us assume that a set of exact linear restrictions $r_1 = R_1\beta$ with $\mathrm{rank}(R_1) = J_1$ is available. Now if we add another independent set $r_2 = R_2\beta$ with $\mathrm{rank}(R_2) = J_2$, and $\mathrm{rank}(R) = J = J_1 + J_2$, then the restricted LSEs $b(R_1)$ and $b(R)$ are unbiased with*

$$\mathrm{V}\left(b(R_1)\right) - \mathrm{V}\left(b(R)\right) \geq 0. \tag{5.33}$$

Hence, a stepwise increase of a set of exact restrictions by adding independent restrictions results in a stepwise decrease of variance in the sense of relation (5.33).

Remark: The proof may be given, alternatively, as follows.

The matrices R_1 and R are connected by the following linear transform:

$$R_1 = PR \quad \text{with} \quad P = (I, 0). \tag{5.34}$$

Using the partitioned matrix A from (5.25), the difference of the covariance matrices may be written as

$$\begin{aligned}\sigma^{-2}&\Big[\mathrm{V}\left(b(R_1)\right) - \mathrm{V}\left(b(R)\right)\Big] \\ &= S^{-1}R'(RS^{-1}R')^{-1}RS^{-1} - S^{-1}R_1'(R_1S^{-1}R_1')^{-1}R_1S^{-1} \\ &= S^{-1}R'(A^{-1} - P'(PAP')^{-1}P)RS^{-1}. \qquad (5.35)\end{aligned}$$

By assumption we have $\mathrm{rank}(R) = J$. Then (see Theorem A.46) this difference becomes nonnegative definite if and only if $A^{-1} - P'(PAP')^{-1}P \geq 0$ or, equivalently (Theorem A.67), if

$$\mathcal{R}(P'PA^{-1}) \subset \mathcal{R}(A^{-1}), \qquad (5.36)$$

which holds trivially.

Comparison of $b(R_1)$ and $b(R_2)$

Let us now investigate the relationship between the restricted least squares estimators for the two sets of restrictions

$$r_j = R_j\beta, \quad \mathrm{rank}(R_j) = J_j \quad (j = 1, 2). \qquad (5.37)$$

The corresponding estimators are $(j = 1, 2)$

$$b(R_j) = b + S^{-1}R_j'(R_jS^{-1}R_j')^{-1}(r_j - R_jb). \qquad (5.38)$$

With the abbreviations

$$\begin{aligned} A_j &= R_jS^{-1}R_j', \qquad (5.39)\\ G_j &= S^{-1}R_j'A_j^{-1}R_jS^{-1}, \qquad (5.40)\end{aligned}$$

we get (cf. (5.13))

$$\mathrm{V}\left(b(R_j)\right) = \sigma^2(S^{-1} - G_j). \qquad (5.41)$$

The restricted LSE $b(R_2)$ is better than $b(R_1)$ if

$$\begin{aligned} C &= \mathrm{V}\left(b(R_1)\right) - \mathrm{V}\left(b(R_2)\right) \\ &= \sigma^2(G_2 - G_1) \\ &= \sigma^2S^{-1}(R_2'A_2^{-1}R_2 - R_1'A_1^{-1}R_1)S^{-1} \geq 0 \qquad (5.42)\end{aligned}$$

or, equivalently, if $R_2'A_2^{-1}R_2 - R_1'A_1^{-1}R_1 \geq 0$.

Theorem 5.2 (Trenkler, 1987) *Under the assumptions (5.37) we have*

$$R_2'A_2^{-1}R_2 - R_1'A_1^{-1}R_1 \geq 0 \qquad (5.43)$$

if and only if there exists a $J_1 \times J_2$-matrix P such that

$$R_1 = PR_2. \qquad (5.44)$$

Proof: Use Theorem A.58 and define $M = R_2' A_2^{-\frac{1}{2}}$ and $N = R_1' A_1^{-\frac{1}{2}}$.
(i) Assume (5.43) and use Theorem A.58. Hence, there exists a matrix H such that

$$N = MH\,.$$

Therefore, we have

$$R_1' A_1^{-\frac{1}{2}} = R_2' A_2^{-\frac{1}{2}} H\,,$$

or, equivalently,

$$R_1 = A_1^{\frac{1}{2}} H' A_2^{-\frac{1}{2}} R_2 = PR_2$$

with the $J_1 \times J_2$-matrix

$$P = A_1^{\frac{1}{2}} H' A_2^{-\frac{1}{2}}\,.$$

(ii) Assume $R_1 = PR_2$. Then we may write the difference (5.43) as

$$R_2' A_2^{-\frac{1}{2}} (I - F) A_2^{-\frac{1}{2}} R_2\,, \tag{5.45}$$

where the matrix F is defined by

$$F = A_2^{\frac{1}{2}} P' (P A_2 P')^{-1} P A_2^{\frac{1}{2}}\,, \tag{5.46}$$

which is symmetric and idempotent. Hence, $I - F$ is idempotent, too. Using the abbreviation $B = R_2' A_2^{-\frac{1}{2}} (I - F)$, the difference (5.45) becomes $BB' \geq 0$ (see Theorem A.41).

Corollary 1 to Theorem 5.2: If $R_1 = PR_2$ with $\operatorname{rank}(R_1) = J_1$ holds, it is necessary that $J_1 \leq J_2$ and $\operatorname{rank}(P) = J_1$. Moreover, we have $r_1 = Pr_2$.

Proof: From Theorem A.23 (iv), we know that in general $\operatorname{rank}(AB) \leq \min(\operatorname{rank}(A), \operatorname{rank}(B))$. By applying this to our problem, we obtain

$$\begin{aligned} \operatorname{rank}(PR_2) &\leq \min(\operatorname{rank}(P), \operatorname{rank}(R_2)) \\ &= \min(\operatorname{rank}(P), J_2)\,. \\ J_1 &= \operatorname{rank}(P) \end{aligned}$$

as $\operatorname{rank}(R_1) = \operatorname{rank}(PR_2) = J_1 \Rightarrow J_1 \leq J_2$. From $r_1 = R_1\beta$ and $R_1 = PR_2$, we may conclude that

$$r_1 = PR_2\beta = Pr_2\,.$$

Note: We may confine ourselves to the case $J_1 < J_2$ since $J_1 = J_2$ entails the identity of the restrictions $r_1 = R_1\beta$ and $r_2 = R_2\beta$ as well as the identity of the corresponding estimators. This fact is seen as follows:

The relation $R_1 = PR_2$ with $\operatorname{rank}(P) = J_1 = J_2$ implies the existence of P^{-1}, so that $R_2 = P^{-1}R_1$ and $r_2 = P^{-1}r_1$ hold. Therefore $r_2 = R_2\beta$ is equivalent to $P^{-1}(r_1 - R_1\beta) = 0$ (i.e., $r_1 = R_1\beta$). For $R_1 = PR_2$ with

$P: J_1 \times J_1$ and $\text{rank}(P) = J_1 = J_2$, we may check the equivalence of the estimators immediately:

$$\begin{aligned} b(R_2) &= b + S^{-1}R_1'P^{-1}(P^{-1}R_1S^{-1}R_1'P^{-1})^{-1} \\ &\quad \times (P^{-1}r_1 - P^{-1}R_1b) \\ &= b(R_1)\,. \end{aligned}$$

The case $J_1 < J_2$: As we have remarked before, any linear restriction is invariant with respect to multiplication by a nonsingular matrix C: $J_2 \times J_2$, that is, the conditions

$$r_2 = R_2\beta \quad \text{and} \quad Cr_2 = CR_2\beta$$

are equivalent. We make use of this equivalence and make a special choice of C. Let us assume that $R_1 = PR_2$ with P a $J_1 \times J_2$-matrix of $\text{rank}(P) = J_1$. We choose a matrix Q of order $(J_2 - J_1) \times J_2$ and $\text{rank}(Q) = J_2 - J_1$ such that $C' = (Q', P')$ has $\text{rank}(C') = J_2$. (The matrix Q is said to be complementary to the matrix P.) Letting $Qr_2 = r_3$ and $QR_2 = R_3$, we have

$$\begin{aligned} Cr_2 &= \begin{pmatrix} Qr_2 \\ Pr_2 \end{pmatrix} = \begin{pmatrix} r_3 \\ r_1 \end{pmatrix}, \\ CR_2 &= \begin{pmatrix} QR_2 \\ PR_2 \end{pmatrix} = \begin{pmatrix} R_3 \\ R_1 \end{pmatrix}. \end{aligned}$$

It is interesting to note that if two linear restrictions $r_1 = R_1\beta$ and $r_2 = R_2\beta$ are connected by a linear transform $R_1 = PR_2$, then we may assume that $r_1 = R_1\beta$ is completely contained in $r_2 = R_2\beta$. Hence, without loss of generality, we may choose $P = (I, 0)$.

Corollary 2 to Theorem 5.2: The set of restrictions

$$\left.\begin{aligned} &r_1 = R_1\beta,\ r_2 = R_2\beta,\ R_1 = PR_2,\ r_1 = Pr_2\,, \\ &\text{rank}(P) = J_1 < J_2 \end{aligned}\right\} \tag{5.47}$$

and

$$r_1 = R_1\beta,\ r_2 = \begin{pmatrix} r_1 \\ r_3 \end{pmatrix} = \begin{pmatrix} R_1 \\ R_3 \end{pmatrix}\beta = R_2\beta, \tag{5.48}$$

with $r_3 = Qr_2$, $R_3 = QR_2$, and Q complementary to P are equivalent.

We may therefore conclude from Theorem 5.2 that two exact linear restrictions are comparable by their corresponding restricted LSEs if and only if $R_1 = PR_2$ and $\text{rank}(P) = J_1 < J_2$. The special case $P = (I, 0)$ describes the nested situation

$$R_2 = \begin{pmatrix} R_1 \\ R_3 \end{pmatrix}, \quad r_2 = \begin{pmatrix} r_1 \\ r_3 \end{pmatrix}. \tag{5.49}$$

5.4 Biased Linear Restrictions and MDE Comparison with the OLSE

If, in addition to the sample information, a linear restriction $r = R\beta$ is included in the analysis, it is often imperative to check this restriction by the F-test for the hypothesis H_0: $R\beta = r$ (see Section 3.7). A rejection of this hypothesis may be caused either by a nonstochastic bias δ,

$$r = R\beta + \delta \quad \text{with} \quad \delta \neq 0\,, \tag{5.50}$$

or by a nonstochastic bias and a stochastic effect,

$$r = R\beta + \delta + \Phi, \quad \Phi \sim (0, \sigma^2 V). \tag{5.51}$$

If there is a bias vector $\delta \neq 0$ in the restriction, then the restricted LSE $b(R)$ becomes biased, too. On the other hand, the covariance matrix of $b(R)$ is not affected by δ, and in any case $b(R)$ continues to have smaller variance than the OLSE b (see (5.14)). Therefore, we need to investigate the influence of δ on the restricted LSE $b(R)$ by using its mean dispersion error.

Under assumption (5.50), we have

$$\mathrm{E}\left(b(R)\right) = \beta + S^{-1}R'(RS^{-1}R')^{-1}\delta\,. \tag{5.52}$$

Using the abbreviations

$$A = RS^{-1}R' = A^{\frac{1}{2}}A^{\frac{1}{2}} \tag{5.53}$$

and

$$H = S^{-1}R'A^{-1}\,, \tag{5.54}$$

we may write

$$\begin{aligned} \mathrm{Bias}(b(R), \beta) &= H\delta\,, & (5.55)\\ \mathrm{V}\left(b(R)\right) &= \mathrm{V}(b) - \sigma^2 HAH'\,, & (5.56)\\ \mathrm{M}(b(R), \beta) &= \mathrm{V}(b) - \sigma^2 HAH' + H\delta\delta'H'\,. & (5.57) \end{aligned}$$

We study the MDE comparison according to the following criteria.

MDE-I Criterion

From Definition 3.10, we know that the biased estimator $b(R)$ is MDE-I-better than the unbiased estimator b if

$$\begin{aligned} \Delta\left(b, b(R)\right) &= \mathrm{V}(b) - \mathrm{V}\left(b(R)\right) - \left(\mathrm{Bias}(b(R), \beta)\right)\left(\mathrm{Bias}(b(R), \beta)\right)' \\ &= \sigma^2 H(A - \sigma^2\delta\delta')H' \geq 0 & (5.58) \end{aligned}$$

or, as $\mathrm{rank}(R) = J$ according to Theorem A.46, if and only if

$$A - \sigma^{-2}\delta\delta' \geq 0. \tag{5.59}$$

This is seen to be equivalent (Theorem A.57, Theorem 5.7) to the following condition:

$$\lambda = \sigma^{-2}\delta' A^{-1}\delta = \sigma^{-2}\delta'(RS^{-1}R')^{-1}\delta \leq 1 \,. \tag{5.60}$$

(Toro-Vizcarrondo and Wallace (1968,1969) give an alternative proof.)

Definition 5.3 (MDE-II criterion; first weak MDE criterion)
Let $\hat\beta_1$ and $\hat\beta_2$ be two competing estimators. The estimator $\hat\beta_2$ is said to be MDE-II-better than the estimator $\hat\beta_1$ if

$$\mathrm{E}(\hat\beta_1 - \beta)'(\hat\beta_1 - \beta) - \mathrm{E}(\hat\beta_2 - \beta)'(\hat\beta_2 - \beta) = \mathrm{tr}\left\{\Delta(\hat\beta_1, \hat\beta_2)\right\} \geq 0 \,. \tag{5.61}$$

If $\hat\beta_2$ is MDE-I-better than $\hat\beta_1$, then $\hat\beta_2$ is also MDE-II-better than $\hat\beta_1$, since $\Delta \geq 0$ entails $\mathrm{tr}\{\Delta\} \geq 0$. The reverse conclusion does not necessarily hold true. Therefore, the MDE-II criterion is weaker than the MDE-I criterion.

Direct application of the MDE-II criterion to the comparison of $b(R)$ and b gives (cf. (5.58))

$$\mathrm{tr}\left\{\Delta\big(b, b(R)\big)\right\} = \sigma^2\,\mathrm{tr}\{HAH'\} - \delta' H' H \delta \geq 0$$

if and only if

$$\begin{aligned} \delta' H' H \delta &\leq \sigma^2\,\mathrm{tr}\{HAH'\} \\ &= \mathrm{tr}\{\mathrm{V}(b) - \mathrm{V}\left(b(R)\right)\} \,. \end{aligned} \tag{5.62}$$

Hence, the biased estimator $b(R)$ is MDE-II-better than the unbiased OLSE b if and only if the squared length of the bias vector of $b(R)$ is less than the total decrease of variance of $b(R)$.

With the abbreviation $X'X = S$, we have $H'SH = A^{-1}$, and therefore $\delta' H' S H \delta = \delta' A^{-1}\delta = \sigma^2\lambda$ with λ from (5.60). Using Theorem A.56 and assuming $\delta \neq 0$, we may conclude that

$$d_K \leq \frac{\delta' H' S H \delta}{\delta' H' H \delta} \leq d_1 \tag{5.63}$$

where $d_1 \geq \ldots \geq d_K > 0$ are the eigenvalues of $S > 0$.

Then we have the following upper bound for the left-hand side of (5.62):

$$\delta' H' H \delta \leq d_K^{-1}\delta' A^{-1}\delta = d_K^{-1}\sigma^2\lambda \,. \tag{5.64}$$

Therefore, a sufficient condition for (5.62) to hold is (cf. Wallace, 1972)

$$\begin{aligned} \lambda &\leq d_K\ \mathrm{tr}\{HAH'\} \\ &= d_K\ \mathrm{tr}\{S^{-1}R'(RS^{-1}R')^{-1}RS^{-1}\} \\ &= \lambda_0 \quad (\text{say}) \,. \end{aligned} \tag{5.65}$$

Definition 5.4 (MDE-III criterion; second weak MDE criterion) *$\hat\beta_2$ is said to be MDE-III-better than $\hat\beta_1$ if*

$$\mathrm{E}(X\hat\beta_1 - X\beta)'(X\hat\beta_1 - X\beta) - \mathrm{E}(X\hat\beta_2 - X\beta)'(X\hat\beta_2 - X\beta)$$

$$\begin{aligned} &= \mathrm{E}(\hat{\beta}_1 - \beta)'S(\hat{\beta}_1 - \beta) - \mathrm{E}(\hat{\beta}_2 - \beta)'S(\hat{\beta}_2 - \beta) \\ &= \mathrm{tr}\{S\Delta(\hat{\beta}_1, \hat{\beta}_2)\} \geq 0 \,. \end{aligned} \tag{5.66}$$

Note: According to Definition 3.9 we see that MDE-III superiority is equivalent to $R(S)$ superiority.

Applying criterion (5.66) to $b(R)$ and b, we see that $b(R)$ is MDE-III-better than b if

$$\begin{aligned} \mathrm{tr}\{S\Delta(b, b(R))\} &= \sigma^2 \,\mathrm{tr}\{SS^{-1}R'(RS^{-1}R')^{-1}RS^{-1}\} - \delta' A^{-1}\delta \\ &= \sigma^2(\mathrm{tr}\{I_J\} - \lambda) \\ &= \sigma^2(J - \lambda) \geq 0 \,; \end{aligned}$$

that is, $b(R)$ is preferred if

$$\lambda \leq J \,. \tag{5.67}$$

It may be observed that for $J \geq 2$ the MDE-III criterion is weaker than the MDE-I criterion. If $J = 1$, both criteria become equivalent.

Theorem 5.5 *Let us suppose that we are given a biased linear restriction* $(r - R\beta = \delta)$. *Then the biased RLSE* $b(R)$ *is better than the unbiased OLSE* b *by*

(i) *MDE-I criterion if*
$\lambda \leq 1$ *(necessary and sufficient),*

(ii) *MDE-II criterion if*
$\lambda \leq \lambda_0$ (λ_0 *from (5.65)) (sufficient), and*

(iii) *MDE-III criterion if*
$\lambda \leq J$ *(necessary and sufficient),*
where $\lambda = \sigma^{-2}(r - R\beta)'(RS^{-1}R')^{-1}(r - R\beta)$.

To test the conditions $\lambda \leq 1$ (or λ_0 or J), we assume $\epsilon \sim N(0, \sigma^2 I)$ and use the test statistic

$$F = \frac{1}{Js^2}(r - Rb)'(RS^{-1}R')^{-1}(r - Rb) \,, \tag{5.68}$$

which has a noncentral $F_{J,T-K}(\lambda)$-distribution. The test statistic F provides a uniformly most powerful test for the MDE criteria (Lehmann, 1986). We test the null hypothesis

$$H_0\colon \lambda \leq 1 \quad (\text{or } \leq \lambda_0 \text{ or } \leq J)$$

against the alternative

$$H_1\colon \lambda > 1 \quad (\text{or } > \lambda_0 \text{ or } > J)$$

based on the decision rule

$$\begin{aligned} \text{do not reject } H_0 \text{ if } F &\leq F_{J,T-K,1-\alpha}(1) \,, \\ \text{or } F &\leq F_{J,T-K,1-\alpha}(\lambda_0) \,, \\ \text{or } F &\leq F_{J,T-K,1-\alpha}(J) \,, \end{aligned}$$

respectively, and reject otherwise.

5.5 MDE Matrix Comparisons of Two Biased Estimators

Up to now we have investigated the relationship between two unbiased RSLEs (Section 5.3) and the relationship between a biased and an unbiased estimator (Section 5.4).

The problem of the MDE comparison of any two estimators is of central interest in statistics. Therefore, we now present a systematic overview of the situations that are to be expected, especially if any two biased estimators have to be compared. This overview comprises the development during the past decade. One of the main results is a matrix theorem of Baksalary and Kala (1983). In this context the investigations of Teräsvirta (1982, 1986) and Trenkler (1985) should also be mentioned. In the following we use the general framework developed in Trenkler and Toutenburg (1990).

Suppose we have available an estimator t for a parameter vector $\theta \in \mathcal{R}^p$. We do not assume that t is necessarily unbiased for θ, that is, $\mathrm{E}(t)$ may be different from θ for some θ.

We denote by

$$\mathrm{D}(t) = \mathrm{E}\big(t - E(t)\big)\big(t - \mathrm{E}(t)\big)' = \mathrm{V}(t) \tag{5.69}$$

the dispersion matrix of t and by

$$d = \mathrm{Bias}(t, \theta) = \mathrm{E}(t) - \theta \tag{5.70}$$

the bias vector of t.

Then the mean dispersion error matrix of t is (cf. (3.45)) given by

$$\mathrm{M}(t, \theta) = \mathrm{D}(t) + dd'. \tag{5.71}$$

Let us consider two competing estimators, t_1 and t_2, of θ. We say that t_2 is superior to t_1 (i.e., t_2 is MDE-I-better than t_1; cf. Definition 3.4) if

$$\Delta(t_1, t_2) = \mathrm{M}(t_1, \theta) - \mathrm{M}(t_2, \theta) \tag{5.72}$$

is a nonnegative-definite (n.n.d.) matrix, that is, $\Delta(t_1, t_2) \geq 0$.

In case the matrix $\Delta(t_1, t_2)$ is positive definite (p.d.), we may give the following definition.

Definition 5.6 *t_2 is said to be strongly MDE-better (or strongly MDE-I-better) than t_1 if $\Delta(t_1, t_2) > 0$ (positive definite).*

For notational convenience, let us define

$$\begin{aligned} d_i &= \mathrm{Bias}(t_i, \theta) \quad (i = 1, 2)\,, & (5.73)\\ \mathrm{D}(t_i) &= \mathrm{V}(t_i) \quad (i = 1, 2)\,, & (5.74)\\ D &= \mathrm{D}(t_1) - \mathrm{D}(t_2)\,. & (5.75) \end{aligned}$$

Then (5.72) becomes

$$\Delta(t_1, t_2) = D + d_1 d_1' - d_2 d_2' \,. \tag{5.76}$$

In order to inspect whether $\Delta(t_1, t_2)$ is n.n.d. or p.d., we may confine ourselves to two cases:

$$\text{Condition 1:} \qquad D > 0 \,,$$
$$\text{Condition 2:} \qquad D \geq 0 \,.$$

Note that it is possible that $\Delta(t_1, t_2) \geq 0$ although condition 1 or condition 2 has not been satisfied; however, this is very rarely the case. Hence, we shall concentrate on these two realistic situations.

As $d_1 d_1' \geq 0$, it is easy to see that

$$D > 0 \quad \Rightarrow \quad D + d_1 d_1' > 0 \,,$$
$$D \geq 0 \quad \Rightarrow \quad D + d_1 d_1' \geq 0 \,.$$

Hence the problem of deciding whether $\Delta(t_1, t_2) > 0$ or $\Delta(t_1, t_2) \geq 0$ reduces to that of deciding whether a matrix of type

$$A - aa' \tag{5.77}$$

is positive or nonnegative definite when A is positive or nonnegative definite.

Condition 1: $D > 0$

Let $A > 0$. Then we have (cf. A.57) the following result.

Theorem 5.7 (Farebrother, 1976) *Suppose that A is p.d. and a is a compatible column vector. Then $A - aa' > (\geq) 0$ if and only if $a'A^{-1}a < (\leq) 1$.*

Direct application of Theorem 5.7 to the matrix $\Delta(t_1, t_2)$ specified by (5.76) gives the following result:

Theorem 5.8 *Suppose that the difference $D = \mathrm{D}(t_1) - \mathrm{D}(t_2)$ of the dispersion matrices of the estimators t_1 and t_2 is positive definite. Then t_2 is strongly MDE-I-superior to t_1 if and only if*

$$d_2'(D + d_1 d_1')^{-1} d_2 < 1 \,, \tag{5.78}$$

and t_2 is MDE-I-better than t_1 if and only if

$$d_2'(D + d_1 d_1')^{-1} d_2 \leq 1 \,. \tag{5.79}$$

By Theorem A.18 (iv) (Rao, 1973a, p. 33) we may write

$$(D + d_1 d_1')^{-1} = D^{-1} - \frac{D^{-1} d_1 d_1' D^{-1}}{1 + d_1' D^{-1} d_1} \,.$$

Setting

$$d_{ij} = d_i' D^{-1} d_j \quad (i, j = 1, 2), \tag{5.80}$$

we get from (5.78) and (5.79)

$$d_2'(D + d_1 d_1')^{-1} d_2 = d_{22} - d_{12}^2 (1 + d_{11})^{-1}.$$

Corollary 1 to Theorem 5.8 (see also Trenkler and Trenkler, 1983): *Under the assumption $D > 0$ we have $\Delta(t_1, t_2) > (\geq) 0$ if and only if*

$$(1 + d_{11})(d_{22} - 1) < (\leq) d_{12}^2 \,. \tag{5.81}$$

Furthermore, each of the two conditions is sufficient for $\Delta(t_1, t_2) > (\geq) 0$:

(i) $(1 + d_{11}) d_{22} < (\leq) 1$,

(ii) $d_{22} < (\leq) 1$.

Corollary 2 to Theorem 5.8: Let $D > 0$ and suppose that d_1 and d_2 are linearly dependent, that is, $d_{12}^2 = d_{11} d_{22}$. Then we have $\Delta(t_1, t_2) > (\geq) 0$ if and only if

$$d_{22} - d_{11} < (\leq) 1. \tag{5.82}$$

Corollary 3 to Theorem 5.8: Let $D > 0$ and suppose that t_1 is unbiased for θ, that is, $d_1 = 0$ and $d_{11} = d_{12} = 0$. Then we have $\Delta(t_1, t_2) > (\geq) 0$ if and only if

$$d_{22} < (\leq) 1. \tag{5.83}$$

Example 5.1 ((Perlman, 1972)): Let t be an estimator of θ. As a competitor to $t_1 = t$, consider $t_2 = \alpha t_1$ with $0 \leq \alpha \leq 1$ so that t_2 is of the shrinkage type. Then $D = (1 - \alpha^2)\,\mathrm{D}(t_1)$, and we have

$$\begin{aligned} D > 0 &\text{ if and only if } \mathrm{D}(t_1) > 0\,, \\ D \geq 0 &\text{ if and only if } \mathrm{D}(t_1) \geq 0\,. \end{aligned}$$

Let us suppose that t is unbiased for θ and $\mathrm{D}(t) > 0$. Consider $t_1 = \alpha_1 t$ and $t_2 = \alpha_2 t$, where $0 \leq \alpha_2 < \alpha_1 < 1$. Then $\mathrm{D}(t_i) = \alpha_i^2\,\mathrm{D}(t)$ and $D = \mathrm{D}(t_1) - \mathrm{D}(t_2) = (\alpha_1^2 - \alpha_2^2)\,\mathrm{D}(t) > 0$. Furthermore, $d_i = \mathrm{Bias}(t_i, \theta) = -(1 - \alpha_i)\theta$, $(i = 1, 2)$, showing the linear dependence of d_1 and d_2. Using definition (5.80), we get

$$d_{ii} = \frac{(1 - \alpha_i)^2}{\alpha_1^2 - \alpha_2^2} \theta' \big(\mathrm{D}(t)\big)^{-1} \theta\,,$$

which yields (cf. (5.82))

$$d_{22} - d_{11} = \frac{2 - \alpha_1 - \alpha_2}{\alpha_1 + \alpha_2} \theta' \big(\mathrm{D}(t)\big)^{-1} \theta\,.$$

Hence, from Corollary 2 to Theorem 5.8 we may conclude that

$$\Delta(\alpha_1 t, \alpha_2 t) > (\geq) 0$$

if and only if

$$\theta'\left(\mathrm{D}(t)\right)^{-1}\theta < (\leq)\,\frac{\alpha_1+\alpha_2}{2-\alpha_1-\alpha_2}\,.$$

If $\alpha_1 = 1$, then $t_1 = t$ is unbiased and $\Delta(t, \alpha_2 t) > (\geq)\, 0$ holds according to (5.83) if and only if

$$d_{22} = \frac{1-\alpha_2}{1+\alpha_2}\theta'\left(\mathrm{D}(t)\right)^{-1}\theta < (\leq)\, 1.$$

Note: The case $D = \mathrm{D}(t_1) - \mathrm{D}(t_2) > 0$ rarely occurs in practice (except in very special situations, as described in the above example). It is more realistic to assume $D \geq 0$.

Condition 2: $D \geq 0$

MDE matrix comparisons of two biased estimators under this condition may be based on the definiteness of a difference of matrices of type $A - aa'$ where $A \geq 0$. Here we state a basic result.

Theorem 5.9 (Baksalary and Kala, 1983) *Let A be an n.n.d. matrix and let a be a column vector. Then $A - aa' \geq 0$ if and only if*

$$a \in \mathcal{R}(A) \quad \textit{and} \quad a'A^- a \leq 1\,, \tag{5.84}$$

where A^- is any g-inverse of A, that is, $AA^-A = A$.

Note: Observe that the requirement $a \in \mathcal{R}(A)$ is equivalent to $a = Ac$ for some vector c. Hence, $a'A^-a = c'AA^-Ac = c'Ac$, and $a'A^-a$ is therefore seen to be invariant to the choice of the g-inverse A^-.

An application of this theorem gives the following result.

Theorem 5.10 *Suppose that the difference $D = \mathrm{D}(t_1) - \mathrm{D}(t_2)$ of the dispersion matrices of two competing estimators t_1 and t_2 is n.n.d. Then t_2 is MDE-better than t_1 if and only if*

$$(i) \qquad d_2 \in \mathcal{R}(D + d_1 d_1')\,, \tag{5.85}$$

$$(ii) \qquad d_2'(D + d_1 d_1')^- d_2 \leq 1\,, \tag{5.86}$$

where d_i is the bias in t_i, $i = 1, 2$, $(D + d_1 d_1')^-$ is any g-inverse of $D + d_1 d_1'$.

To determine a g-inverse of $D + d_1 d_1'$, let us now consider two possibilities:

(i) $d_1 \in \mathcal{R}(D)$,

(ii) $d_1 \notin \mathcal{R}(D)$.

If $d_1 \in \mathcal{R}(D)$, a g-inverse of $D + d_1 d_1'$ is given by (cf. Theorem A.68)

$$(D + d_1 d_1')^- = D^- - \frac{D^- d_1 d_1' D^-}{1 + d_1' D^- d_1}\,. \tag{5.87}$$

Because $d_1 \in \mathcal{R}(D)$, we have $d_1 = Df_1$ with a suitable vector f_1. Since we have assumed $D \geq 0$, it follows that $d_1' D^- d_1 = f_1' D f_1 \geq 0$ and $1 + d_1' D^- d_1 > 0$.

Since $D \geq 0$ and $d_1 d_1' \geq 0$, we get

$$\begin{aligned} \mathcal{R}(D + d_1 d_1') &= \mathcal{R}(D) + \mathcal{R}(d_1 d_1') \\ &= \mathcal{R}(D) + \mathcal{R}(d_1) . \end{aligned}$$

Now $d_1 \in \mathcal{R}(D)$ implies

$$\mathcal{R}(D + d_1 d_1') = \mathcal{R}(D) \tag{5.88}$$

(cf. Theorem A.76 (ii)). Based on (5.87) and (5.88), we may state the next result.

Corollary 1 to Theorem 5.10: Assume that $d_1 \in \mathcal{R}(D)$ and $d_2 \in \mathcal{R}(D + d_1 d_1') = \mathcal{R}(D)$, and let $d_{ij} = d_i' D^- d_j (i, j = 1, 2)$, where D^- is any g-inverse of D. Then we have

$$\Delta(t_1, t_2) \geq 0 \quad \textit{if and only if} \quad (1 + d_{11})(d_{22} - 1) \leq d_{12}^2 . \tag{5.89}$$

Furthermore, each of the following conditions is sufficient for $\Delta(t_1, t_2) \geq 0$:

$$(1 + d_{11}) d_{22} \leq 1 , \tag{5.90}$$

$$d_{22} \leq 1 . \tag{5.91}$$

Since both d_1 and d_2 belong to the range of D, there exist vectors f_i with $d_i = Df_i$ $(i = 1, 2)$ such that $d_{ij} = d_i' D^- d_j = f_i' D f_j$; that is, d_{ij} is invariant to the choice of D^- (cf. Theorem A.69).

It is easily seen that $d_{12}^2 = d_{11} d_{22}$ if d_1 and d_2 are linearly dependent.

Corollary 2 to Theorem 5.10: Let the assumptions of Corollary 1 be valid, and assume that d_1 and d_2 are linearly dependent. Then we have $\Delta(t_1, t_2) \geq 0$ if and only if

$$d_{22} - d_{11} \leq 1 . \tag{5.92}$$

Corollary 3 to Theorem 5.10: Suppose t_1 is unbiased (i.e., $d_1 = 0$) and $d_2 \in \mathcal{R}(D)$. Then we have $\Delta(t_1, t_2) \geq 0$ if and only if

$$d_{22} \leq 1 . \tag{5.93}$$

Case $d_1 \notin \mathcal{R}(D)$

In order to obtain the explicit formulation of condition (5.86), we need a g-inverse of $D + d_1 d_1'$. Applying Theorem A.70 gives the following result.

Corollary 4 to Theorem 5.10: Suppose that $d_1 \notin \mathcal{R}(D)$ and $d_2 \in \mathcal{R}(D + d_1 d_1')$. Then $\Delta(t_1, t_2) \geq 0$ if and only if

$$d_2' D^+ d_2 - 2\phi(d_2' v)(d_2' u) + \gamma \phi^2 (d_2' u)^2 \leq 1\,, \tag{5.94}$$

with the notation

$$\begin{aligned} u &= (I - DD^+) d_1, \quad v = D^+ d_1\,, \\ \gamma &= 1 + d_1' D^+ d_1\,, \\ \phi &= (u'u)^{-1}. \end{aligned}$$

Moreover, if $d_2 \in \mathcal{R}(D)$, we immediately get

$$\begin{aligned} d_2' u &= f_2' D(I - DD^+) d_1 = f_2'(D - DDD^+) d_1 \\ &= f_2'(D - DD^+ D) d_1 = 0 \end{aligned}$$

using $(DD^+)' = D^+ D$ since D is symmetric.

Corollary 5 to Theorem 5.10: Assume that $d_1 \notin \mathcal{R}(D)$ and $d_2 \in \mathcal{R}(D)$. Then we have $\Delta(t_1, t_2) \geq 0$ if and only if

$$d_2' D^+ d_2 \leq 1\,. \tag{5.95}$$

We have thus investigated conditions under which the matrix $D + d_1 d_1' - d_2 d_2'$ is n.n.d. in various situations concerning the relationship between d_1 and d_2 and the range $\mathcal{R}(D + d_1 d_1')$. These conditions may also be presented in equivalent alternative forms. In Bekker and Neudecker (1989), one may find an overview of such characterizations (cf. also Theorems A.74–A.78).

5.6 MDE Matrix Comparison of Two Linear Biased Estimators

In Section 5.5, we investigated the MDE matrix superiority of an estimator t_2 with respect to any other estimator t_1. In this section we wish to apply these results for the case of two *linear* estimators b_1 and b_2, which is of central interest in linear models.

Consider the standard regression model $y = X\beta + \epsilon$, $\epsilon \sim (0, \sigma^2 I)$ and $\text{rank}(X) = K$. Let us consider two competing heterogeneous linear estimators

$$b_i = C_i y + c_i \quad (i = 1, 2), \tag{5.96}$$

where $C_i : K \times T$ and $c_i : K \times 1$ are nonstochastic. Then it is easy to see that

$$\text{V}(b_i) = \sigma^2 C_i C_i' \tag{5.97}$$

$$\begin{aligned} d_i &= \text{Bias}(b_i, \beta) = (C_i X - I)\beta + c_i \,, && (5.98) \\ \text{M}(b_i, \beta) &= \sigma^2 C_i C_i' + d_i d_i' \quad (i = 1, 2) \,, && (5.99) \end{aligned}$$

from which the difference of the dispersion matrices of b_1 and b_2 becomes

$$D = \sigma^2 (C_1 C_1' - C_2 C_2') \,, \tag{5.100}$$

which is symmetric.

As we have seen in Section 5.5, the definiteness of the matrix D has main impact on the MDE superiority of b_2 over b_1 according to the condition

$$\Delta(b_1, b_2) = D + d_1 d_1' - d_2 d_2' \geq 0 \,. \tag{5.101}$$

Since we are interested in the case for which the matrix D is n.n.d. or p.d., the following characterization may be very useful.

Theorem 5.11 (Baksalary, Liski, and Trenkler, 1989) *The matrix D (5.100) is n.n.d. if and only if*

$$(i) \quad \mathcal{R}(C_2) \subset \mathcal{R}(C_1) \tag{5.102}$$

and

$$(ii) \quad \lambda_{\max}(C_2'(C_1 C_1')^- C_2) \leq 1, \tag{5.103}$$

where $\lambda_{\max}(\cdot)$ denotes the maximal eigenvalue of the matrix inside the parantheses. This eigenvalue is invariant to the choice of the g-inverse $(C_1 C_1')^-$.

Theorem 5.12 *We have $D > 0$ if and only if*

$$C_1 C_1' > 0 \tag{5.104}$$

and

$$\lambda_{\max}(C_2'(C_1 C_1')^{-1} C_2) < 1. \tag{5.105}$$

Proof: Assume that $D > 0$. Because $C_2 C_2' \geq 0$ always holds, we get

$$C_1 C_1' = D + C_2 C_2' > 0 \,,$$

which is regular, and we may write its inverse in the form

$$(C_1 C_1')^{-1} = (C_1 C_1')^{-\frac{1}{2}} (C_1 C_1')^{-\frac{1}{2}} .$$

Applying Theorem A.39, we get

$$(C_1 C_1')^{-\frac{1}{2}} D (C_1 C_1')^{-\frac{1}{2}} = I - (C_1 C_1')^{-\frac{1}{2}} C_2 C_2' (C_1 C_1')^{-\frac{1}{2}} > 0 \,. \tag{5.106}$$

The eigenvalues of the p.d. matrix $(C_1 C_1')^{-\frac{1}{2}} D (C_1 C_1')^{-\frac{1}{2}}$ are positive. Using the properties of eigenvalues, $\lambda(I - A) = 1 - \lambda(A)$ and $\lambda(PP') = \lambda(P'P)$, we find

$$\lambda\big((C_1 C_1')^{-\frac{1}{2}} C_2 C_2' (C_1 C_1')^{-\frac{1}{2}}\big) = \lambda(C_2'(C_1 C_1')^{-1} C_2).$$

This holds for all eigenvalues and in particular for the maximal eigenvalue. Therefore, we have proved the necessity of (5.104) and (5.105). The proof of the sufficiency is trivial, as (5.104) and (5.105) immediately imply relationship (5.106) and hence $D > 0$.

5.7 MDE Comparison of Two (Biased) Restricted Estimators

Suppose that we have two competing restrictions on β $(i = 1, 2)$,

$$r_i = R_i\beta + \delta_i \,, \tag{5.107}$$

where R_i is a $J_i \times K$-matrix of full row rank J_i.

The corresponding linearly restricted least-squares estimators are given by

$$b(R_i) = b + S^{-1}R_i'(R_iS^{-1}R_i')^{-1}(r_i - R_ib). \tag{5.108}$$

Let $S^{-\frac{1}{2}}$ denote the unique p.d. square root of the matrix $S^{-1} = (X'X)^{-1}$. As we have assumed that $\operatorname{rank}(R_i) = J_i$, we see that the $J_i \times K$-matrix $R_iS^{-\frac{1}{2}}$ is of rank J_i. Therefore (cf. Theorem A.66), its Moore-Penrose inverse is

$$(R_iS^{-\frac{1}{2}})^+ = S^{-\frac{1}{2}}R_i'(R_iS^{-1}R_i')^{-1}. \tag{5.109}$$

Noticing that the matrix $(i = 1, 2)$

$$P_i = (R_iS^{-\frac{1}{2}})^+R_iS^{-\frac{1}{2}} \tag{5.110}$$

is idempotent of rank $J_i < K$ and an orthogonal projector on the column space $\mathcal{R}(S^{-\frac{1}{2}}R_i')$, we observe that (cf. (5.55) and (5.56))

$$\begin{aligned} d_i = \operatorname{Bias}(b(R_i), \beta) &= S^{-1}R_i'(R_iS^{-1}R_i')^{-1}\delta_i \\ &= S^{-\frac{1}{2}}(R_iS^{-\frac{1}{2}})^+\delta_i \,, & (5.111) \\ \operatorname{V}\left(b(R_i)\right) &= \sigma^2S^{-\frac{1}{2}}(I - P_i)S^{-\frac{1}{2}} \,, & (5.112) \end{aligned}$$

where $\delta_i = R_i\beta - r_i$, $i = 1, 2$. Denoting $P_{21} = P_2 - P_1$, the difference of the dispersion matrices can be written as

$$D = \operatorname{V}\left(b(R_1)\right) - \operatorname{V}\left(b(R_2)\right) = \sigma^2S^{-\frac{1}{2}}P_{21}S^{-\frac{1}{2}} \tag{5.113}$$

and hence we have the following equivalence:

$$D \geq 0 \quad \text{if and only if} \quad P_{21} \geq 0 \,. \tag{5.114}$$

Note: If we use the notation

$$c_i = S^{\frac{1}{2}}d_i = (R_iS^{-\frac{1}{2}})^+\delta_i \,, \tag{5.115}$$

we may conclude that $b(R_2)$ is MDE-I-better than $b(R_1)$ if

$$\Delta\big(b(R_1), b(R_2)\big) = S^{-\frac{1}{2}}(\sigma^2 P_{21} + c_1 c_1' - c_2 c_2')S^{-\frac{1}{2}} \geq 0$$

or, equivalently, if

$$P_{21} + c_1 c_1' - c_2 c_2' \geq 0\,. \tag{5.116}$$

According to Theorem 5.12, we see that the symmetric $K \times K$-matrix P_2 cannot be p.d., because $P_2 = S^{-\frac{1}{2}} R_2'(R_2 S^{-1} R_2')^{-1} R_2 S^{-\frac{1}{2}}$ is of rank $J_2 < K$ and hence P_2 is singular. Therefore, condition (5.104) does not hold.

We have to confine ourselves to the case $P_{21} \geq 0$. According to a result by Ben-Israel and Greville (1974, p. 71) we have the following equivalence.

Theorem 5.13 *Let $P_{21} = P_2 - P_1$ with P_1, P_2 from (5.110). Then the following statements are equivalent:*

(i) $P_{21} \geq 0$,

(ii) $\mathcal{R}(S^{-\frac{1}{2}} R_1') \subset \mathcal{R}(S^{-\frac{1}{2}} R_2')$,

(iii) There exists a matrix F such that $R_1 = F R_2$,

(iv) $P_2 P_1 = P_1$,

(v) $P_1 P_2 = P_1$,

(vi) P_{21} is an orthogonal projector.

Note: The equivalence of $P_{21} \geq 0$ and condition (iii) has been proved in Theorem 5.2.

Let us assume that $D \geq 0$ (which is equivalent to conditions (i)–(vi)). As in the discussion following Theorem 5.10, let us consider two cases:

(i) $\quad c_1 \in \mathcal{R}(P_{21})$,

(ii) $\quad c_1 \notin \mathcal{R}(P_{21})$.

Case (i): $c_1 \in \mathcal{R}(P_{21})$.

Since P_{21} is an orthogonal projector, the condition $c_1 \in \mathcal{R}(P_{21})$ is equivalent to

$$P_{21} c_1 = c_1\,. \tag{5.117}$$

We have the following relationships for c_i and P_i, $i = 1, 2$:

$$\left.\begin{array}{ll} P_1 c_1 = c_1, & P_2 c_2 = c_2\,, \\ P_1 c_2 = c_1, & P_2 c_1 = c_1\,. \end{array}\right\} \tag{5.118}$$

Proof:

$$\begin{aligned} P_1 c_1 &= S^{-\frac{1}{2}} R_1'(R_1 S^{-1} R_1')^{-1} R_1 S^{-\frac{1}{2}} S^{-\frac{1}{2}} R_1'(R_1 S^{-1} R_1')^{-1} \delta_1 \\ &= c_1 \\ P_2 c_2 &= c_2 \quad \text{(using the above procedure)} \\ P_2 c_1 &= P_2 P_1 c_1 = P_1 c_1 = c_1 \quad \text{(cf. (iv))} \\ P_1 c_2 &= S^{-\frac{1}{2}} R_1'(R_1 S^{-1} R_1')^{-1} F \delta_2 = c_1 \quad \text{(cf. (iii))} \end{aligned}$$

as $R_1 = F R_2$ implies $r_1 = F r_2$ and

$$\delta_1 = r_1 - R_1\beta = F(r_2 - R_2\beta) = F\delta_2 .$$

Thus we obtain the following result: Suppose that $D \geq 0$ and $c_1 \in \mathcal{R}(P_{21})$ or, equivalently (cf. (5.118)),

$$\begin{aligned} c_1 = P_{21} c_1 &= P_2 c_1 - P_1 c_1 \\ &= c_1 - c_1 = 0 , \end{aligned}$$

which implies that $\delta_1 = 0$ and $b(R_1)$ unbiased. Relation (5.118) implies $P_{21} c_2 = P_2 c_2 = c_2$ and, hence, $c_2 \in \mathcal{R}(P_{21})$ and

$$\begin{aligned} c_2' P_{21}^- c_2 &= c_2' c_2 \\ &= \delta_2'(R_2 S^{-1} R_2')^{-1} \delta_2 . \end{aligned}$$

Applying Theorem 5.9 leads to the following theorem.

Theorem 5.14 *Suppose that the linear restrictions $r_1 = R_1\beta$ and $r_2 = R_2\beta + \delta_2$ are given, and assume that*

$$D = \mathrm{V}\left(b(R_1)\right) - \mathrm{V}\left(b(R_2)\right) \geq 0 .$$

Then the biased estimator $b(R_2)$ is MDE-superior to the unbiased estimator $b(R_1)$ if and only if

$$\sigma^{-2} \delta_2'(R_2 S^{-1} R_2')^{-1} \delta_2 \leq 1 . \tag{5.119}$$

Case (ii): $c_1 \notin \mathcal{R}(P_{21})$.

The case $c_1 \notin \mathcal{R}(P_{21})$ is equivalent to $c_1 \neq 0$. Assuming $D \geq 0$, we have $\Delta\big(b(R_1), b(R_2)\big) \geq 0$ if and only if (5.84) is fulfilled (cf. Theorem 5.9), that is, if and only if

$$c_2'(\sigma^2 P_{21} + c_1 c_1')^+ c_2 = \sigma^{-2} c_2' P_{21} c_2 + 1 \leq 1$$

or, equivalently, if

$$P_{21} c_2 = 0 , \tag{5.120}$$

that is (cf.(5.117)), if

$$c_1 = c_2 . \tag{5.121}$$

This way we have prooved the following

Theorem 5.15 *Assume that $\delta_i = r_i - R_i\beta \neq 0$. Then we have the following equivalence:*

$$\Delta\big(b(R_1), b(R_2)\big) \geq 0 \quad \textit{if and only if} \quad \text{Bias}(b(R_1), \beta) = \text{Bias}(b(R_2), \beta). \tag{5.122}$$

Note: An alternative proof is given in Toutenburg (1989b).

Summary: The results given in Theorems 5.14 and 5.15 may be summed up as follows. Suppose that we are given two linear restrictions $r_i = R_i\beta + \delta_i$, $i = 1, 2$. Let $b(R_i)$, $i = 1, 2$, denote the corresponding RLSEs. Assume that the difference of their dispersion matrices is n.n.d. (i.e., $\text{V}\left(b(R_1)\right) - \text{V}\left(b(R_2)\right) \geq 0$). Then both linear restrictions are comparable under the MDE-I criterion if

(i) $\delta_1 = 0$ (i.e., $b(R_1)$ is unbiased) or

(ii) $\text{Bias}(b(R_1), \beta) = \text{Bias}(b(R_2), \beta)$.

If (ii) holds, then the difference of the MDE matrices of both estimators reduces to the difference of their dispersion matrices:

$$\Delta\big(b(R_1), b(R_2)\big) = \text{V}\left(b(R_1)\right) - \text{V}\left(b(R_2)\right).$$

We consider now the special case of stepwise biased restrictions. The preceding comparisons of two RLSEs have shown the necessity of $\text{V}\left(b(R_1)\right) - \text{V}\left(b(R_2)\right)$ being nonnegative definite. This condition is equivalent to $R_1 = PR_2$ (cf. Theorems 5.2 and 5.13 (iii)). According to Corollary 2 of Theorem 5.2, we may assume without loss of generality that $P = (I, 0)$.

Therefore, assuming $\text{V}\left(b(R_1)\right) - \text{V}\left(b(R_2)\right) \geq 0$, we may specify the competing linear restrictions as follows:

$$\left.\begin{array}{rclrcl}
r_1 & = & R_1\beta\,, & \operatorname{rank}\underset{J_1\times K}{(R_1)} & = & J_1\,, \\
r_3 & = & R_3\beta + \delta_3, & \operatorname{rank}\underset{J_3\times K}{(R_3)} & = & J_3\,, \\
r_2 & = & R_2\beta + \delta_2, & \operatorname{rank}\underset{J_2\times K}{(R_2)} & = & J_2\,, \\
\text{where} & & & & & \\
r_2 & = & \begin{pmatrix} r_1 \\ r_3 \end{pmatrix}, & R_2 & = & \begin{pmatrix} R_1 \\ R_3 \end{pmatrix}, \\
\delta_2 & = & \begin{pmatrix} 0 \\ \delta_3 \end{pmatrix}, & J_1 + J_3 & = & J_2\,.
\end{array}\right\} \tag{5.123}$$

Furthermore, from Theorems 5.14 and 5.15, we know that we may confine our attention to the case $r_1 - R_1\beta = \delta_1 = 0$.

In the following we investigate the structure of the parameter condition (5.119) for the MDE superiority of $b(R_2)$ in comparison to $b(R_1)$. We are especially interested in the relationships among the competing estimators

$$\begin{aligned} b &= S^{-1}X'y \quad \text{(unbiased)} \\ b(R_1) &= b + S^{-1}R_1'(R_1S^{-1}R_1')^{-1}(r_1 - R_1b) \\ &\quad \text{(unbiased)} \end{aligned} \tag{5.124}$$

$$\begin{aligned} b(R_3) &= b + S^{-1}R_3'(R_3S^{-1}R_3')^{-1}(r_3 - R_3b) \\ &\quad \text{(biased in case } \delta_3 \neq 0) \end{aligned} \tag{5.125}$$

$$\begin{aligned} b\begin{pmatrix} R_1 \\ R_3 \end{pmatrix} &= b + S^{-1}(R_1' R_3')\left(\begin{pmatrix} R_1 \\ R_3 \end{pmatrix} S^{-1}(R_1' \, R_3')\right)^{-1} \\ &\quad \times \left(\begin{pmatrix} r_1 \\ r_3 \end{pmatrix} - \begin{pmatrix} R_1 \\ R_3 \end{pmatrix} b\right) \\ &\quad \text{(biased in case } \delta_3 \neq 0). \end{aligned} \tag{5.126}$$

We again use the notation (cf. (5.53) and (5.54))

$$\begin{aligned} A_i &= R_iS^{-1}R_i', \quad A_i > 0 \quad (i = 1, 2, 3) \\ H_i &= S^{-1}R_i'A_i^{-1} \quad (i = 1, 2, 3). \end{aligned}$$

Additionally, we may write (cf. (5.55))

$$\text{Bias } b(R_i, \beta) = H_i\delta_i \quad (i = 1, 2, 3). \tag{5.127}$$

Comparison of $b(R_1)$ and b

Each of these estimators is unbiased and so $b(R_1)$ is always MDE-better than b according to relationship (5.14):

$$\begin{aligned} \Delta\big(b, b(R_1)\big) &= \mathrm{V}(b) - \mathrm{V}\big(b(R_1)\big) \\ &= \sigma^2 H_1 A_1 H_1' \\ &= \sigma^2 S^{-1}R_1'A_1^{-1}R_1S^{-1} \geq 0\,. \end{aligned} \tag{5.128}$$

Comparison of $b(R_3)$ and b

We have

$$\begin{aligned} \Delta\big(b, b(R_3)\big) &= S^{-\frac{1}{2}}[\sigma^2 S^{-\frac{1}{2}}R_3'A_3^{-1}R_3S^{-\frac{1}{2}} \\ &\quad - S^{-\frac{1}{2}}R_3'A_3^{-1}\delta_3\delta_3'A_3^{-1}R_3S^{-\frac{1}{2}}]S^{-\frac{1}{2}} \\ S^{-\frac{1}{2}}R_3'A_3^{-1}\delta_3 &= \sigma^2 S^{-\frac{1}{2}}R_3'A_3^{-1}R_3S^{-\frac{1}{2}}[\sigma^{-2}(R_3S^{-\frac{1}{2}})^+]\delta_3 \\ \sigma^2 S^{-\frac{1}{2}}R_3'A_3^{-1}R_3S^{-\frac{1}{2}} &\geq 0\,. \end{aligned}$$

Therefore, we may apply Theorem 5.9 and arrive at the equivalence

$$\Delta\big(b, b(R_3)\big) \geq 0 \quad \text{if and only if} \quad \lambda_3 = \sigma^{-2}\delta_3'A_3^{-1}\delta_3 \leq 1\,. \tag{5.129}$$

This condition was already deduced in (5.60) using an alternative set of arguments.

Comparison of $b\begin{pmatrix} R_1 \\ R_3 \end{pmatrix}$ and $b(R_1)$

Using $R_1 = PR_2$, $P = (I, 0)$, $R_2 = \begin{pmatrix} R_1 \\ R_3 \end{pmatrix}$, $\delta_2 = \begin{pmatrix} 0 \\ \delta_3 \end{pmatrix}$, and Theorem A.19 allows condition (5.119) to be expressed in the form

$$\begin{aligned} \sigma^{-2}\delta_2'(R_2S^{-1}R_2')^{-1}\delta_2 &= \sigma^{-2}(0, \delta_3')\begin{pmatrix} R_1S^{-1}R_1' & R_1S^{-1}R_3' \\ R_3S^{-1}R_1' & R_3S^{-1}R_3' \end{pmatrix}^{-1}\begin{pmatrix} 0 \\ \delta_3 \end{pmatrix} \\ &= \sigma^{-2}\delta_3'(A_3 - R_3S^{-1}R_1'A_1^{-1}R_1S^{-1}R_3')^{-1}\delta_3 \\ &\leq 1. \end{aligned} \quad (5.130)$$

Comparing conditions (5.129) and (5.130) gives (cf. Theorem 5.7)

$$\sigma^{-2}\delta_3'A_3^{-1}\delta_3 \leq \sigma^{-2}\delta_3'(A_3 - R_3S^{-1}R_1'A_1^{-1}R_1S^{-1}R_3')^{-1}\delta_3. \quad (5.131)$$

Summarizing our results leads us to the following.

Theorem 5.16 (Toutenburg, 1989b) *Suppose that the restrictions (5.123) hold true. Then we have these results:*

(a) The biased linearly restricted estimator $b\begin{pmatrix} R_1 \\ R_3 \end{pmatrix}$ is MDE-better than the unbiased RLSE $b(R_1)$ if and only if $b\begin{pmatrix} R_1 \\ R_3 \end{pmatrix}$ is MDE-better than b, that is, if condition (5.130) holds.

(b) Suppose that $\Delta\left(b\begin{pmatrix} R_1 \\ R_3 \end{pmatrix}, b(R_1)\right) \geq 0$; then necessarily

$$\Delta(b(R_3), b) \geq 0\,.$$

Interpretation: Adding an exact restriction $r_1 = R_1\beta$ to the model $y = X\beta + \epsilon$ in any case leads to an increase in efficiency compared with the OLSE b. Stepwise adding of another restriction $r_3 - R_3\beta = \delta_3$ will further improve the efficiency in the sense of the MDE criterion if and only if the condition (5.130) holds. If the condition (5.130) is fulfilled, then necessarily the biased estimator $b(R_3)$ has to be MDE-superior to b. This fact is necessary but not sufficient.

Remark: The difference of the dispersion matrices of $b(R_1)$ and $b\begin{pmatrix} R_1 \\ R_3 \end{pmatrix}$ is nonnegative definite (cf. Theorem 5.2).

Using $P = (I, 0)$, we may write the matrix F from (5.46) as

$$\begin{aligned} F &= A_2^{\frac{1}{2}}\begin{pmatrix} I \\ 0 \end{pmatrix}\left((I\,0)A_2\begin{pmatrix} I \\ 0 \end{pmatrix}\right)^{-1}(I, 0)A_2^{\frac{1}{2}} \\ &= A_2^{\frac{1}{2}}\begin{pmatrix} A_1^{-1} & 0 \\ 0 & 0 \end{pmatrix}A_2^{\frac{1}{2}}. \end{aligned} \quad (5.132)$$

Thus we arrive at the following interesting relationship:

$$
\begin{aligned}
& \mathrm{V}\left(b(R_1)\right) - \mathrm{V}\left(b\begin{pmatrix} R_1 \\ R_3 \end{pmatrix}\right) \\
&\quad = \sigma^2 S^{-1} R_2' A_2^{-\frac{1}{2}} (I - F) A_2^{-\frac{1}{2}} R_2 S^{-1} \\
&\quad = \sigma^2 S^{-1} R_2' A_2^{-1} R_2 S^{-1} - \sigma^2 S^{-1} R_1' A_1^{-1} R_1 S^{-1} \\
&\quad = \left[\mathrm{V}(b) - \mathrm{V}\left(b\begin{pmatrix} R_1 \\ R_3 \end{pmatrix}\right)\right] - \left[\mathrm{V}(b) - \mathrm{V}\left(b(R_1)\right)\right], \qquad (5.133)
\end{aligned}
$$

which may be interpreted as follows: A decrease in variance by using the restrictions $r_1 = R_1\beta$ and $r_3 = R_3\beta$ in comparison to $\mathrm{V}(b)$ equals a decrease in variance by using $r_1 = R_1\beta$ in comparison to $\mathrm{V}(b)$ plus a decrease in variance by using $r_3 = R_3\beta$ in comparison to $\mathrm{V}\left(b(R_1)\right)$.

Let us now apply Theorem A.19 to the partitioned matrix A_2 using the notation

$$
\begin{aligned}
U &= R_3 - R_3 S^{-1} R_1' A_1^{-1} R_1 \\
&= R_3 S^{-\frac{1}{2}} (I_K - S^{-\frac{1}{2}} R_1' A_1^{-1} R_1 S^{-\frac{1}{2}}) S^{\frac{1}{2}} . \qquad (5.134)
\end{aligned}
$$

We see that the matrix $S^{-\frac{1}{2}} R_1' A_1^{-1} R_1 S^{-\frac{1}{2}}$ is idempotent of rank J_1. Then (cf. Theorem A.61 (vi)) the matrix $I_K - S^{-\frac{1}{2}} R_1' A_1^{-1} R_1 S^{-\frac{1}{2}}$ is idempotent of rank $K - J_1$. To show $\mathrm{rank}(U) = J_3$, we note that $J_3 \leq K - J_1$, that is, $J_1 + J_3 \leq K$ is a necessary condition.

Let us use the abbreviation

$$
\underset{J_3 \times J_3}{Z} = A_3 - R_3 S^{-1} R_1' A_1^{-1} R_1 S^{-1} R_3' \qquad (5.135)
$$

so that Z is regular. Now we exchange the submatrices in the matrix A_2, call this matrix $\tilde{A}_2$, and apply Albert's theorem (A.74) to $\tilde{A}_2$:

$$
\tilde{A}_2 = \begin{pmatrix} R_3' S^{-1} R_3 & R_3' S^{-1} R_1 \\ R_1' S^{-1} R_3 & R_1' S^{-1} R_1 \end{pmatrix} = \begin{pmatrix} A_3 & R_3' S^{-1} R_1 \\ R_1' S^{-1} R_3 & A_1 \end{pmatrix},
$$

which shows that $\tilde{A}_2 > 0$ is equivalent to $Z > 0$ (see Theorem A.74 (ii)(b)).

By straightforward calculation, we get

$$
\begin{aligned}
\mathrm{V}\left(b(R_1)\right) - \mathrm{V}\left(b\begin{pmatrix} R_1 \\ R_3 \end{pmatrix}\right) &= \sigma^2 S^{-1} U' Z^{-1} U S^{-1}, \\
\mathrm{Bias}\; b\left(\begin{pmatrix} R_1 \\ R_3 \end{pmatrix}, \beta\right) &= -S^{-1} U' Z^{-1} \delta_3 ,
\end{aligned}
$$

from which the following difference of the MDE matrices becomes n.n.d., that is,

$$
\Delta\left(b(R_1), b\begin{pmatrix} R_1 \\ R_3 \end{pmatrix}\right) = S^{-1} U' Z^{-\frac{1}{2}} [\sigma^2 I - Z^{-\frac{1}{2}} \delta_3 \delta_3' Z'^{-\frac{1}{2}}] Z'^{-\frac{1}{2}} U S^{-1} \geq 0 \qquad (5.136)
$$

when $\text{rank}(U) = J_3$ if and only if (cf. (5.130))

$$\lambda = \sigma^{-2}\delta_3' Z^{-1}\delta_3 \leq 1 \tag{5.137}$$

(see also Theorem 5.7).

Thus we have found an explicit presentation of the necessary and sufficient condition (5.119). This result is based on the special structure of the restrictions (5.123). A test of hypothesis for condition (5.130) can be conducted employing the test statistic

$$F = \frac{(r_3 - R_3 b)' Z^{-1} (r_3 - R_3 b)}{J_3 s^2} \sim F_{J_3, T-K}(\lambda), \tag{5.138}$$

where λ is the parameter defined by (5.137). The decision rule is as follows:

$$\begin{aligned} \text{do not reject } H_0\text{: } \lambda \leq 1 \text{ if } F \leq F_{J_3,T-K}(1), \\ \text{reject } H_0\text{: } \lambda \leq 1 \text{ if } F > F_{J_3,T-K}(1) \end{aligned}$$

Note: Based on this decision rule, we may define a so-called pretest estimator

$$b^* = \begin{cases} b\begin{pmatrix} R_1 \\ R_3 \end{pmatrix} & \text{if } F \leq F_{J_3,T-K}(\lambda), \\ b(R_1) & \text{if } F > F_{J_3,T-K}(\lambda). \end{cases}$$

The MDE matrix of this estimator is not of a simple structure. The theory of pretest estimators is discussed in full detail in Judge and Bock (1978). Applications of pretest procedures to problems of model choice are given in Trenkler and Pordzik (1988) and Trenkler and Toutenburg (1992). Dube, Srivastava, Toutenburg, and Wijekoon (1991) discuss model choice problems under linear restrictions by using Stein-type estimators.

5.8 Stochastic Linear Restrictions

5.8.1 *Mixed Estimator*

In many models of practical interest, in addition to the sample information of the matrix (y, X), supplementary information is available that often may be expressed (or, at least, approximated) by a linear stochastic restriction of the type

$$r = R\beta + \phi, \quad \phi \sim (0, \sigma^2 V), \tag{5.139}$$

where $r : J \times 1$, $R : J \times K$, $\text{rank}(R) = J$, and R and V may be assumed to be known. Let us at first suppose $V > 0$ and, hence, is regular. The vector r may be interpreted as a random variable with expectation $\text{E}(r) = R\beta$. Therefore the restriction (5.139) does not hold exactly except in the mean. We assume r to be known (i.e., to be a realized value of the random vector) so that all the expectations are conditional on r as, for example, $\text{E}(\hat{\beta}|r)$.

In the following we do not mention this separately. Examples for linear stochastic restrictions of type (5.139) are unbiased preestimates of β from models with smaller sample size or from comparable designs. As an example of practical interest, we may mention the imputation of missing values by unbiased estimates (such as sample means). This problem will be discussed in more detail in Chapter 8.

Durbin (1953) was one of the first who used sample and auxiliary information simultaneously, by developing a stepwise estimator for the parameters. Theil and Goldberger (1961) and Theil (1963) introduced the *mixed estimation technique* by unifying the sample and the prior information (5.139) in a common model

$$\begin{pmatrix} y \\ r \end{pmatrix} = \begin{pmatrix} X \\ R \end{pmatrix} \beta + \begin{pmatrix} \epsilon \\ \phi \end{pmatrix}. \tag{5.140}$$

An essential assumption is to suppose that both random errors are uncorrelated:

$$\mathrm{E}(\epsilon\phi') = 0\,. \tag{5.141}$$

This assumption underlines the external character of the auxiliary information. In contrast to the preceding parts of Chapter 5, we now assume the generalized regression model, that is, $\mathrm{E}(\epsilon\epsilon') = \sigma^2 W$. With (5.141) the matrix of variance-covariance becomes

$$\mathrm{E}\begin{pmatrix} \epsilon \\ \phi \end{pmatrix}(\epsilon, \phi)' = \sigma^2 \begin{pmatrix} W & 0 \\ 0 & V \end{pmatrix}. \tag{5.142}$$

Calling the augmented matrices and vectors in the mixed model (5.140) $\tilde{y}$, $\tilde{X}$, and $\tilde{\epsilon}$, that is,

$$\tilde{y} = \begin{pmatrix} y \\ r \end{pmatrix}, \quad \tilde{X} = \begin{pmatrix} X \\ R \end{pmatrix}, \quad \tilde{\epsilon} = \begin{pmatrix} \epsilon \\ \phi \end{pmatrix}, \tag{5.143}$$

we may write

$$\tilde{y} = \tilde{X}\beta + \tilde{\epsilon}, \quad \epsilon \sim (0, \sigma^2 \tilde{W}), \tag{5.144}$$

where

$$\tilde{W} = \begin{pmatrix} W & 0 \\ 0 & V \end{pmatrix} > 0\,. \tag{5.145}$$

As $\mathrm{rank}(\tilde{X}) = \mathrm{rank}(X) = K$ holds, model (5.144) is seen to be a generalized linear model. Therefore, we may apply Theorem 4.4 (using the notation $S = X'W^{-1}X$).

Theorem 5.17 *In the mixed model (5.140) the best linear unbiased estimator of β is*

$$\begin{aligned} \hat{\beta}(R) &= (S + R'V^{-1}R)^{-1}(X'W^{-1}y + R'V^{-1}r) & (5.146) \\ &= b + S^{-1}R'(V + RS^{-1}R')^{-1}(r - Rb)\,, & (5.147) \end{aligned}$$

and $\hat{\beta}(R)$ has the dispersion matrix

$$\mathrm{V}\left(\hat{\beta}(R)\right) = \sigma^2(S + R'V^{-1}R)^{-1}. \tag{5.148}$$

The estimator $\hat{\beta}(R)$ is called the *mixed estimator* for β.

Proof: Straightforward application of Theorem 4.4 to model (5.144) gives the GLSE of β:

$$\begin{aligned}\hat{\beta} &= (\tilde{X}'\tilde{W}^{-1}\tilde{X})^{-1}\tilde{X}'\tilde{W}^{-1}\tilde{y} \\ &= (X'W^{-1}X + R'V^{-1}R)^{-1}(X'W^{-1}y + R'V^{-1}r). \end{aligned} \tag{5.149}$$

Again using the notation $S = X'W^{-1}X$ and applying Theorem A.18 (iii), we get

$$(S + R'V^{-1}R)^{-1} = S^{-1} - S^{-1}R'(V + RS^{-1}R')^{-1}RS^{-1}. \tag{5.150}$$

If we insert this formula in relationship (5.146), then identity (5.147) follows.

Note: The relationship (5.147) yields a representation of the mixed estimator as the GLSE b plus a linear term adjusting b such that $\mathrm{E}\left(b(R)\right) = R\beta$ holds. The form (5.147) was first derived in the paper of Toutenburg (1975a) in connection with optimal prediction under stochastic restrictions with $\mathrm{rank}(V) < J$ (see also Schaffrin, 1987). In contrast to (5.146), presentation (5.147) no longer requires regularity of the dispersion matrix V. Therefore, formula (5.147) allows the simultaneous use of exact and stochastic restrictions. In particular, we have the following convergence result:

$$\lim_{V \to 0} \hat{\beta}(R) = b(R)\,, \tag{5.151}$$

where $b(R)$ is the RLSE (5.11) under the exact restriction $r = R\beta$.

Comparison of $\hat{\beta}(R)$ and the GLSE

The mixed estimator is unbiased and has a smaller dispersion matrix than GLSE b in the sense that

$$\mathrm{V}(b) - \mathrm{V}\left(\hat{\beta}(R)\right) = \sigma^2 S^{-1}R'(V + RS^{-1}R')^{-1}RS^{-1} \geq 0 \tag{5.152}$$

(cf. (5.148) and (5.150)). This gain in efficiency is apparently independent of whether $\mathrm{E}(r) = R\beta$ holds.

5.8.2 *Assumptions about the Dispersion Matrix*

In model (5.141), we have assumed the structure of the dispersion matrix of ϕ as $\mathrm{E}(\phi\phi') = \sigma^2 V$, that is, with the same factor of proportionality σ^2 as that occurring in the sample model. But in practice it may happen that this is not the adequate parameterization. Therefore, it may sometimes be more realistic to suppose that $\mathrm{E}(\phi\phi') = V$, with the consequence that the mixed estimator involves the unknown σ^2:

$$\hat{\beta}(R, \sigma^2) = (\sigma^{-2}S + R'V^{-1}R)^{-1}(\sigma^{-2}X'W^{-1}y + R'V^{-1}r). \tag{5.153}$$

There are some proposals to overcome this problem:

Using the Sample Variance s^2 to Estimate σ^2 in $\hat{\beta}(R, \sigma^2)$

One possibility is to estimate σ^2 by s^2, as proposed by Theil (1963). The resulting estimator $\hat{\beta}(R, s^2)$ is no longer unbiased in general. If certain conditions hold ($s^{-2} - \sigma^{-2} = O(T^{-\frac{1}{2}})$ in probability), then $\hat{\beta}(R, s^2)$ is asymptotically unbiased and has asymptotically the same dispersion matrix as $\hat{\beta}(R, \sigma^2)$. Properties of this estimator have been analyzed by Giles and Srivastava (1991); Kakwani (1968, 1974); Mehta and Swamy (1970); Nagar and Kakwani (1964); Srivastava, Chandra, and Chandra (1985); Srivastava and Upadhyaha (1975); and Swamy and Mehta (1969) to cite a few.

Using a Constant

Theil (1963), Hartung (1978), Teräsvirta and Toutenburg (1980), and Toutenburg (1982, pp. 53–60) have investigated an estimator $\hat{\beta}(R, c)$, where c is a nonstochastic constant that has to be chosen such that the unbiased estimator $\hat{\beta}(R, c)$ has a smaller covariance matrix than the GLSE b.

With the notation $M_c = (cS + R'V^{-1}R)$, we get

$$\begin{aligned} \hat{\beta}(R, c) &= M_c^{-1}(cX'W^{-1}y + R'V^{-1}r), & (5.154)\\ \mathrm{V}\left(\hat{\beta}(R, c)\right) &= M_c^{-1}(c^2\sigma^2 S + R'V^{-1}R)M_c^{-1}. & (5.155) \end{aligned}$$

If we define the matrix

$$B(c, \sigma^2) = \sigma^2 S^{-1} + (2c\sigma^2 - 1)(R'V^{-1}R)^{-1}, \tag{5.156}$$

then the difference of the dispersion matrices becomes n.n.d., that is,

$$\begin{aligned} \Delta(c) &= \mathrm{V}(b) - \mathrm{V}\left(\hat{\beta}(R, c)\right) \\ &= M_c^{-1}(R'V^{-1}R)B(c, \sigma^2)(R'V^{-1}R)M_c^{-1} \geq 0 \qquad (5.157) \end{aligned}$$

if and only if $B(c, \sigma^2) \geq 0$ as $M_c^{-1} > 0$ and $(R'V^{-1}R) > 0$.

We now discuss two possibilities.

Case (a): With $B(0, \sigma^2) = \sigma^2 S^{-1} - (R'V^{-1}R)^{-1} < 0$ (negative definite), $B(\sigma^{-2}/2, \sigma^2) = \sigma^2 S^{-1}$ (positive definite), and $a'B(c, \sigma^2)a$ ($a \neq 0$ a fixed K-vector) being a continuous function of c, there exists a critical value $c_0(a)$ such that

$$\left.\begin{aligned} a'B(c_0(a), \sigma^2)a &= 0, \quad 0 < c_0(a) < \tfrac{1}{2}\sigma^{-2}, \\ a'B(c, \sigma^2)a &> 0, \quad \text{for } c > c_0(a). \end{aligned}\right\} \tag{5.158}$$

Solving $a'B(c_0(a), \sigma^2)a = 0$ for $c_0(a)$ gives the critical value as

$$c_0(a) = (2\sigma^2)^{-1} - \frac{a'S^{-1}a}{2a'(R'V^{-1}R)^{-1}a}, \tag{5.159}$$

which clearly is unknown as a function of σ^2.

Using prior information on σ^2 helps to remove this difficulty.

Theorem 5.18 *Suppose that we are given a lower and an upper bound for σ^2 such that*

(i) $0 < \sigma_1^2 < \sigma^2 < \sigma_2^2 < \infty$, *and*

(ii) $B(0, \sigma_2^2) < 0$ *is negative definite.*

Then the family of estimators $\hat{\beta}(R, c)$ having a smaller dispersion matrix than the GLSE b is specified by $\mathcal{F}_c = \{\hat{\beta}(R, c) : c \geq \sigma_1^{-2}\}$.

Proof: From $B(0, \sigma_2^2) < 0$ it follows that $B(0, \sigma^2) < 0$ too. Now, $\sigma_1^{-2} > \frac{1}{2}\sigma^{-2}$ and thus $\sigma_1^{-2} > c_0(a)$ is fulfilled (cf. (5.159)), that is, $\Delta(c) \geq 0$ for $c \geq \sigma_1^{-2}$.

Case (b): $B(0, \sigma^2)$ is nonnegative definite. Then $B(c, \sigma^2) \geq 0$, and therefore $\Delta(c) \geq 0$ for all $c > 0$. To examine the condition $B(0, \sigma^2) \geq 0$, we assume a lower bound $0 < \sigma_1^2 < \sigma^2$ with $B(0, \sigma_1^2) \geq 0$. Therefore, the corresponding family of estimators is $\mathcal{F}_c = \{\hat{\beta}(R, c) : c \geq 0\}$.

Summarizing, we may state that prior information about σ^2 in the form of $\sigma_1^2 < \sigma^2 < \sigma_2^2$ in any case will make it possible to find a constant c such that the estimator $\hat{\beta}(R, c)$ has a smaller variance compared to b in the sense that $\Delta(c) \geq 0$ (cf. (5.157)).

Measuring the Gain in Efficiency

The fact that $\Delta(c)$ is nonnegative definite is qualitative. In order to quantify the gain in efficiency by using the estimator $\hat{\beta}(R, c)$ instead of the GLSE b, we define a scalar measure. We choose the risk $R_1(\hat{\beta}, \beta, A)$ from (4.3) and specify $A = S = X'W^{-1}X$. Then the measure for the gain in efficiency is defined by

$$\delta(c) = \frac{R_1(b, \cdot, S) - R_1(\hat{\beta}(R, c), \cdot, S)}{R_1(b, \cdot, S)} \tag{5.160}$$

$$= \frac{\operatorname{tr}\{S\Delta(c)\}}{\sigma^2 K}, \tag{5.161}$$

since

$$R_1(b, \cdot, S) = \sigma^2 \operatorname{tr}\{SS^{-1}\} = \sigma^2 K.$$

In any case, we have $0 \leq \delta(c) \leq 1$. Suppose c to be a suitable choice for σ^{-2} in the sense that approximately $c\sigma^2 = 1$ and, therefore, $\mathrm{V}\left(\hat{\beta}(R, c)\right) \approx M_c^{-1}$.

Then we get

$$\begin{aligned}\delta(c) &\approx 1-\frac{\operatorname{tr}\{SM_c^{-1}\}}{\sigma^2K}\\ &= 1-\frac{\operatorname{tr}(S(S+c^{-1}R'V^{-1}R)^{-1})}{c\sigma^2K}\\ &\approx 1-\frac{\operatorname{tr}\{S(S+c^{-1}R'V^{-1}R)^{-1}\}}{K}\,. \end{aligned} \tag{5.162}$$

The closer $\delta(c)$ is to 1, the more important the auxiliary information becomes. The closer $\delta(c)$ is to 0, the less important is its influence on the estimator compared with the sample information. This balance has led Theil (1963) to the definition of the so-called posterior precision of both types of information:

$$\begin{aligned}\lambda(c,\text{sample}) &= \frac{1}{K}\operatorname{tr}\{S(S+c^{-1}R'V^{-1}R)^{-1}\}\,,\\ \lambda(c,\text{prior information}) &= \frac{1}{K}\operatorname{tr}\{c^{-1}R'V^{-1}R(S+c^{-1}R'V^{-1}R)^{-1}\}\,,\end{aligned}$$

with

$$\lambda(c,\text{sample})+\lambda(c,\text{prior information})=1\,. \tag{5.163}$$

In the following, we shall confine ourselves to stochastic variables ϕ such that $\mathrm{E}(\phi\phi')=\sigma^2V$.

5.8.3 Biased Stochastic Restrictions

Analogous to Section 5.4, we assume that $\mathrm{E}(r)-R\beta=\delta$ with $\delta\neq 0$. Then the stochastic restriction (5.141) becomes

$$r=R\beta+\delta+\phi,\quad \phi\sim(0,\sigma^2V). \tag{5.164}$$

Examples for this type of prior information are given in Teräsvirta (1979a) for the so-called one-input distributed lag model, and in Hill and Ziemer (1983) and in Toutenburg (1989b) for models with incomplete design matrices that are filled up by imputation. If assumption (5.164) holds, the mixed estimator (5.147) becomes biased:

$$\mathrm{E}\left(\hat\beta(R)\right)=\beta+S^{-1}R'(V+RS^{-1}R')^{-1}\delta\,. \tag{5.165}$$

MDE-I Superiority of $\hat\beta(R)$ over b

Denoting the difference of the covariance matrices by D, we get:

$$\begin{aligned}\mathrm{V}(b)-\mathrm{V}\left(\hat\beta(R)\right) &= D\\ &= \sigma^2S^{-1}R'(V+RS^{-1}R')^{-1}RS^{-1}\geq 0\,, \qquad (5.166)\\ \operatorname{Bias}(\hat\beta(R),\beta) &= S^{-1}R'(V+RS^{-1}R')^{-1}\delta\\ &= Dd\,, \qquad (5.167)\end{aligned}$$

with

$$d = SR^{+}\delta\sigma^{-2} \quad \text{and} \quad R^{+} = R'(RR')^{-1}. \tag{5.168}$$

Therefore, $\text{Bias}(\hat{\beta}(R), \beta) \in \mathcal{R}(D)$ and we may apply Theorem 5.6.

Theorem 5.19 *The biased estimator $\hat{\beta}(R)$ is MDE-I-superior over the GLSE b if and only if*

$$\lambda = \sigma^{-2}\delta'(V + RS^{-1}R')^{-1}\delta \le 1. \tag{5.169}$$

If ϵ and ϕ are independently normally distributed, then λ is the noncentrality parameter of the statistic

$$F = \frac{1}{Js^2}(r - Rb)'(V + RS^{-1}R')^{-1}(r - Rb), \tag{5.170}$$

which follows a noncentral $F_{J,T-K}(\lambda)$-distribution under H_0: $\lambda \le 1$.

Remark: Comparing conditions (5.169) and (5.60) for the MDE-I superiority of the mixed estimator $\hat{\beta}(R)$ and the RLSE $b(R)$, respectively, over the LSE b, we see from the fact

$$(RS^{-1}R')^{-1} - (V + RS^{-1}R')^{-1} \ge 0$$

that condition (5.169) is weaker than condition (5.60). Therefore, introducing a stochastic term ϕ in the restriction $r = R\beta$ leads to an increase of the region of parameters, ensuring the estimator based on auxiliary information to be better than b.

Let us now discuss the converse problem; that is, we want to derive the parameter conditions under which the GLSE b becomes MDE-I-superior over $\hat{\beta}(R)$.

MDE-I Superiority of b over $\hat{\beta}(R)$

The following difference of the MDE matrices is nonnegative definite:

$$\begin{aligned}
\Delta(\hat{\beta}(R), b) &= \text{M}(\hat{\beta}(R), \beta) - \text{V}(b) \\
&= -\sigma^2 S^{-1}R'(V + RS^{-1}R')^{-1}RS^{-1} \\
&\quad + \text{Bias}(\hat{\beta}(R), \beta)\,\text{Bias}(\hat{\beta}(R), \beta)' \ge 0
\end{aligned} \tag{5.171}$$

if and only if (see Theorem A.46)

$$-I_J + (V + RS^{-1}R')^{-\frac{1}{2}}\delta\delta'(V + RS^{-1}R')^{-\frac{1}{2}} \ge 0. \tag{5.172}$$

According to Theorem A.59, this matrix is never nonnegative definite if $J \ge 2$. For $J = 1$, the restriction becomes

$$\underset{1,1}{r} = \underset{1,K}{R'}\,\beta + \underset{1,1}{\delta} + \underset{1,1}{\phi}, \quad \phi \sim (0, \sigma^2 \underset{1,1}{v}). \tag{5.173}$$

Then for the matrix (5.172), we have

$$-1 + \delta^2(v + R'S^{-1}R)^{-1} \ge 0$$

if and only if

$$\lambda = \frac{\delta^2}{(v + R'S^{-1}R)} \geq 1\,. \tag{5.174}$$

The following theorem summarizes our findings.

Theorem 5.20 *The biased estimator $\hat{\beta}(R)$ is MDE-I-superior over the GLSE b if and only if (cf. (5.169))*

$$\lambda = \sigma^{-2}\delta'(V + RS^{-1}R')^{-1}\delta \leq 1\,.$$

Conversely, b is MDE-I-better than $\hat{\beta}(R)$

(i) for $J = 1$ if and only if $\lambda \geq 1$, and
(ii) for $J \geq 2$ in no case.

Interpretation: Suppose that $J = 1$; then the region of parameters λ is divided in two disjoint subregions $\{\lambda < 1\}$ and $\{\lambda > 1\}$, respectively, such that in each subregion one of the estimators $\hat{\beta}(R)$ and b is superior to the other. For $\lambda = 1$, both estimators become equivalent. For $J \geq 2$, there exists a region ($\lambda \leq 1$) where $\hat{\beta}(R)$ is better than b, but there exists no region where b is better than $\hat{\beta}(R)$.

This theorem holds analogously for the restricted LSE $b(R)$ (use $V = 0$ in the proof).

MDE-II Superiority of $\hat{\beta}(R)$ over b

We want to extend the conditions of acceptance of the biased mixed estimator by employing the weaker MDE criteria of Section 5.4.

According to Definition 5.3, the mixed estimator $\hat{\beta}(R)$ is MDE-II-better than the GLSE b if

$$\operatorname{tr}\{\Delta(b, \hat{\beta}(R)\} = \operatorname{tr}\{\mathrm{V}(b) - \mathrm{V}(\hat{\beta}(R)\} - \operatorname{Bias}(\hat{\beta}(R), \beta)' \operatorname{Bias}(\hat{\beta}(R), \beta) \geq 0\,. \tag{5.175}$$

Applying (5.166) and (5.167) and using the notation

$$A = V + RS^{-1}R', \tag{5.176}$$

(5.175) is found to be equivalent to

$$Q(\delta) = \sigma^{-2}\delta'A^{-1}RS^{-1}S^{-1}R'A^{-1}\delta \leq \operatorname{tr}(S^{-1}R'A^{-1}RS^{-1})\,. \tag{5.177}$$

This condition is not testable in the sense that there does not exist a statistic having $Q(\delta)$ as noncentrality parameter. Based on an idea of Wallace (1972) we search for a condition that is sufficient for (5.177) to hold. Let us assume that there is a symmetric $K \times K$-matrix G such that

$$\sigma^{-2}\delta'A^{-1}RS^{-1}GS^{-1}R'A^{-1}\delta = \sigma^{-2}\delta'A^{-1}\delta = \lambda\,. \tag{5.178}$$

Such a matrix is given by

$$G = S + SR^{+}VR^{+\prime}S\,, \tag{5.179}$$

where $R^+ = R'(RR')^{-1}$ (Theorem A.66 (vi)). Then we get the identity

$$RS^{-1}GS^{-1}R' = A\,.$$

By Theorem A.44, we have

$$\lambda_{\min}(G) \leq \frac{\sigma^{-2}\delta' A^{-1}RS^{-1}GS^{-1}R'A^{-1}\delta}{Q(\delta)} = \frac{\lambda}{Q(\delta)} \tag{5.180}$$

or, equivalently,

$$Q(\delta) \leq \frac{\lambda}{\lambda_{\min}(G)}\,. \tag{5.181}$$

Therefore, we may state that

$$\lambda \leq \lambda_{\min}(G)\,\mathrm{tr}(S^{-1}R'(V + RS^{-1}R')^{-1}RS^{-1}) = \lambda_2\,, \tag{5.182}$$

for instance, is sufficient for condition (5.177) to hold. Moreover, condition (5.182) is testable. Under H_0: $\lambda \leq \lambda_2$, the statistic F (5.170) has an $F_{J,T-K}(\lambda_2)$-distribution.

Remark: In the case of exact linear restrictions, we have $V = 0$ and hence $G = S$. For $W = I$, condition (5.182) will coincide with condition (5.65) for the MDE-II superiority of the RLSE $b(R)$ to b.

MDE-III Comparison of $\hat{\beta}(R)$ and b

According to Definition 5.4 (cf. (5.66)), the estimator $\hat{\beta}(R)$ is MDE-III-better than b if (with A from (5.176))

$$\mathrm{tr}\left\{S\Delta\big(b, \hat{\beta}(R)\big)\right\}\sigma^2\,\mathrm{tr}\{A^{-1}RS^{-1}R'\} - \delta' A^{-1}RS^{-1}R'A^{-1}\delta \geq 0 \tag{5.183}$$

or, equivalently, if

$$\begin{aligned}
\sigma^{-2}\delta' A^{-1}RS^{-1}R'A^{-1}\delta &\leq J - \mathrm{tr}(A^{-1}V) \\
&= J - \sum_{j=1}^{J}(1+\lambda_j)^{-1} \\
&= \sum \lambda_j(1+\lambda_j)^{-1},
\end{aligned} \tag{5.184}$$

where $\lambda_1 \geq \ldots \geq \lambda_J > 0$ are the eigenvalues of $V^{-\frac{1}{2}}RS^{-1}R'V^{-\frac{1}{2}}$.

This may be shown as follows:

$$\begin{aligned}
\mathrm{tr}(A^{-1}V) &= \mathrm{tr}(V^{\frac{1}{2}}A^{-1}V^{\frac{1}{2}}) && \text{[Theorem A.13]} \\
&= \mathrm{tr}\left((V^{-\frac{1}{2}}AV^{-\frac{1}{2}})^{-1}\right) && \text{[Theorem A.18]} \\
&= \mathrm{tr}\left((I + V^{-\frac{1}{2}}RS^{-1}R'V^{-\frac{1}{2}})^{-1}\right) \\
&= \mathrm{tr}\left((I+\Lambda)^{-1}\right) && \text{[Theorem A.27 (v)]} \\
&= \sum_{j=1}^{J}(1+\lambda_j)^{-1} && \text{[Theorem A.27 (iii)].}
\end{aligned}$$

The left-hand side of (5.184) may be bounded by λ from (5.169):

$$\begin{aligned}\sigma^{-2}\delta' A^{-1}(RS^{-1}R' + V - V)A^{-1}\delta \\ = \sigma^{-2}\delta' A^{-1}\delta - \sigma^{-2}\delta' A^{-1} V A^{-1}\delta \\ \le \sigma^{-2}\delta' A^{-1}\delta = \lambda\,. \end{aligned} \tag{5.185}$$

Then the condition

$$\lambda \le \sum \lambda_j (1+\lambda_j)^{-1} = \lambda_3\,, \tag{5.186}$$

for instance, is sufficient for (5.183) to hold. Condition (5.186) may be tested using F from (5.170), since the statistic F has an $F_{J,T-K}(\lambda_3)$-distribution under H_0: $\lambda \le \lambda_3$.

Remark: From $\lambda_1 \ge \ldots \ge \lambda_J > 0$, it follows that

$$J\frac{\lambda_J}{1+\lambda_J} \le \sum_{j=1}^{J} \frac{\lambda_j}{1+\lambda_j} \le J\frac{\lambda_1}{1+\lambda_1}\,. \tag{5.187}$$

Suppose that $\lambda_J > (J-1)^{-1}$ and $J \ge 2$, then $\lambda_3 > 1$ and the MDE-III criterion indeed leads to a weaker condition than the MDE-I criterion. For $J = 1$, we get $\lambda_3 = \lambda_1/(1+\lambda_1) < 1$.

Further problems such as

- MDE-I comparison of two biased mixed estimators
- stepwise procedures for adapting biased stochastic restrictions

are discussed in papers by Freund and Trenkler (1986), Teräsvirta (1979b, 1981, 1982, 1986), and Toutenburg (1989a, 1989b).

5.9 Weakened Linear Restrictions

5.9.1 Weakly (R, r)-Unbiasedness

In the context of modeling and testing a linear relationship, it may happen that some auxiliary information is available, such as prior estimates, natural restrictions on the parameters ($\beta_i < 0$, etc.), analysis of submodels, or estimates by experts. A very popular and flexible approach is to incorporate auxiliary information in the form of a linear stochastic restriction ($r : J \times 1$, $R : J \times K$)

$$r = R\beta + \phi, \quad \phi \sim (0, V). \tag{5.188}$$

However, this information heavily depends on the knowledge of the dispersion matrix V of ϕ. In statistical practice, unfortunately, the matrix V is rarely known, and consequently $\hat{\beta}(R)$ cannot be computed. Nevertheless, we should still be interested in extracting the remaining applicable part of the information contained in (5.188). In the following, we may look for a concept that leads to the use of the auxiliary information (5.188). Note that (5.188) implies

$$\mathrm{E}(r) = R\beta\,. \tag{5.189}$$

In order to take the information (5.188) into account while constructing estimators $\hat{\beta}$ for β, we require that

$$\mathrm{E}(R\hat{\beta}|r) = r\,. \tag{5.190}$$

Definition 5.21 *An estimator $\hat{\beta}$ for β is said to be weakly (R, r)-unbiased with respect to the stochastic linear restriction $r = R\beta + \phi$ if $\mathrm{E}(R\hat{\beta}|r) = r$.*

This definition was first introduced by Toutenburg, Trenkler, and Liski (1992).

5.9.2 *Optimal Weakly (R, r)-Unbiased Estimators*

Heterogeneous Estimator

First we choose a linear heterogeneous function for the estimator, that is, $\hat{\beta} = Cy + d$. Then the requirement of weakly (R, r)-unbiasedness is equivalent to

$$\mathrm{E}(R\hat{\beta}) = RCX\beta + Rd = r\,. \tag{5.191}$$

If we use the risk function $R_1(\hat{\beta}, \beta, A)$ from (4.39) where $A > 0$, we have to consider the following optimization problem:

$$\min_{C,d,\lambda} \{R_1(\hat{\beta}, \beta, A) - 2\lambda'(RCX\beta + Rd - r)\} = \min_{C,d,\lambda} g(C, d, \lambda) \tag{5.192}$$

where λ is a J-vector of Lagrangian multipliers.

Differentiating the function $g(C, d, \lambda)$ with respect to C, d, and λ gives the first-order equations for an optimum (Theorems A.91, A.92)

$$\frac{1}{2}\frac{\partial g}{\partial d} = Ad + A(CX - I)\beta - R'\lambda = 0\,, \tag{5.193}$$

$$\begin{aligned}\frac{1}{2}\frac{\partial g}{\partial C} &= ACX\beta'\beta X' - A\beta\beta' X' + Ad\beta' X' \\ &\quad + \sigma^2 ACW - R'\lambda\beta' X' = 0\,.\end{aligned} \tag{5.194}$$

$$\frac{1}{2}\frac{\partial g}{\partial \lambda} = RCX\beta + Rd - r = 0\,. \tag{5.195}$$

Solving (5.193) for Ad gives

$$Ad = -A(CX - I)\beta + R'\lambda \tag{5.196}$$

and inserting in (5.194) yields

$$\sigma^2 ACW = 0\,.$$

As A and W are positive definite, we conclude $C = 0$. Now using (5.195) again, we obtain

$$\hat{d} = \beta + A^{-1}R'\lambda . \tag{5.197}$$

Premultiplying (5.194) by R, we get

$$R\hat{d} = r = R\beta + (RA^{-1}R')\lambda ,$$

from which we find

$$\hat{\lambda} = (RA^{-1}R')^{-1}(r - R\beta)$$

and (cf. (5.197))

$$\hat{d} = \beta + A^{-1}R'(RA^{-1}R')^{-1}(r - R\beta) .$$

The following theorem summarizes our findings.

Theorem 5.22 *In the regression model $y = X\beta + \epsilon$, the heterogeneous R_1-optimal weakly (R, r)-unbiased estimator for β is given by*

$$\hat{\beta}_1(\beta, A) = \beta + A^{-1}R'(RA^{-1}R')^{-1}(r - R\beta) , \tag{5.198}$$

and its risk conditional on r is

$$R_1[\hat{\beta}_1(\beta, A), \beta, A] = (r - R\beta)'(RA^{-1}R')^{-1}(r - R\beta) . \tag{5.199}$$

Interpretation:

(i) $\hat{\beta}_1(\beta, A)$ is the sum of the R_1-optimal heterogeneous estimator $\hat{\beta}_1 = \beta$ and a correction term adjusting for the weakly (R, r)-unbiasedness:

$$\mathrm{E}[R\hat{\beta}_1(\beta, A)] = R\beta + (RA^{-1}R')(RA^{-1}R')^{-1}(r - R\beta) = r . \tag{5.200}$$

(ii) The estimator $\hat{\beta}_1(\beta, A)$ depends on the unknown parameter vector β and thus is not of direct use. However, if β is replaced by an unbiased estimator $\tilde{\beta}$, the resulting feasible estimator $\hat{\beta}(\tilde{\beta}, A)$ becomes weakly (R, r)-unbiased:

$$\mathrm{E}[R\hat{\beta}_1(\tilde{\beta}, A)] = R\,\mathrm{E}(\tilde{\beta}) + (RA^{-1}R')(RA^{-1}R')^{-1}(r - R\,\mathrm{E}\,(\tilde{\beta})) = r . \tag{5.201}$$

Although $\hat{\beta}_1(\beta, A)$ involves the unknown β, it characterizes the structure of operational estimators being weakly (R, r)-unbiased and indicates that this class of estimators may have better statistical properties.

(iii) Since $R_1(\hat{\beta}, \beta, A)$ is a convex function of C, our solution $\hat{d} = \hat{\beta}_1(\beta, A)$ from (5.198) yields a minimum.

(iv) Formula (5.199) for the minimal risk is an easy consequence of (4.39) and (5.198).

(v) As $\hat{\beta}_1(\beta, A)$ explicitly depends on the weight matrix A, variation with respect to A defines a new class of estimators. Hence, the matrix A may be interpreted to be an additional parameter. For instance, let β be replaced by the OLSE $b_0 = (X'X)^{-1}X'y$. Then the choice $A = X'X = S$ results in the restricted LSE $b(R)$ (cf. (5.11))

$$\hat{\beta}_1(b, S) = b + S^{-1}R'(RS^{-1}R')^{-1}(r - Rb).$$

Homogeneous Estimator

If $\hat{\beta} = Cy$, then the requirement of weakly (R, r)-unbiasedness is equivalent to

$$RCX\beta = r\,. \tag{5.202}$$

If we set $d = 0$ in (5.192) and differentiate, we obtain the following first-order equations for an optimum:

$$\frac{1}{2}\frac{\partial g}{\partial C} = ACB - A\beta\beta'X' - R\lambda'\beta'X' = 0\,, \tag{5.203}$$

$$\frac{1}{2}\frac{\partial g}{\partial \lambda} = RCX\beta - r = 0, \tag{5.204}$$

where the matrix B is defined as

$$B = X\beta\beta'X' + \sigma^2 W\,. \tag{5.205}$$

Obviously B is positive definite and its inverse is (cf. Theorem A.18, (iv))

$$B^{-1} = \sigma^{-2}\left(W^{-1} - \frac{W^{-1}X\beta\beta'X'W^{-1}}{\sigma^2 + \beta'X'W^{-1}X\beta}\right)\,. \tag{5.206}$$

Solving (5.203) for C yields

$$C = \beta\beta'X'B^{-1} + A^{-1}R'\lambda'\beta'X'B^{-1}. \tag{5.207}$$

Combining this with equation (5.204)

$$RCX\beta = r = [R\beta + (RA^{-1}R')\lambda']\alpha(\beta) \tag{5.208}$$

leads to the optimal λ, which is

$$\hat{\lambda}' = (RA^{-1}R')^{-1}\left(\frac{r}{\alpha(\beta)} - R\beta\right), \tag{5.209}$$

where $\alpha(\beta)$ is defined in (4.21). Inserting $\hat{\lambda}$ in (5.207), we obtain the solution for C as

$$\hat{C} = \beta\beta'X'B^{-1} + A^{-1}R'(RA^{-1}R')^{-1}\left([\alpha(\beta)]^{-1}r - R\beta\right)\beta'X'B^{-1}. \tag{5.210}$$

Summarizing our derivations, we may state that the R_1-optimal, homogeneous, weakly (R, r)-unbiased estimator is

$$\hat{\beta}_2(\beta, A) = \beta\alpha(y) + A^{-1}R'(RA^{-1}R')^{-1}\left(\frac{r}{\alpha(\beta)} - R\beta\right)\alpha(y)\,, \tag{5.211}$$

where

$$\alpha(y) = \beta' X' B^{-1} y = \frac{\beta' X' W^{-1} y}{\sigma^2 + \beta' S \beta} \tag{5.212}$$

is used for abbreviation (cf. (4.18)–(4.21)).

It should be emphasized that $\hat{\beta}_2 = \beta\alpha(y)$ is the R_1-optimal homogeneous estimator for β (cf. (4.20)).

With $\mathrm{E}\left(\alpha(y)\right) = \alpha(\beta)$, we see that $\hat{\beta}_2(\beta, A)$ is weakly (R, r)-unbiased:

$$\mathrm{E}[R\hat{\beta}_2(\beta, A)] = R\beta\alpha(\beta) + \frac{r}{\alpha(\beta)}\alpha(\beta) - R\beta\alpha(\beta) = r\,. \tag{5.213}$$

With respect to β, this estimator is biased:

$$\mathrm{Bias}[\hat{\beta}_2(\beta, A), \beta] = \beta(\alpha(\beta) - 1) + z\alpha(\beta)\,, \tag{5.214}$$

where

$$z = A^{-1}R'(RA^{-1}R')^{-1}\left(\frac{r}{(\alpha(\beta))} - R\beta\right). \tag{5.215}$$

Obviously, the dispersion matrix is

$$\mathrm{V}\big(\hat{\beta}_2(\beta, A)\big) = \mathrm{V}(\hat{\beta}_2) + zz'\frac{\sigma^2\alpha(\beta)}{\sigma^2 + \beta' S\beta} + 2z'\beta\frac{\sigma^2\alpha(\beta)}{\sigma^2 + \beta' S\beta} \tag{5.216}$$

with $\mathrm{V}(\hat{\beta}_2)$ from (4.24). This implies that the MDE matrix of $\hat{\beta}_2(\beta, A)$ is

$$\mathrm{M}(\hat{\beta}_2(\beta, A), \beta) = \mathrm{M}(\hat{\beta}_2, \beta) + zz'\alpha(\beta)\,, \tag{5.217}$$

where $\mathrm{M}(\hat{\beta}_2, \beta)$ is the mean dispersion error matrix from (4.25). Obviously, we have

$$\Delta(\hat{\beta}_2(\beta, A), \hat{\beta}_2) = zz'\alpha(\beta) \geq 0\,. \tag{5.218}$$

Theorem 5.23 *The R_1-optimal, homogeneous, weakly (R, r)-unbiased estimator for β is given by $\hat{\beta}_2(\beta, A)$ (5.211). This estimator has the R_1-risk*

$$\begin{aligned} R_1(\hat{\beta}_2(\beta, A), \beta, A) &= R_1(\hat{\beta}_2, \beta, A) \\ &+ \alpha(\beta)\Big(\frac{r}{\alpha(\beta)} - R\beta\Big)'\Big(RA^{-1}R'\Big)^{-1}\Big(\frac{r}{\alpha(\beta)} - R\beta\Big), \end{aligned} \tag{5.219}$$

where $R_1(\hat{\beta}_2, \beta, A) = \mathrm{tr}\left(A\,\mathrm{M}(\hat{\beta}_2, \beta)\right)$ is the R_1-risk of $\hat{\beta}_2$ (4.20).

5.9.3 Feasible Estimators—Optimal Substitution of β in $\hat{\beta}_1(\beta, A)$

From the relationship (5.201), we know that any substitution of β by an unbiased estimator $\tilde{\beta}$ leaves $\hat{\beta}_1(\beta, A)$ weakly (R, r)-unbiased. To identify an estimator $\tilde{\beta}$ such that the feasible version $\hat{\beta}_1(\tilde{\beta}, A)$ is optimal with respect to the quadratic risk, we confine ourselves to well-defined classes of estimators. Let us demonstrate this for the class $\{\tilde{\beta} = \tilde{C}y | \tilde{C}X = I\}$ of linear homogeneous estimators.

With the notation

$$\tilde{A} = A^{-1}R'(RA^{-1}R')^{-1}, \tag{5.220}$$

we obtain

$$\hat{\beta}_1(\tilde{C}y, A) = \tilde{C}y + \tilde{A}(r - \tilde{C}y), \tag{5.221}$$

which is unbiased for β:

$$\mathrm{E}(\hat{\beta}_1(\tilde{C}y, A)) = \tilde{C}X\beta + \tilde{A}(r - R\tilde{C}X\beta) = \beta \tag{5.222}$$

and has the dispersion matrix

$$\mathrm{V}\left(\hat{\beta}_1(\tilde{C}y, A)\right) = \sigma^2(I - \tilde{A}R)\tilde{C}W\tilde{C}'(I - \tilde{A}\tilde{R})'. \tag{5.223}$$

Furthermore, the matrix

$$Q = I - A^{-\frac{1}{2}}R'(RA^{-1}R')^{-1}RA^{-\frac{1}{2}}, \tag{5.224}$$

is idempotent of rank $K - J$, and it is readily seen that

$$(I - R'\tilde{A}')A(I - \tilde{A}R) = A^{\frac{1}{2}}QA^{\frac{1}{2}}. \tag{5.225}$$

Let $\Lambda = (\lambda_1, \ldots, \lambda_K)$ denote a $K \times K$-matrix of K-vectors λ_i of Lagrangian multipliers. Then the R_1-optimal, unbiased version $\tilde{\beta} = \tilde{C}y$ of the estimator $\hat{\beta}(\tilde{\beta}, A)$ is the solution to the following optimization problem

$$\begin{aligned} &\min_{\tilde{C},\Lambda}\left\{\operatorname{tr}[A\,\mathrm{V}\left(\hat{\beta}_1(\tilde{C}y, A)\right)] - 2\sum_{i=1}^{K}\lambda_i'(\tilde{C}X - I)_{(i)}\right\} \\ &= \min_{\tilde{C},\Lambda}\left\{\sigma^2\operatorname{tr}[A^{\frac{1}{2}}QA^{\frac{1}{2}}\tilde{C}W\tilde{C}'] - 2\sum_{i=1}^{K}\lambda_i'(\tilde{C}X - I)_{(i)}\right\} \\ &= \min_{\tilde{C},\Lambda} g(\tilde{C}, \Lambda). \end{aligned} \tag{5.226}$$

Differentiating with respect to $\tilde{C}$ and Λ, respectively, gives the necessary conditions for a minimum:

$$\frac{1}{2}\frac{\partial g(\tilde{C}, \Lambda)}{\partial \tilde{C}} = A^{\frac{1}{2}}QA^{\frac{1}{2}}\tilde{C}W - \Lambda X' = 0 \tag{5.227}$$

$$\frac{1}{2}\frac{\partial g(\tilde{C}, \Lambda)}{\partial \Lambda} = \tilde{C}X - I = 0. \tag{5.228}$$

Postmultiplying (5.227) by $W^{-1}X$ and using (5.228) give

$$\hat{\Lambda} = A^{\frac{1}{2}}QA^{\frac{1}{2}}S^{-1} \tag{5.229}$$

and consequently from (5.227)

$$A^{\frac{1}{2}}QA^{\frac{1}{2}}[\tilde{C} - S^{-1}X'W^{-1}] = 0. \tag{5.230}$$

The principal solution of (5.230) is then given by

$$\tilde{C}_* = S^{-1}X'W^{-1} \tag{5.231}$$

with the corresponding estimator $\tilde{\beta} = b$ being the GLSE, and

$$\hat{\beta}_1(\tilde{C}_* y, A) = b + A^{-1}R'(RA^{-1}R')^{-1}(r - Rb). \tag{5.232}$$

An interesting special case is to choose $A = S$, transforming the risk $R_1(\hat{\beta}, \beta, S)$ to the R_3-risk (cf. (4.5)). Hence we may state the following theorem, by using the convexity argument again.

Theorem 5.24 *Let $\hat{\beta}_1(\tilde{C}y, S)$ be the class of weakly (R, r)-unbiased estimators with $\hat{\beta} = \tilde{C}y$ being an unbiased estimator for β. Then in this class the estimator $\hat{\beta}_1(b, A)$ minimizes the risk $R_1(\hat{\beta}, \beta, A)$. Choosing $A = S$ then makes the optimal estimator $\hat{\beta}_1(b, S)$ equal to the restricted least-squares estimator*

$$b(R) = b + S^{-1}R'(RS^{-1}R')^{-1}(r - Rb)\,, \tag{5.233}$$

which is R_3-optimal.

Remark: To get feasible weakly (R, r)-unbiased estimators, one may use the idea of incorporating a prior guess for β (cf. Toutenburg et al., 1992). Alternatively, in Chapter 8 we shall discuss the method of weighted mixed regression, which values sample information more highly than auxiliary information.

5.9.4 RLSE instead of the Mixed Estimator

The correct prior information (5.137) is operational if the dispersion matrix V is known. If V is unknown, we may use the methods of Section 5.8.2 to estimate V.

An alternative idea would be to use the restricted least-squares estimator $b(R)$, which may be interpreted as a misspecified mixed estimator mistakenly using dispersion matrix $V_m = 0$ instead of V. To highlight this fact, we use the notation

$$b(R) = b(R, V_m) = b + S^{-1}R'(RS^{-1}R' + V_m)^{-1}(r - Rb). \tag{5.234}$$

With respect to the correct specification of the stochastic restriction

$$r = R\beta + \phi, \quad \phi \sim (0, V)\,,$$

the estimator $b(R, V_m)$ is unbiased for β:

$$\mathrm{E}\left(b(R, V_m)\right) = \beta \tag{5.235}$$

but has the covariance matrix

$$\mathrm{V}\left(b(R, V_m)\right) = \mathrm{V}\left(b(R)\right) + \sigma^2 S^{-1}R'(RS^{-1}R')^{-1}V(RS^{-1}R')^{-1}RS^{-1} \tag{5.236}$$

where $\mathrm{V}\big(b(R)\big)$ is the covariance matrix of the RLSE from (5.13).

Because of the unbiasedness of the competing estimators $b(R, V_m)$ and $\hat{\beta}(R)$, the MDE comparison is reduced to the comparison of their covariance matrices. Letting

$$A = S^{-1}R'(RS^{-1}R')^{-1}V^{\frac{1}{2}}, \tag{5.237}$$

we get the following expression for the difference of the covariance matrices:

$$\Delta\big(b(R, V_m), \hat{\beta}(R)\big) = \sigma^2 A[I - (I + V^{\frac{1}{2}}(RS^{-1}R')^{-1}V^{\frac{1}{2}})^{-1}]A'. \tag{5.238}$$

Based on the optimality of the mixed estimator $\hat{\beta}(R)$, it is seen that the estimator $b(R, V_m)$ has to be less efficient; that is, in any case it holds that

$$\Delta\big(b(R, V_m), \hat{\beta}(R)\big) \geq 0. \tag{5.239}$$

Since V is unknown, we cannot estimate the extent of this loss.

Comparing the estimators $b(R, V_m)$ and the GLSE b, the misspecified estimator $b(R, V_m)$ is MDE-superior to b if

$$\Delta\big(b, b(R, V_m)\big) = \sigma^2 A[V^{-\frac{1}{2}}RS^{-1}R'V^{-\frac{1}{2}} - I]A' \geq 0 \tag{5.240}$$

or, equivalently, if

$$\lambda_{\min}(V^{-\frac{1}{2}}RS^{-1}R'V^{-\frac{1}{2}}) \geq 1.$$

Again this condition is not operational because V is unknown in this set-up.

5.10 Exercises

Exercise 1. Assuming that $k = R_1\beta_1 + R_2\beta_2$ with R_1 as a nonsingular matrix, show that the restricted regression estimator of β_2 in the model $y = X_1\beta_1 + X_2\beta_2$ is equal to the least-squares estimator of β_2 in the model $(y - X_1R_1^{-1}k) = (X_2 - X_1R_1^{-1}R_2)\beta_2 + \epsilon$.

Exercise 2. Compare the least-squares and restricted least squares estimators with respect to the risk under a general quadratic loss function defined by $\mathrm{E}(\hat{\beta} - \beta)'Q(\hat{\beta} - \beta)$ for any estimator $\hat{\beta}$ of β where Q is a nonsingular matrix with nonstochastic elements.

Exercise 3. Consider the estimation of β by $\theta\, b(R)$ with θ as a scalar and $b(R)$ as the restricted least-squares estimator. Determine the value of θ that minimizes the trace of mean dispersion error matrix of $\theta\, b(R)$. Comment on the utility of the estimator thus obtained.

Exercise 4. Find an unbiased estimator of σ^2 based on residuals obtained from restricted least-squares estimation.

Exercise 5. Obtain an estimator of β in the bivariate model $y_t = \alpha + \beta x_t + \epsilon_t$ with $\mathrm{E}(\epsilon_t) = 0$ and $\mathrm{E}(\epsilon_t^2) = \sigma^2$ (known) for all t when an unbiased estimate b_0 with standard error σc is available from some extraneous source. Find the variance of this estimator and examine its efficiency with respect to the conventional unbiased estimator.

Exercise 6. Consider the model $y = \alpha \mathbf{1} + X\beta + \epsilon$ with $\mathrm{E}(\epsilon) = 0$, $\mathrm{E}(\epsilon\epsilon') = \sigma^2 I$, and $\mathbf{1}$ denoting a column vector having all elements unity. Find the mixed estimator of α when $k = R\beta + v$ with $\mathrm{E}(v) = 0$ and $E(vv') = V$ is available and σ^2 is known. What are its properties?

Exercise 7. Show that the least-squares estimator ignoring the stochastic linear restrictions has the same asymptotic properties as the mixed estimator. Does this kind of result carry over if we compare the asymptotic properties of least-squares and restricted least-squares estimators?

Exercise 8. Formulate the inequality restrictions on the regression coefficients in the form of a set of stochastic linear restrictions and obtain the mixed estimator assuming σ^2 to be known. Derive expressions for the bias vector and mean dispersion error matrix of the estimator.

Exercise 9. Discuss the estimation of β when both $k_1 = R_1\beta$ and $k_2 = R_2\beta + v$ are to be utilized simultaneously.

Exercise 10. When unbiased estimates of a set of linear combinations of the regression coefficients are available from some extraneous source, present a procedure for testing the compatibility of the sample and extraneous information.

6

Prediction Problems in the Generalized Regression Model

6.1 Introduction

The problem of prediction in linear models has been discussed in the monograph by Bibby and Toutenburg (1977) and also in the papers by Toutenburg (1968, 1970a, 1970b, 1970c). One of the main aims of the above publications is to examine the conditions under which biased estimators can lead to an improvement over conventional unbiased procedures. In the following, we will concentrate on recent results connected with alternative superiority criteria.

6.2 Some Simple Linear Models

To demonstrate the development of statistical prediction in regression we will first present some illustrative examples of linear models.

6.2.1 The Constant Mean Model

The simplest "regression" may be described by

$$y_t = \mu + \epsilon_t \quad (t = 1, \dots, T),$$

where $\epsilon = (\epsilon_1, \dots, \epsilon_T)' \sim (0, \sigma^2 I)$ and μ is a scalar constant. T denotes the index (time) of the last observation of the random process $\{y_t\}$. We assume that a prediction of a future observation $y_{T+\tau}$ is required. Extrapolation gives

$$y_{T+\tau} = \mu + \epsilon_{T+\tau}\,.$$

One would expect to estimate $y_{T+\tau}$ by adding the estimators of μ and $\epsilon_{T+\tau}$. The actual value of the random variable $\epsilon_{T+\tau}$ cannot be predicted, as it is uncorrelated with the past values $\epsilon_1, \ldots, \epsilon_T$; thus we simply forecast $\epsilon_{T+\tau}$ by its expected value, that is, $\mathrm{E}(\epsilon_{T+\tau}) = 0$. The quantity μ is a constant over time, so its estimate from the past will give a predictor for the future.

Thus we are led to the predictor

$$\hat{y}_{T+\tau} = T^{-1}\sum_{t=1}^{T} y_t = \bar{y}\,,$$

which is unbiased:

$$\mathrm{E}\,\hat{y}_{T+\tau} = \mathrm{E}\,\bar{y} = \mu \quad \Rightarrow \quad \mathrm{E}(\hat{y}_{T+\tau} - y_{T+\tau}) = 0$$

and has variance

$$\operatorname{var}(\hat{y}_{T+\tau}) = \frac{\sigma^2}{T} \quad \Rightarrow \quad \mathrm{E}(\hat{y}_{T+\tau} - y_{T+\tau})^2 = \sigma^2\left(1 + \frac{1}{T}\right).$$

The precision of the predictor, as indicated by the mean square error $\sigma^2(1+T^{-1})$, will improve with an increase in the sample size T.

6.2.2 The Linear Trend Model

If the mean μ has a linear trend with time, we have the model

$$y_t = \alpha + \beta t + \epsilon_t \quad (t = 1, \ldots, T)\,,$$

where α is the expectation of y_0, β is the slope, and ϵ_t is the added random variation (see Figure 6.1).

If we transform t to $\tilde{t} = t - \bar{t}$, then the predictor of any future value $y_{\tilde{T}+\tau}$ with $\tilde{T} = T - \bar{t}$ is simply obtained by

$$\hat{y}_{T+\tau} = \hat{\alpha} + \hat{\beta}(\tilde{T} + \tau)\,,$$

where $\hat{\alpha}$ and $\hat{\beta}$ are the unbiased, ordinary least-squares estimates of α and β (see Chapter 3):

$$\begin{aligned}
\hat{\alpha} &= \bar{y}, \quad \hat{\beta} = \frac{\sum_t \tilde{t}(y_t - \bar{y})}{\sum_t \tilde{t}^2}\,,\\
\operatorname{var}(\hat{\alpha}) &= \frac{\sigma^2}{T}\,, \quad \operatorname{var}(\hat{\beta}) = \frac{\sigma^2}{\sum_{t=1}^{T} \tilde{t}^2}\,.
\end{aligned}$$

Due to the transformation of t to $\tilde{t}$, $\hat{\alpha}$ and $\hat{\beta}$ are independent.

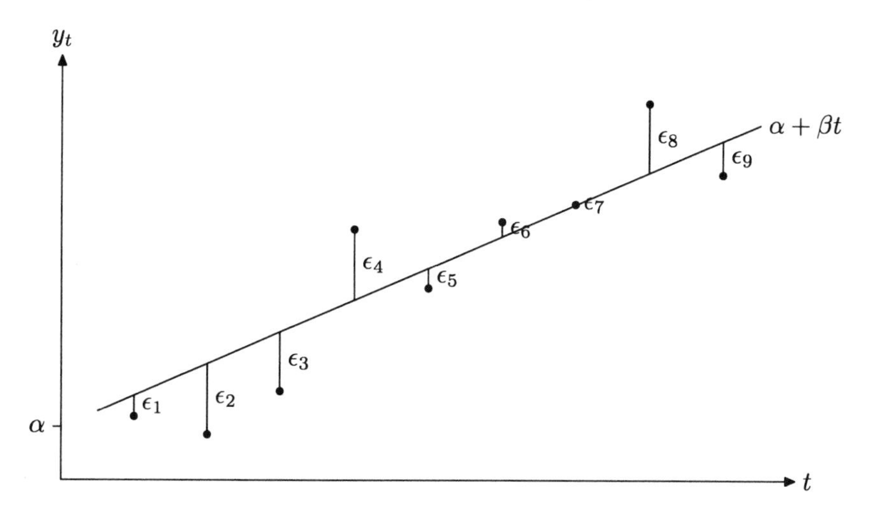

FIGURE 6.1. A linear trend model

Denoting the forecast error by $e_{\tilde{T}+\tau}$ we have

$$\begin{aligned} e_{\tilde{T}+\tau} &= y_{\tilde{T}+\tau} - \hat{y}_{\tilde{T}+\tau} \\ &= [\alpha + \beta(\tilde{T}+\tau) + \epsilon_{\tilde{T}+\tau}] - [\hat{\alpha} + \hat{\beta}(\tilde{T}+\tau)] \\ &= (\alpha - \hat{\alpha}) + (\beta - \hat{\beta})(\tilde{T}+\tau) + \epsilon_{\tilde{T}+\tau}\,. \end{aligned}$$

Hence, $\mathrm{E}(e_{\tilde{T}+\tau}) = 0$ and the predictor $\hat{y}_{\tilde{T}+\tau}$ is unbiased. This leads to the following expression for the mean dispersion error:

$$\begin{aligned} \mathrm{MDE}(\hat{y}_{\tilde{T}+\tau}) &= \mathrm{E}(e_{\tilde{T}+\tau})^2 \\ &= \mathrm{var}(\hat{\alpha}) + \mathrm{var}(\hat{\beta}) + \sigma^2 \\ &= \sigma^2 \left(\frac{1}{T} + \frac{(\tilde{T}+\tau)^2}{\sum \tilde{t}^2} + 1 \right). \end{aligned}$$

From this it is seen that increasing the predictor's horizon (i.e., τ) will decrease the expected precision of the forecast.

6.2.3 Polynomial Models

The polynomial trend model of order K is of the form

$$y_t = \alpha + \beta_1 t + \beta_2 t^2 + \cdots + \beta_K t^K + \epsilon_t\,,$$

and its forecast again is based on the OLSE of $\alpha, \beta_1, \ldots, \beta_K$:

$$\hat{y}_{T+\tau} = \hat{\alpha} + \hat{\beta}_1(T+\tau) + \cdots + \hat{\beta}_K(T+\tau)^K\,.$$

Using a high-degree polynomial trend does not necessarily improve prediction. In any given problem an appropriate degree of the polynomial has to be determined (cf. Rao, 1967; Gilchrist, 1976). The examples discussed above are special cases of the general regression model described in the next section.

6.3 The Prediction Model

The statistical investigations of the preceding chapters concentrated on the problem of fitting the model

$$y = X\beta + \epsilon, \quad \epsilon \sim (0, \sigma^2 W), \quad \text{rank}(X) = K \tag{6.1}$$

to a matrix of data (y, X) in an optimal manner, where optimality was related to the choice of an estimator of β. Another important task is to adopt the model to not-yet-realized values of the endogeneous variable Y. Henceforth we assume X to be nonstochastic.

Let $\{\Upsilon\}$ denote a set of indices and y_τ, $\tau \in \{\Upsilon\}$ a set of y-values, partially or completely unknown. A basic requirement for the prediction of y_τ is the assumption that the y_τ follow the same model as the vector y, that is,

$$y_{\tau *} = x'_{\tau *}\beta + \epsilon_{\tau *} \tag{6.2}$$

with the same β as in the sample model (6.1).

In matrix form, the n values $y_{1*}, \ldots, y_{n*}$ to be predicted may be summarized in the model

$$\underset{n,1}{y_*} = \underset{n,K}{X_*}\,\beta + \underset{n,1}{\epsilon_*}\,, \quad \epsilon_* \sim (0, \sigma^2 \underset{n,n}{W_*})\,. \tag{6.3}$$

The index $*$ relates to future observations.

In a general situation, we assume that

$$E(\epsilon\epsilon'_*) = \sigma^2 \underset{T,n}{W_0} = \sigma^2(\underset{T,1}{w_1}, \ldots, \underset{T,1}{w_n}) \neq 0\,. \tag{6.4}$$

This assumption is the main source for an improvement of the prediction compared to the classical prediction based on the corollary to the Gauss-Markov-Aitken Theorem (Theorem 4.4). In the following we assume the matrix X_* is known. Restrictions on the rank of X_* are generally not necessary. If we have $\text{rank}(X_*) = K \leq n$, then the predictors can be improved (cf. Section 6.5).

Classical Prediction

In a classical set-up for prediction of y_*, we consider the estimation of the conditional expectation $E(y_*|X_*) = X_*\beta$. By Theorem 4.5 we obtain for any component $x'_{\tau *}\beta$ of $X_*\beta$ that the best linear unbiased estimator is (p stands for predictor)

$$\hat{p}_\tau = x'_{\tau *}b\,, \tag{6.5}$$

where $b = S^{-1}X'W^{-1}y$ is the Gauss-Markov-Aitken estimator of β from the model (6.1), with

$$\text{var}(\hat{p}_{\tau}) = x'_{\tau *}V(b)x_{\tau *}\,. \tag{6.6}$$

Then the classical prediction $\hat{p}_{\text{classical}} = \hat{p}_0$ for the whole vector y_* becomes

$$\hat{p}_0 = X_* b \tag{6.7}$$

with

$$V(\hat{p}_0) = X_* V(b) X'_* \tag{6.8}$$

and

$$V(b) = \sigma^2 S^{-1} \quad \text{with} \quad S = X'W^{-1}X\,.$$

Remarks: (i) As we will see in the following sections, possible improvements of the classical prediction in the generalized model (6.1) depend only on the correlation of the disturbances ϵ and ϵ_*. This fundamental result is due to Goldberger (1962). We shall use this information to derive optimal linear predictors for y_*.

(ii) If X is stochastic and/or β becomes stochastic, then the results of this chapter remain valid for conditional distributions (cf. Toutenburg, 1970d).

6.4 Optimal Heterogeneous Prediction

Here we shall derive some optimal predictors for the random variable y_*. This may be seen as an alternative to the classical prediction.

The prediction p of y_* will be based on the sample information given by y; that is, we choose the predictor p as a function of y, namely, $p = f(y)$. In view of the linearity of the models (6.1) and (6.3), and because of the simplicity of a linear statistic, we confine ourselves to predictions that are linear in y.

The linear heterogeneous set-up is

$$p = Cy + d\,, \tag{6.9}$$

where $C : n \times T$ and $d : n \times 1$ are nonstochastic. For the risk function, we choose the quadratic form $(A > 0)$

$$R_A(p, y_*) = \text{E}(p - y_*)'A(p - y_*)\,. \tag{6.10}$$

The matrix A gives different weights to errors of prediciton of different components of $y_{\tau *}$ and is at the choice of the customer.

Example 6.1: Suppose that t is an ordered time indicator (e.g., years) such that $t = 1, \ldots, T$ corresponds to the sample and $\{\Upsilon\} = (T+1, T+2, \ldots,$ $T+n)$ denotes the periods of forecasting. For the prediction of an economic variable it may be reasonable to have maximum goodness of fit in the period

$T+1$ and decreasing fit in the periods $T+i$, $i=2,\ldots,n$. The appropriate choice of A would be:

$$A = \mathrm{diag}(a_1,\ldots,a_n) \quad \text{with} \quad a_1 > \cdots > a_n > 0$$

and

$$\sum a_i = 1\,.$$

If no prior weights are available, it is reasonable to choose $A = I$.

Using set-up (6.9), we have

$$p - y_* = [(CX - X_*)\beta + d] + (C\epsilon - \epsilon_*)\,, \tag{6.11}$$

and the quadratic risk becomes

$$\begin{aligned} R_A(p, y_*) &= \mathrm{tr}A[(CX - X_*)\beta + d][(CX - X_*)\beta + d]' \\ &\quad + \sigma^2 \mathrm{tr}[A(CWC' + W_* - 2CW_0)] \\ &= u^2 + v^2\,. \end{aligned} \tag{6.12}$$

If β is known, the first expression u^2 depends only on d, and the minimization of $R_A(p, y_*)$ with respect to C and d may be carried out separately for u^2 and v^2 (cf. Section 4.1). With

$$\hat{d} = -(CX - X_*)\beta\,, \tag{6.13}$$

the minimum value of u^2 as 0 is attained. The minimization of v^2 with respect to C results in the necessary condition for C (Theorems A.91–A.95)

$$\frac{1}{2}\frac{\partial v^2}{\partial C} = ACW - AW_0' = 0\,. \tag{6.14}$$

From this relationship we obtain the solution to our problem:

$$\hat{C}_1 = W_0' W^{-1} \tag{6.15}$$

and

$$\hat{d} = X_*\beta - W_0' W^{-1} X\beta\,. \tag{6.16}$$

Theorem 6.1 *If β is known, the $R_A(p, y_*)$-optimal, heterogeneous prediction of y_* is*

$$\hat{p}_1 = X_*\beta + W_0' W^{-1}(y - X\beta) \tag{6.17}$$

with

$$\mathrm{E}(\hat{p}_1) = X_*\beta \tag{6.18}$$

and

$$R_A(\hat{p}_1, y_*) = \sigma^2\, \mathrm{tr}[A(W_* - W_0' W^{-1} W_0)]\,. \tag{6.19}$$

Remark: $\hat{p}_1$ is the optimal linear prediction generally. Furthermore, $\hat{p}_1$ is unbiased for the conditional expectation $X_*\beta$ of y_*.

As $\hat{p}_1$ depends on the unknown parameter β itself, this prediction—as well as the R_1-optimal estimation $\hat{\beta}_1 = \beta$—is not operational.

Nevertheless, Theorem 6.1 yields two remarkable results: the structure (6.17) of an optimal prediction and the lower bound (6.19) of the $R_A(p, y_*)$-risk in the class of linear predictors. Similar to the problems related to the optimal linear estimator $\hat{\beta}_1 = \beta$ (cf. Section 4.1), we have to restrict the set of linear predictors $\{Cy + d\}$.

6.5 Optimal Homogeneous Prediction

Letting $d = 0$ in (6.9) and in $R_A(p, y_*)$ defined in (6.12), similar to (4.13)–(4.16), we obtain by differentiating and equating to the null matrix

$$\frac{1}{2}\frac{\partial R_A(p, y_*)}{\partial C} = AC(X\beta\beta'X' + \sigma^2 W) - A(\sigma^2 W_0' + X_*\beta\beta'X') = 0 .$$

A solution to this is given by the matrix

$$\hat{C}_2 = (\sigma^2 W_0' + X_*\beta\beta'X')(X\beta\beta'X' + \sigma^2 W)^{-1}.$$

Applying Theorem A.18 (iv), we derive the optimal homogeneous predictor

$$\hat{p}_2 = \hat{C}_2 y = X_*\hat{\beta}_2 + W_0' W^{-1}(y - X\hat{\beta}_2) , \tag{6.20}$$

where $\hat{\beta}_2 = \beta\left[\frac{\beta'X'W^{-1}y}{\sigma^2+\beta'S\beta}\right]$ is the optimal homogeneous estimator of β (cf. (4.20)).

Define

$$Z = X_* - W_0' W^{-1} X . \tag{6.21}$$

Then, with $R_A(\hat{p}_1, y_*)$ from (6.19) and $M(\hat{\beta}_2, \beta)$ from (4.25), we may conclude that

$$R_A(\hat{p}_2, y_*) = \text{tr}\{AZM(\hat{\beta}_2, \beta)Z'\} + R_A(\hat{p}_1, y_*) . \tag{6.22}$$

Hint: Because of its dependence on $\hat{\beta}_2$ and, hence, on $\sigma^{-1}\beta$, the optimal homogeneous prediction again is not operational. Using prior information of the form

$$\sigma^{-2}(\beta - \beta_0)'\text{diag}(c_1^2, \cdots, c_K^2)(\beta - \beta_0) \le 1 \tag{6.23}$$

may help in finding feasible operational solutions that might have a smaller risk than $\hat{p}_2$. These investigations are given in full detail in Toutenburg (1968, 1975b).

Condition of Unbiasedness

To have operational solutions to our prediction problem when β is unknown, we confine ourselves to the class of homogeneous unbiased predictors (cf. arguments in Section 4.1). Letting $d = 0$ it follows immediately from (6.11) that $\mathrm{E}(p) = \mathrm{E}(y_*) = X_*\beta$; that is,

$$\mathrm{E}(p - y_*) = (CX - X_*)\beta = 0$$

is valid for all vectors β if and only if

$$CX = X_* \,. \tag{6.24}$$

Under this condition we obtain (cf. (6.12))

$$R_A(p, y_*) = \sigma^2 \mathrm{tr}\{A(CWC' + W_* - 2CW_0)\} = v^2. \tag{6.25}$$

Therefore, we are led to the following linearly restrained optimization problem:

$$\min_{C,\Lambda} \tilde{R}_A = \min_{C,\Lambda} \left\{ \sigma^{-2} R_A(p, y_*) - 2\sum_{\tau=1}^{n} \lambda'_\tau (CX - X_*)'_\tau \right\} \tag{6.26}$$

with $(CX - X_*)'_\tau$ as the τth column of $(CX - X_*)$ and

$$\underset{K,n}{\Lambda'} = (\underset{K,1}{\lambda_1}, \ldots, \underset{K,1}{\lambda_n})$$

a matrix of Lagrangian multipliers, where each λ_i is a K-vector.

The optimal matrices $\hat{C}_3$ and $\hat{\Lambda}$ are solutions to the normal equations

$$\frac{1}{2}\frac{\partial \tilde{R}_A}{\partial C} = ACW - AW_0' - \Lambda X' = 0 \tag{6.27}$$

and

$$\frac{1}{2}\frac{\partial \tilde{R}_A}{\partial \Lambda} = CX - X_* = 0 \,. \tag{6.28}$$

Because of the regularity of $A > 0$, it follows from (6.27) that

$$C = W_0' W^{-1} + \Lambda X' W^{-1}.$$

Using (6.28) and setting $S = X'W^{-1}X$, we obtain

$$CX = W_0' W^{-1} X + \Lambda S = X_* \,,$$

and hence we find

$$\hat{\Lambda} = (X_* - W_0' W^{-1} X) S^{-1}.$$

Combining these expressions gives the optimal matrix $\hat{C}_3$:

$$\hat{C}_3 = W_0' W^{-1} + X_* S^{-1} X' W^{-1} - W_0' W^{-1} X S^{-1} X' W^{-1}$$

and, finally, the optimal predictor $\hat{p}_3 = \hat{C}_3 y$:

$$\hat{p}_3 = X_* b + W_0' W^{-1}(y - Xb). \tag{6.29}$$

Theorem 6.2 *The $R_A(p, y_*)$-optimal, homogeneous unbiased predictor of y_* is of the form $\hat{p}_3$ (6.29) with $b = S^{-1}X'W^{-1}y$, the GLSE. Using the notation Z from (6.21), we get the risk*

$$R_A(\hat{p}_3, y_*) = \operatorname{tr}\{AZV(b)Z'\} + R_A(\hat{p}_1, y_*) . \tag{6.30}$$

Comparison of the Optimal Predictors

From (6.30) we may conclude that

$$R_A(\hat{p}_3, y_*) - R_A(\hat{p}_1, y_*) = \operatorname{tr}\{A^{\frac{1}{2}}ZV(b)Z'A^{\frac{1}{2}}\} \geq 0 \tag{6.31}$$

and, analogously (cf. (6.22))

$$R_A(\hat{p}_2, y_*) - R_A(\hat{p}_1, y_*) = \operatorname{tr}\{A^{\frac{1}{2}}ZM(\hat{\beta}_2, \beta)Z'A^{\frac{1}{2}}\} \geq 0 , \tag{6.32}$$

as the matrices in brackets are nonnegative definite.

For the comparison of $\hat{p}_3$ and $\hat{p}_2$, we see that the following difference is nonnegative definite:

$$R_A(\hat{p}_3, y_*) - R_A(\hat{p}_2, y_*) = \operatorname{tr}\{A^{\frac{1}{2}}Z[V(b) - M(\hat{\beta}_2, \beta)]Z'A^{\frac{1}{2}}\} \geq 0 , \tag{6.33}$$

if, as a sufficient condition,

$$V(b) - M(\hat{\beta}_2, \beta) = \sigma^2 S^{-1} - \frac{\sigma^2 \beta\beta'}{\sigma^2 + \beta' S \beta} \geq 0 . \tag{6.34}$$

But this is seen to be equivalent to the following condition

$$\beta' S \beta \leq \sigma^2 + \beta' S \beta ,$$

which trivially holds.

Corollary to Theorems 6.1 and 6.2: Consider the three classes of heterogeneous, homogeneous, and homogeneous unbiased linear predictors. Then the optimal predictors of each class are $\hat{p}_1$, $\hat{p}_2$, and $\hat{p}_3$, respectively, with their risks ordered in the following manner:

$$R_A(\hat{p}_1, y_*) \leq R_A(\hat{p}_2, y_*) \leq R_A(\hat{p}_3, y_*) . \tag{6.35}$$

Convention: Analogous to the theory of estimation, we say that the best linear unbiased predictor $\hat{p}_3$ is the Gauss-Markov (GM) predictor or the BLUP (best linear unbiased predictor) of y_*.

Example 6.2 (One-step-ahead prediction): An important special case of prediction arises when $n = 1$ and $\tau = T + 1$, that is, with the scalar model

$$y_* = y_{T+1} = x'_{T+1}\beta + \epsilon_{T+1} , \tag{6.36}$$

where $\epsilon_{T+1} \sim (0, \sigma^2 w_*) = (0, \sigma_*^2)$. The covariance vector of ϵ and ϵ_{T+1} is the first column of $\sigma^2 W_0$ (6.4):

$$\operatorname{E}(\epsilon\, \epsilon_{T+1}) = \sigma^2 w . \tag{6.37}$$

Then the GM predictor of $y_* = y_{T+1}$ is (cf. (6.29)) of the form

$$\hat{p}_3 = x'_{T+1}b + w'W^{-1}(y - Xb)\,. \tag{6.38}$$

As a particular case, let us assume that W is the dispersion matrix of the first-order autoregressive process. Then we have $\sigma_*^2 = \sigma^2$ and the structure of the vector w as

$$w = \mathrm{E}(\epsilon\, \epsilon_{T+1}) = \sigma^2 \begin{pmatrix} \rho^T \\ \rho^{T-1} \\ \vdots \\ \rho \end{pmatrix}. \tag{6.39}$$

Postmultiplying by the matrix W^{-1} (4.103) gives

$$w'W^{-1} = \rho(0, \cdots, 0, 1) \tag{6.40}$$

so that

$$w'W^{-1}w = \rho^2.$$

Therefore, the one-step-ahead GM predictor of y_* becomes

$$\hat{p}_3 = x'_{T+1}b + \rho\hat{\epsilon}_T\,. \tag{6.41}$$

Here $\hat{\epsilon}_T$ is the last component of the estimated residual vector $y - Xb = \hat{\epsilon}$.

For $n = 1$, the (n, n)-matrix A becomes a positive scalar, which may be fixed as 1. Then the predictor $\hat{p}_3$ (6.41) has the risk

$$R(\hat{p}_3, y_{T+1}) = (x'_{T+1} - \rho x'_T)V(b)(x_{T+1} - \rho x_t) + \sigma^2(1 - \rho^2) \tag{6.42}$$

(cf. Goldberger, 1962) .

6.6 MDE Matrix Comparisons between Optimal and Classical Predictors

Predicting future values of the dependent variable in the generalized linear regression model is essentially based on two alternative methods: the classical one, which estimates the expected value of the regressand to be predicted; and the optimal one, which minimizes some quadratic risk over a chosen class of predictors. We now present some characterizations of the interrelationships of these two types of predictors and the involved estimators of β. These investigations are mainly based on the results derived in Toutenburg and Trenkler (1990).

The classical predictor estimates the conditional expectation $X_*\beta$ of y_* by $X_*\hat{\beta}$, where $\hat{\beta}$ is an estimator of β. Since X_* is known, classical predictors $X_*\hat{\beta}$ vary with respect to the chosen estimator $\hat{\beta}$. Hence, optimality or superiority of classical predictors may be expected to be strongly related to the superiority of estimators.

Let us first give the following definition concerning the superiority of classical predictors.

Definition 6.3 ($X_*\beta$-superiority) *Consider two estimators $\hat\beta_1$ and $\hat\beta_2$. Then the classical predictor $X_*\hat\beta_2$ of y_* is said to be $X_*\beta$-superior to the predictor $X_*\hat\beta_1$ if*

$$M(X_*\hat\beta_1, X_*\beta) - M(X_*\hat\beta_2, X_*\beta) \geq 0 \,. \tag{6.43}$$

Using $M(X_*\hat\beta_i, X_*\beta) = \mathrm{E}(X_*\hat\beta_i - X_*\beta)(X_*\hat\beta_i - X_*\beta)'$, we have

$$\begin{aligned} M(X_*\hat\beta_1, X_*\beta) - M(X_*\hat\beta_2, X_*\beta) &= X_*[M(\hat\beta_1, \beta) - M(\hat\beta_2, \beta)]X'_* \\ &= X_*\Delta(\hat\beta_1, \hat\beta_2)X'_*\,, \end{aligned} \tag{6.44}$$

where $\Delta(\hat\beta_1, \hat\beta_2)$ is the difference between the MDE matrices of the estimators $\hat\beta_1$ and $\hat\beta_2$ (cf. (3.46)).

It follows that superiority of the estimator $\hat\beta_2$ over $\hat\beta_1$ implies the $X_*\beta$-superiority of the predictor $X_*\hat\beta_2$ over $X_*\hat\beta_1$. Therefore, the semi-ordering (in the Loewner sense) of estimators implies the same semi-ordering of the corresponding classical predictors. The superiority condition for estimators, (i.e., $\Delta(\hat\beta_1, \hat\beta_2) \geq 0$) and that for classical predictors (i.e., condition (6.44)) become equivalent if the (n, K)-matrix X_* has full column rank K (see Theorem A.46), which, however, may rarely be the case.

Both criteria also become equivalent in any case if we admit *all* matrices X_* in Definition 6.3, so that $X_*\beta$ superiority reduces to the MDE-I superiority of estimators.

If we are mainly interested in predicting the random vector y_* itself, then we should introduce an alternative mean dispersion error criterion for a predictor p by defining the following matrix:

$$M(p, y_*) = \mathrm{E}(p - y_*)(p - y_*)'. \tag{6.45}$$

Observe that

$$M(p, y_*) = V(p - y_*) + d_*d'_*\,, \tag{6.46}$$

where

$$d_* = \mathrm{E}(p) - X_*\beta \tag{6.47}$$

denotes the bias of p with respect to $X_*\beta$.

On the other hand,

$$M(p, X_*\beta) = V(p) + d_*d'_* \tag{6.48}$$

and since

$$V(p - y_*) = V(p) - \mathrm{cov}(p, y_*) - \mathrm{cov}(y_*, p) + V(y_*)\,, \tag{6.49}$$

we have in general

$$M(p, y_*) \neq M(p, X_*\beta)\,. \tag{6.50}$$

Example 6.3: If $p = Cy + d$ is a linear predictor, we have

$$\begin{aligned} M(p, y_*) &= \sigma^2[CWC' - CW_0 - W_0'C' + W_*] + d_*d_*' \,, & (6.51) \\ M(p, X_*\beta) &= \sigma^2 CWC' + d_*d_*' \,, & (6.52) \end{aligned}$$

where the bias of p with respect to $X_*\beta$ is given by

$$d_* = (CX - X_*)\beta + d \,. \tag{6.53}$$

Definition 6.4 **(y_* superiority)** *Consider two predictors p_1 and p_2 of y_*. The predictor p_2 is said to be y_*-superior to p_1 if*

$$M(p_1, y_*) - M(p_2, y_*) \geq 0 \,. \tag{6.54}$$

Let us now pose the question as to when $X_*\beta$ superiority implies y_* superiority, and vice versa, that is, when

$$M(p_1, y_*) - M(p_2, y_*) = M(p_1, X_*\beta) - M(p_2, X_*\beta) \tag{6.55}$$

holds.

From (6.46) and (6.49) it becomes clear that this will be the case if $\text{cov}(p, y_*) = 0$. For linear predictors, this means that W_0 should be zero. We may state the following result (Toutenburg and Trenkler, 1990):

Theorem 6.5 *Suppose that $\sigma^{-2}\,\text{E}(\epsilon\, \epsilon_*') = W_0 = 0$, and let p_1 and p_2 be two predictors. Then the following conditions are equivalent for competing predictors:*

(i) $M(p_1, y_*) - M(p_2, y_*) \geq 0$,

(ii) $M(p_1, X_*\beta) - M(p_2, X_*\beta) \geq 0$,

(iii) $R_A(p_1, y_*) - R_A(p_2, y_*) \geq 0$ *for all* $A \geq 0$,

(iv) $R_A(p_1, X_*\beta) - M(p_2, X_*\beta) \geq 0$ *for all* $A \geq 0$,

where (cf. (6.10))

$$\begin{aligned} R_A(p_i, X_*\beta) &= \text{E}[(p_i - X_*\beta)'A(p_i - X_*\beta)] \,, \\ R_A(p_i, y_*) &= \text{E}[(p_i - y_*)'A(p_i - y_*)] \,, \quad i = 1, 2. \end{aligned}$$

Now assume $\hat\beta$ to be any estimator of β, and let

$$p(\hat\beta) = X_*\hat\beta + W_0'W^{-1}(y - X\hat\beta) \tag{6.56}$$

be the predictor. With the (n, K)-matrix Z from (6.21), we get

$$p(\hat\beta) - y_* = Z(\hat\beta - \beta) + W_0'W^{-1}\epsilon - \epsilon_* \,. \tag{6.57}$$

If $\hat\beta = Dy + d$ is a linear estimator of β, it immediately follows that

$$\begin{aligned} \text{E}[(\hat\beta - \beta)(W_0'W^{-1}\epsilon - \epsilon_*)'] &= D\,\text{E}[\epsilon(\epsilon'W^{-1}W_0 - \epsilon_*')] \\ &= \sigma^2 D(WW^{-1}W_0 - W_0) \\ &= 0 \,, & (6.58) \end{aligned}$$

and from this (cf. (6.51)) we obtain the MDE matrix (6.45) of $p(\hat{\beta})$ as

$$M(p(\hat{\beta}), y_*) = ZM(\hat{\beta}, \beta)Z' + \sigma^2(W_* - W_0'W^{-1}W_0). \tag{6.59}$$

6.6.1 *Comparison of Classical and Optimal Prediction with Respect to the y_* Superiority*

Consider linear heterogeneous estimators for β given by $\hat{\beta} = Dy + d$, which are not necessarily unbiased. It might be expected that the classical predictor

$$\hat{p}_0 = X_*\hat{\beta} \tag{6.60}$$

for y_* is outperformed with respect to the MDE matrix criterion (6.54) by the predictor $p(\hat{\beta})$ given in (6.56), since the latter uses more information. This, however, does not seem always to be the case.

Let

$$b_{*o} = X_*[(DX - I)\beta + d] \tag{6.61}$$

denote the bias of $\hat{p}_0$ with respect to $X_*\beta$. Then we have (cf. (6.51))

$$\begin{aligned} M(\hat{p}_0, y_*) &= \sigma^2 X_* DWD'X_*' - \sigma^2 X_* DW_0 \\ &\quad - \sigma^2 W_0'D'X_*' + \sigma^2 W_* + b_{*0}b_{*0}', \end{aligned} \tag{6.62}$$

and with (6.58) and (6.59) we obtain

$$\begin{aligned} M(p(\hat{\beta}), y_*) &= \sigma^2 ZDWD'Z' - \sigma^2 W_0'W^{-1}W_0 \\ &\quad + \sigma^2 W_* + b_{*1}b_{*1}', \end{aligned} \tag{6.63}$$

where

$$\begin{aligned} b_{*1} &= Z[(DX - I)\beta + d] \\ &= b_{*0} - W_0'W^{-1}X[(DX - I)\beta + d] \end{aligned} \tag{6.64}$$

is the bias of $p(\hat{\beta})$ with respect to $X_*\beta$.

Introducing the notation

$$P = W^{-\frac{1}{2}}XDWD'X'W^{-\frac{1}{2}}, \tag{6.65}$$

$$G = W_0'W^{-\frac{1}{2}}(I - P)W^{-\frac{1}{2}}W_0, \tag{6.66}$$

$$E = DWD'X'W^{-\frac{1}{2}} - DW^{-\frac{1}{2}}, \tag{6.67}$$

we obtain the following representation for the difference of the MDE matrices of $\hat{p}_0$ and $p(\hat{\beta})$:

$$\begin{aligned} M(\hat{p}_0, y_*) - M(p(\hat{\beta}), y_*) &= \sigma^2 G + \sigma^2 X_* EW^{-\frac{1}{2}}W_0 \\ &\quad + \sigma^2 W_0'W^{-\frac{1}{2}}E'X_*' \\ &\quad + b_{*0}b_{*0}' - b_{*1}b_{*1}'. \end{aligned} \tag{6.68}$$

Now the crucial problem is to find the conditions under which the difference (6.68) is nonnegative definite. As indicated above, it turns out that there is no general solution to this problem. Nevertheless, we are able to find some simplifications in some special cases.

Assume that $E = 0$. This condition is equivalent to the equation

$$DW(D'X' - I) = 0\,, \tag{6.69}$$

which is satisfied, for example, for the so-called guess prediction using $D = 0$. An important case is given by $\hat{\beta}_1 = \beta$. Furthermore, we notice that (6.69) is sufficient for P to be a projector, which implies that $G \geq 0$:

$$\begin{aligned}
P &= W^{-\frac{1}{2}}XDWD'X'W^{-\frac{1}{2}} = W^{-\frac{1}{2}}XDW^{\frac{1}{2}} && \text{(use (6.69))}\\
P^2 &= (W^{-\frac{1}{2}}XDW^{\frac{1}{2}})(W^{-\frac{1}{2}}XDWD'X'W^{-\frac{1}{2}})\\
&= W^{-\frac{1}{2}}XD(WD')X'W^{-\frac{1}{2}} && \text{(use (6.69))}\\
&= P\,,
\end{aligned}$$

so that P is idempotent, and hence $I - P$ is also idempotent, implying $G \geq 0$.

Theorem 6.6 *Assume that (6.69) is satisfied. Then the predictor $p(\hat{\beta})$ (from (6.56)) is y_*-superior to the classical predictor $\hat{p}_0 = X_*\hat{\beta}$ if and only if*

$$(i) \quad b_{*1} \in \mathcal{R}(\sigma^2 G + b_{*0}b'_{*0}) \tag{6.70}$$

and

$$(ii) \quad b'_{*1}(\sigma^2 G\,, + b_{*0}b'_{*0})^- b_{*1} \leq 1 \tag{6.71}$$

where the choice of the g-inverse is arbitrary.

Proof: Use Theorem A.71.

Examples:

(a) Let $D = S^{-1}X'W^{-1}$ and $d = 0$, so that $\hat{\beta} = Dy = b$ is the GLSE. Then it is easily seen that (6.69) is satisfied:

$$S^{-1}X'W^{-1}W(W^{-1}XS^{-1}X' - I) = 0.$$

Since b is unbiased, both $p(b)$ ($= \hat{p}_3$ (6.29)) and $\hat{p}_0 = X_*b$ are unbiased, and by Theorem 6.6 we get

$$M(X_*b, y_*) - M(p(b), y_*) \geq 0\,. \tag{6.72}$$

This result was first derived by Goldberger (1962).

(b) Consider the case where we have an additional linear restriction $r = R\beta + \delta$ with $\text{rank}(R) = J$. Then the corresponding linearly restricted least-squares estimator is given by

$$\begin{aligned}
b(R) &= b + S^{-1}R'(RS^{-1}R')^{-1}(r - Rb)\\
&= \bar{D}y + \bar{d}
\end{aligned} \tag{6.73}$$

with

$$\bar{D} = (I - S^{-1}R'(RS^{-1}R')^{-1}R)S^{-1}X'W^{-1} \tag{6.74}$$

and

$$\bar{d} = S^{-1}R'(RS^{-1}R')^{-1}r. \tag{6.75}$$

After some straightforward calculations, it is easily seen that the matrix $\bar{D}$ (6.74) belonging to the heterogeneous estimator (6.73) satisfies condition (6.69), not depending on whether the restrictions $r = R\beta$ are valid. Now consider the predictors

$$\hat{p}_0 = X_* b(R)$$

and

$$p(b(R)) = X_* b(R) + W_0' W^{-1}(y - Xb(R)).$$

With the notation

$$\begin{aligned} \bar{G} &= W_0' W^{-\frac{1}{2}}(I - \bar{P})W^{-\frac{1}{2}}W_0 \geq 0, \\ \bar{P} &= W^{-\frac{1}{2}}X\bar{D}W\bar{D}'X'W^{-\frac{1}{2}} \quad \text{(cf. (6.65), (6.66))}, \end{aligned}$$

and defining

$$b_{*0} = X_* S^{-1}R'(RS^{-1}R')^{-1}\delta, \tag{6.76}$$

$$b_{*1} = ZS^{-1}R'(RS^{-1}R')^{-1}\delta, \tag{6.77}$$

with

$$\delta = r - R\beta, \tag{6.78}$$

we finally obtain

$$M(\hat{p}_0, y_*) - M(p(b(R)), y_*) = \sigma^2 \bar{G} + b_{*0}b_{*0}' - b_{*1}b_{*1}'. \tag{6.79}$$

In order to decide if this difference is nonnegative definite, we have to use Theorem 6.6. As a conclusion, we may state that the predictor $\hat{p}(b(R))$ is y_*-superior to the classical predictor $\hat{p}_0 = X_* b(R)$ if and only if conditions (6.70) and (6.71) are satisfied. If $\delta = 0$ (i.e., if the linear restrictions are satisfied exactly), then $b_{*0} = b_{*1} = 0$ and $M(\hat{p}_0, y_*) - M(p(b(R), y_*) = \sigma^2 \bar{G} \geq 0$.

6.6.2 Comparison of Classical and Optimal Predictors with Respect to the $X_*\beta$ Superiority

We now compare the predictors $\hat{p}_0 = X_*\hat{\beta}$ and $p(\hat{\beta})$ (cf. (6.56)) for a linear heterogeneous estimator $\hat{\beta} = Dy + d$ with respect to criterion (6.43). Different from the y_* optimality of $p(\beta)$, it might be expected that $\hat{p}_0$ is a more efficient predictor according to the $X_*\beta$ criterion when compared with $p(\hat{\beta})$. Hence, let us investigate the conditions for the classical predictor $\hat{p}_0 = X_* \hat{\beta}$ to be superior to the predictor $p(\hat{\beta})$, according to Definition 6.3; that is, let us find when (see (6.43))

$$M(p(\hat{\beta}), X_*\beta) - M(\hat{p}_0, X_*\beta) \geq 0\,. \tag{6.80}$$

Using (6.48) we get

$$M(\hat{p}_0, X_*\beta) = \sigma^2 X_* D W D' X_*' + b_{*0} b_{*0}' \tag{6.81}$$

with b_{*0} from (6.61) and

$$\begin{aligned} M(p(\hat{\beta}), X_*\beta) &= \sigma^2 X_* D W D' X_*' + \sigma^2 W_0' W^{-1} W_0 \\ &+ \sigma^2 W_0' W^{-1} X D W D' X' W^{-1} W_0 \\ &+ \sigma^2 X_* D W_0 + \sigma^2 W_0' D' X_*' - \sigma^2 X_* D W D' X' W^{-1} W_0 \\ &- \sigma^2 W_0' W^{-1} X D W D' X_*' - \sigma^2 W_0' W^{-1} X D W_0 \\ &- \sigma^2 W_0' D' X' W^{-1} W_0 + b_{*1} b_{*1}' \end{aligned} \tag{6.82}$$

with b_{*1} from (6.64).

Hence the difference (6.80) between the MDE matrices becomes

$$\begin{aligned} & M(p(\hat{\beta}), X_*\beta) - M(\hat{p}_0, X_*\beta) \\ &= -\sigma^2 G - b_{*0} b_{*0}' + b_{*1} b_{*1}' - \sigma^2 X_* E W^{-\frac{1}{2}} W_0 \\ &\quad - \sigma^2 W_0' W^{-\frac{1}{2}} E' X_*' + \sigma^2 W_0' W^{-1} [I - XD] W_0 \\ &\quad + \sigma^2 W_0' [I - D'X'] W^{-1} W_0 \end{aligned} \tag{6.83}$$

with G from (6.66) and E from (6.67).

Similar to the problem discussed before, it is not an easy task to decide whether this difference is nonnegative definite. Therefore we confine ourselves again to situations for which this difference assumes a simple structure. This occurs, for example, if condition (6.69) is satisfied such that after some calculations condition (6.83) reduces to

$$M(p(\hat{\beta}), X_*\beta) - M(\hat{p}_0, X_*\beta) = \sigma^2 G + b_{*1} b_{*1}' - b_{*0} b_{*0}'\,. \tag{6.84}$$

Theorem 6.7 *Let $\hat{\beta} = Dy + d$ be a linear estimator such that the matrix D satisfies condition (6.69) (which is equivalent to $E = 0$). Then the classical predictor $\hat{p}_0 = X_* \hat{\beta}$ is $X_*\beta$-superior to the predictor $p(\hat{\beta}) = X_*\beta + W_0' W^{-1}(y - X\hat{\beta})$ if and only if*

$$(i) \quad b_{*0} \in \mathcal{R}(\sigma^2 G + b_{*1} b_{*1}') \tag{6.85}$$

and

$$(ii) \quad b_{*0}' (\sigma^2 G + b_{*1} b_{*1}')^- b_{*0} \leq 1\,. \tag{6.86}$$

Example 6.4: Let $\hat{\beta} = b$. Then $\hat{p}_0 = X_* b$ is $X_*\beta$-superior to $p(b)$ in accordance with the extended Gauss-Markov-Aitken theorem.

This may be seen as follows:

$$\begin{aligned}
M(X_*b, X_*\beta) &= \sigma^2 X_* S^{-1} X_*', && (6.87)\\
p(b) - X_*\beta &= Z S^{-1} X' W^{-1} \epsilon + W_0' W^{-1} \epsilon,\\
M(p(b), X_*\beta) &= \sigma^2 Z S^{-1} Z' + \sigma^2 W_0' W^{-1} W_0\\
&\quad + \sigma^2 Z S^{-1} X' W^{-1} W_0 + \sigma^2 W_0' W^{-1} X S^{-1} Z'\\
&= \sigma^2 X_* S^{-1} X_*' + \sigma^2 W_0' W^{-1} W_0\\
&\quad - \sigma^2 W_0' W^{-1} X S^{-1} X' W^{-1} W_0\\
&= \sigma^2 X_* S^{-1} X_*' + \sigma^2 G && (6.88)
\end{aligned}$$

with

$$G = W_0'(W^{-\frac{1}{2}} - W^{-1} X S^{-1} X' W^{-\frac{1}{2}})(W^{-\frac{1}{2}} - W^{-\frac{1}{2}} X S^{-1} X' W^{-1}) W_0 \geq 0 .$$

Therefore, we obtain

$$M(p(b), X_*\beta) - M(X_*b, X_*\beta) = \sigma^2 G \geq 0 . \tag{6.89}$$

Interpretation: The investigations of this section have shown very clearly that optimality is strongly dependent on the chosen criterion and/or its respective parameters. If we consider the two predictors X_*b (classical) and $p(b) = \hat{p}_3$ (R_A-optimal), we notice that $p(b)$ is y_*-superior to $X_*\beta$ (cf. (6.72)):

$$M(X_*b, y_*) - M(p(b), y_*) \geq 0$$

with respect to the R_A optimality of $\hat{p}_3 = p(b)$. If we change the criterion, that is, if we compare both predictors with respect to the $X_*\beta$ superiority, we obtain

$$M(p(b), X_*\beta) - M(X_*b, X_*\beta) \geq 0 ,$$

which is the reverse relationship.

6.7 Prediction Regions

In Sections 3.8.1 and 3.8.2, we derived confidence intervals and ellipsoids for the parameter β and its components.

The related problem in this section consists of the derivation of prediction regions for the random variable y_*.

In addition to (6.3), we assume a joint normal distribution, that is,

$$(\epsilon_*, \epsilon) \sim N_{n+T}\left((0,0), \sigma^2 \begin{pmatrix} W_* & W_0' \\ W_0 & W \end{pmatrix}\right), \tag{6.90}$$

where the joint dispersion matrix is assumed to be regular. This is seen to be equivalent (cf. Theorem A.74 (ii)(b)) to

$$W_* - W_0' W^{-1} W_0 > 0 \,. \tag{6.91}$$

We choose the R_A-optimal homogeneous predictor as

$$\hat{p}_3 = X_* b + W_0' W^{-1}(y - Xb) \,.$$

Using (6.90) and (6.30), this predictor is normally distributed:

$$\hat{p}_3 - y_* \sim N_n(0, \sigma^2 \Sigma_b) \,, \tag{6.92}$$

with $Z = X_* - W_0' W^{-1} X$ from (6.21) and

$$\Sigma_b = Z S^{-1} Z' + W_* - W_0' W^{-1} W_0 \,. \tag{6.93}$$

Since $\hat{p}_3$ is unbiased, we have $\sigma^2 \Sigma_b = M(\hat{p}_3, y_*)$ (cf. (6.45)). Thus it follows from Theorem A.85 (ii) that

$$(\hat{p}_3 - y_*)' \Sigma_b^{-1} (\hat{p}_3 - y_*) \sim \sigma^2 \chi_n^2 \,. \tag{6.94}$$

This quadratic form describes a random ellipsoid with center $\hat{p}_3$. Its distribution depends on the unknown parameter σ^2, which has to be estimated.

Theorem 6.8 *Let $s^2 = (y - Xb)' W^{-1} (y - Xb)(T - K)^{-1}$ be the estimator of σ^2. Then*

$$n^{-1} s^{-2} (\hat{p}_3 - y_*)' \Sigma_b^{-1} (\hat{p}_3 - y_*) \sim F_{n, T-K} \,. \tag{6.95}$$

Proof: Consider the standardized vector of disturbances

$$\Phi = \begin{pmatrix} W^{-\frac{1}{2}} \epsilon \\ W_*^{-\frac{1}{2}} \epsilon_* \end{pmatrix} . \tag{6.96}$$

Then, by using (6.90), we obtain

$$\Phi \sim N_{n+T}(0, \sigma^2 V) \,, \tag{6.97}$$

with

$$V = \begin{pmatrix} I_T & W^{-\frac{1}{2}} W_0 W_*^{-\frac{1}{2}} \\ W_*^{-\frac{1}{2}} W_0' W^{-\frac{1}{2}} & I_n \end{pmatrix} . \tag{6.98}$$

From this we get the representation

$$\begin{aligned} \hat{p}_3 - y_* &= [Z S^{-1} X' W^{-\frac{1}{2}} + W_0' W^{-\frac{1}{2}}, -W_*^{\frac{1}{2}}] \Phi & (6.99) \\ &= (A_1, A_2) \Phi \,, & (6.100) \end{aligned}$$

and with (6.92) we have

$$\Sigma_b = (A_1, A_2) V \begin{pmatrix} A_1' \\ A_2' \end{pmatrix} . \tag{6.101}$$

The following matrix is seen to be symmetric and idempotent:

$$V^{\frac{1}{2}}\begin{pmatrix} A_1' \\ A_2' \end{pmatrix}\Sigma_b^{-1}(A_1\ A_2)V^{\frac{1}{2}}. \tag{6.102}$$

By using

$$V^{-\frac{1}{2}}\Phi \sim N(0,\sigma^2 I). \tag{6.103}$$

and (6.99), (6.101), and (6.103), we may apply Theorem A.87 to show that

$$\begin{aligned}
&(\hat{p}_3 - y_*)'\Sigma_b^{-1}(\hat{p}_3 - y_*) \\
&= (\Phi' V^{-1/2})[V^{1/2}\begin{pmatrix} A_1' \\ A_2' \end{pmatrix}]\Sigma_b^{-1}[(A_1, A_2)V^{1/2}](V^{-1/2}\Phi) \\
&\sim \sigma^2\chi^2_n.
\end{aligned} \tag{6.104}$$

The estimator $s^2 = (y - Xb)'W^{-1}(y - Xb)(T - K)^{-1}$ (cf. (4.66)) may be rewritten in the following manner:

$$\begin{aligned}
W^{-\frac{1}{2}}(y - Xb) &= (I - W^{-\frac{1}{2}}XS^{-1}X'W^{-\frac{1}{2}})W^{-\frac{1}{2}}\epsilon \\
&= (I - M)W^{-\frac{1}{2}}\epsilon.
\end{aligned} \tag{6.105}$$

The matrix

$$M = W^{\frac{1}{2}}XS^{-1}X'W^{-\frac{1}{2}} \tag{6.106}$$

is idempotent of $\text{rank}(M) = \text{tr}(M) = K$ and $I - M$ is idempotent of rank $T - K$. Therefore, we obtain

$$\begin{aligned}
(T - K)s^2 &= \epsilon' W^{-\frac{1}{2}}(I - M)W^{-\frac{1}{2}}\epsilon \\
&= \Phi'\begin{pmatrix} I - M & 0 \\ 0 & 0 \end{pmatrix}\Phi = \Phi' M_1 \Phi \\
&= (\Phi' V^{-\frac{1}{2}})V^{\frac{1}{2}}M_1 V^{\frac{1}{2}}(V^{-\frac{1}{2}}\Phi),
\end{aligned} \tag{6.107}$$

where $M_1 = \begin{pmatrix} I - M & 0 \\ 0 & 0 \end{pmatrix}$ is idempotent of rank $T - K$, and, hence, $\Phi' M_1 \Phi \sim \sigma^2\chi^2_{T-K}$.

As a consequence of these calculations, we have found a representation of $(\hat{p}_3 - y_*)'\Sigma_b^{-1}(\hat{p}_3 - y_*)$ and of s^2 as quadratic forms involving the same vector $V^{-1/2}\Phi$. Therefore, we may use Theorem A.89 to check the independence of these quadratic forms. The necessary condition for this to hold is

$$V^{\frac{1}{2}}M_1 V^{\frac{1}{2}}V^{\frac{1}{2}}\begin{pmatrix} A_1' \\ A_2' \end{pmatrix}\Sigma_b^{-1}(A_1, A_2)V^{\frac{1}{2}} = 0. \tag{6.108}$$

Therefore, the condition

$$M_1 V\begin{pmatrix} A_1' \\ A_2' \end{pmatrix} = 0$$

would be sufficient for (6.108) to hold. But this condition is fulfilled as

$$\begin{aligned}
M_1 V \begin{pmatrix} A_1' \\ A_2' \end{pmatrix}
&= \begin{pmatrix} I-M & 0 \\ 0 & 0 \end{pmatrix} \begin{pmatrix} I & W^{-\frac{1}{2}} W_0 W_*^{-\frac{1}{2}} \\ W_*^{-\frac{1}{2}} W_0' W^{-\frac{1}{2}} & I \end{pmatrix} \begin{pmatrix} A_1' \\ A_2' \end{pmatrix} \\
&= (I-M)(A_1' + W^{-\frac{1}{2}} W_0 W_*^{-\frac{1}{2}} A_2') \\
&= (I-M)(W^{-\frac{1}{2}} X S^{-1} Z' + + W^{-\frac{1}{2}} W_0 - W^{-\frac{1}{2}} W_0) \quad \text{[cf. 6.99)]} \\
&= (I - W^{-\frac{1}{2}} X S^{-1} X' W^{-\frac{1}{2}}) W^{-\frac{1}{2}} X S^{-1} Z' \quad \text{[cf. (6.106)]} \\
&= W^{-\frac{1}{2}} X S^{-1} Z' - W^{-\frac{1}{2}} X S^{-1} Z' = 0 .
\end{aligned} \tag{6.109}$$

The F-distribution (6.95) is a consequence of Theorem A.86, and this completes the proof.

The result of Theorem 6.8 provides the basis to construct prediction regions in the sense of the following definition.

Definition 6.9 *A compact set $B(p(\hat{\beta}))$ is called a region with expected coverage q $(0 \le q \le 1)$ for the unknown random vector y_* centered around $p(\hat{\beta})$ if*

$$E_y P_{y_*} \{ y_* \in B(p(\hat{\beta})) \} = q . \tag{6.110}$$

From this definition and Theorem 6.8, we immediately obtain the following result.

Theorem 6.10 *The ellipsoid*

$$B(\hat{p}_3) = \{ y_* : n^{-1} s^{-2} (y_* - \hat{p}_3)' \Sigma_b^{-1} (y_* - \hat{p}_3) \le F_{n,T-K,1-\alpha} \} \tag{6.111}$$

is a region with expected coverage $(1-\alpha)$ for the vector y_*.

Comparing the Efficiency of Prediction Ellipsoids

Similar to point estimators and point predictors, we may pose the question of which prediction region should be regarded as optimal. If the predictor $p(\hat{\beta})$ is unbiased, then as a measure of optimality we choose a quantity related to the volume of a prediction ellipsoid.

Let V_n denote the volume of the n-dimensional unit sphere, and let $a'Aa = 1$ with $A : n \times n$ positive definite be any ellipsoid. Then its volume is given by

$$V_A = V_n |A|^{-\frac{1}{2}} , \tag{6.112}$$

and its squared volume by

$$V_A^2 = V_n^2 |A^{-1}| . \tag{6.113}$$

Applying this rule, we may calculate the squared volume of the ellipsoid $B(\hat{p}_3)$ (6.111) as follows:

$$\begin{aligned} A^{-1} &= nsF_{n,T-K,1-\alpha}\Sigma_b, \\ |A^{-1}| &= (ns^2F_{n,T-K,1-\alpha})^n|\Sigma_b| \end{aligned}$$

(cf. Theorem A.16 (ii)). Taking expectation with respect to the random variable $(s^2)^n$, we obtain the mean of the squared volume:

$$\bar{V}(B(\hat{p}_3)) = V_n^2\,\mathrm{E}(s^{2n})(nF_{n,T-K,1-\alpha})^n|ZS^{-1}Z' + W_* - W_0'W^{-1}W_0|\,. \tag{6.114}$$

Theorem 6.11 *Suppose that there are two unbiased estimators $\hat{\beta}_1$ and $\hat{\beta}_2$ for β having dispersion matrices $V(\hat{\beta}_1)$ and $V(\hat{\beta}_2)$, respectively, and the corresponding predictors*

$$p(\hat{\beta}_i) = X_*\hat{\beta}_i + W_0'W^{-1}(y - X\hat{\beta}_i)\,, \quad i = 1, 2\,.$$

Assume further that $p(\hat{\beta}_1)$ and $p(\hat{\beta}_2)$ satisfy the necessary conditions for F-distribution in the sense of (6.95). Then we have the result

$$\begin{aligned} V(\hat{\beta}_1) - V(\hat{\beta}_2) &\geq 0 \\ \Rightarrow \quad \bar{V}(B(\hat{p}(\hat{\beta}_1))) - \bar{V}(B(p(\hat{\beta}_2))) &\geq 0\,. \end{aligned} \tag{6.115}$$

Proof: Let

$$V_n^2\,\mathrm{E}(s^{2n})(nF_{n,T-K,1-\alpha})^n = c_n$$

denote the constant term of (6.114). Then the means of the squared volume of the prediction ellipsoids $B(p(\hat{\beta}_i))$, $i = 1, 2$, is

$$\bar{V}(B(p(\hat{\beta}_i))) = c_n|\sigma^{-2}ZV(\hat{\beta}_i)Z' + W_* - W_0'W^{-1}W_0|\,.$$

Assume $V(\hat{\beta}_1) - V(\hat{\beta}_2) \geq 0$. Then we obtain

$$\begin{aligned} \Sigma_1 &= \sigma^2 ZV(\hat{\beta}_1)Z' + W_* - W_0W^{-1}W_0 \\ &\geq \sigma^{-2}ZV(\hat{\beta}_2)Z' + W_* - W_0'W^{-1}W_0 = \Sigma_2\,, \end{aligned}$$

that is, $\Sigma_1 = \Sigma_2 + B$, where B is nonnegative definite. Therefore, by Theorem A.40 we have $|\Sigma_2| \leq |\Sigma_1|$.

Hint: For more detailed discussions of prediction regions, the reader is referred to Aitchison (1966), Aitchison and Dunsmore (1968), Toutenburg (1970d, 1971, 1975b), and Guttmann (1970).

For literature on some other aspects of prediction with special reference to growth curve models, the reader is referred to papers by Rao (1962,1964, 1977, 1984, 1987), and Rao and Boudreau (1985).

6.8 Simultaneous Prediction of Actual and Average Values of Y

Generally, predictions from a linear regression model are made either for the actual values of the study variable or for the average values at a time. However, situations may occur in which one may be required to consider the predictions of both the actual and average values simultaneously. For example, consider the installation of an artificial tooth in patients through a specific device. Here a dentist would like to know the life of a restoration, on the average. On the other hand, a patient would be more interested in knowing the actual life of restoration in his/her case. Thus a dentist is interested in the prediction of average value but he may not completely ignore the interest of patients in the prediction of actual value. The dentist may assign higher weight to prediction of average values in comparison to the prediction of actual values. Similarly, a patient may give more weight to prediction of actual values in comparison to that of average values.

This section considers the problem of simultaneous prediction of actual and average values of the study variable in a linear regression model when a set of linear restrictions binding the regression coefficients is available, and analyzes the performance properties of predictors arising from the methods of restricted regression and mixed regression in addition to least squares.

6.8.1 Specification of Target Function

Let us postulate the classical linear regression model (2.45). If $\hat{\beta}$ denotes an estimator of β, then the predictor for the values of study variables within the sample is generally formulated as $\hat{T} = X\hat{\beta}$, which is used for predicting either the actual values y or the average values $\text{E}(y) = X\beta$ at a time.

For situations demanding prediction of both the actual and average values together, Toutenburg and Shalabh (1996) defined the following stochastic target function

$$T(y) = \lambda y + (1 - \lambda)\,\text{E}(y) = T \tag{6.116}$$

and used $\hat{T} = X\hat{\beta}$ for predicting it where $0 \leq \lambda \leq 1$ is a nonstochastic scalar specifying the weight to be assigned to the prediction of actual and average values of the study variable (see, e. g. , Shalabh, 1995).

Remark (i). In cases for which $\lambda = 0$, we have $T = \text{E}(y) = X\beta$ and then optimal prediction coincides with optimal estimation of β, whereas optimality may be defined, for example, by minimal variance in the class of linear unbiased estimators or by some mean dispersion error criterion if biased estimators are considered. The other extreme case, $\lambda = 1$, leads to $T = y$. Optimal prediction of y is then equivalent to optimal estimation of $X\beta + \epsilon$. If the disturbances are uncorrelated, this coincides again with optimal estimation of $X\beta$, that is, of β itself. If the disturbances are correlated according to $\mathrm{E}(\epsilon\epsilon') = \sigma^2 W$, then this information leads to solutions $\hat{y} = X\hat{\beta} + \hat{\epsilon}$ (cf. (6.56) and Goldberger, 1962).

Remark (ii). The two alternative prediction problems—$X\beta$ superiority and the y superiority, respectively—are discussed in full detail in Section 6.6. As a central result, we have the fact that the superiority (in the Loewner ordering of definite matrices) of one predictor over another predictor can change if the criterion is changed. This was one of the motivations to define a target as in (6.116), which combines these two risks.

In the following we consider this problem but with the nonstochastic scalar λ replaced by a nonstochastic matrix Λ. The target function is therefore

$$T(y) = \Lambda y + (I - \Lambda)\,\mathrm{E}(y) = T\,,\text{say}. \tag{6.117}$$

Our derivation of the results makes no assumption about Λ, but one may have in mind Λ as a diagonal matrix with elements $0 \leq \lambda_i \leq 1$, $i = 1, \ldots, T$.

6.8.2 Exact Linear Restrictions

Let us suppose that we are given a set of J exact linear restrictions binding the regression coefficients $r = R\beta$ (see (5.1)).

If these restrictions are ignored, the least squares estimator of β is $b = (X'X)^{-1}X'y$, which may not necessarily obey $r = R\beta$. Such is, however, not the case with the restricted regression estimator given by (see (5.11))

$$b(R) = b + (X'X)^{-1}R'[R(X'X)^{-1}R']^{-1}(r - Rb).$$

By employing these estimators, we get the following two predictors for the values of the study variable within the sample:

$$\hat{T} = Xb\,, \tag{6.118}$$

$$\hat{T}(R) = Xb(R)\,. \tag{6.119}$$

In the following, we compare the estimators b and $b(R)$ with respect to the predictive mean-dispersion error (MDEP) of their corresponding predictions $\hat{T} = Xb$ and $\hat{T}(R) = Xb(R)$ for the target function T.

From (6.117), and the fact that the ordinary least-squares estimator and the restricted estimator are both unbiased, we see that

$$\mathrm{E}_\Lambda(T) = \mathrm{E}(y)\,, \tag{6.120}$$

$$\mathrm{E}_\Lambda(\hat{T}) = X\beta = \mathrm{E}(y)\,, \tag{6.121}$$

$$\mathrm{E}_\Lambda(\hat{T}(R)) = X\beta = \mathrm{E}(y)\,, \tag{6.122}$$

but

$$\mathrm{E}(\hat{T}) = \mathrm{E}(\hat{T}(R)) \neq T\,. \tag{6.123}$$

Equation (6.123) reflects the stochastic nature of the target function T, a problem that differs from the common problem of unbiasedness of a statistic for a fixed but unknown (possibly matrix-valued) parameter. Therefore, both the predictors are only "weakly unbiased" in the sense that

$$\begin{aligned} \mathrm{E}_\Lambda(\hat{T} - T) &= 0\,, &(6.124)\\ \mathrm{E}_\Lambda(\hat{T}(R) - T) &= 0\,. &(6.125)\end{aligned}$$

6.8.3 MDEP Using Ordinary Least Squares Estimator

To compare alternative predictors, we use the matrix-valued mean-dispersion error for $\tilde{T} = X\hat{\beta}$ as follows:

$$\mathrm{MDEP}_\Lambda(\tilde{T}) = \mathrm{E}(\tilde{T} - T)(\tilde{T} - T)'\,. \tag{6.126}$$

First we note that

$$\begin{aligned} T &= \Lambda y + (I - \Lambda)\,\mathrm{E}(y)\\ &= X\beta + \Lambda\epsilon\,, &(6.127)\\ \hat{T} &= Xb\\ &= X\beta + P\epsilon\,, &(6.128)\end{aligned}$$

with the symmetric and idempotent projection matrix $P = X(X'X)^{-1}X'$. Hence we get

$$\begin{aligned} \mathrm{MDEP}_\Lambda(\hat{T}) &= \mathrm{E}(P - \Lambda)\epsilon\epsilon'(P - \Lambda)'\\ &= \sigma^2(P - \Lambda)(P - \Lambda)'\,, &(6.129)\end{aligned}$$

using our previously made assumptions on ϵ.

6.8.4 MDEP Using Restricted Estimator

The problem is now solved by the calculation of

$$\mathrm{MDEP}_\Lambda(\hat{T}(R)) = \mathrm{E}(\hat{T}(R) - T)(\hat{T}(R) - T)'\,. \tag{6.130}$$

Using the abbreviation

$$F = X(X'X)^{-1}R'[R(X'X)^{-1}R']^{-1}R(X'X)^{-1}X' \tag{6.131}$$

and

$$r - Rb = -R(X'X)^{-1}X'\epsilon\,, \tag{6.132}$$

we get from (5.11), (6.119), (6.127), and (6.128) the following

$$\begin{aligned} \hat{T}(R) - T &= Xb(R) - T\\ &= (P - F - \Lambda)\epsilon\,. &(6.133)\end{aligned}$$

As $F = F'$, $P = P'$, and $PF = FP = F$, we have

$$\begin{aligned} \text{MDEP}_\Lambda(\hat{T}(R)) &= \sigma^2(P - F - \Lambda)(P - F - \Lambda)' \\ &= \sigma^2[(P - \Lambda)(P - \Lambda)' - (F - \Lambda F - F\Lambda')] \end{aligned} \quad (6.134)$$

6.8.5 *MDEP Matrix Comparison*

Using results (6.129) and (6.134), the difference of the MDEP-matrices can be written as

$$\begin{aligned} \Delta_\Lambda(\hat{T}; \hat{T}(R)) &= \text{MDEP}_\Lambda(\hat{T}) - \text{MDEP}_\Lambda(\hat{T}(R)) \\ &= \sigma^2(F - \Lambda F - F\Lambda') \\ &= \sigma^2\left[(I - \Lambda)F(I - \Lambda)' - \Lambda F \Lambda'\right] . \end{aligned} \quad (6.135)$$

Then $\hat{T}(R)$ becomes MDEP-superior to $\hat{T}$ if $\Delta_\Lambda(\hat{T}; \hat{T}(R)) \geq 0$.

For $\Delta_\Lambda(\hat{T}; \hat{T}(R))$ to be nonnegative definite, it follows from Baksalary, Schipp, and Trenkler (1992) that necessary and sufficient conditions are

$$\begin{aligned} &\text{(i)} \quad \mathcal{R}(\Lambda F) \subset (\mathcal{R}(I - \Lambda)F) \\ &\text{(ii)} \quad \lambda_1 \leq 1 \end{aligned}$$

where λ_1 denotes the largest characteristic root of the matrix

$$[(I - \Lambda)F(I - \Lambda')]^+ \Lambda F \Lambda' .$$

For the simple special case of $\Lambda = \theta I$, the conditions reduce to $\theta \leq \frac{1}{2}$. Further applications of this target function approach are given in Toutenburg, Fieger, and Heumann (1999).

6.9 Kalman Filter

The Kalman filter (KF) commonly employed by control engineers and other physical scientists has been successfully used in such diverse areas as the processing of signals in aerospace tracking and underwater sonar, and statistical quality control. More recently, it has been used in some nonengineering applications such as short-term forecasting and analysis of life lengths from dose-response experiments. Unfortunately, much of the published work on KF is in engineering literature and uses a language, notation, and style that is not familiar to statisticians. The original papers on the subject are Kalman (1960) and Kalman and Bucy (1961). We believe that KF can be discussed under the general theory of linear models and linear prediction. We first mention the problem and some lemmas used in the solution of the problem. All the results in this section are discussed in a paper by Rao (1994).

6.9.1 *Dynamical and Observational Equations*

Consider a time sequence of p- and q-vector random variables $\{x(t), y(t)\}$, $t = 1, 2, \dots$ with the structural equations

$$x(t) = Fx(t-1) + \xi(t) \tag{6.136}$$

$$y(t) = Hx(t) + \eta(t) \tag{6.137}$$

where F and H are matrices of order $p \times p$ and $q \times p$, respectively, and the following stochastic relationships hold.

1. $\{\xi(t)\}$ and $\{\eta(t)\}$ are independent sequences of p and q random vectors with zero means and covariance matrices V_t and W_t, respectively.

2. $\xi(t)$ and $x(u)$ are independent for $t > u$, and $\eta(t)$ and $x(u)$ are independent for $t \geq u$.

We can observe only $y(t)$, and not $x(t)$ and the problem is to predict $x(t)$ given $y(1), \dots, y(t)$. Generally the covariance matrices V_t and W_t are independent of t. In the sequal we take $V_t = V$ and $W_t = W$, noting that the theory applies even if V_t and W_t are time dependent.

6.9.2 *Some Theorems*

Consider the linear model

$$x = A\beta + \xi \tag{6.138}$$

where $x : p \times 1$, $A : p \times K$, $\beta : K \times 1$, and $\xi : p \times 1$,

$$y = B\beta + \eta \tag{6.139}$$

where $y : q \times 1$, $B : q \times K$, $\eta : q \times 1$ with

$$\mathrm{E}\begin{pmatrix} \xi \\ \eta \end{pmatrix} = \begin{pmatrix} 0 \\ 0 \end{pmatrix}, \mathrm{D}\begin{pmatrix} \xi \\ \eta \end{pmatrix} = \begin{pmatrix} V_{11} & V_{12} \\ V_{21} & V_{22} \end{pmatrix} \tag{6.140}$$

Note that we write $\mathrm{D}(x) = \mathrm{E}[(x - \mathrm{E}(x))(x - \mathrm{E}(x))']$ for the variance-covariance matrix of a vector variable x and $\mathrm{cov}(x, y) = \mathrm{E}\left[(x - \mathrm{E}(x))(y - \mathrm{E}(y))'\right]$. We wish to predict x given y under different assumptions on the unknown β.

Assumption A_1: β has a prior distribution with $\mathrm{E}(\beta) = \beta_0$ and $\mathrm{D}(\beta) = \Gamma$. (This is sometimes possible using technological considerations as in aerospace tracking problems.)

Assumption A_2: We may choose a noninformative prior for β.

Assumption A_3: We may consider β as an unknown but fixed parameter.

Theorem 6.12 (Rao, 1994) *Under A_1, the minimum mean-dispersion linear predictor (MDLP) of x given y is*

$$\hat{x} = A\beta_0 + C(y - B\beta_o) \tag{6.141}$$

where $C = (A\Gamma B' + V_{12})(B\Gamma B' + V_{22})^{-1}$ with the mean dispersion (MD)

$$A\Gamma A' + V_{11} + C(B\Gamma B' + V_{22})C' - (A\Gamma B' + V_{12})C' - C(B\Gamma A' + V_{21}) \tag{6.142}$$

The proof follows on standard lines of finding the linear regression of one vector variable on another, observing that

$$\begin{aligned} \mathrm{E}(\eta) &= A\beta_0\,, \mathrm{E}(y) = B\beta_0 & (6.143)\\ \mathrm{D}(y) &= B\Gamma B' + V_{22}\,, \mathrm{cov}(x,y) = A\Gamma B' + V_{12}\,. & (6.144) \end{aligned}$$

The solution for the noninformative prior is obtained by taking the limit of (6.141) as $\Gamma^{-1} \to 0$.

The case of $\Gamma = 0$ and a known value of β occurs in economic applications. The solution in such a case is obtained by putting $\Gamma = 0$ and $\beta = \beta_0$ (known value) in (6.141) and (6.142).

If β is a fixed unknown parameter or a random variable with an unknown prior distribution, we may find predictions independent of β as in Theorem 6.12.

Theorem 6.13 (Rao, 1994) *Let $\mathcal{R}(A') \subset \mathcal{R}(B')$. Then a linear predictor of x whose error is independent of the unknown β is (6.147) with the MDLP (6.148) as given below.*

Proof. Let $L'y$ be a predictor of x. The condition that the error $x - L'y$ is independent of β implies

$$A - L'B = 0\,. \tag{6.145}$$

Subject to condition (6.145) we minimize the MDLP

$$\mathrm{D}(x - L'y) = \mathrm{D}(\xi - D'\eta) = V_{11} - L'V_{21} - V_{12}L + L'V_{22}L\,.$$

The minimum is attained when

$$\begin{aligned} V_{22}L - B\Lambda &= V_{21}\\ B'L &= A' \end{aligned} \tag{6.146}$$

where Λ is a Lagrangian matrix multiplier. The optimum L is

$$L = V_{22}^{-1}(V_{21} + BG)$$

where

$$G = (B'V_{22}^{-1}B)^{-1}(A' - B'V_{22}^{-1}V_{21})\,.$$

The predictor, which we call a constrained linear predictor (CLP), of x given y is

$$(V_{12} + G'B')V_{22}^{-1}y \tag{6.147}$$

with the MDLP

$$V_{11} - V_{12}V_{22}^{-1}V_{21} + G'B'V_{22}^{-1}BG\,. \tag{6.148}$$

Note that if β is known, then the second terms in (6.146) and (6.147) are zero, which is the classical case of unconstrained linear prediction.

Theorem 6.14 (Rao 1994) *Suppose that the vector y in (6.139) has the partitioned form*

$$y = \begin{pmatrix} y_1 \\ y_2 \end{pmatrix} = \begin{pmatrix} B_1\beta + \eta_1 \\ B_2\beta + \eta_2 \end{pmatrix}$$

with $\mathcal{R}(A') \subset \mathcal{R}(B_1')$ and $\mathcal{R}(B_2') \subset \mathcal{R}(B_1')$. Let $L_1'y_1$ be the CLP of x given y_1 and $L_2'y_1$ be the CLP of y_2 given y_1. Then the CLP of x given y is

$$D_1y_1 + K(y_2 - L_2'y_1) \tag{6.149}$$

where

$$K = \operatorname{cov}(x - L_1'y_1, y_2 - L_2'y_1)[D(y_2 - L_2'y_1)]^{-1}$$

Proof. Observe that a linear predictor of x on y_1 and y_2 is of the form

$$\hat{x} = L'y_1 + M'(y_2 - L_2'y_1) = L_1'y_1 + (L - L_1)'y_1 + M'(y_2 - L_2'y_1) \tag{6.150}$$

where L and M are arbitrary matrices. Note that if the linear predictor (6.150) is unbiased for β, that is, $\mathrm{E}(\hat{x} - x) = 0$, then $\mathrm{E}(L - L_1)'y_1 = 0$, since $\mathrm{E}(x - L_1'y_1') = 0$ and $\mathrm{E}\, M'(y_2 - L_2'y_1) = 0$. Further, it is easy to verify that

$$\begin{aligned} \operatorname{cov}((L - L_1)'y_1, x - L_1'y_1) &= 0 \\ \operatorname{cov}((L - L_1)'y_1, M(y_2 - L_2'y_1)) &= 0\,. \end{aligned}$$

In such a case

$$\operatorname{cov}(\hat{x} - x, \hat{x} - x) = M'AM + (L - L_1)'C(L - L_1) - DM' - MD' \tag{6.151}$$

where

$$\begin{aligned} A &= \operatorname{cov}(y_2 - L_2'y_1, y_2 - L_2y_1) \\ C &= \operatorname{cov}(y_1, y_1) \\ D &= \operatorname{cov}(y_1 - L_1'y_1, y_2 - L_2'y_1) \end{aligned}$$

Now, by minimizing (6.151) with respect to L and M, we have

$$M' = DA^{-1},\ L = L_1$$

giving the optimum CLP as

$$L_1y_1 + DA^{-1}(y_2 - L_1y_1)\,.$$

6.9.3 Kalman Model

Consider the Kalman model introduced in (6.136) and (6.137),

$$\begin{aligned} x(t) &= Fx(t-1) + \xi(t)\, t = 1, 2, \ldots && (6.152) \\ y(t) &= Hx(t) + \eta(t)\,. && (6.153) \end{aligned}$$

From (6.152), we have

$$\begin{aligned} x(1) &= Fx(0) + \xi(1) \\ x(2) &= Fx(1) + \xi(2) = F^2x(0) + \epsilon(2)\,, \quad \epsilon(2) = F\xi(1) + \xi(2) \\ &\vdots \\ x(t) &= F^t x(0) + \epsilon(t)\,, && (6.154) \end{aligned}$$

where $\epsilon(t) = F^{t-1}\xi(1) + \ldots + \xi(t)$. Similarly,

$$y(t) = HF^t x(0) + \delta(t)\,, \tag{6.155}$$

where $\delta(t) = H\epsilon(t) + \eta(t)$. Writing

$$Y(t) = \begin{pmatrix} y(1) \\ \vdots \\ y(t) \end{pmatrix}, Z(t) = \begin{pmatrix} HF \\ \vdots \\ HF^t \end{pmatrix}, \Delta(t) = \begin{pmatrix} \delta(1) \\ \vdots \\ \delta(t) \end{pmatrix}$$

we have the observational equation

$$Y(t) = Z(t)x(0) + \Delta(t)\, t = 1, 2, \ldots\,. \tag{6.156}$$

Equations (6.154) and (6.156) are of the form (6.138) and (6.139) with $x(0)$ in the place of β; consequently, the results of Theorems 6.12, 6.13 and 6.14 can be used to predict $x(s)$ given $Y(t)$, depending on the assumptions made on $x(0)$. We write such a predictor as $x(s|t)$, and its MDLP by $P(s|t)$. We are seeking $\hat{x}(t|t)$ and its MDLP,

$$P(t|t) = \mathrm{D}\left(x(t) - \hat{x}(t|t)\right). \tag{6.157}$$

We will now show how $\hat{x}(t+1|t+1)$ can be derived knowing $\hat{x}(t|t)$ and its MDLP. From equation (6.152),

$$\begin{aligned} \hat{x}(t+1|t) &= F\hat{x}(t|t) \\ \mathrm{D}(\hat{x}(t+1|t)) &= FP(t|t)F' + V = P(t+1|t)\,. && (6.158) \end{aligned}$$

From the equation (6.153)

$$\begin{aligned} \hat{y}(t+1|t) &= HF\hat{x}(t|t) \\ \mathrm{D}[\hat{y}(t+1|t)] &= HP(t+1|t)H' + W = S(t+1) \\ \mathrm{cov}[\hat{x}(t+1|t), \hat{y}(t+1|t)] &= P(t+1|t)H' = C(t+1)\,. \end{aligned}$$

Then

$$\hat{x}(t+1|t+1) = \hat{x}(t+1|t) + K\hat{y}(t+1|t) \tag{6.159}$$

where

$$\begin{aligned} K &= C(t+1)[S(t+1)]^{-1} \\ D[\hat{x}(t+1|t+1)] &= P(t+1|t) - C(t+1)[S(t+1)]^{-1}C(t+1)'. \end{aligned} \tag{6.160}$$

Following the terminology in the KF theory, we call the second expression on the right-hand side of (6.159) the Kalman gain in prediction, which brings about the reduction in the MDLP by the second term in (6.160). Thus, starting with $\hat{x}(t|t)$, we can derive $\hat{x}(t+1|t+1)$. We begin with $\hat{x}(s|t)$ making an appropriate assumption on $x(0)$ and build up successively $\hat{x}(s+1|t), \ldots, \hat{x}(t|t)$.

6.10 Exercises

Exercise 1. Derive optimal homogeneous and heterogeneous predictors for $X\beta$ and comment on their usefulness.

Exercise 2. If we use $\theta X_* b$ with θ as a fixed scalar to predict $X_*\beta$, find the value of θ that ensures minimum risk under quadratic loss function with A as the loss matrix.

Exercise 3. Discuss the main results related to y_* superiority and $X_*\beta$ superiority of classical and optimal predictors when the disturbances in the model are independently and identically distributed.

Exercise 4. In a classical linear regression model $y = X\beta + \epsilon$, the predictor $\hat{p}_0$ can be used for y_* as well as $E(y_*) = X_*\beta$. Compare the quantities $E(\hat{p}_0 - y_*)'(\hat{p}_0 - y_*)$ and $E(\hat{p}_0 - X_*\beta)'(\hat{p}_0 - X_*\beta)$, and interpret the outcome.

Exercise 5. Let the performance criterion be given as

$$E\left[\lambda(y_* - \hat{p})'W_*^{-1}(y_* - \hat{p}) + (1-\lambda)(X_*\beta - \hat{p})'W_*^{-1}(X_*\beta - \hat{p})\right]$$

$(0 < \lambda < 1)$ for any predictor $\hat{p}$. With respect to it, compare the predictors $\hat{p}_0$ and $\hat{p}_3$.

Exercise 6. Suppose that the predicted values for y are $\hat{y}$ from model $y = X_1\beta_1 + \epsilon_1$ and $\tilde{y}$ from model $y = X_1\beta_1 + X_2\beta_2 + \epsilon_2$. Compare $\hat{y}$ and $\tilde{y}$ with respect to the criteria of unbiasedness and dispersion matrix.

7 Sensitivity Analysis

7.1 Introduction

This chapter discusses the influence of individual observations on the estimated values of parameters and prediction of the dependent variable for given values of regressor variables. Methods for detecting outliers and deviation from normality of the distribution of errors are given in some detail. The material of this chapter is drawn mainly from the excellent book by Chatterjee and Hadi (1988).

7.2 Prediction Matrix

We consider the classical linear model

$$y = X\beta + \epsilon, \quad \epsilon \sim (0, \sigma^2 I)$$

with the usual assumptions. In particular, we assume that the matrix X of order $T \times K$ has the full rank K. The quality of the classical ex-post predictor $\hat{p} = Xb_0 = \hat{y}$ of y with $b_0 = (X'X)^{-1}X'y$, the OLSE (ordinary least-squares estimator), is strongly determined by the $T \times T$-matrix

$$P = X(X'X)^{-1}X' = (p_{ij}), \tag{7.1}$$

which is symmetric and idempotent of $\text{rank}(P) = \text{tr}(P) = \text{tr}(I_K) = K$. The matrix $M = I - P$ is also symmetric and idempotent and has $\text{rank}(M) =$

$T - K$. The estimated residuals are defined by

$$\begin{aligned} \hat{\epsilon} = (I - P)y &= y - Xb_0 \\ &= y - \hat{y} = (I - P)\epsilon \,. \end{aligned} \tag{7.2}$$

Definition 7.1 (Chatterjee and Hadi, 1988) *The matrix P given in (7.1) is called the prediction matrix, and the matrix $I - P$ is called the residuals matrix.*

Remark: The matrix P is sometimes called the *hat matrix* because it maps y onto $\hat{y}$.

The (i, j)th element of the matrix P is denoted by p_{ij} where

$$p_{ij} = p_{ji} = x_j'(X'X)^{-1}x_i \quad (i, j = 1, \ldots, T) \,. \tag{7.3}$$

The ex-post predictor $\hat{y} = Xb_0 = Py$ has the dispersion matrix

$$V(\hat{y}) = \sigma^2 P \,. \tag{7.4}$$

Therefore, we obtain (denoting the ith component of $\hat{y}$ by $\hat{y}_i$ and the ith component of $\hat{\epsilon}$ by $\hat{\epsilon}_i$)

$$\text{var}(\hat{y}_i) = \sigma^2 p_{ii} \,, \tag{7.5}$$

$$\text{V}(\hat{\epsilon}) = \text{V}\left((I - P)y\right) = \sigma^2(I - P) \,, \tag{7.6}$$

$$\text{var}(\hat{\epsilon}_i) = \sigma^2(1 - p_{ii}) \tag{7.7}$$

and for $i \neq j$

$$\text{cov}(\hat{\epsilon}_i, \hat{\epsilon}_j) = -\sigma^2 p_{ij} \,. \tag{7.8}$$

The correlation coefficient between $\hat{\epsilon}_i$ and $\hat{\epsilon}_j$ then becomes

$$\rho_{ij} = \text{corr}(\hat{\epsilon}_i, \hat{\epsilon}_j) = \frac{-p_{ij}}{\sqrt{1 - p_{ii}}\sqrt{1 - p_{jj}}} \,. \tag{7.9}$$

Thus the covariance matrices of the predictor Xb_0 and the estimator of error $\hat{\epsilon}$ are entirely determined by P. Although the disturbances ϵ_i of the model are i.i.d., the estimated residuals $\hat{\epsilon}_i$ are not identically distributed and, moreover, they are correlated. Observe that

$$\hat{y}_i = \sum_{j=1}^{T} p_{ij} y_i = p_{ii} y_i + \sum_{j \neq i} p_{ij} y_j \quad (i = 1, \ldots, T) \,, \tag{7.10}$$

implying that

$$\frac{\partial \hat{y}_i}{\partial y_i} = p_{ii} \quad \text{and} \quad \frac{\partial \hat{y}_i}{\partial y_j} = p_{ij} \,. \tag{7.11}$$

Therefore, p_{ii} can be interpreted as the amount of *leverage* each value y_i has in determining $\hat{y}_i$ regardless of the realized value y_i. The second relation of (7.11) may be interpreted, analogously, as the influence of y_j in determining $\hat{y}_i$.

Decomposition of P

Assume that X is partitioned as $X = (X_1, X_2)$ with $X_1 : T \times p$ and $\text{rank}(X_1) = p$, $X_2 : T \times (K-p)$ and $\text{rank}(X_2) = K - p$. Let $P_1 = X_1(X_1'X_1)^{-1}X_1'$ be the (idempotent) prediction matrix for X_1, and let $W = (I - P_1)X_2$ be the projection of the columns of X_2 onto the orthogonal complement of X_1. Then the matrix $P_2 = W(W'W)^{-1}W'$ is the prediction matrix for W, and P can be expressed as (using Theorem A.45)

$$P = P_1 + P_2 \tag{7.12}$$

or

$$X(X'X)^{-1}X' = X_1(X_1'X_1)^{-1}X_1' + (I - P_1)X_2[X_2'(I-P_1)X_2]^{-1}X_2'(I-P_1). \tag{7.13}$$

Equation (7.12) shows that the prediction matrix P can be decomposed into the sum of two (or more) prediction matrices. Applying the decomposition (7.13) to the linear model including a dummy variable, that is, $y = 1\alpha + X\beta + \epsilon$, we obtain

$$P = \frac{11'}{T} + \tilde{X}(\tilde{X}'\tilde{X})^{-1}\tilde{X}' = P_1 + P_2 \tag{7.14}$$

and

$$p_{ii} = \frac{1}{T} + \tilde{x}_i'(\tilde{X}'\tilde{X})^{-1}\tilde{x}_i\,, \tag{7.15}$$

where $\tilde{X} = (x_{ij} - \bar{x}_i)$ is the matrix of the mean-corrected x-values. This is seen as follows. Application of (7.13) to $(1, X)$ gives

$$P_1 = 1(1'1)^{-1}1' = \frac{11'}{T} \tag{7.16}$$

and

$$\begin{aligned} W = (I - P_1)X &= X - 1\left(\frac{1}{T}1'X\right) \\ &= X - (1\bar{x}_1, 1\bar{x}_2, \ldots, 1\bar{x}_K) \\ &= (x_1 - \bar{x}_1, \ldots, x_K - \bar{x}_K)\,. \end{aligned} \tag{7.17}$$

The size and the range of the elements of P are measures for the influence of data on the predicted values $\hat{y}_t$. Because of the symmetry of P, we have $p_{ij} = p_{ji}$, and the idempotence of P implies

$$p_{ii} = \sum_{j=1}^{n} p_{ij}^2 = p_{ii}^2 + \sum_{j \neq i} p_{ij}^2\,. \tag{7.18}$$

From this equation we obtain the important property

$$0 \le p_{ii} \le 1\,. \tag{7.19}$$

Reformulating (7.18):

$$p_{ii} = p_{ii}^2 + p_{ij}^2 + \sum_{k \neq i,j} p_{ik}^2 \quad (j \text{ fixed}), \tag{7.20}$$

which implies that $p_{ij}^2 \leq p_{ii}(1 - p_{ii})$, and therefore, using (7.19), we obtain

$$-0.5 \leq p_{ij} \leq 0.5 \quad (i \neq j). \tag{7.21}$$

If X contains a column of constants (1 or $c1$), then in addition to (7.19) we obtain

$$p_{ii} \geq T^{-1} \quad (\text{for all } i) \tag{7.22}$$

and

$$P1 = 1. \tag{7.23}$$

Relationship (7.22) is a direct consequence of (7.15). Since $\tilde{X}'1 = 0$ and hence $P_2 1 = 0$, we get from (7.14)

$$P1 = 1\frac{T}{T} + 0 = 1. \tag{7.24}$$

The diagonal elements p_{ii} and the off-diagonal elements p_{ij} $(i \neq j)$ are interrelated according to properties (i)–(iv) as follows (Chatterjee and Hadi, 1988, p. 19):

(i) If $p_{ii} = 1$ or $p_{ii} = 0$, then $p_{ij} = 0$.

Proof: Use (7.18).

(ii) We have

$$(p_{ii}p_{jj} - p_{ij}^2) \geq 0. \tag{7.25}$$

Proof: Since P is nonnegative definite, we have $x'Px \geq 0$ for all x, and especially for $x_{ij} = (0, \ldots, 0, x_i, 0, x_j, 0, \ldots, 0)'$, where x_i and x_j occur at the ith and jth positions $(i \neq j)$. This gives

$$x_{ij}'Px_{ij} = (x_i, x_j)\begin{pmatrix} p_{ii} & p_{ij} \\ p_{ji} & p_{jj} \end{pmatrix}\begin{pmatrix} x_i \\ x_j \end{pmatrix} \geq 0.$$

Therefore, $P_{ij} = \begin{pmatrix} p_{ii} & p_{ij} \\ p_{ji} & p_{jj} \end{pmatrix}$ is nonnegative definite, and hence its determinant is nonnegative:

$$|P_{ij}| = p_{ii}p_{jj} - p_{ij}^2 \geq 0.$$

(iii) We have

$$(1 - p_{ii})(1 - p_{jj}) - p_{ij}^2 \geq 0. \tag{7.26}$$

Proof: Analogous to (ii), using $I - P$ instead of P leads to (7.26).

(iv) We have

$$p_{ii} + \frac{\hat{\epsilon}_i^2}{\hat{\epsilon}'\hat{\epsilon}} \leq 1 \,. \tag{7.27}$$

Proof: Let $Z = (X, y)$, $P_X = X(X'X)^{-1}X'$ and $P_Z = Z(Z'Z)^{-1}Z'$. Then (7.13) and (7.2) imply

$$\begin{aligned} P_Z &= P_X + \frac{(I - P_X)yy'(I - P_X)}{y'(I - P_X)y} \\ &= P_X + \frac{\hat{\epsilon}\hat{\epsilon}'}{\hat{\epsilon}'\hat{\epsilon}} \,. \end{aligned} \tag{7.28}$$

Hence we find that the ith diagonal element of P_Z is equal to $p_{ii} + \hat{\epsilon}_i^2/\hat{\epsilon}'\hat{\epsilon}$. If we now use (7.19), then (7.27) follows.

Interpretation: If a diagonal element p_{ii} is close to either 1 or 0, then the elements p_{ij} (for all $j \neq i$) are close to 0.

The classical predictor of y is given by $\hat{y} = Xb_0 = Py$, and its first component is $\hat{y}_1 = \sum p_{1j}y_j$. If, for instance, $p_{11} = 1$, then $\hat{y}_1$ is fully determined by the observation y_1. On the other hand, if p_{11} is close to 0, then y_1 itself and all the other observations $y_2, \ldots, y_T$ have low influence on $\hat{y}_1$.

Relationship (7.27) indicates that if p_{ii} is large, then the standardized residual $\hat{\epsilon}_i/\hat{\epsilon}'\hat{\epsilon}$ becomes small.

Conditions for p_{ii} to be Large

If we assume the simple linear model

$$y_t = \alpha + \beta x_t + \epsilon_t, \quad t = 1, \ldots, T \,,$$

then we obtain from (7.15)

$$p_{ii} = \frac{1}{T} + \frac{(x_i - \bar{x})^2}{\sum_{t=1}^{T}(x_t - \bar{x})^2} \,. \tag{7.29}$$

The size of p_{ii} is dependent on the distance $|x_i - \bar{x}|$. Therefore, the influence of any observation (y_i, x_i) on $\hat{y}_i$ will be increasing with increasing distance $|x_i - \bar{x}|$.

In the case of multiple regression we have a similar relationship. Let λ_i denote the eigenvalues and γ_i $(i = 1, \ldots, K)$ the orthonormal eigenvectors of the matrix $X'X$. Furthermore, let θ_{ij} be the angle between the column vector x_i and the eigenvector γ_j $(i, j = 1, \ldots, K)$. Then we have

$$p_{ij} = \|x_i\| \, \|x_j\| \sum_{r=1}^{K} \lambda_r^{-1} \cos\theta_{ir} \cos\theta_{rj} \tag{7.30}$$

and

$$p_{ii} = x_i' x_i \sum_{r=1}^{K} \lambda_r^{-1} (\cos \theta_{ir})^2. \tag{7.31}$$

The proof is straightforward by using the spectral decomposition of $X'X = \Gamma \Lambda \Gamma'$ and the definition of p_{ij} and p_{ii} (cf. (7.3)), that is,

$$\begin{aligned} p_{ij} &= x_i'(X'X)^{-1}x_j = x_i' \Gamma \Lambda^{-1} \Gamma' x_j \\ &= \sum_{r=1}^{K} \lambda_r^{-1} x_i' \gamma_r x_j' \gamma_r \\ &= \|x_i\| \, \|x_j\| \sum \lambda_r^{-1} \cos \theta_{ir} \cos \theta_{jr} , \end{aligned}$$

where $\|x_i\| = (x_i' x_i)^{\frac{1}{2}}$ is the norm of the vector x_i.

Therefore, p_{ii} tends to be large if

(i) $x_i' x_i$ is large in relation to the square of the vector norm $x_j' x_j$ of the other vectors x_j (i.e., x_i is far from the other vectors x_j) or

(ii) x_i is parallel (or almost parallel) to the eigenvector corresponding to the smallest eigenvalue. For instance, let λ_K be the smallest eigenvalue of $X'X$, and assume x_i to be parallel to the corresponding eigenvector γ_K. Then we have $\cos \theta_{iK} = 1$, and this is multiplied by λ_K^{-1}, resulting in a large value of p_{ii} (cf. Cook and Weisberg, 1982, p. 13).

Multiple X-Rows

In the statistical analysis of linear models there are designs (as, e.g., in the analysis of variance of factorial experiments) that allow a repeated response y_t for the same fixed x-vector. Let us assume that the ith row $(x_{i1}, \ldots, x_{iK})$ occurs a times in X. Then it holds that

$$p_{ii} \le a^{-1}. \tag{7.32}$$

This property is a direct consequence of (7.20). Let $J = \{j : x_i = x_j\}$ denote the set of indices of rows identical to the ith row. This implies $p_{ij} = p_{ii}$ for $j \in J$, and hence (7.20) becomes

$$p_{ii} = a p_{ii}^2 + \sum_{j \notin J} p_{ij}^2 \ge a p_{ii}^2 ,$$

including (7.32).

Example 7.1: We consider the matrix

$$X = \begin{pmatrix} 1 & 2 \\ 1 & 2 \\ 1 & 1 \end{pmatrix}$$

with $K = 2$ and $T = 3$, and calculate

$$\begin{aligned} X'X &= \begin{pmatrix} 3 & 5 \\ 5 & 9 \end{pmatrix}, \quad |X'X| = 2, \quad (X'X)^{-1} = \frac{1}{2}\begin{pmatrix} 9 & -5 \\ -5 & 3 \end{pmatrix}, \\ P &= X(X'X)^{-1}X' = \begin{pmatrix} 0.5 & 0.5 & 0 \\ 0.5 & 0.5 & 0 \\ 0 & 0 & 1 \end{pmatrix}. \end{aligned}$$

The first row and the second row of P coincide. Therefore we have $p_{11} \le \frac{1}{2}$. Inserting $\bar{x} = \frac{5}{3}$ and $\sum_{t=1}^{3}(x_t - \bar{x})^2 = \frac{6}{9}$ in (7.29) results in

$$p_{ii} = \frac{1}{3} + \frac{(x_i - \bar{x})^2}{\sum(x_t - \bar{x}^2)},$$

that is, $p_{11} = p_{22} = \frac{1}{3} + \frac{1/9}{6/9} = \frac{1}{2}$ and $p_{33} = \frac{1}{3} + \frac{4/9}{6/9} = 1$.

7.3 The Effect of a Single Observation on the Estimation of Parameters

In Chapter 3 we investigated the effect of one variable X_i (or sets of variables) on the fit of the model. The effect of including or excluding columns of X is measured and tested by the statistic F.

In this section we wish to investigate the effect of rows (y_t, x_t') instead of columns x_t on the estimation of β. Usually, not all observations (y_t, x_t') have equal influence in a least-squares fit and on the estimator $(X'X)^{-1}X'y$. It is important for the data analyst to be able to identify observations that individually or collectively have excessive influence compared to other observations. Such rows of the data matrix (y, X) will be called *influential observations*.

The measures for the goodness of fit of a model are mainly based on the residual sum of squares

$$\begin{aligned} \hat{\epsilon}'\hat{\epsilon} &= (y - Xb)'(y - Xb) \\ &= y'(I - P)y = \epsilon'(I - P)\epsilon. \end{aligned} \tag{7.33}$$

This quadratic form and the residual vector $\hat{\epsilon} = (I - P)\epsilon$ itself may change considerably if an observation is excluded or added. Depending on the change in $\hat{\epsilon}$ or $\hat{\epsilon}'\hat{\epsilon}$, an observation may be identified as influential or not. In the literature, a large number of statistical measures have been proposed for diagnosing influential observations. We describe some of them and focus attention on the detection of a single influential observation. A more detailed presentation is given by Chatterjee and Hadi (1988, Chapter 4).

7.3.1 Measures Based on Residuals

Residuals play an important role in regression diagnostics, since the ith residual $\hat{\epsilon}_i$ may be regarded as an appropriate guess for the unknown random error ϵ_i.

The relationship $\hat{\epsilon} = (I - P)\epsilon$ implies that $\hat{\epsilon}$ would even be a good estimator for ϵ if $(I - P) \approx I$, that is, if all p_{ij} are sufficiently small and if the diagonal elements p_{ii} are of the same size. Furthermore, even if the random errors ϵ_i are i.i.d. (i.e., $\mathrm{E}\,\epsilon\epsilon' = \sigma^2 I$), the identity $\hat{\epsilon} = (I - P)\epsilon$ indicates that the residuals are not independent (unless P is diagonal) and do not have the same variance (unless the diagonal elements of P are equal). Consequently, the residuals can be expected to be reasonable substitutes for the random errors if

(i) the diagonal elements p_{ii} of the matrix P are almost equal, that is, the rows of X are almost homogeneous, implying homogeneity of variances of the $\hat{\epsilon}_t$, and

(ii) the off-diagonal elements p_{ij} $(i \neq j)$ are sufficiently small, implying uncorrelated residuals.

Hence it is preferable to use transformed residuals for diagnostic purposes. That is, instead of $\hat{\epsilon}_i$ we may use a transformed standardized residual $\tilde{\epsilon}_i = \hat{\epsilon}_i/\sigma_i$, where σ_i is the standard deviation of the ith residual. Several standardized residuals with specific diagnostic power are obtained by different choices of $\hat{\sigma}_i$ (Chatterjee and Hadi, 1988, p. 73).

(i) *Normalized Residual.* Replacing σ_i by $(\hat{\epsilon}'\hat{\epsilon})^{\frac{1}{2}}$ gives

$$a_i = \frac{\hat{\epsilon}_i}{\sqrt{\hat{\epsilon}'\hat{\epsilon}}} \quad (i = 1, \ldots, T). \tag{7.34}$$

(ii) *Standardized Residual.* Replacing σ_i by $s = \sqrt{\hat{\epsilon}'\hat{\epsilon}/(T - K)}$, we obtain

$$b_i = \frac{\hat{\epsilon}_i}{s} \quad (i = 1, \ldots, T). \tag{7.35}$$

(iii) *Internally Studentized Residual.* With $\hat{\sigma}_i = s\sqrt{1 - p_{ii}}$ we obtain

$$r_i = \frac{\hat{\epsilon}_i}{s\sqrt{1 - p_{ii}}} \quad (i = 1, \ldots, T). \tag{7.36}$$

(iv) *Externally Studentized Residual.* Let us assume that the ith observation is omitted. This fact is indicated by writing the index (i) in brackets. Using this indicator, we may define the estimator of σ_i^2 when the ith row (y_i, x_i') is omitted as

$$s_{(i)}^2 = \frac{y_{(i)}'(I - P_{(i)})y_{(i)}}{T - K - 1}, \quad (i = 1, \ldots, T). \tag{7.37}$$

If we take $\hat{\sigma}_i = s_{(i)}\sqrt{1-p_{ii}}$, the ith externally Studentized residual is defined as

$$r_i^* = \frac{\hat{\epsilon}_i}{s_{(i)}\sqrt{1-p_{ii}}} \quad (i = 1, \ldots, T). \tag{7.38}$$

7.3.2 Algebraic Consequences of Omitting an Observation

Let $(y_{(i)}, X_{(i)})$ denote the remaining data matrix when the ith observation vector $(y_i, x_{i1}, \ldots, x_{iK})$ is omitted.

Using the rowwise representation of the matrix $X' = (x_1, \ldots, x_T)$, we obtain

$$X'X = \sum_{t=1}^{T} x_t x_t' = X_{(i)}'X_{(i)} + x_i x_i'. \tag{7.39}$$

Assume that $\text{rank}(X_{(i)}) = K$. Then the inverse of $X_{(i)}'X_{(i)}$ may be calculated using Theorem A.18 (iv) (if $x_i'(X'X)^{-1}x_i \neq 1$ holds) as

$$(X_{(i)}'X_{(i)})^{-1} = (X'X)^{-1} + \frac{(X'X)^{-1}x_i x_i'(X'X)^{-1}}{1 - x_i'(X'X)^{-1}x_i}. \tag{7.40}$$

This implies that the following bilinear forms become functions of the elements of the matrix P:

$$x_r'(X_{(i)}'X_{(i)})^{-1}x_k = p_{rk} + \frac{p_{ri}p_{ik}}{1 - p_{ii}} \quad (r, k \neq i). \tag{7.41}$$

The rth diagonal element of the prediction matrix

$$P_{(i)} = X_{(i)}(X_{(i)}'X_{(i)})^{-1}X_{(i)}'$$

then is

$$p_{rr(i)} = p_{rr} + \frac{p_{ri}^2}{1 - p_{ii}} \quad (r \neq i). \tag{7.42}$$

From (7.42), we observe that $p_{rr(i)}$ may be large if either p_{rr} or p_{ii} is large and/or if p_{ri} is large. Let us look at the case where the ith row of X occurs twice. If the rth row and the ith row are identical, then (7.42) reduces to

$$p_{rr(i)} = \frac{p_{ii}}{1 - p_{ii}}. \tag{7.43}$$

If the ith row is identical to the rth row, then (cf. (7.32)) we get $p_{ii} \leq 0.5$. If p_{ii} $(= p_{rr})$ is near 0.5, this implies that $p_{rr(i)}$ $(= p_{ii(r)})$ will be close to 1 and the influence of the ith observation on $\hat{y}_r$ will be undetected. This is called the *masking effect*.

When the ith observation is omitted, then in the reduced data set the OLSE for β may be written as

$$\hat{\beta}_{(i)} = (X_{(i)}'X_{(i)})^{-1}X_{(i)}'y_{(i)}. \tag{7.44}$$

Therefore, the ith residual is of the form

$$\begin{aligned}
\hat{\epsilon}_{i(i)} &= y_i - x_i'\hat{\beta}_{(i)} = y_i - x_i'(X_{(i)}'X_{(i)})^{-1}X_{(i)}'y_{(i)} \\
&= y_i - x_i'\left[(X'X)^{-1} + \frac{(X'X)^{-1}x_ix_i'(X'X)^{-1}}{1-p_{ii}}\right](X'y - x_iy_i) \\
&= y_i - x_ib + p_{ii}y_i - \frac{p_{ii}x_i'b}{1-p_{ii}} + \frac{p_{ii}^2y_i}{1-p_{ii}} \\
&= y_i - \hat{y}_i + p_{ii}y_i - \frac{p_{ii}\hat{y}_i}{1-p_{ii}} + \frac{p_{ii}^2y_i}{1-p_{ii}} \\
&= \frac{y_i - \hat{y}_i}{1-p_{ii}} = \frac{\hat{\epsilon}_i}{1-p_{ii}}\,.
\end{aligned} \tag{7.45}$$

Hence, the difference between the OLSEs in the full and the reduced data sets, respectively, is seen to be

$$b - \hat{\beta}_{(i)} = \frac{(X'X)^{-1}x_i\hat{\epsilon}_i}{1-p_{ii}}\,, \tag{7.46}$$

which can be easily deduced by combining equations (7.44) and (7.40). Based on formula (7.46) we may investigate the interrelationships among the four types of residuals defined before. Equations (7.34) and (7.35) provide us with the relationship between the ith standardized residual b_i and the ith normalized residual a_i:

$$b_i = a_i\sqrt{T-K}\,. \tag{7.47}$$

In the same manner it is proved that the ith internally Studentized residual r_i is proportional to b_i, and hence to a_i, in the following manner:

$$r_i = \frac{b_i}{\sqrt{1-p_{ii}}} = a_i\sqrt{\frac{T-K}{1-p_{ii}}}\,. \tag{7.48}$$

7.3.3 Detection of Outliers

To find the relationships between the ith internally and externally Studentized residuals, we need to write $(T-K)s^2 = y'(I-P)y$ as a function of $s_{(i)}^2$, that is, as $(T-K-1)s_{(i)}^2 = y_{(i)}'(I-P_{(i)})y_{(i)}$. This is done by noting that omitting the ith observation is equivalent to fitting the *mean-shift outlier model*

$$y = X\beta + e_i\delta + \epsilon\,, \tag{7.49}$$

where e_i (see Definition A.8) is the ith unit vector; that is, $e_i' = (0,\dots,0,1,0,\dots,0)$. The argument is as follows. Suppose that either y_i or $x_i'\beta$ deviates systematically by δ from the model $y_i = x_i'\beta + \epsilon_i$. Then the ith observation $(y_i, x_i'\beta)$ would have a different intercept than the remaining observations and $(y_i, x_i'\beta)$ would hence be an outlier. To check this fact, we test the hypothesis

$$H_0\colon \delta = 0 \quad (\text{i.e., } \mathrm{E}(y) = X\beta)$$

against the alternative

$$H_1\colon \delta \neq 0 \quad (\text{i.e., } \mathrm{E}(y) = X\beta + e_i\delta)$$

using the likelihood-ratio test statistic

$$F_i = \frac{(SSE(H_0) - SSE(H_1))/1}{SSE(H_1)/(T - K - 1)}, \tag{7.50}$$

where $SSE(H_0)$ is the residual sum of squares in the model $y = X\beta + \epsilon$ containing all the T observations:

$$SSE(H_0) = y'(I - P)y = (T - K)s^2.$$

$SSE(H_1)$ is the residual sum of squares in the model $y = X\beta + e_i\delta + \epsilon$. Applying relationship (7.13), we obtain

$$(X, e_i)[(X, e_i)'(X, e_i)]^{-1}(X, e_i)' = P + \frac{(I - P)e_i e_i'(I - P)}{e_i'(I - P)e_i}. \tag{7.51}$$

The left-hand side may be interpreted as the prediction matrix $P_{(i)}$ when the ith observation is omitted. Therefore, we may conclude that

$$\begin{aligned} SSE(H_1) &= (T - K - 1)s_{(i)}^2 = y_{(i)}'(I - P_{(i)})y_{(i)} \\ &= y'\left(I - P - \frac{(I - P)e_i e_i'(I - P)}{e_i'(I - P)e_i}\right)y \\ &= SSE(H_0) - \frac{\hat{\epsilon}_i^2}{1 - p_{ii}} \end{aligned} \tag{7.52}$$

holds, where we have made use of the following relationships: $(I - P)y = \hat{\epsilon}$ and $e_i'\hat{\epsilon} = \hat{\epsilon}_i$ and, moreover, $e_i'Ie_i = 1$ and $e_i'Pe_i = p_{ii}$.

Therefore, the test statistic (7.50) may be written as

$$F_i = \frac{\hat{\epsilon}_i^2}{(1 - p_{ii})s_{(i)}^2} = (r_i^*)^2, \tag{7.53}$$

where r_i^* is the ith externally Studentized residual.

Theorem 7.2 (Beckman and Trussel, 1974) *Assume the design matrix X is of full column rank K.*

(i) *If* $\mathrm{rank}(X_{(i)}) = K$ *and* $\epsilon \sim N_T(0, \sigma^2 I)$, *then the externally Studentized residuals* r_i^* $(i = 1, \ldots, T)$ *are* t_{T-K-1}*-distributed.*

(ii) *If* $\mathrm{rank}(X_{(i)}) = K - 1$, *then the residual* r_i^* *is not defined.*

Assume $\mathrm{rank}(X_{(i)}) = K$. Then Theorem 7.2 (i) implies that the test statistic $(r_i^*)^2 = F_i$ from (7.53) is distributed as central $F_{1,T-K-1}$ under H_0 and noncentral $F_{1,T-K-1}(\delta^2(1 - p_{ii})\sigma^2)$ under H_1, respectively. The noncentrality parameter decreases (tending to zero) as p_{ii} increases. That is, the detection of outliers becomes difficult when p_{ii} is large.

Relationships between r_i^* and r_i

Equations (7.52) and (7.36) imply that

$$\begin{aligned} s_{(i)}^2 &= \frac{(T-K)s^2}{T-K-1} - \frac{\hat{\epsilon}_i^2}{(T-K-1)(1-p_{ii})} \\ &= s^2 \left(\frac{T-K-r_i^2}{T-K-1} \right) \end{aligned} \qquad (7.54)$$

and, hence,

$$r_i^* = r_i \sqrt{\frac{T-K-1}{T-K-r_i^2}} \,. \qquad (7.55)$$

Inspecting the Four Types of Residuals

The normalized, standardized, and internally and externally Studentized residuals are transformations of the OLS residuals $\hat{\epsilon}_i$ according to $\hat{\epsilon}_i/\sigma_i$, where σ_i is estimated by the corresponding statistics defined in (7.34) to (7.37), respectively. The normalized as well as the standardized residuals a_i and b_i, respectively, are easy to calculate but they do not measure the variability of the variances of the $\hat{\epsilon}_i$. Therefore, in the case of large differences in the diagonal elements p_{ii} of P or, equivalently (cf. (7.7)), of the variances of $\hat{\epsilon}_i$, application of the Studentized residuals r_i or r_i^* is well recommended. The externally Studentized residuals r_i^* are advantageous in the following sense:

(i) $(r_i^*)^2$ may be interpreted as the F-statistic for testing the significance of the unit vector e_i in the mean-shift outlier model (7.49).

(ii) The internally Studentized residual r_i follows a beta distribution (cf. Chatterjee and Hadi, 1988, p. 76) whose quantiles are not included in standard textbooks.

(iii) If $r_i^2 \to T-K$ then $r_i^{*2} \to \infty$ (cf. (7.55)). Hence, compared to r_i, the residual r_i^* is more sensitive to outliers.

Example 7.2: We go back to Section 3.8.3 and consider the following data set including the response vector y and the variable X_4 (which was detected to be the most important variable compared to X_1, X_2, and X_3):

$$\begin{pmatrix} y \\ X_4 \end{pmatrix}' = \begin{pmatrix} 18 & 47 & 125 & 40 & 37 & 20 & 24 & 35 & 59 & 50 \\ -10 & 19 & 100 & 17 & 13 & 10 & 5 & 22 & 35 & 20 \end{pmatrix}.$$

TABLE 7.1. Internally and externally Studentized residuals

i	$1-p_{ii}$	$\hat{y}_i$	$\hat{\epsilon}_i$	r_i^2	$r_i^{*2}=F_i$
1	0.76	11.55	6.45	1.15	1.18
2	0.90	41.29	5.71	0.76	0.74
3	0.14	124.38	0.62	0.06	0.05
4	0.90	39.24	0.76	0.01	0.01
5	0.89	35.14	1.86	0.08	0.07
6	0.88	32.06	-12.06	3.48	5.38
7	0.86	26.93	-2.93	0.21	0.19
8	0.90	44.37	-9.37	2.05	2.41
9	0.88	57.71	1.29	0.04	0.03
10	0.90	42.32	7.68	1.38	1.46

Including the dummy variable 1, the matrix $X = (1, X_4)$ gives ($T = 10, K = 2$)

$$\begin{aligned} X'X &= \begin{pmatrix} 10 & 231 \\ 231 & 13153 \end{pmatrix}, \quad |X'X| = 78169 \\ (X'X)^{-1} &= \frac{1}{78169}\begin{pmatrix} 13153 & -231 \\ -231 & 10 \end{pmatrix}. \end{aligned}$$

The diagonal elements of $P = X(X'X)^{-1}X'$ are

$$\begin{aligned} p_{11} &= 0.24\,, & p_{66} &= 0.12\,, \\ p_{22} &= 0.10\,, & p_{77} &= 0.14\,, \\ p_{33} &= 0.86\,, & p_{88} &= 0.10\,, \\ p_{44} &= 0.10\,, & p_{99} &= 0.12\,, \\ p_{55} &= 0.11\,, & p_{1010} &= 0.11\,, \end{aligned}$$

where $\sum p_{ii} = 2 = K = \operatorname{tr} P$ and $p_{ii} \geq \frac{1}{10}$ (cf. (7.22)). The value p_{33} differs considerably from the other p_{ii}. To calculate the test statistic F_i (7.53), we have to find the residuals $\hat{\epsilon}_i = y_i - \hat{y}_i = y_i - x_i'b_0$, where $\hat{\beta} = (21.80; 1.03)$ (cf. Section 3.8.3, first step of the procedure). The results are summarized in Table 7.1.

The residuals r_i^2 and r_i^{*2} are calculated according to (7.36) and (7.55), respectively. The standard deviation was found to be $s = 6.9$.

From Table B.2 (Appendix B) we have the quantile $F_{1,7,0.95} = 5.59$, implying that the null hypothesis H_0: "ith observation $(y_i, 1, x_{4i})$ is not an outlier" is not rejected for all $i = 1, \ldots, 10$. The third observation may be identified as a high-leverage point having remarkable influence on the regression line. Taking $\bar{x}_4 = 23.1$ and $s^2(x_4) = 868.544$ from Section 3.8.3 and applying formula (7.29), we obtain

$$p_{33} = \frac{1}{10} + \frac{(100-23.1)^2}{\sum_{t=1}^{10}(x_t - \bar{x})^2} = \frac{1}{10} + \frac{76.9^2}{9 \cdot 868.544}$$

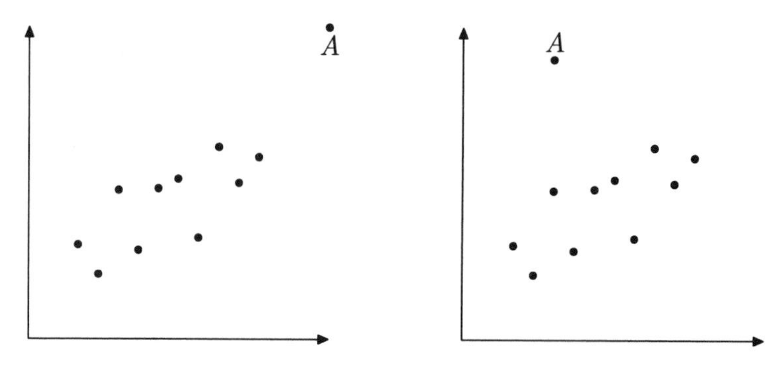

FIGURE 7.1. High-leverage point A

FIGURE 7.2. Outlier A

$$= 0.10 + 0.76 = 0.86.$$

Therefore, the large value of $p_{33} = 0.86$ is mainly caused by the large distance between x_{43} and the mean value $\bar{x}_4 = 23.1$.

Figures 7.1 and 7.2 show typical situations for points that are very far from the others. Outliers correspond to extremely large residuals, but high-leverage points correspond to extremely small residuals in each case when compared with other residuals.

7.4 Diagnostic Plots for Testing the Model Assumptions

Many graphical methods make use of the residuals to detect deviations from the stated assumptions. From experience one may prefer graphical methods over numerical tests based on residuals. The most common residual plots are

(i) empirical distribution of the residuals, stem-and-leaf diagrams, Box-Whisker plots;

(ii) normal probability plots;

(iii) residuals versus fitted values or residuals versus x_i plots (see Figures 7.3 and 7.4).

These plots are useful in detecting deviations from assumptions made on the linear model.

The externally Studentized residuals also may be used to detect violation of normality. If normality is present, then approximately 68% of the residuals r_i^* will be in the interval $[-1, 1]$. As a rule of thumb, one may identify the ith observation as an outlier if $|r_i^*| > 3$.

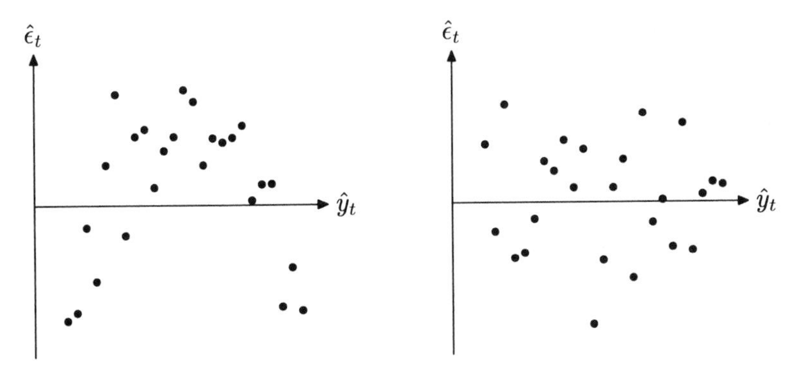

FIGURE 7.3. Plot of the residuals $\hat{\epsilon}_t$ versus the fitted values $\hat{y}_t$ (suggests deviation from linearity)

FIGURE 7.4. No violation of linearity

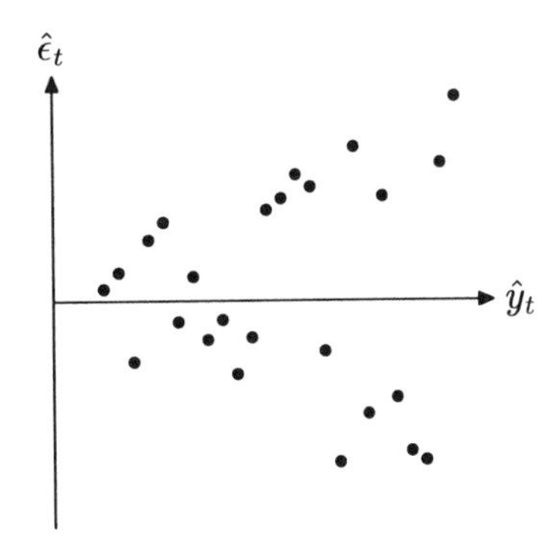

FIGURE 7.5. Signals for heteroscedasticity

If the assumptions of the model are correctly specified, then we have

$$\operatorname{cov}(\hat{\epsilon}, \hat{y}') = \mathrm{E}\left((I - P)\epsilon\epsilon' P\right) = 0 \,. \tag{7.56}$$

Therefore, plotting $\hat{\epsilon}_t$ versus $\hat{y}_t$ (Figures 7.3 and 7.4) exhibits a random scatter of points. A situation as in Figure 7.4 is called a null plot. A plot as in Figure 7.5 indicates heteroscedasticity of the covariance matrix.

7.5 Measures Based on the Confidence Ellipsoid

Under the assumption of normally distributed disturbances, that is, $\epsilon \sim N(0, \sigma^2 I)$, we have $b_0 = (X'X)^{-1}X'y \sim N(\beta, \sigma^2(X'X)^{-1})$ and

$$\frac{(\beta - b_0)'(X'X)(\beta - b_0)}{Ks^2} \sim F_{K,T-K} \,. \tag{7.57}$$

Then the inequality

$$\frac{(\beta - b_0)'(X'X)(\beta - b_0)}{Ks^2} \leq F_{K,T-K,1-\alpha} \tag{7.58}$$

defines a $100(1-\alpha)\%$ confidence ellipsoid for β centered at b_0. The influence of the ith observation (y_i, x_i') can be measured by the change of various parameters of the ellipsoid when the ith observation is omitted. Strong influence of the ith observation would be equivalent to significant change of the corresponding measure.

Cook's Distance

Cook (1977) suggested the index

$$C_i = \frac{(b - \hat{\beta}_{(i)})'X'X(b - \hat{\beta}_{(i)})}{Ks^2} \tag{7.59}$$

$$= \frac{(\hat{y} - \hat{y}_{(i)})'(\hat{y} - \hat{y}_{(i)})}{Ks^2} \quad (i = 1, \ldots, T) \tag{7.60}$$

to measure the influence of the ith observation on the center of the confidence ellipsoid or, equivalently, on the estimated coefficients $\hat{\beta}_{(i)}$ (7.44) or the predictors $\hat{y}_{(i)} = X\hat{\beta}_{(i)}$. The measure C_i can be thought of as the scaled distance between b and $\hat{\beta}_{(i)}$ or $\hat{y}$ and $\hat{y}_{(i)}$, respectively. Using (7.46), we immediately obtain the following relationship:

$$C_i = \frac{1}{K}\frac{p_{ii}}{1 - p_{ii}} r_i^2, \tag{7.61}$$

where r_i is the ith internally Studentized residual. C_i becomes large if p_{ii} and/or r_i^2 are large. Furthermore C_i is proportional to r_i^2. Applying (7.53) and (7.55), we get

$$\frac{r_i^2(T - K - 1)}{T - K - r_i^2} \sim F_{1,T-K-1},$$

indicating that C_i is not exactly F-distributed. To inspect the relative size of C_i for all the observations, Cook (1977), by analogy of (7.58) and (7.59), suggests comparing C_i with the $F_{K,T-K}$-percentiles. The greater the percentile corresponding to C_i, the more influential is the ith observation.

Let for example $K = 2$ and $T = 32$, that is, $(T - K) = 30$. The 95% and the 99% quantiles of $F_{2,30}$ are 3.32 and 5.59, respectively. When $C_i = 3.32$, $\hat{\beta}_{(i)}$ lies on the surface of the 95% confidence ellipsoid. If $C_j = 5.59$ for $j \neq i$, then $\hat{\beta}_{(j)}$ lies on the surface of the 99% confidence ellipsoid, and hence the jth observation would be more influential than the ith observation.

Welsch-Kuh's Distance

The influence of the ith observation on the predicted value $\hat{y}_i$ can be measured by the scaled difference $(\hat{y}_i - \hat{y}_{i(i)})$—by the change in predicting y_i

when the ith observation is omitted. The scaling factor is the standard deviation of $\hat{y}_i$ (cf. (7.5)):

$$\frac{|\hat{y}_i - \hat{y}_{i(i)}|}{\sigma\sqrt{p_{ii}}} = \frac{|x_i'(b - \hat{\beta}_{(i)})|}{\sigma\sqrt{p_{ii}}}\,. \tag{7.62}$$

Welsch and Kuh (1977) suggest the use of $s_{(i)}$ (7.37) as an estimate of σ in (7.63). Using (7.46) and (7.38), (7.63) can be written as

$$\begin{aligned} WK_i &= \frac{|\frac{\hat{\epsilon}_i}{1-p_{ii}} x_i'(X'X)^{-1}x_i|}{s_{(i)}\sqrt{p_{ii}}} \\ &= |r_i^*|\sqrt{\frac{p_{ii}}{1-p_{ii}}}\,. \end{aligned} \tag{7.63}$$

WK_i is called the Welsch-Kuh statistic. When $r_i^* \sim t_{T-K-1}$ (see Theorem 7.2), we can judge the size of WK_i by comparing it to the quantiles of the t_{T-K-1}-distribution. For sufficiently large sample sizes, one may use $2\sqrt{K/(T-K)}$ as a cutoff point for WK_i, signaling an influential ith observation.

Remark: The literature contains various modifications of Cook's distance (cf. Chatterjee and Hadi, 1988, pp. 122–135).

Measures Based on the Volume of Confidence Ellipsoids

Let $x'Ax \leq 1$ define an ellipsoid and assume A to be a symmetric (positive-definite or nonnegative-definite) matrix. From spectral decomposition (Theorem A.30), we have $A = \Gamma\Lambda\Gamma'$, $\Gamma\Gamma' = I$. The volume of the ellipsoid $x'Ax = (x'\Gamma)\Lambda(\Gamma'x) = 1$ is then seen to be

$$V = c_K \prod_{i=1}^{K} \lambda_i^{-\frac{1}{2}} = c_K\sqrt{|\Lambda^{-1}|}\,,$$

that is, inversely proportional to the root of $|A|$. Applying these arguments to (7.58), we may conclude that the volume of the confidence ellipsoid (7.58) is inversely proportional to $|X'X|$. Large values of $|X'X|$ indicate an informative design. If we take the confidence ellipsoid when the ith observation is omitted, namely,

$$\frac{(\beta - \hat{\beta}_{(i)})'(X_{(i)}'X_{(i)})(\beta - \hat{\beta}_{(i)})}{Ks_{(i)}^2} \leq F_{K,T-K-1,1-\alpha}\,, \tag{7.64}$$

then its volume is inversely proportional to $|X_{(i)}'X_{(i)}|$. Therefore, omitting an influential (informative) observation would decrease $|X_{(i)}'X_{(i)}|$ relative to $|X'X|$. On the other hand, omitting an observation having a large residual will decrease the residual sum of squares $s_{(i)}^2$ relative to s^2. These two ideas can be combined in one measure.

Andrews-Pregibon Statistic

Andrews and Pregibon (1978) have compared the volume of the ellipsoids (7.58) and (7.64) according to the ratio

$$\frac{(T-K-1)s_{(i)}^2|X_{(i)}'X_{(i)}|}{(T-K)s^2|X'X|}. \tag{7.65}$$

Let us find an equivalent representation. Define $Z = (X, y)$ and consider the partitioned matrix

$$Z'Z = \begin{pmatrix} X'X & X'y \\ y'X & y'y \end{pmatrix}. \tag{7.66}$$

Since $\text{rank}(X'X) = K$, we get (cf. Theorem A.16 (vii))

$$\begin{aligned} |Z'Z| &= |X'X||y'y - y'X(X'X)^{-1}X'y| \\ &= |X'X|(y'(I-P)y) \\ &= |X'X|(T-K)s^2 . \end{aligned} \tag{7.67}$$

Analogously, defining $Z_{(i)} = (X_{(i)}, y_{(i)})$, we get

$$|Z_{(i)}'Z_{(i)}| = |X_{(i)}'X_{(i)}|(T-K-1)s_{(i)}^2. \tag{7.68}$$

Therefore the ratio (7.65) becomes

$$\frac{|Z_{(i)}'Z_{(i)}|}{|Z'Z|}. \tag{7.69}$$

Omitting an observation that is far from the center of data will result in a large reduction in the determinant and consequently a large increase in volume. Hence, small values of (7.69) correspond to this fact. For the sake of convenience, we define

$$AP_i = 1 - \frac{|Z_{(i)}'Z_{(i)}|}{|Z'Z|}, \tag{7.70}$$

so that large values will indicate influential observations. AP_i is called the Andrews-Pregibon statistic.

Using $Z_{(i)}'Z_{(i)} = Z'Z - z_i z_i'$ with $z_i = (x_i', y_i)$ and Theorem A.16 (x), we obtain

$$\begin{aligned} |Z_{(i)}'Z_{(i)}| &= |Z'Z - z_i z_i'| \\ &= |Z'Z|(1 - z_i'(Z'Z)^{-1}z_i) \\ &= |Z'Z|(1 - p_{zii}), \end{aligned}$$

implying that

$$AP_i = p_{zii}, \tag{7.71}$$

where p_{zii} is the ith diagonal element of the prediction matrix $P_Z = Z(Z'Z)^{-1}Z'$. From (7.28) we get

$$p_{zii} = p_{ii} + \frac{\hat{\epsilon}_i^2}{\hat{\epsilon}'\hat{\epsilon}}\,. \tag{7.72}$$

Thus AP_i does not distinguish between high-leverage points in the X-space and outliers in the Z-space. Since $0 \le p_{zii} \le 1$ (cf. (7.19)), we get

$$0 \le AP_i \le 1\,. \tag{7.73}$$

If we apply the definition (7.36) of the internally Studentized residuals r_i and use $s^2 = \hat{\epsilon}'\hat{\epsilon}/(T-K)$, (7.73) implies

$$AP_i = p_{ii} + (1-p_{ii})\frac{r_i^2}{T-K} \tag{7.74}$$

or

$$(1-AP_i) = (1-p_{ii})\left(1-\frac{r_i^2}{T-K}\right). \tag{7.75}$$

The first quantity of (7.75) identifies high-leverage points and the second identifies outliers. Small values of $(1-AP_i)$ indicate influential points (high-leverage points or outliers), whereas independent examination of the single factors in (7.75) is necessary to identify the nature of influence.

Variance Ratio

As an alternative to the Andrews-Pregibon statistic and the other measures, one can identify the influence of the ith observation by comparing the estimated dispersion matrices of b_0 and $\hat{\beta}_{(i)}$:

$$V(b_0) = s^2(X'X)^{-1} \quad \text{and} \quad V(\hat{\beta}_{(i)}) = s_{(i)}^2(X_{(i)}'X_{(i)})^{-1}$$

by using measures based on the determinant or the trace of these matrices. If $(X_{(i)}'X_i)$ and $(X'X)$ are positive definite, one may apply the following variance ratio suggested by Belsley et al. (1980):

$$VR_i = \frac{|s_{(i)}^2(X_{(i)}'X_{(i)})^{-1}|}{|s^2(X'X)^{-1}|} \tag{7.76}$$

$$= \left(\frac{s_{(i)}^2}{s^2}\right)^K \frac{|X'X|}{|X_{(i)}'X_{(i)}|}\,. \tag{7.77}$$

Applying Theorem A.16 (x), we obtain

$$\begin{aligned} |X_{(i)}'X_{(i)}| &= |X'X - x_i x_i'| \\ &= |X'X|(1 - x_i'(X'X)^{-1}x_i) \\ &= |X'X|(1-p_{ii})\,. \end{aligned}$$

TABLE 7.2. Cook's C_i; Welsch-Kuh, WK_i; Andrews-Pregibon, AP_i; variance ratio VR_i, for the data set of Table 7.1

i	C_i	WK_i	AP_i	VR_i
1	0.182	0.610	0.349	1.260
2	0.043	0.289	0.188	1.191
3	0.166	0.541	0.858	8.967
4	0.001	0.037	0.106	1.455
5	0.005	0.096	0.122	1.443
6	0.241	0.864	0.504	0.475
7	0.017	0.177	0.164	1.443
8	0.114	0.518	0.331	0.803
9	0.003	0.068	0.123	1.466
10	0.078	0.405	0.256	0.995

With this relationship and using (7.54), we may conclude that

$$VR_i = \left(\frac{T-K-r_i^2}{T-K-1}\right)^K \frac{1}{1-p_{ii}}. \tag{7.78}$$

Therefore, VR_i will exceed 1 when r_i^2 is small (no outliers) and p_{ii} is large (high-leverage point), and it will be smaller than 1 whenever r_i^2 is large and p_{ii} is small. But if both r_i^2 and p_{ii} are large (or small), then VR_i tends toward 1. When all observations have equal influence on the dispersion matrix, VR_i is approximately equal to 1. Deviation from unity then will signal that the ith observation has more influence than the others. Belsley et al. (1980) propose the approximate cut-off "quantile"

$$|VR_i - 1| \geq \frac{3K}{T}. \tag{7.79}$$

Example 7.3 (Example 7.2 continued): We calculate the measures defined before for the data of Example 7.2 (cf. Table 7.1). Examining Table 7.2, we see that Cook's C_i has identified the sixth data point to be the most influential one. The cutoff quantile $2\sqrt{K/T-K} = 1$ for the Welsch-Kuh distance is not exceeded, but the sixth data point has the largest indication, again.

In calculating the Andrews-Pregibon statistic AP_i (cf. (7.71) and (7.72)), we insert $\hat{\epsilon}'\hat{\epsilon} = (T-K)s^2 = 8 \cdot (6.9)^2 = 380.88$. The smallest value $(1 - AP_i) = 0.14$ corresponds to the third observation, and we obtain

$$\begin{aligned}(1 - AP_3) = 0.14 &= (1-p_{33})\left(1 - \frac{r_3^2}{8}\right) \\ &= 0.14 \cdot (1 - 0.000387),\end{aligned}$$

indicating that (y_3, x_3) is a high-leverage point, as we have noted already. The sixth observation has an AP_i value next to that of the third observation. An inspection of the factors of $(1 - AP_6)$ indicates that (y_6, x_6) tends to be an outlier:

$$(1 - AP_6) = 0.496 = 0.88 \cdot (1 - 0.437).$$

These conclusions hold for the variance ratio also. Condition (7.79), namely, $|VR_i - 1| \geq \frac{6}{10}$, is fulfilled for the third observation, indicating significance in the sense of (7.79).

Remark: In the literature one may find many variants and generalizations of the measures discussed here. A suitable recommendation is the monograph of Chatterjee and Hadi (1988).

7.6 Partial Regression Plots

Plotting the residuals against a fixed independent variable can be used to check the assumption that this regression has a linear effect on Y. If the residual plot shows the inadequacy of a linear relation between Y and some fixed X_i, it does not display the true (nonlinear) relation between Y and X_i. *Partial regression plots* are refined residual plots to represent the correct relation for a regressor in a multiple model under consideration. Suppose that we want to investigate the nature of the marginal effect of a variable X_k, say, on Y in case the other independent variables under consideration are already included in the model. Thus partial regression plots may provide information about the marginal importance of the variable X_k that may be added to the regression model.

Let us assume that one variable X_1 is included and that we wish to add a second variable X_2 to the model (cf. Neter, Wassermann, and Kutner, 1990, p. 387). Regressing Y on X_1, we obtain the fitted values

$$\hat{y}_i(X_1) = \hat{\beta}_0 + x_{1i}\hat{\beta}_1 = \tilde{x}'_{1i}\tilde{\beta}_1 , \tag{7.80}$$

where

$$\tilde{\beta}_1 = (\hat{\beta}_0, \hat{\beta}_1)' = (\tilde{X}_1'\tilde{X}_1)^{-1}\tilde{X}_1'y \tag{7.81}$$

and $\tilde{X}_1 = (1, x_1)$.

Hence, we may define the residuals

$$e_i(Y|X_1) = y_i - \hat{y}_i(X_1) . \tag{7.82}$$

Regressing X_2 on $\tilde{X}_1$, we obtain the fitted values

$$\hat{x}_{2i}(X_1) = \tilde{x}_{1i}'b_1^* \tag{7.83}$$

with $b_1^* = (\tilde{X}_1'\tilde{X}_1)^{-1}\tilde{X}_1'x_2$ and the residuals

$$e_i(X_2|X_1) = x_{2i} - \hat{x}_{2i}(X_1) . \tag{7.84}$$

Analogously, in the full model $y = \beta_0 + X_1\beta_1 + X_2\beta_2 + \epsilon$, we have

$$e_i(Y|X_1, X_2) = y_i - \hat{y}_i(X_1, X_2), \tag{7.85}$$

where

$$\hat{y}_i(X_1, X_2) = \tilde{X}_1 b_1 + X_2 b_2 \tag{7.86}$$

and b_1 and b_2 are as defined in (3.98) (replace X_1 by $\tilde{X}_1$). Then we have

$$e(Y|X_1, X_2) = e(Y|X_1) - b_2 e(X_2|X_1). \tag{7.87}$$

The proof is straightforward. Writing (7.87) explicitly gives

$$\begin{aligned} y - \tilde{X}_1 b_1 - X_2 b_2 &= [y - \tilde{X}_1(\tilde{X}_1'\tilde{X}_1)^{-1}\tilde{X}_1' y] \\ &\quad - [X_2 - \tilde{X}_1(\tilde{X}_1'\tilde{X}_1)^{-1}\tilde{X}_1']b_2 \\ &= \tilde{M}_1(y - X_2 b_2) \end{aligned} \tag{7.88}$$

with the symmetric idempotent matrix

$$\tilde{M}_1 = I - \tilde{X}_1(\tilde{X}_1'\tilde{X}_1)^{-1}\tilde{X}_1. \tag{7.89}$$

Consequently, (7.88) may be rewritten as

$$\tilde{X}_1(\tilde{X}_1'\tilde{X}_1)^{-1}\tilde{X}_1'(y - X_2 b_2 - b_1) = 0. \tag{7.90}$$

Using the second relation in (3.99), we see that (7.90) holds, and hence (7.87) is proved.

The partial regression plot is obtained by plotting the residuals $e_i(Y|X_1)$ against the residuals $e_i(X_2|X_1)$. Figures 7.6 and 7.7 present some standard partial regression plots. If the vertical deviations of the plotted points around the line $e(Y|X_1) = 0$ are squared and summed, we obtain the residual sum of squares

$$\begin{aligned} RSS_{\tilde{X}_1} &= (y - \tilde{X}_1(\tilde{X}_1'\tilde{X}_1)^{-1}\tilde{X}_1' y)'(y - \tilde{X}_1(\tilde{X}_1'\tilde{X}_1)^{-1}\tilde{X}_1' y) \\ &= y'\tilde{M}_1 y \\ &= \left[e(y|X_1)\right]'\left[e(Y|X_1)\right]. \end{aligned} \tag{7.91}$$

The vertical deviations of the plotted points in Figure 7.6 taken with respect to the line through the origin with slope b_1 are the estimated residuals $e(Y|X_1, X_2)$. Using relation (3.169), we get from (7.86) the extra sum of squares relationship

$$SS_{\text{Reg}}(X_2|\tilde{X}_1) = RSS_{\tilde{X}_1} - RSS_{\tilde{X}_1, X_2}. \tag{7.92}$$

This relation is the basis for the interpretation of the partial regression plot: If the scatter of the points around the line with slope b_2 is much less than the scatter around the horizontal line, then adding an additional independent variable X_2 to the regression model will lead to a substantial reduction of the error sum of squares and, hence, will substantially increase the fit of the model.

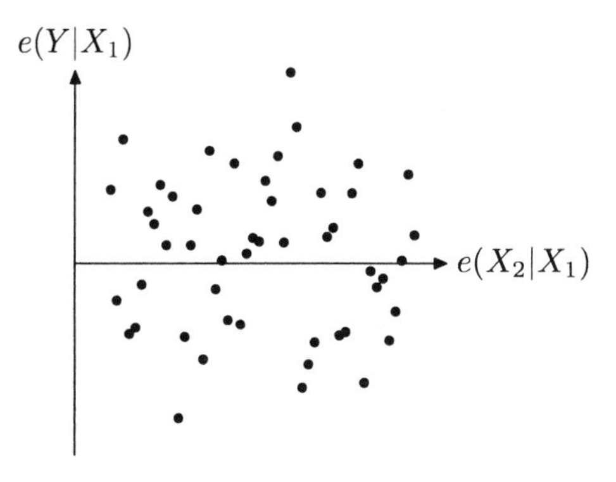

FIGURE 7.6. Partial regression plot (of $e(X_2|X_1)$ versus $e(Y|X_1)$) indicating no additional influence of X_2 compared to the model $y = \beta_0 + X_1\beta_1 + \epsilon$

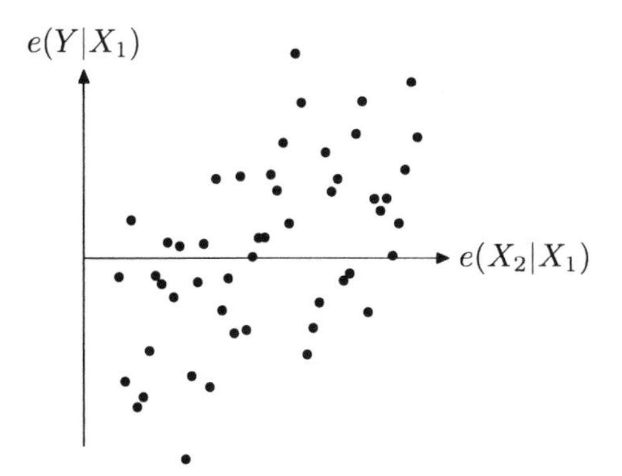

FIGURE 7.7. Partial regression plot (of $e(X_2|X_1)$ versus. $e(Y|X_1)$) indicating additional linear influence of X_2

7.7 Regression Diagnostics for Removing an Observation with Animating Graphics

Graphical techniques are an essential part of statistical methodology. One of the important graphics in regression analysis is the residual plot. In regression analysis the plotting of residuals versus the independent variable or predicted values has been recommended by Draper and Smith (1966) and Cox and Snell (1968). These plots help to detect outliers, to assess the presence of inhomogenity of variance, and to check model adequacy. Larsen and McCleary (1972) introduced partial residual plots, which can detect the importance of each independent variable and assess some nonlinearity or necessary transformation of variables.

For the purpose of regression diagnostics Cook and Weisberg (1989) introduced dynamic statistical graphics. They considered interpretation of two proposed types of dynamic displays, rotation and animation, in regression diagnostics. Some of the issues that they addressed by using dynamic graphics include adding predictors to a model, assessing the need to transform, and checking for interactions and normality. They used animation to show dynamic effects of adding a variable to a model and provided methods for simultaneously adding variables to a model.

Assume the classical linear, normal model:

$$\begin{aligned} y &= X\beta + \epsilon \\ &= X_1\beta_1 + X_2\beta_2 + \epsilon, \quad \epsilon \sim N(0, \sigma^2 I). \end{aligned} \tag{7.93}$$

X consists of X_1 and X_2 where X_1 is a $T \times (K-1)$-matrix, and X_2 is a $T \times 1$-matrix, that is, $X = (X_1, X_2)$. The basic idea of Cook and Weisberg (1989) is to begin with the model $y = X_1\beta_1 + \epsilon$ and then smoothly add X_2, ending with a fit of the full model $y = X_1\beta_1 + X_2\beta_2 + \epsilon$, where β_1 is a $(K-1) \times 1$-vector and β_2 is an unknown scalar. Since the animated plot that they proposed involves only fitted values and residuals, they worked in terms of a modified version of the full model (7.93) given by

$$\begin{aligned} y &= Z\beta^* + \epsilon \\ &= X_1\beta_1^* + \tilde{X}_2\beta_2^* + \epsilon \end{aligned} \tag{7.94}$$

where $\tilde{X}_2 = Q_1X_2/||Q_1X_2||$ is the part of X_2 orthogonal to X_1, normalized to unit length, $Q_1 = I - P_1$, $P_1 = X_1(X_1'X_1)^{-1}X_1'$, $Z = (X_1, \tilde{X}_2)$, and $\beta^* = (\beta_1^{*\prime}, \beta_2^{*\prime})'$.

Next, for each $0 < \lambda \leq 1$, they estimate β^* by

$$\hat{\beta}_\lambda = \left(Z'Z + \frac{1-\lambda}{\lambda}ee'\right)^{-1} Z'y \tag{7.95}$$

where e is a $K \times 1$-vector of zeros except for single 1 corresponding to X_2. Since

$$\begin{aligned} \left(Z'Z + \frac{1-\lambda}{\lambda}ee'\right)^{-1} &= \begin{pmatrix} X_1'X_1 & 0 \\ 0' & \tilde{X}_2'\tilde{X}_2 + \frac{1-\lambda}{\lambda} \end{pmatrix}^{-1} \\ &= \begin{pmatrix} X_1'X_1 & 0 \\ 0' & \frac{1}{\lambda} \end{pmatrix}^{-1}, \end{aligned}$$

we obtain

$$\hat{\beta}_\lambda = \begin{pmatrix} (X_1'X_1)^{-1}X_1'y \\ \lambda\tilde{X}_2'y \end{pmatrix}.$$

So as λ tends to 0, (7.95) corresponds to the regression of y on X_1 alone. And if $\lambda = 1$, then (7.95) corresponds to the ordinary least-squares regression of y on X_1 and X_2. Thus as λ increases from 0 to 1, $\hat{\beta}_\lambda$ represents a continuous change of estimators that add X_2 to the model, and an animated plot of $\hat{\epsilon}(\lambda)$ versus $\hat{y}(\lambda)$, where $\hat{\epsilon}(\lambda) = y - \hat{y}(\lambda)$ and $\hat{y}(\lambda) = Z\hat{\beta}_\lambda$, gives a dynamic view of the effects of adding X_2 to the model that already includes X_1. This idea corresponds to the weighted mixed regression estimator (8.46).

Using Cook and Weisberg's idea of animation, Park, Kim, and Toutenburg (1992) proposed an animating graphical method to display the effects of removing an outlier from a model for regression diagnostic purpose.

We want to view dynamic effects of removing the ith observation from the model (7.93). First, we consider the mean shift model $y = X\beta + \gamma_i e_i + \epsilon$ (see (7.49)) where e_i is the vector of zeros except for single a 1 corresponding to the ith observation. We can work in terms of a modified version of the mean shift model given by

$$\begin{aligned} y &= Z\beta^* + \epsilon \\ &= X\tilde{\beta} + \gamma_i^* \tilde{e} + \epsilon \end{aligned} \tag{7.96}$$

where $\tilde{e}_i = Q_x e_i / ||Q_x e_i||$ is the orthogonal part of e_i to X normalized to unit length, $Q = I - P$, $P = X(X'X)^{-1}X'$, $Z = (X, \tilde{e}_i)$, and $\beta^* = \begin{pmatrix} \tilde{\beta} \\ \gamma_i^* \end{pmatrix}$. And then for each $0 < \lambda \leq 1$, we estimate β^* by

$$\hat{\beta}_\lambda = \left(Z'Z + \frac{1-\lambda}{\lambda} ee' \right)^{-1} Z'y\,, \tag{7.97}$$

where e is the $(K+1) \times 1$-vector of zeros except for single a 1 for the $(K+1)$th element. Now we can think of some properties of $\hat{\beta}_\lambda$. First, without loss of generality, we take X and y of the forms $X = \begin{pmatrix} X_{(i)} \\ x_i' \end{pmatrix}$ and $y = \begin{pmatrix} y_{(i)} \\ y_i \end{pmatrix}$, where x_i' is the ith row vector of X, $X_{(i)}$ is the matrix X without the ith row, and $y_{(i)}$ is the vector y without y_i. That is, place the ith observation to the bottom and so e_i and e become vectors of zeros except for the last 1. Then since

$$\left(Z'Z + \frac{1-\lambda}{\lambda} ee' \right)^{-1} = \begin{pmatrix} X'X & 0 \\ 0' & \frac{1}{\lambda} \end{pmatrix}^{-1} = \begin{pmatrix} (X'X)^{-1} & 0 \\ 0' & \lambda \end{pmatrix}$$

and

$$Z'y = \begin{pmatrix} X'y \\ \tilde{e}_i' y \end{pmatrix}$$

we obtain

$$\hat{\beta}_\lambda = \begin{pmatrix} \hat{\tilde{\beta}} \\ \hat{\gamma}_i^* \end{pmatrix} = \begin{pmatrix} (X'X)^{-1}X'y \\ \lambda \tilde{e}_i^* y \end{pmatrix}$$

and

$$\hat{y}(\lambda) = Z\hat{\beta}_\lambda = X(X'X)^{-1}X'y + \lambda \tilde{e}\tilde{e}'y\,.$$

Hence at $\lambda = 0$, $\hat{y}(\lambda) = (X'X)^{-1}X'y$ is the predicted vector of observed values for the full model by the method of ordinary least squares. And at $\lambda = 1$, we can get the following lemma, where $\hat{\beta}_{(i)} = (X_{(i)}' X_{(i)})^{-1} X_{(i)} y_{(i)}$.

Lemma 7.3

$$\hat{y}(1) = \begin{pmatrix} X_{(i)}\hat{\beta}_{(i)} \\ y_{(i)} \end{pmatrix}$$

Proof: Using Theorem A.18 (iv),

$$\begin{aligned} (X'X)^{-1} &= (X'_{(i)}X_{(i)} + x_i x_i')^{-1} \\ &= (X'_{(i)}X_{(i)})^{-1} - \frac{(X'_{(i)}X_{(i)})^{-1}x_i x_i'(X'_{(i)}X_{(i)})^{-1}}{1+t_{ii}}, \end{aligned}$$

where

$$t_{ii} = x_i'(X'_{(i)}X_{(i)})^{-1}x_i .$$

We have

$$\begin{aligned} P &= X(X'X)^{-1}X' \\ &= \begin{pmatrix} X_{(i)} \\ x_i' \end{pmatrix} \left((X'_{(i)}X_{(i)})^{-1} - \frac{(X'_{(i)}X_{(i)})^{-1}x_i x_i'(X'_{(i)}X_{(i)})^{-1}}{1+t_{ii}} \right) (X'_{(i)}x_i) \end{aligned}$$

and

$$\begin{aligned} Py &= X(X'X)^{-1}X'y \\ &= \begin{pmatrix} X_{(i)}\hat{\beta}_{(i)} - \frac{1}{1+t_{ii}}(X'_{(i)}(X'_{(i)}X_{(i)})^{-1}x_i x_i'\hat{\beta}_{(i)} - X'_{(i)}(X'_{(i)}X_{(i)})^{-1}x_i y_i) \\ \frac{1}{1+t_{ii}}(x_i'\hat{\beta}_{(i)} + t_{ii}y_i) \end{pmatrix} . \end{aligned}$$

Since

$$(I-P)e_i = \frac{1}{1+t_{ii}} \begin{pmatrix} -X_{(i)}(X'_{(i)}X_{(i)})^{-1}x_i \\ 1 \end{pmatrix}$$

and

$$||(I-P)e_i||^2 = \frac{1}{1+t_{ii}},$$

we get

$$\tilde{e}_i\tilde{e}_i{}'y = \frac{1}{1+t_{ii}} \begin{pmatrix} X'_{(i)}(X'_{(i)}X_{(i)})^{-1}x_i x_i'\hat{\beta}_{(i)} - X'_{(i)}(X'_{(i)}X_{(i)})^{-1}x_i y_i \\ -x_i'\hat{\beta}_{(i)} + y_i \end{pmatrix} .$$

Therefore,

$$X(X'X)^{-1}X'y + \tilde{e}_i\tilde{e}_i{}'y = \begin{pmatrix} X_{(i)}\hat{\beta}_{(i)} \\ y_i \end{pmatrix} .$$

Thus as λ increases from 0 to 1, an animated plot of $\hat{\epsilon}(\lambda)$ versus $\hat{\lambda}$ gives a dynamic view of the effects of removing the ith observation from model (7.93).

The following lemma shows that the residuals $\hat{\epsilon}(\lambda)$ and fitted values $\hat{y}(\lambda)$ can be computed from the residuals $\hat{\epsilon}$, fitted values $\hat{y} = \hat{y}(0)$ from the full model, and the fitted values $\hat{y}(1)$ from the model that does not contain the ith observation.

Lemma 7.4

(i) $\hat{y}(\lambda) = \lambda\hat{y}(1) + (1-\lambda)\hat{y}(0)$

(ii) $\hat{\epsilon}(\lambda) = \hat{\epsilon} - \lambda(\hat{y}(1) - \hat{y}(0))$

Proof: Using the fact

$$\begin{pmatrix} X'X & X'e_i \\ e_i'X & e_i'e_i \end{pmatrix} - 1$$

$$= \begin{pmatrix} (X'X)^{-1} + (X'X)^{-1}X'e_iHe_i'He_i'X(X'X)^{-1} & -(X'X)^{-1}X'e_iH \\ -He_i'X(X'X)^{-1} & H \end{pmatrix}$$

where

$$\begin{aligned} H &= (e_i'e_i - e_i'X(X'X)^{-1}Xe_i)^{-1} \\ &= (e_i'(I-P)e_i)^{-1} \\ &= \frac{1}{||Qe_i||^2}, \end{aligned}$$

we can show that $P(X, e_i)$, the projection matrix onto the column space of (X, e_i), becomes

$$\begin{aligned} P(X,e_i) &= (X \quad e_i)\begin{pmatrix} X'X & X'e_i \\ e_i'X & e_i'e_i \end{pmatrix}^{-1}\begin{pmatrix} X' \\ e_i' \end{pmatrix} \\ &= P + \frac{(I-P)e_ie_i'(I-P)}{||Qe_i||^2} \\ &= P + \tilde{e}_i\tilde{e}_i'. \end{aligned}$$

Therefore

$$\begin{aligned} \hat{y}(\lambda) &= X(X'X)^{-1}X'y + \lambda e_ie_i'y \\ &= \hat{y}(0) + \lambda(P(X,e_i) - P)y \\ &= \hat{y}(0) + \lambda(\hat{y}(1) - \hat{y}(0)) \\ &= \lambda\hat{y}(1) + (1-\lambda)\hat{y}(0) \end{aligned}$$

and property (ii) can be proved by the fact that

$$\begin{aligned} \hat{\epsilon}(\lambda) &= y - \hat{y}(\lambda) \\ &= y - \hat{y}(0) - \lambda(\hat{y}(1) - \hat{y}(0)) \\ &= \hat{\epsilon} - \lambda(\hat{y}(1) - \hat{y}(0)). \end{aligned}$$

Because of the simplicity of Lemma 7.4, an animated plot of $\hat{\epsilon}(\lambda)$ versus $\hat{y}(\lambda)$ as λ is varied between 0 and 1 can be easily computed.

The appropriate number of frames (values of λ) for an animated residual plot depends on the speed with which the computer screen can be refreshed and thus on the hardware being used. With too many frames, changes often become too small to be noticed, and as consequence the overall trend can be missed. With too few frames, smoothness and the behavior of individual points cannot be detected.

When there are too many observations, and it is difficult to check all animated plots, it is advisable to select several suspicious observations based on nonanimated diagnostic measures such as Studentized residuals, Cook's distance, and so on.

From animated residual plots for individual observations, $i = 1, 2, \ldots, n$, it would be possible to diagnose which observation is most influential in changing the residuals $\hat{\epsilon}$, and the fitted values of y, $\hat{y}(\lambda)$, as λ changes from 0 to 1. Thus, it may be possible to formulate a measure to reflect which observation is most influential, and which kind of influential points can be diagnosed in addition to those that can already be diagnosed by well-known diagnostics. However, our primary intent is only to provide a graphical tool to display and see the effects of continuously removing a single observation from a model. For this reason, we do not develop a new diagnostic measure that could give a criterion when an animated plot of removing an observation is significant or not. Hence, development of a new measure based on such animated plots remains open to further research.

Example 7.4 (Phosphorus Data): In this example, we illustrate the use of $\hat{\epsilon}(\lambda)$ versus $\hat{y}(\lambda)$ as an aid to understanding the dynamic effects of removing an observation from a model. Our illustration is based on the phosphorus data reported in Snedecor and Cochran (1967, p. 384). An investigation of the source from which corn plants obtain their phosphorus was carried out. Concentrations of phosphorus in parts per millions in each of 18 soils was measured. The variables are

$$\begin{aligned} X_1 &= \text{concentrations of inorganic phosphorus in the soil,} \\ X_2 &= \text{concentrations of organic phosphorus in the soil, and} \\ y &= \text{phosphorus content of corn grown in the soil at } 20^\circ\text{C.} \end{aligned}$$

The data set together with the ordinary residuals e_i, the diagonal terms h_{ii} of hat matrix $H = X(X'X)^{-1}X'$, the Studentized residuals r_i, and Cook's distances C_i are shown in Table 7.3 under the linear model assumption. We developed computer software, that plots the animated residuals and some related regression results. The plot for the 17th observation shows the most significant changes in residuals among 18 plots. In fact, the 17th observation has the largest target residual e_i, Studentized residuals r_{ii}, and Cook's distances C_i, as shown in Table 7.3.

Figure 7.8 shows four frames of an animated plot of $\hat{\epsilon}(\lambda)$ versus $\hat{y}(\lambda)$ for removing the 17th observation. The first frame (a) is for $\lambda = 0$ and thus corresponds to the usual plot of residuals versus fitted values from

Soil	X_1	X_2	y	e_i	h_{ii}	r_i	C_i
1	0.4	53	64	2.44	0.26	0.14	0.002243
2	0.4	23	60	1.04	0.19	0.06	0.000243
3	3.1	19	71	7.55	0.23	0.42	0.016711
4	0.6	34	61	0.73	0.13	0.04	0.000071
5	4.7	24	54	−12.74	0.16	−0.67	0.028762
6	1.7	65	77	12.07	0.46	0.79	0.178790
7	9.4	44	81	4.11	0.06	0.21	0.000965
8	10.1	31	93	15.99	0.10	0.81	0.023851
9	11.6	29	93	13.47	0.12	0.70	0.022543
10	12.6	58	51	−32.83	0.15	−1.72	0.178095
11	10.9	37	76	−2.97	0.06	−0.15	0.000503
12	23.1	46	96	−5.58	0.13	−0.29	0.004179
13	23.1	50	77	−24.93	0.13	−1.29	0.080664
14	21.6	44	93	−5.72	0.12	−0.29	0.003768
15	23.1	56	95	−7.45	0.15	−0.39	0.008668
16	1.9	36	54	−8.77	0.11	−0.45	0.008624
17	26.8	58	168	58.76	0.20	3.18	0.837675
18	29.9	51	99	−15.18	0.24	−0.84	0.075463

TABLE 7.3. Data, ordinary residuals e_i, diagonal terms h_{ii} of hat matrix $H = X(X'X)^{-1}X'$, studentized residuals r_i, and Cook's distances C_i from Example 7.4

the regression of y on $X = (X_1, X_2)$, and we can see in (a) the 17th observation is located on the upper right corner. The second (b), third (c), and fourth (d) frames correspond to $\lambda = \frac{1}{2}, \frac{2}{3}$, and 1, respectively. So the fourth frame (d) is the usual plot of the residuals versus the fitted values from the regression of $y_{(17)}$ on $X_{(17)}$ where the subscript represents omission of the corresponding observation. We can see that as λ increases from 0 to 1, the 17th observation moves to the right and down, becoming the rightmost point in (b), (c), and (d). Cconsidering the plotting form, the residual plot in (a) has an undesirable form because it does not have a random form in a band between -60 and $+60$, but in (d) its form has randomness in a band between -20 and $+20$.

7.8 Exercises

Exercise 1. Examine the impact of an influential observation on the coefficient of determination.

Exercise 2. Obtain an expression for the change in residual sum of squares when one observation is deleted. Can it be used for studying the change in residual sum of squares when one observation is added to the data set?

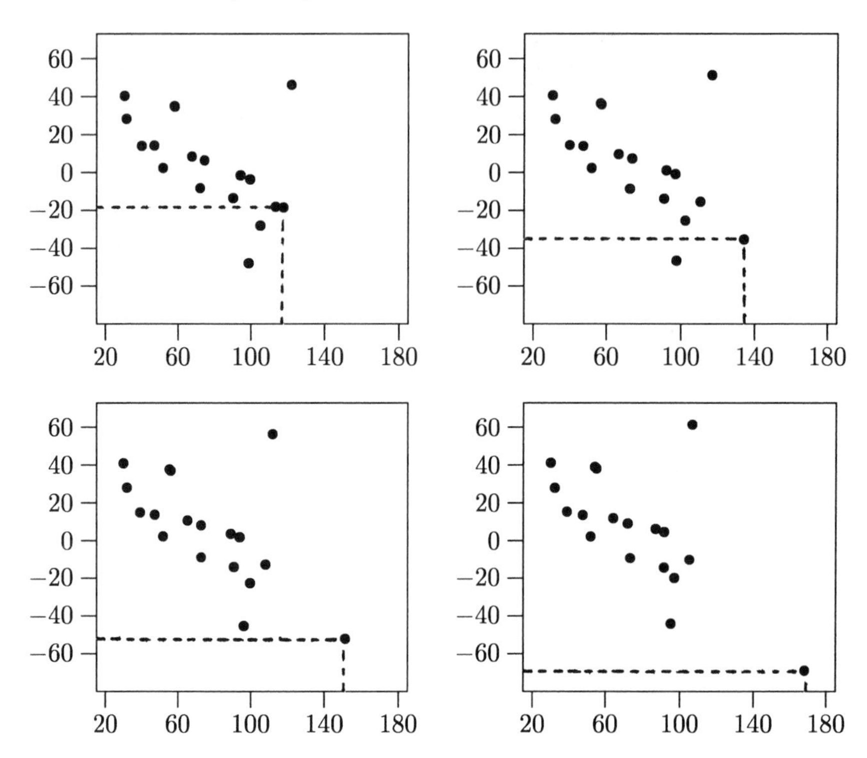

FIGURE 7.8. Four frames $\lambda = 0$ (a), $\lambda = \frac{1}{3}$ (b), $\lambda = \frac{2}{3}$ (c) and $\lambda = 1$ (d) (left to right, top down) of an animated plot of $\hat{\epsilon}(\lambda)$ versus $\hat{y}(\lambda)$ for data in Example 7.4 when removing the 17th observation (marked by dotted lines).

Exercise 3. If we estimate β by the mean of vectors $\hat{\beta}_{(i)}$, what are its properties? Compare these properties with those of the least-squares estimator.

Exercise 4. Analyze the effect of including an irrelevant variable in the model on the least-squares estimation of regression coefficients and its efficiency properties. How does the inference change when a dominant variable is dropped?

Exercise 5. For examining whether an observation belongs to the model $y_t = x_t'\beta + \epsilon_t$; $t = 1, 2, \ldots, n-1$, it is proposed to test the null hypothesis $\mathrm{E}(y_n) = x_n'\beta$ against the alternative $\mathrm{E}(y_n) \neq x_n'\beta$. Obtain the likelihood ratio test.

8
Analysis of Incomplete Data Sets

Standard statistical procedures assume the availability of complete data sets. In frequent cases, however, not all values are available, and some responses may be missing due to various reasons. Rubin (1976, 1987) and Little and Rubin (1987) have discussed some concepts for handling missing data based on decision theory and models for mechanisms of nonresponse.

Standard statistical methods have been developed to analyze rectangular data sets D of the form

$$D = \begin{pmatrix} d_{11} & \cdots & \cdots & d_{1m} \\ \vdots & * & & \vdots \\ \vdots & & & * \\ \vdots & & * & \vdots \\ d_{T1} & \cdots & \cdots & d_{Tm} \end{pmatrix},$$

where the rows of the matrix D represent units (cases, observations) and the columns represent variables observed on each unit. In practice, some of the observations d_{ij} are missing. This fact is indicated by the symbol "$*$."

Examples:

- Respondents do not answer all items of a questionnaire. Answers can be missing by chance (a question was overlooked) or not by chance (individuals are not willing to give detailed information concerning sensitive items, such as drinking behavior, income, etc.).

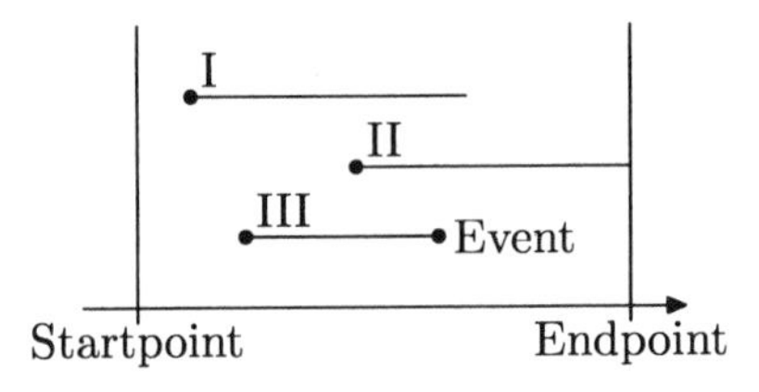

FIGURE 8.1. Censored individuals (I: dropout, II: censored by the endpoint) and an individual with response (event) (III)

- In clinical long-term studies some individuals do not participate over the whole period and drop out. The different situations are indicated in Figure 8.1. In the case of dropout, it is difficult to characterize the stochastic nature of the event.
- Physical experiments in industrial production (quality control) sometimes end with possible destruction of the object being investigated. Further measurements for destructed objects cannot be obtained.
- Censored regression, see Section 3.14.

8.1 Statistical Methods with Missing Data

There are several general approaches to handling the missing-data problem in statistical analysis. We briefly describe the idea behind these approaches in the following sections.

8.1.1 Complete Case Analysis

The simplest approach is to discard all incomplete cases. The analysis is performed using only the complete cases, i.e., those cases for which all t_c observations in a row of the matrix D are available. The advantage of this approach is simplicity, because standard statistical analyses (and statistical software packages) can be applied to the complete part of the data without modification.

Using complete case analysis tends to become inefficient if the percentage of cases with missing values is large. The selection of complete cases can lead to selectivity biases in estimates if selection is heterogeneous with respect to covariates.

8.1.2 Available Case Analysis

Another approach to missing values, that is similar to complete case analysis in some sense is the so-called available case analysis. Again, the analysis is restricted to complete cases. The difference is the definition of "complete." For complete case analysis, only cases having observations for all variables are used. Available case analysis uses all cases that are complete with respect to the variables of the current step in the analysis. If the correlation of D_1 and D_2 is of interest, cases with missing values in variable D_3 can still be used.

8.1.3 *Filling in the Missing Values*

Imputation ("filling in") is a general and flexible alternative to the complete case analysis. The missing values in the data matrix D are replaced by guesses or correlation-based predictors transforming D to a complete matrix. The completed data set then can be analyzed by standard procedures.

However, this method can lead to biases in statistical analyses, as the imputed values in general are different from the true but missing values. We shall discuss this problem in detail in the case of regression (see Section 8.6).

Some of the current practices in imputation are:

Hot-deck imputation. The imputed value for each missing value is selected (drawn) from a distribution, which is estimated from the complete cases in most applications.

Cold deck-imputation. A missing value is replaced by a constant value from external sources, such as an earlier realization of the survey.

Mean imputation. Based on the sample of the responding units, means are substituted for the missing values.

Regression (correlation) imputation. Based on the correlative structure of the subset of complete data, missing values are replaced by predicted values from a regression of the missing item on items observed for the unit.

Multiple imputation. $k \geq 2$ values are imputed for a missing value, giving k completed data sets (cf. Rubin, 1987). The k complete data sets are analyzed, yielding k estimates, which are combined to a final estimate.

8.1.4 *Model-Based Procedures*

Model-based procedures are based on a model for the data and the missing mechanism. The maximum-likelihood methods, as described in Section 8.7.3, factorize the likelihood of the data and the missing mechanism according to missing patterns. Bayesian methods, which operate on the observed data posterior of the unknown parameters (conditioned on the observed quantities), are described in detail in Schafer (1997).

8.2 Missing-Data Mechanisms

Knowledge of the mechanism for nonresponse is a central element in choosing an appropriate statistical analysis. The nature of the missing data mechanism can be described in terms of the distribution $f(R|D)$ of a missing indicator R conditional on the data D.

8.2.1 Missing Indicator Matrix

Rubin (1976) introduced the matrix R consisting of indicator variables r_{ij}, which has the same dimension as the data matrix D. The elements r_{ij} have values $r_{ij} = 1$ if d_{ij} is observed (reported), and $r_{ij} = 0$ if d_{ij} is missing.

8.2.2 Missing Completely at Random

Missing values are said to be missing completely at random (MCAR), if

$$f(R|D) = f(R) \quad \forall D\,. \tag{8.1}$$

The data D cannot be used to specify the distribution of R; the values are missing completely at random.

8.2.3 Missing at Random

Missing values are said to be missing at random (MAR), if

$$f(R|D) = f(R|D_{\text{obs}}) \quad \forall D_{\text{mis}}\,. \tag{8.2}$$

The dependence of the distribution of R on the data D can be specified using the observed data D_{obs} alone. Conditional on the observed values, the unobserved values D_{mis} are missing at random.

8.2.4 Nonignorable Nonresponse

The conditional distribution $f(R|D)$ cannot be simplified as above, that is, even after conditioning on the observed data, the distribution of R still depends on the unobserved data D_{mis}. In this case the missing data mechanism cannot be ignored (see Section 8.7.3).

8.3 Missing Pattern

A pattern of missing values in the data matrix D is called monotone if rows and columns can be rearranged such that the following condition holds. For all $j = 1, \ldots, m-1$: D_{j+1} is observed for all cases, where D_j is observed

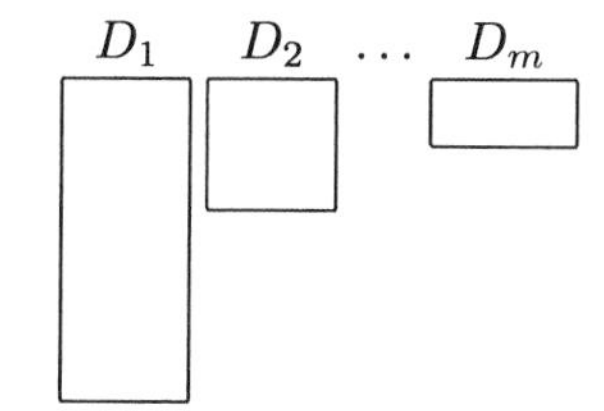

FIGURE 8.2. Monotone Missing Pattern

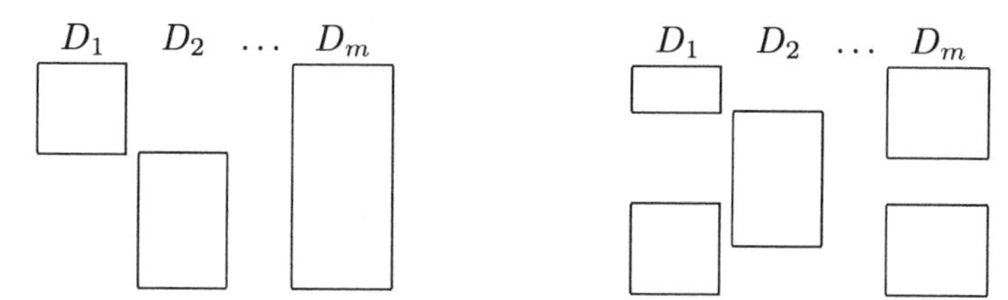

FIGURE 8.3. Special Missing Pattern

FIGURE 8.4. General Missing Pattern

(Figure 8.2). Univariate missingness, that is, missing values in only one variable D_j, is a special case.

Figure 8.3 shows a pattern in which two variables D_j and $D_{j'}$ are never observed together; a situation that might show up when the data of two studies are merged. Figure 8.4 shows a general pattern, in which no specific structure can be described.

8.4 Missing Data in the Response

In controlled experiments, such as clinical trials, the design matrix X is fixed and the response is observed for factor levels of X. In this situation it is realistic to assume that missing values occur in the response Y and not in the design matrix X resulting in unbalanced response. Even if we can assume that MCAR holds, sometimes it may be more advantageous to fill up the vector Y than to confine the analysis to the complete cases. This is the fact, for example, in factorial (cross-classified) designs with few replications. For the following let us assume that the occurrence of missing y values does not depend on the values of y, that is, MAR holds.

Let Y be the response variable and $X\colon T \times K$ be the design matrix, and assume the linear model $y = X\beta + \epsilon$, $\epsilon \sim N(0, \sigma^2 I)$. The OLSE of β for complete data is given by $b = (X'X)^{-1}X'y$, and the unbiased estimator of σ^2 is given by $s^2 = (y - Xb)'(y - Xb)(T - K)^{-1} = \sum_{t=1}^{T}(y_t - \hat{y}_t)^2/(T - K)$.

8.4.1 Least-Squares Analysis for Filled-up Data—Yates Procedure

The following method was proposed by Yates (1933). Assume that $t_* = T - t_c$ responses in y are missing. Reorganize the data matrices according to

$$\begin{pmatrix} y_{obs} \\ y_{mis} \end{pmatrix} = \begin{pmatrix} X_c \\ X_* \end{pmatrix} \beta + \begin{pmatrix} \epsilon_c \\ \epsilon_* \end{pmatrix} . \tag{8.3}$$

The indices c and $*$ indicate the complete and partially incomplete parts of the model, respectively. In the current case, X_* is fully observed; the index $*$ is used to denote the connection to the unobserved responses y_{mis}.

The complete-case estimator of β is given by

$$b_c = (X_c' X_c)^{-1} X_c' y_{obs} \tag{8.4}$$

using the $t_c \times K$-matrix X_c and the observed responses y_{obs} only. The classical predictor of the $(T - t_c)$-vector y_{mis} is given by

$$\hat{y}_{mis} = X_* b_c . \tag{8.5}$$

It is easily seen that inserting this estimator into model (8.3) for y_{mis} and estimating β in the filled-up model is equivalent to minimizing the following function with respect to β (cf. (3.7)):

$$\begin{aligned} S(\beta) &= \left\{ \begin{pmatrix} y_{obs} \\ \hat{y}_{mis} \end{pmatrix} - \begin{pmatrix} X_c \\ X_* \end{pmatrix} \beta \right\}' \left\{ \begin{pmatrix} y_{obs} \\ \hat{y}_{mis} \end{pmatrix} - \begin{pmatrix} X_c \\ X_* \end{pmatrix} \beta \right\} \\ &= \sum_{t=1}^{t_c} (y_t - x_t'\beta)^2 + \sum_{t=t_c+1}^{T} (\hat{y}_t - x_t'\beta)^2 . \end{aligned} \tag{8.6}$$

The first sum is minimized by b_c given in (8.4). Replacing β in the second sum by b_c is equating this sum to zero (cf. (8.5)). Therefore, b_c is seen to be the OLSE of β in the filled-up model.

Estimating σ^2

If the data are complete, then $s^2 = \sum_{t=1}^{T} (y_t - \hat{y}_t)^2 / (T - K)$ is the corresponding estimator of σ^2. If $T - t_c$ cases are incomplete, that is, observations y_{mis} are missing in (8.3), then the variance σ^2 can be estimated using the complete case estimator

$$\hat{\sigma}_c^2 = \frac{\sum_{t=1}^{t_c} (y_t - \hat{y}_t)^2}{(t_c - K)} .$$

On the other hand, if the missing data are filled up according to the method of Yates, then we automatically get the estimator

$$\hat{\sigma}_{Yates}^2 = \frac{1}{(T-K)} \left\{ \sum_{t=1}^{t_c} (y_t - \hat{y}_t)^2 + \sum_{t=t_c+1}^{T} (\hat{y}_t - \hat{y}_t)^2 \right\}$$

$$= \frac{\sum_{t=1}^{t_c}(y_t - \hat{y}_t)^2}{(T-K)}, \tag{8.7}$$

which makes use of t_c observations but has $T-K$ instead of $t_c - K$ degrees of freedom. As

$$\hat{\sigma}^2_{\text{Yates}} = \hat{\sigma}^2_c \frac{t_c - K}{T - K} < \hat{\sigma}^2_c$$

we have to make an adjustment by multiplying by $(T-K)/(t_c - K)$ before using it in tests of significance. It corresponds to the conditional mean imputation as in first-order regression (which is be described in Section 8.8.3). Its main aim is to fill up the data to ensure application of standard procedures existing for balanced designs.

8.4.2 Analysis of Covariance—Bartlett's Method

Bartlett (1937) suggested an improvement of Yates's ANOVA, which is known under the name Bartlett's ANCOVA (analysis of covariance). This procedure is as follows:

(i) The missing values in y_{mis} are replaced by an arbitrary estimator $\hat{y}_{\text{mis}}$ (a guess).

(ii) Define an indicator matrix $Z : T \times (T - t_c)$ as covariable according to

$$Z = \begin{pmatrix} 0 & 0 & 0 & \cdots & 0 \\ \vdots & \vdots & \vdots & & \vdots \\ 0 & 0 & 0 & \cdots & 0 \\ 1 & 0 & 0 & \cdots & 0 \\ 0 & 1 & 0 & \cdots & 0 \\ \vdots & & \ddots & & \vdots \\ \vdots & & & \ddots & \vdots \\ 0 & 0 & 0 & \cdots & 1 \end{pmatrix}.$$

The t_c null vectors indicate the observed cases and the $(T - t_c)$-vectors e_i' indicate the missing values. The covariable Z is incorporated into the linear model by introducing the $(T - t_c)$-vector γ of additional parameters:

$$\begin{pmatrix} y_{\text{obs}} \\ \hat{y}_{\text{mis}} \end{pmatrix} = X\beta + Z\gamma + \epsilon = (X, Z)\begin{pmatrix} \beta \\ \gamma \end{pmatrix} + \epsilon. \tag{8.8}$$

The OLSE of the parameter vector $(\beta', \gamma')'$ is found by minimizing the error sum of squares:

$$S(\beta, \gamma) = \sum_{t=1}^{t_c}(y_t - x_t'\beta - 0'\gamma)^2 + \sum_{t=t_c+1}^{T}(\hat{y}_t - x_t'\beta - e_t'\gamma)^2. \tag{8.9}$$

The first term is minimal for $\hat{\beta} = b_c$ (8.4), whereas the second term becomes minimal (equating to zero) for $\hat{\gamma} = \hat{y}_{\text{mis}} - X_* b_c$. Therefore the solution to $\min_{\beta,\gamma} S(\beta,\gamma)$ is given by

$$\begin{pmatrix} b_c \\ \hat{y}_{\text{mis}} - X_* b_c \end{pmatrix} . \tag{8.10}$$

Choosing the guess $\hat{y}_{\text{mis}} = X_* b_c$ as in Yates's method, we get $\hat{\gamma} = 0$. With both methods we have $\hat{\beta} = b_c$, the complete-case OLSE. Introducing the additional parameter γ (which is without any statistical interest) has one advantage: The degrees of freedom in estimating σ^2 in model (8.8) are now T minus the number of estimated parameters, that is, $T - K - (T - t_c) = t_c - K$. Therefore we get a correct (unbiased) estimator $\hat{\sigma}^2 = \hat{\sigma}_c^2$.

8.5 Shrinkage Estimation by Yates Procedure

8.5.1 Shrinkage Estimators

The procedure of Yates essentially involves first estimating the parameters of the model with the help of the complete observations alone and obtaining the predicted values for the missing observations. These predicted values are then substituted in order to get a repaired or completed data set, which is finally used for the estimation of parameters. This strategy is adopted now using shrinkage estimators.

Now there are two popular ways for obtaining predicted values of the study variable. One is the least-squares method, which gives $\hat{y}_{\text{mis}} = X_* b_c$ as predictions for the missing observations on the study variable, and the other is the Stein-rule method, providing the following predictions:

$$\begin{aligned} \hat{y}_{\text{mis}} &= \left(1 - \frac{kR_c}{(t_c - K + 2) b_c' X_c' X_c b_c}\right) X_* b_c \\ &= \left(1 - \frac{k\hat{\epsilon}_c'\hat{\epsilon}_c}{(t_c - K + 2)\hat{y}_c'\hat{y}_c}\right) X_* b_c \end{aligned} \tag{8.11}$$

where $\hat{y}_c = X_c b_c$ and $R_c = (y_c - X_c b_c)'(y_c - X_c b_c) = \hat{\epsilon}_c'\hat{\epsilon}_c$ is the residual sum of squares and k is a positive nonstochastic scalar.

If we replace y_{mis} in (8.3) by $\hat{y}_{\text{mis}}$ and then apply the least-squares method, the following estimator of β is obtained

$$\begin{aligned} \hat{\beta} &= (X_c' X_c + X_*' X_*)^{-1} (X_c' y_c + X_*' \hat{y}_{\text{mis}}) \\ &= b_c - \frac{kR_c}{(t_c - K + 2) b_c' X_c' X_c b_c} \left(I + (X_*' X_*)^{-1} X_c' X_c\right)^{-1} b_c , \end{aligned} \tag{8.12}$$

which is of shrinkage type (see Section 3.10.3).

8.5.2 Efficiency Properties

It can be easily seen that $\hat{\beta}$ (8.12) is consistent but biased. The exact expressions for its bias vector and mean squared error matrix can be straightforwardly obtained, for example, from Judge and Bock (1978). However, they turn out to be intricate enough and may not lead to some clear inferences regarding the gain/loss in efficiency of $\hat{\beta}$ with respect to b_c. We therefore consider their asymptotic approximations with the first specification that t_c increases but t_* stays fixed.

In order to analyze the asymptotic property of $\hat{\beta}$ when t_c increases but t_* stays fixed, we assume that $V_c = t_c(X_c'X_c)^{-1}$ tends to a finite nonsingular matrix as t_c tends to infinity.

Theorem 8.1 *The asymptotic approximations for the bias vector of $\hat{\beta}$ (8.12) up to order $O(t_c^{-1})$ and the mean squared error matrix up to order $O(t_c^{-2})$ are given by*

$$\text{Bias}(\hat{\beta}) = -\frac{\sigma^2 k}{t_c \beta' V_c^{-1} \beta}\beta \tag{8.13}$$

$$\text{M}(\hat{\beta}) = \frac{\sigma^2}{t_c}V_c - \frac{2\sigma^4 k}{t_c^2 \beta' V_c^{-1}\beta}\left(V_c - \left(\frac{4+k}{2\beta' V_c^{-1}\beta}\right)\beta\beta'\right). \tag{8.14}$$

From (8.13) we observe that the bias vector has sign opposite to β. Further, the magnitude of the bias declines as k tends to be small and/or t_c grows large.

Comparing $\text{V}(b_c) = (\sigma^2/t_c)V_c$ and (8.14), we notice that the expression $[\text{V}(b_c) - \text{M}(\hat{\beta})]$ cannot be positive definite for positive values of k. Similarly, the expression $[\text{M}(\hat{\beta}) - \text{V}(b_c)]$ cannot be positive definite except in the trivial case of $K = 1$. We thus find that none of the two estimators b_c and $\hat{\beta}$ dominates over the other with respect to the criterion of mean dispersion error matrix, at least to the order of our approximation.

Next, let us compare b_c and $\hat{\beta}$ with respect to a weaker criterion of risk. If we choose the MDE-III criterion, then $\hat{\beta}$ is superior to b_c if

$$k < 2(K-2) \tag{8.15}$$

provided that K exceeds 2.

Let us now consider a second specification of more practical interest when both t_c and t_* increase. Let us define t as t_c for t_c less than t_* and as t_* for t_c greater than t_* so that $t \to \infty$ is equivalent to $t_c \to \infty$ and $t_* \to \infty$.

Assuming the asymptotic cooperativeness of explanatory variables, that is, $V_c = t_c(X_c'X_c)^{-1}$ and $V_* = t_*(X_*'X_*)^{-1}$ tend to finite nonsingular matrices as t_c and t_* grow large, we have the following results for $\hat{\beta}$.

Theorem 8.2 *For the estimator $\hat{\beta}$, the asymptotic approximations for the bias vector to order $O(t^{-1})$ and the mean squared error matrix to order*

$O(t^{-2})$ are given by

$$B(\hat{\beta}) = -\frac{\sigma^2 k}{t_c \beta' V_c^{-1} \beta} G\beta \tag{8.16}$$

$$\begin{aligned} M(\hat{\beta}) &= \frac{\sigma^2}{t_c} V_c \\ &\quad - \frac{2\sigma^2 k}{t_c^2 \beta' V_c^{-1} \beta} \left(GV_c - \frac{1}{\beta' V_c^{-1} \beta} (G\beta\beta' + \beta\beta' G' + \frac{k}{2} G\beta\beta' G') \right) \end{aligned} \tag{8.17}$$

where

$$G = V_c \left(V_c + \frac{t_c}{t_*} V_* \right)^{-1} . \tag{8.18}$$

Choosing the performance criterion to be the risk under weighted squared error loss function specified by weight matrix Q of order $O(1)$, we find from (8.17) that $\hat{\beta}$ is superior to b_c when

$$k < 2 \left(\frac{\beta' V_c^{-1} \beta}{\beta' G' Q G \beta} \operatorname{tr} QGV_c - \frac{2\beta' G' Q \beta}{\beta' G' Q G \beta} \right) \tag{8.19}$$

provided that the quantity on the right-hand side of the inequality is positive.

Let δ be the largest characteristic root of QV_* or $Q^{\frac{1}{2}} V_* Q^{\frac{1}{2}}$ in the metric of QV_c or $Q^{\frac{1}{2}} V_c Q^{\frac{1}{2}}$. Now we observe that

$$\begin{aligned} \operatorname{tr} QGV_c &\geq \left(\frac{t_*}{t_* + \delta t_c} \right) \operatorname{tr} QV_c \\ \frac{\beta' G' Q G \beta}{\beta' V_c^{-1} \beta} &\leq \left(\frac{t_*}{t_* + \delta t_c} \right)^2 \lambda_p \\ \frac{\beta' G' Q \beta}{\beta' G' Q G \beta} &\leq \left(\frac{t_*}{t_* + \delta t_c} \right) \end{aligned} \tag{8.20}$$

and hence we see that condition (8.19) is satisfied as long as

$$k < 2 \left(1 + \frac{\delta t_c}{t_*} \right) \left[T - 2 \left(\frac{t_*}{t_* + \delta t_*} \right)^{-1} \right] ; \quad T > \left(\frac{t_*}{t_* + \delta t_*} \right)^2 , \tag{8.21}$$

which is easy to check in any given application owing to the absence of β.

Similar to (8.21), one can derive various sufficient versions of the condition (8.19).

For the proof of the theorem and for further results, the reader is referred to Toutenburg, Srivastava, and Fieger (1997).

8.6 Missing Values in the X-Matrix

In econometric models, other than in experimental designs in biology or pharmacy, the matrix X does not have a fixed design but contains observations of exogeneous variables, which may be random, including the possibility that some data are missing. In general, we may assume the following structure of data:

$$\begin{pmatrix} y_{\text{obs}} \\ y_{\text{mis}} \\ y_{\text{obs}} \end{pmatrix} = \begin{pmatrix} X_{\text{obs}} \\ X_{\text{obs}} \\ X_{\text{mis}} \end{pmatrix} \beta + \epsilon \,, \quad \epsilon \sim (0, \sigma^2 I) \,.$$

Estimation of y_{mis} corresponds to the prediction problem, which is discussed in Chapter 6 in full detail. The classical prediction of y_{mis} using X_{obs} is equivalent to the method of Yates.

8.6.1 *General Model*

Based on the above arguments, we may drop the cases in $(y_{\text{mis}}, X_{\text{obs}})$ and now confine ourselves to the structure

$$y_{\text{obs}} = \begin{pmatrix} X_{\text{obs}} \\ X_{\text{mis}} \end{pmatrix} \beta + \epsilon \,.$$

We change the notation as follows:

$$\begin{pmatrix} y_c \\ y_* \end{pmatrix} = \begin{pmatrix} X_c \\ X_* \end{pmatrix} \beta + \begin{pmatrix} \epsilon_c \\ \epsilon_* \end{pmatrix}, \quad \begin{pmatrix} \epsilon_c \\ \epsilon_* \end{pmatrix} \sim (0, \sigma^2 I) \,. \tag{8.22}$$

The submodel

$$y_c = X_c \beta + \epsilon_c \tag{8.23}$$

represents the completely observed data, where we have $y_c : t_c \times 1$, $X_c : t_c \times K$ and assume $\text{rank}(X_c) = K$. Let us further assume that X is nonstochastic (if X is stochastic, unconditional expectations have to be replaced by conditional expectations).

The remaining part of (8.22), that is,

$$y_* = X_* \beta + \epsilon_* \,, \tag{8.24}$$

is of dimension $T - t_c = t_*$. The vector y_* is completely observed. The notation X_* shall underline that X_* is partially incomplete (whereas X_{mis} stands for a completely missing matrix). Combining both submodels (8.23) and (8.24) in model (8.22) corresponds to investigating the mixed model (5.140). Therefore, it seems to be a natural idea to use the mixed model estimators for handling nonresponse in X_* by imputation methods.

The optimal, but due to the unknown elements in X_*, nonoperational estimator is given by the mixed estimator of β in the model (8.22) according to Theorem 5.17 as

$$\begin{aligned} \hat{\beta}(X_*) &= (X_c'X_c + X_*'X_*)^{-1}(X_c'y_c + X_*'y_*) \\ &= b_c + S_c^{-1}X_*'(I_{t_*} + X_*S_c^{-1}X_*')^{-1}(y_* - X_*b_c), \end{aligned} \tag{8.25}$$

where $b_c = (X_c'X_c)^{-1}X_c'y_c$ is the OLSE of β in the complete-case model (8.23) and $S_c = X_c'X_c$.

The estimator $\hat{\beta}(X_*)$ is unbiased for β and has the dispersion matrix (cf. (5.148))

$$\mathrm{V}\left(\hat{\beta}(X_*)\right) = \sigma^2(S_c + S_*)^{-1}, \tag{8.26}$$

where $S_* = X_*'X_*$ is used for abbreviation.

8.6.2 *Missing Values and Loss in Efficiency*

We now discuss the consequences of confining the analysis to the complete-case model (8.23), assuming that the selection of complete cases is free of selectivity bias. Our measure to compare $\hat{\beta}_c$ and $\hat{\beta}(X_*)$ is the scalar risk

$$R(\hat{\beta}, \beta, S_c) = \mathrm{tr}\{S_c \mathrm{V}(\hat{\beta})\},$$

which coincides with the MDE-III risk (cf. (5.66)). From Theorem A.18 (iii) we have the identity

$$(S_c + X_*'X_*)^{-1} = S_c^{-1} - S_c^{-1}X_*'(I_{t_*} + X_*S_c^{-1}X_*')^{-1}X_*S_c^{-1}.$$

Applying this, we get the risk of $\hat{\beta}(X_*)$ as

$$\begin{aligned} \sigma^{-2}R(\hat{\beta}(X_*), \beta, S_c) &= \mathrm{tr}\{S_c(S_c + S_*)^{-1}\} \\ &= K - \mathrm{tr}\{(I_{t_*} + B'B)^{-1}B'B\}, \end{aligned} \tag{8.27}$$

where $B = S_c^{-1/2}X_*'$.

The $t_* \times t_*$-matrix $B'B$ is nonnegative definite of $\mathrm{rank}(B'B) = J^*$. If $\mathrm{rank}(X_*) = t_* < K$ holds, then $J^* = t_*$ and $B'B > 0$ follow.

Let $\lambda_1 \geq \ldots \geq \lambda_{t_*} \geq 0$ denote the eigenvalues of $B'B$, $\Lambda = \mathrm{diag}(\lambda_1, \ldots, \lambda_{t_*})$, and let P be the matrix of orthogonal eigenvectors. Then we have $B'B = P\Lambda P'$ (cf. Theorem A.30) and

$$\begin{aligned} \mathrm{tr}\{(I_{t_*} + B'B)^{-1}B'B\} &= \mathrm{tr}\{P(I_{t_*} + \Lambda)^{-1}P'P\Lambda P'\} \\ &= \mathrm{tr}\{(I_{t_*} + \Lambda)^{-1}\Lambda\} = \sum_{i=1}^{t_*} \frac{\lambda_i}{1 + \lambda_i}. \end{aligned} \tag{8.28}$$

Assuming MCAR and stochastic X, the MDE-III risk of the complete-case estimator b_c is

$$\sigma^{-2}R(b_c, \beta, S_c) = \mathrm{tr}\{S_c S_c^{-1}\} = K. \tag{8.29}$$

Using the MDE-III criterion for the comparison of b_c and $\hat{\beta}(X_*)$, we may conclude that

$$R(b_c, \beta, S_c) - R(\hat{\beta}(X_*), \beta, S_c) = \sum_{i=1}^{t_*} \frac{\lambda_i}{1+\lambda_i} \geq 0$$

holds, and, hence, $\hat{\beta}(X_*)$ in any case is superior to b_c. This result is expected. To have more insight into this relationship, let us apply another criterion by comparing the size of the risks instead of their differences.

Definition 8.3 *The relative efficiency of an estimator $\hat{\beta}_1$ compared to another estimator $\hat{\beta}_2$ is defined by the ratio*

$$\text{eff}(\hat{\beta}_1, \hat{\beta}_2, A) = \frac{R(\hat{\beta}_2, \beta, A)}{R(\hat{\beta}_1, \beta, A)}. \tag{8.30}$$

$\hat{\beta}_1$ is said to be less efficient than $\hat{\beta}_2$ if

$$\text{eff}(\hat{\beta}_1, \hat{\beta}_2, A) \leq 1.$$

Using (8.27)–(8.29), the efficiency of b_c compared to $\hat{\beta}(X_*)$ is

$$\text{eff}(b_c, \hat{\beta}(X_*), S_c) = 1 - \frac{1}{K} \sum_{i=1}^{t_*} \frac{\lambda_i}{1+\lambda_i} \leq 1. \tag{8.31}$$

The relative efficiency of the estimator b_c compared to the mixed estimator in the full model (8.22) falls in the interval

$$\max\left\{0, 1 - \frac{t_*}{K} \frac{\lambda_1}{1+\lambda_1}\right\} \leq \text{eff}(b_c, \hat{\beta}(X_*), S_c) \leq 1 - \frac{t_*}{K} \frac{\lambda_{t_*}}{1+\lambda_{t_*}} \leq 1. \tag{8.32}$$

Examples:

(i) Let $X_* = X_c$, so that the matrix X_c is used twice. Then $B'B = X_c S_c^{-1} X_c'$ is idempotent of rank K. Therefore, we have $\lambda_i = 1$ for $i = 1, ..., K$; $\lambda_i = 0$ else (cf. Theorem A.61 (i)) and

$$\text{eff}(b_c, \hat{\beta}(X_c), S_c) = \frac{1}{2}.$$

(ii) $t_* = 1$ (one row of X is incomplete). Then $X_* = x_*'$ becomes a K-vector and $B'B = x_*' S_c^{-1} x_*$ a scalar. Let $\mu_1 \geq \ldots \geq \mu_K > 0$ be the eigenvalues of S_c and let $\Gamma = (\gamma_1, \ldots, \gamma_K)$ be the matrix of the corresponding orthogonal eigenvectors.
Therefore, we may write

$$\hat{\beta}(x_*) = (S_c + x_* x_*')^{-1}(X_c' y_c + x_* y_*).$$

Using Theorem A.44, we have

$$\mu_1^{-1} x_*' x_* \leq x_*' S_c^{-1} x_* = \sum_{j=1}^{K} \mu_j^{-1} (x_*' \gamma_j)^2 \leq \mu_K^{-1} x_*' x_*,$$

and according to (8.31), $\text{eff}(b_c, \hat{\beta}(x_*), S_c)$ becomes

$$\begin{aligned}\text{eff}(b_c, \hat{\beta}(x_*), S_c) &= 1 - \frac{1}{K}\frac{x_*' S_c^{-1} x_*}{1 + x_*' S_c^{-1} x_*} \\ &= 1 - \frac{1}{K}\frac{\sum_{j=1}^K \mu_j^{-1}(x_*'\gamma_j)^2}{1 + \sum_{j=1}^K \mu_j^{-1}(x_*'\gamma_j)^2} \le 1\,.\end{aligned}$$

The interval (8.32) has the form

$$1 - \frac{\mu_1 \mu_K^{-1} x_*' x_*}{K(\mu_1 + x_*' x_*)} \le \text{eff}(b_c, \hat{\beta}(x_*), S_c) \le 1 - \frac{x_*' x_*}{K(\mu_1 \mu_K^{-1})(\mu_K + x_*' x_*)}\,.$$

The relative efficiency of b_c compared to $\hat{\beta}(x_*)$ is dependent on the norm $(x_*' x_*)$ of the vector x_* as well as on the eigenvalues of the matrix S_c, that is, on their condition number μ_1/μ_K and the span $\mu_1 - \mu_k$.
Let $x_* = g\gamma_j \quad (j = 1, \ldots, K)$ and $\mu = (\mu_1, \ldots, \mu_K)'$, where g is a scalar and γ_i is the jth orthonormal eigenvector of S_c corresponding to the eigenvalue μ_j. Then for these vectors $x_* = g\gamma_j$, which are parallel to the eigenvectors of S_c, the quadratic risk of the estimators $\hat{\beta}(g\gamma_j)$ $(j = 1, \ldots, K)$ becomes

$$\sigma^{-2} R(\hat{\beta}(g\gamma_j), \beta, S_c) = \text{sp}\{\Gamma\mu\Gamma'(\Gamma\mu\Gamma' + g^2\gamma_j\gamma_j')^{-1}\} = K - 1 + \frac{\mu_j}{\mu_j + g^2}\,.$$

Inspecting this equation, we note that $\text{eff}(b_c, \hat{\beta}(g\gamma_j)$ has its maximum for $j = 1$ (i.e., if x_* is parallel to the eigenvector corresponding to the maximal eigenvalue μ_1). Therefore, the loss in efficiency by leaving out one incomplete row is minimal for $x_* = g\gamma_1$ and maximal for $x_* = g\gamma_K$. This fact corresponds to the result of Silvey (1969), who proved that the goodness of fit of the OLSE may be improved optimally if additional observations are taken in the direction that is the most imprecise. But this is just the direction of the eigenvector γ_K, corresponding to the minimal eigenvalue μ_K of S_c.

8.7 Methods for Incomplete X-Matrices

8.7.1 Complete Case Analysis

The technically easiest method is to confine the analysis to the completely observed submodel (8.23); the partially incomplete cases are not used at all. The corresponding estimator of β is $b_c = S_c^{-1} X_c' y_c$, with covariance matrix $\text{V}(b_c) = \sigma^2 S_c^{-1}$. This estimator is unbiased as long as missingness is independent of y, that is, if

$$f(y|R,X) = \frac{f(y,R|X)}{f(R|X)} = f(y|X)$$

holds, and the model is correctly specified. Here R is the missing indicator matrix introduced in Section 8.2.1.

8.7.2 Available Case Analysis

Suppose that the regressors $X_1, \ldots, X_K$ (or $X_2, \ldots, X_K$ if $X_1 = 1$) are stochastic. Then the $(X_1, \ldots, X_K, y)$ have a joint distribution with mean $\mu = (\mu_1, \ldots, \mu_K, \mu_y)$ and covariance matrix

$$\Sigma = \begin{pmatrix} \Sigma_{xx} & \Sigma_{xy} \\ \Sigma_{yx} & \sigma_{yy} \end{pmatrix}.$$

Then β can be estimated by solving the normal equations

$$\hat{\Sigma}_{xx}\hat{\beta} = \hat{\Sigma}_{yx}\,, \tag{8.33}$$

where $\hat{\Sigma}_{xx}$ is the $K \times K$-sample covariance matrix. The solutions are

$$\hat{\beta} = \hat{\Sigma}_{yx}\hat{\Sigma}_{xx}^{-1}\,,$$

with

$$\hat{\beta}_0 = \hat{\mu}_y - \sum_{j=1}^{K} \hat{\beta}_j \hat{\mu}_j\,,$$

the term for the intercept or constant variable $X_1 = (1, \ldots, 1)'$.

The (i,j)th element of $\hat{\Sigma}_{xx}$ is computed from the pairwise observed elements of the variables x_i and x_j. Similarly, $\hat{\Sigma}_{yx}$ makes use of the pairwise observed elements of x_i and y. Based on simulation studies, Haitovsky (1968) has investigated the performance of this method and has concluded that in many situations the complete-case estimator b_c is superior to the estimator $\hat{\beta}$ from this method.

8.7.3 Maximum-Likelihood Methods

Having a monotone pattern (cf. Figure 8.2), the common distribution of the data D (given some parameter ϕ) can be factorized as follows:

$$\begin{aligned}
&\prod_{i=1}^{T} f(d_{i1}, d_{i2}, \ldots, d_{iK}|\phi) \\
&\quad = \prod_{i=1}^{T} f(d_{i1}|\phi_1) \prod_{i=1}^{t_2} f(d_{i2}|d_{i1}, \phi_2) \cdots \prod_{i=1}^{t_K} f(d_{iK}|d_{i1}, \ldots, d_{i,K-1}, \phi_K)\,,
\end{aligned}$$

where $t_2, \ldots, t_K$ are the number of observations for variables $2, \ldots, K$, respectively.

Consider a model $y = X\beta + \epsilon$, where the joint distribution of y and X is a multivariate normal distribution with mean μ and covariance matrix Σ. Without missing values, ML estimates of μ and Σ are used as in Section 8.7.2 to obtain the estimates of the regression parameters.

For the case of $X = (X_1, \ldots, X_K)$ with missing values in X_1 only, the joint distribution of Y and X_1 conditional on the remaining X's can be factored as

$$f(y, X_1|X_2, \ldots, X_K, \phi) = f(y|X_2, \ldots, X_K, \phi_1) f(X_1|X_2, \ldots, X_K, y, \phi_2).$$

The corresponding likelihood of ϕ_1 and ϕ_2 can be maximized seperately, as ϕ_1 and ϕ_2 are distinct sets of parameters. The results are two complete data problems, which can be solved using standard techniques. The results can be combined to obtain estimates of the regression of interest (cf. Little, 1992):

$$\hat{\beta}_{y1|1,\ldots,K} = \frac{\tilde{\beta}_{1y|2,\ldots,K,y}\hat{\sigma}_{yy|2,\ldots,K}}{\tilde{\sigma}_{11|2,\ldots,K,y} + \tilde{\beta}^2_{1y|2,\ldots,K,y}\hat{\sigma}_{yy|2,\ldots,K}}$$

$$\hat{\beta}_{yj|1,\ldots,K} = -\frac{\hat{\beta}_{yj|2,\ldots,K}\tilde{\sigma}_{11|2,\ldots,K,y} - \tilde{\beta}_{1y|2,\ldots,K,y}\tilde{\beta}_{1j|2,\ldots,K,y}\hat{\sigma}_{yy|2,\ldots,K}}{\tilde{\sigma}_{11|2,\ldots,K,y} + \tilde{\beta}^2_{1y|2,\ldots,K,y}\hat{\sigma}_{yy|2,\ldots,K}}$$

where parameters with a tilde ($\tilde{}$) belong to ϕ_2 (the regression of X_1 on $X_2, \ldots, X_K, y$, from the t_c complete cases) and parameters with a hat ($\hat{}$) belong to ϕ_1 (the regression of y on $X_2, \ldots, X_K$, estimated from all T cases).

In this case the assumption of joint normality has to hold only for (y, X_1); covariates $X_2, \ldots, X_K$ may also be categorical variables. General patterns of missing data require iterative approaches such as the EM algorithm by Dempster, Laird, and Rubin (1977). A detailed discussion of likelihood-based approaches can be found in Little and Rubin (1987).

8.8 Imputation Methods for Incomplete X-Matrices

This section gives an overview of methods that impute values for missing observations. Most of the methods presented here are based on the assumption that the variables in X are continuous.

The conditions under which the respective procedures yield consistent estimates of the regression parameters are discussed in Section 8.9.

8.8.1 Maximum-Likelihood Estimates of Missing Values

Suppose that the errors are normally distributed (i.e., $\epsilon \sim N(0, \sigma^2 I_T)$) and, moreover, assume a monotone pattern of missing values. Then the likelihood can be factorized with one component for the observed data and one for the missing data (cf. Little and Rubin, 1987). We confine ourselves to the simplest case of a completely nonobserved matrix X_*. Therefore, X_* may be interpreted as an unknown parameter to be estimated. The loglikelihood of model (8.22) may be written as

$$\begin{aligned}\ln L(\beta, \sigma^2, X_*) &= -\frac{n}{2}\ln(2\pi) - \frac{n}{2}\ln(\sigma^2) \\ &\quad - \frac{1}{2\sigma^2}(y_c - X_c\beta, y_* - X_*\beta)' \begin{pmatrix} y_c - X_c\beta \\ y_* - X_*\beta \end{pmatrix}.\end{aligned}$$

Differentiating with respect to β, σ^2, and X_* gives

$$\begin{aligned}\frac{\partial \ln L}{\partial \beta} &= \frac{1}{2\sigma^2}\{X_c'(y_c - X_c\beta) + X_*'(y_* - X_*\beta)\} = 0\,, \\ \frac{\partial \ln L}{\partial \sigma^2} &= \frac{1}{2\sigma^2}\Big\{ -n + \frac{1}{\sigma^2}(y_c - X_c\beta)'(y_c - X_c\beta) \\ &\quad + \frac{1}{\sigma^2}(y_* - X_*\beta)'(y_* - X_*\beta)\Big\} = 0 \\ \frac{\partial \ln L}{\partial X_*} &= \frac{1}{2\sigma^2}(y_* - X_*\beta)\beta' = 0\,.\end{aligned}$$

Solving for β and σ^2,

$$\hat\beta = b_c = S_c^{-1}X_c'y_c\,, \tag{8.34}$$

$$\hat\sigma^2 = \frac{1}{m}(y_c - X_c b_c)'(y_c - X_c b_c)\,, \tag{8.35}$$

results in ML estimators that are based on the data of the complete-case submodel (8.23). The ML estimate $\hat X_*$ is the solution of the equation

$$y_* = \hat X_* b_c\,. \tag{8.36}$$

In the one-regressor model (i.e., $K = 1$) the solution is unique:

$$\hat x_* = \frac{y_*}{b_c}\,,$$

where $b_c = (x_c'x_c)^{-1}x_c'y_c$ (cf. Kmenta, 1971). For $K > 1$ there exists a $t_* \times (K-1)$-fold set of solutions $\hat X_*$. If any solution $\hat X_*$ is substituted for X_* in the mixed model, that is,

$$\begin{pmatrix} y_c \\ y_* \end{pmatrix} = \begin{pmatrix} X_c \\ \hat X_* \end{pmatrix}\beta + \begin{pmatrix} \epsilon_c \\ \epsilon_* \end{pmatrix},$$

then we are led to the following identity:

$$\hat\beta(\hat X_*) = (S_c + \hat X_*'\hat X_*)^{-1}(X_c'y_c + \hat X_*'y_*)$$

$$\begin{aligned} &= (S_c + \hat{X}_*'\hat{X}_*)^{-1}(S_c\beta + X_c'\epsilon_c + \hat{X}_*'\hat{X}_*\beta + \hat{X}_*'\hat{X}_* S_c^{-1} X_c'\epsilon_c) \\ &= \beta + (S_c + \hat{X}_*'\hat{X}_*)^{-1}(S_c + \hat{X}_*'\hat{X}_*)S_c^{-1}X_c'\epsilon_c \\ &= \beta + S_c^{-1}X_c'\epsilon_c \\ &= b_c \,. \end{aligned} \tag{8.37}$$

This corresponds to the results of Section 8.4.1. Therefore, filling up missing values X_* by their ML estimators $\hat{X}_*$ and calculating the mixed estimator $\hat{\beta}(\hat{X}_*)$ gives $\hat{\beta}(\hat{X}_*) = b_c$.

On the other hand, if we don't have a monotone pattern, the ML equations have to be solved by iterative procedures as, for example, the EM algorithm (Dempster et al., 1977) or other procedures (cf. Oberhofer and Kmenta, 1974).

8.8.2 Zero-Order Regression

The zero-order regression (ZOR) method is due to Wilks (1938) and is also called the method of sample means. A missing value x_{ij} of the jth regressor X_j is replaced by the sample mean of the observed values of X_j computed from the complete cases or the available cases.

Let

$$\Phi_j = \{i : x_{ij} \text{ missing}\}, \quad j = 1, \ldots, K \tag{8.38}$$

denote the index sets of the missing values of X_j, and let M_j be the number of elements in Φ_j. Then for j fixed, any missing value x_{ij} in X_* is replaced by

$$\hat{x}_{ij} = \bar{x}_j = \frac{1}{T - M_j} \sum_{i \notin \Phi_j} x_{ij} \,, \tag{8.39}$$

using all available cases, or

$$\hat{x}_{ij} = \bar{x}_j = \frac{1}{T - t_c} \sum_{i=1}^{t_c} x_{ij} \,,$$

using the complete cases only.

If the sample mean can be expected to be a good estimator for the unknown mean μ_j of the jth column, then this method may be recommended. If, on the other hand, the data in the jth column have a trend or follow a growth curve, then $\bar{x}_j$ is not a good estimator and its use can cause some bias. If all the missing values in the matrix X_* are replaced by their corresponding column means $\bar{x}_j$ $(j = 1, \ldots, K)$, this results in a filled-up matrix $X_{(1)}$, say, and in an operationalized version of the mixed model (8.22), that is,

$$\begin{pmatrix} y_c \\ y_* \end{pmatrix} = \begin{pmatrix} X_c \\ X_{(1)} \end{pmatrix} \beta + \begin{pmatrix} \epsilon \\ \epsilon_{(1)} \end{pmatrix} .$$

Inspecting the vector of errors $\epsilon_{(1)}$, namely,

$$\epsilon_{(1)} = (X_* - X_{(1)})\beta + \epsilon_* ,$$

we have

$$\epsilon_{(1)} \sim \{(X_* - X_{(1)})\beta, \sigma^2 I_{t_*}\} ,$$

where again $t_* = T - t_c$.

In general, replacing missing values results in a biased mixed estimator unless $X_* - X_{(1)} = 0$ holds. If X is a matrix of stochastic regressor variables, then one may expect that at least $\mathrm{E}(X_* - X_{(1)}) = 0$ holds.

8.8.3 *First-Order Regression*

The notation "first-order regression" (FOR) is used for a set of methods that make use of the correlative structure of the covariate matrix X. Based on the index sets Φ_j of (8.38), the dependence of any column X_j ($j = 1, \dots, K$, j fixed) with the remaining columns is modeled by additional regressions, that is,

$$x_{ij} = \theta_{0j} + \sum_{\substack{l=1 \\ l \neq j}}^{K} x_{il}\theta_{lj} + u_{ij} , \quad i \notin \Phi = \bigcup_{j=1}^{K} \Phi_j = t_c + 1, \dots, T , \qquad (8.40)$$

with parameters estimated from the complete cases only. Alternatively, the parameters could be estimated from all cases $i \notin \Phi_j$, but then the auxiliary regressions would again involve incomplete data.

The missing values x_{ij} of X_* are estimated and replaced by

$$\hat{x}_{ij} = \hat{\theta}_{0j} + \sum_{\substack{l=1 \\ l \neq j}}^{K} x_{il}\hat{\theta}_{lj} . \qquad (8.41)$$

Example 8.1 (Disjoint sets Φ_j of indices): Let X_c be an $t_c \times K$-matrix and X_* the following $2 \times K$-matrix:

$$X_* = \begin{pmatrix} * & x_{t_c+1,2} & x_{t_c+1,3} & \cdots & x_{t_c+1,K} \\ x_{t_c+2,1} & * & x_{t_c+2,3} & \cdots & x_{t_c+2,K} \end{pmatrix} ,$$

where "*" indicates missing values. The corresponding index sets are

$$\Phi_1 = \{t_c + 1\} , \Phi_2 = \{t_c + 2\} , \Phi_3 = \cdots = \Phi_K = \emptyset ,$$
$$\Phi = \bigcup_{j=1}^{K} \Phi_j = \{t_c + 1, t_c + 2\} .$$

Then we have the following two additional regressions:

$$x_{1i} = \theta_{01} + \sum_{l=2}^{K} x_{il}\theta_{l1} + u_{i1} , \quad i = 1, \dots, t_c ,$$

$$x_{i2} = \theta_{02} + x_{i1}\theta_{12} + \sum_{l=3}^{K} x_{il}\theta_{l2} + u_{i2}, \quad i = 1, \ldots, t_c .$$

The parameters in the above two equations are estimated by their corresponding OLSEs $\hat{\theta}_1$ and $\hat{\theta}_2$, respectively, and x_{1i} and x_{2i} are estimated by their respective classical predictors, that is,

$$\hat{x}_{t_c+1,1} = \hat{\theta}_{01} + \sum_{l=2}^{K} x_{t_c+1,l}\hat{\theta}_{l1}$$

and

$$\hat{x}_{t_c+2,2} = \hat{\theta}_{02} + \sum_{\substack{l=1 \\ l\neq 2}}^{K} x_{t_c+2,l}\hat{\theta}_{l2} .$$

This procedure gives the filled-up matrix

$$\hat{X}_* = \begin{pmatrix} \hat{x}_{t_c+1,1} & x_{t_c+1,2} & x_{t_c+1,3} & \cdots & x_{t_c+1,K} \\ x_{t_c+2,1} & \hat{x}_{t_c+2,2} & x_{t_c+2,3} & \cdots & x_{t_c+2,K} \end{pmatrix} = X_{(2)} .$$

Thus, the operationalized mixed model is

$$\begin{pmatrix} y_c \\ y_* \end{pmatrix} = \begin{pmatrix} X_c \\ X_{(2)} \end{pmatrix} \beta + \begin{pmatrix} \epsilon_c \\ \epsilon_{(2)} \end{pmatrix}$$

with the vector of errors $\epsilon_{(2)}$:

$$\begin{aligned} \epsilon_{(2)} &= (X_* - X_{(2)})\beta + \epsilon_* \\ &= \begin{pmatrix} x_{t_c+1,1} - \hat{x}_{t_c+1,1} & 0 & 0 & \cdots & 0 \\ 0 & x_{t_c+2,2} - \hat{x}_{t_c+2,2} & 0 & \cdots & 0 \end{pmatrix} \beta + \epsilon_* \\ &= \begin{pmatrix} (x_{t_c+1,1} - \hat{x}_{t_c+1,1})\beta_1 \\ (x_{t_c+2,2} - \hat{x}_{t_c+2,2})\beta_2 \end{pmatrix} + \begin{pmatrix} \epsilon_{t_c+1} \\ \epsilon_{t_c+2} \end{pmatrix} . \end{aligned}$$

Example 8.2 (Nondisjoint sets of indices Φ_j): Let $t_* = 1$ and

$$x_* = (*, *, x_{t_c+1,3}, \ldots, x_{t_c+1,K})' .$$

Then we have $\Phi_1 = \Phi_2 = \{t_c + 1\}, \Phi_3 = \cdots = \Phi_K = \emptyset$. We calculate the estimators $\hat{\theta}_1$ and $\hat{\theta}_2$ analogously to the previous example. To calculate $\hat{x}_{t_c+1,1}$, we need $\hat{x}_{t_c+1,2}$ and vice versa. Many suggestions have been made to overcome this problem in the case of nondisjoint sets of indices. Afifi and Elashoff (1967) proposed specific means (cf. Buck, 1960, also). Dagenais (1973) described a generalized least-squares procedure using first-order approximations to impute for missing values in X_*. Alternatively, one takes that additional regression model having the largest coefficient of determination. All other missing values are replaced by column means. This way, one can combine ZOR and FOR procedures.

This procedure can be extended, in that the values of the response variable y are also used in the estimation of the missing values in $\hat{X}_*$. Toutenburg, Srivastava, and Fieger (1996) have presented some results on the asymptotic properties of this procedure. Generally biased estimators result in additionally using y in the auxillary regressions.

8.8.4 Multiple Imputation

Single imputations for missing values as described in Sections 8.8.2 and 8.8.3 underestimate the standard errors, because imputation errors are not taken into account. Multiple imputation was proposed by Rubin and is described in full detail in Rubin (1987).

The idea is to impute more than one value, drawn from the predictive distribution, for each missing observation. The I imputations result in I complete data problems with estimates $\hat{\theta}_i$, that can be combined to the final estimates by

$$\hat{\theta} = \frac{1}{I} \sum_{i=1}^{I} \hat{\theta}_i \,.$$

The corresponding variance can be estimated by

$$\hat{s}^2 = s_{\text{w}}^2 + \left(1 + \frac{1}{I}\right) s_{\text{b}}^2 \,,$$

where s_{w}^2 is the average variance within the I repeated imputation steps ($s_{\text{w}}^2 = 1/I \sum \hat{s}_i^2$), and $s_{\text{b}}^2 = \sum (\hat{\theta}_i - \hat{\theta})^2/(I-1)$ is the variance between the imputation steps (which takes care of the imputation error).

The draws are from the predictive distribution of the missing values conditioned on the observed data and the responses y. For example, consider $X = (X_1, X_2)$ with missing values in X_1. Imputations for missing X_1 values are then drawn from the conditional distribution of X_1 given X_2 and y.

8.8.5 Weighted Mixed Regression

Imputation for missing values in X_* in any case gives a filled-up matrix X_R, say, where X_R equals $X_{(1)}$ for ZOR, $X_{(2)}$ for FOR, and $\hat{X}_*$ for ML estimation. The operationalized mixed model may be written as

$$\begin{pmatrix} y_{\text{c}} \\ y_* \end{pmatrix} = \begin{pmatrix} X_{\text{c}} \\ X_R \end{pmatrix} \beta + \begin{pmatrix} \epsilon_{\text{c}} \\ \epsilon_R \end{pmatrix} \tag{8.42}$$

with

$$\epsilon_R = (X_* - X_R)\beta + \epsilon_* \,.$$

Let

$$\delta = (X_* - X_R)\beta \,;$$

then in general we may expect that $\delta \neq 0$. The least-squares estimator of β in the model (8.42) is given by the mixed estimator

$$\hat{\beta}(X_R) = (S_c + S_R)^{-1}(X_c'y_c + X_R'y_*), \tag{8.43}$$

which is a solution to the minimization problem

$$\min_\beta S(\beta) = \min_\beta\{(y_c - X_c\beta)'(y_c - X_c\beta) + (y_* - X_R\beta)'(y_* - X_R\beta)\},$$

where

$$S_R = X_R'X_R .$$

The mixed estimator has

$$\mathrm{E}\left(\hat{\beta}(X_R)\right) = \beta + (S_c + S_R)^{-1}X_R'\delta \tag{8.44}$$

and hence $\hat{\beta}(X_R)$ is biased if $\delta \neq 0$.

The decision to apply either complete-case analysis or to work with some imputed values depends on the comparison of the unbiased estimator b_c and the biased mixed estimator for $\hat{\beta}(X_R)$. If one of the mean dispersion error criteria is used, then the results of Section 5.8.3 give the appropriate conditions.

The scalar MDE-II and MDE-III criteria (cf. Section 5.4) were introduced to weaken the conditions for superiority of a biased estimator over an unbiased estimator. We now propose an alternative method that, analogous to weaker MDE superiority, shall weaken the superiority conditions for the biased mixed estimator. The idea is to give the completely observed submodel (8.23) a higher weight than the filled-up submodel $y_* = X_R\beta + \epsilon_R$.

To give the observed "sample" matrix X_c a different weight than the nonobserved matrix X_R in estimating β, Toutenburg (1989b) and Toutenburg and Schaffrin (1989) suggested solving

$$\min_\beta\{(y_c - X_c\beta)'(y_c - X_c\beta) + \lambda(y_* - X_R\beta)'(y_* - X_R\beta)\}, \tag{8.45}$$

where λ is a scalar factor. Differentiating (8.45) with respect to β and equating to zero gives the normal equation

$$(S_c + \lambda S_R)\beta - (X_c'y_c + \lambda X_R'y_*) = 0 .$$

The solution defined by

$$b(\lambda) = (S_c + \lambda S_R)^{-1}(X_c'y_c + \lambda X_R'y_*) . \tag{8.46}$$

may be called the weighted mixed-regression estimator (WMRE). This estimator may be interpreted as the familiar mixed estimator in the model

$$\begin{pmatrix} y_c \\ \sqrt{\lambda}y_* \end{pmatrix} = \begin{pmatrix} X_c \\ \sqrt{\lambda}X_R \end{pmatrix}\beta + \begin{pmatrix} \epsilon_c \\ \sqrt{\lambda}v_* \end{pmatrix} .$$

If $Z = Z(\lambda) = (S_c + \lambda S_R)$, and $\delta = (X_* - X_R)\beta$, we have

$$b(\lambda) \quad = \quad Z^{-1}(X_c'X_c\beta + X_c'\epsilon_c + \lambda X_R'X_*\beta + \lambda X_R'\epsilon_*)$$

$$= \beta + \lambda Z^{-1} X_R'(X_* - X_R)\beta + Z^{-1}(X_c'\epsilon_c + \lambda X_R'\epsilon_*) , \quad (8.47)$$

from which it follows that the WMRE is biased:

$$\text{Bias}\left(b(\lambda)\right) = \lambda Z^{-1} X_R' \delta \quad (8.48)$$

and has the covariance matrix

$$\text{V}\left(b(\lambda)\right) = \sigma^2 Z^{-1}(S_c + \lambda^2 S_R) Z^{-1} . \quad (8.49)$$

Note: Instead of weighting the approximation matrix X_R by a uniform factor $\sqrt{\lambda}$, one may give each of the t_* rows of X_R a different weight $\sqrt{\lambda_j}$ and solve

$$\min_\beta \left\{ (y_c - X_c\beta)'(y_c - X_c\beta) + \sum_{i=1}^{t_*} \lambda_i (y_*^{(i)} - x_R'^{(i)}\beta)^2 \right\}$$

or, equivalently,

$$\min_\beta \left\{ (y_c - X_c\beta)'(y_c - X_c\beta) + (y_* - X_R\beta)'\Lambda(y_* - X_R\beta) \right\},$$

where $\Lambda = \text{diag}(\lambda_1, \ldots, \lambda_{t_*})$. The solution of this optimization problem is seen to be of the form

$$b(\lambda_1, \ldots, \lambda_{t_*}) = \left(S_c + \sum_{i=1}^{t_*} \lambda_i x_R^{(i)} x_R'^{(i)} \right)^{-1} \left(X_c' y_c + \sum_{i=1}^{t_*} \lambda_i x_R^{(i)} y_*^{(i)} \right)$$

or, equivalently,

$$b(\Lambda) = (S_c + X_R'\Lambda X_R)^{-1}(X_c' y_c + X_R'\Lambda y_*) ,$$

which may be interpreted as the familiar mixed estimator in the model

$$\begin{pmatrix} y_c \\ \sqrt{\lambda_1} y_*^{(1)} \\ \vdots \\ \sqrt{\lambda_{t_*}} y_*^{(t_*)} \end{pmatrix} = \begin{pmatrix} X_c \\ \sqrt{\lambda_1} x_R'^{(1)} \\ \vdots \\ \sqrt{\lambda_{t_*}} x_R'^{(t_*)} \end{pmatrix} + \begin{pmatrix} \epsilon_c \\ \sqrt{\lambda_1} v_*^{(1)} \\ \vdots \\ \sqrt{\lambda_{t_*}} v_*^{(t_*)} \end{pmatrix}$$

or, equivalently written,

$$\begin{pmatrix} y_c \\ \sqrt{\Lambda} y_* \end{pmatrix} = \begin{pmatrix} X_c \\ \sqrt{\Lambda} X_R \end{pmatrix} + \begin{pmatrix} \epsilon_c \\ \sqrt{\Lambda} v_* \end{pmatrix},$$

where $\sqrt{\Lambda} = \text{diag}(\sqrt{\Lambda_1}, \ldots, \sqrt{\Lambda_{t_*}})$..

Minimizing the MDEP

In this section we concentrate on the first problem of a uniform weight λ. A reliable criterion to choose λ is to minimize the mean dispersion error of prediction (MDEP) with respect to λ. Let

$$\tilde{y} = \tilde{x}'\beta + \tilde{\epsilon}, \quad \tilde{\epsilon} \sim (0, \sigma^2),$$

be a nonobserved (future) realization of the regression model that is to be predicted by

$$p = \tilde{x}'b(\lambda)\,.$$

The MDEP of p is

$$\begin{aligned} \mathrm{E}(p-\tilde{y})^2 &= \mathrm{E}\left(\tilde{x}'(b(\lambda)-\beta)-\tilde{\epsilon}\right)^2 \\ &= \left(\tilde{x}'\,\mathrm{Bias}\,(b(\lambda))\right)^2 + \tilde{x}'\,\mathrm{V}\,(b(\lambda))\tilde{x} + \sigma^2\,. \end{aligned} \tag{8.50}$$

Using (8.48) and (8.49), we obtain

$$\begin{aligned} \mathrm{E}(p-\tilde{y})^2 &= g(\lambda) \\ &= \lambda^2(\tilde{x}'Z^{-1}X_R'\delta\delta'X_RZ^{-1}\tilde{x}) \\ &\quad + \sigma^2\tilde{x}'Z^{-1}(S_c+\lambda^2S_R)Z^{-1}\tilde{x}+\sigma^2\,. \end{aligned} \tag{8.51}$$

Using the relations

$$\begin{aligned} \frac{\partial\,\mathrm{tr}\,AZ^{-1}}{\partial\lambda} &= \mathrm{tr}\,\frac{\partial\,\mathrm{tr}\,AZ^{-1}}{\partial Z^{-1}}\frac{\partial Z^{-1}}{\partial\lambda}\,, \\ \frac{\partial\,\mathrm{tr}\,AZ^{-1}}{\partial Z^{-1}} &= A'\,, \\ \frac{\partial Z^{-1}}{\partial\lambda} &= -Z^{-1}\frac{\partial Z}{\partial\lambda}Z^{-1} \end{aligned}$$

(cf.Theorems A.94, A.95 (i), and A.96, respectively) gives

$$\frac{\partial}{\partial\lambda}\,\mathrm{tr}\,AZ^{-1} = -\,\mathrm{tr}\,Z^{-1}A'Z^{-1}\frac{\partial Z}{\partial\lambda}\,.$$

Now for $Z = Z(\lambda)$ we get

$$\frac{\partial Z}{\partial\lambda} = S_R\,.$$

Differentiating $g(\lambda)$ in (8.51) with respect to λ and equating to zero then gives

$$\begin{aligned} \frac{1}{2}\frac{\partial g(\lambda)}{\partial\lambda} &= \lambda(\tilde{x}'Z^{-1}X_R'\delta)^2 - \lambda^2\tilde{x}'Z^{-1}S_RZ^{-1}X_R'\delta\delta'X_RZ^{-1}\tilde{x} \\ &\quad + \sigma^2\lambda\tilde{x}'Z^{-1}S_RZ^{-1}\tilde{x} - \sigma^2\tilde{x}'Z^{-1}S_RZ^{-1}(S_c+\lambda^2S_R)Z^{-1}\tilde{x} \\ &= 0\,, \end{aligned}$$

from which we get the relation

$$\lambda = \frac{1}{1+\sigma^{-2}\rho_1(\lambda)\rho_2^{-1}(\lambda)}\,,\quad 0\le\lambda\le 1\,, \tag{8.52}$$

where

$$\begin{aligned} \rho_1(\lambda) &= \tilde{x}'Z^{-1}S_cZ^{-1}X_R'\delta\delta'X_RZ^{-1}\tilde{x}\,, \\ \rho_2(\lambda) &= \tilde{x}'Z^{-1}S_RZ^{-1}S_cZ^{-1}\tilde{x}\,. \end{aligned}$$

Thus, the optimal λ minimizing the MDEP (8.51) of $p = \tilde{x}'b(\lambda)$ is the solution to relation (8.52). Noting that $Z = Z(\lambda)$ is a function of λ, also, solving (8.52) for λ results in a procedure of iterating the λ-values, whereas σ^2 and δ are estimated by some suitable procedure. This general problem needs further investigation.

The problem becomes somewhat simpler in the case where only one row of the regressor matrix is incompletely observed, that is, $t_* = 1$ in (8.24):

$$y_* = x_*\beta + \epsilon_* , \quad \epsilon_* \sim (0, \sigma^2) .$$

Then we have $S_R = x_R x_R'$, $\delta = (x_*' - x_R')\beta$ (a scalar) and

$$\begin{aligned} \rho_1(\lambda) &= (\tilde{x}' Z^{-1} S_c Z^{-1} x_R)(x_R' Z^{-1} \tilde{x}) \delta^2 , \\ \rho_2(\lambda) &= (\tilde{x}' Z^{-1} x_R)(x_R' Z^{-1} S_c Z^{-1} \tilde{x}) . \end{aligned}$$

So λ becomes

$$\lambda = \frac{1}{1 + \sigma^{-2}\delta^2} . \tag{8.53}$$

Interpretation of the Result

(i) We note that $0 \le \lambda \le 1$, so that λ is, indeed, a weight given to the incompletely observed model.

(ii) $\lambda = 1$ holds for $\sigma^{-2}\delta^2 = 0$. If σ^2 is finite, then the incompletely observed but (by the replacement of x_* by x_R) "repaired" model is given the same weight as the completely observed model when $\delta = 0$. Now, $\delta = (x_*' - x_R')\beta = 0$ implies that the unknown expectation $\mathrm{E}\, y_* = x_*'\beta$ of the dependent variable y_* is estimated exactly by $x_R'\beta$ (for all β). Thus $\delta = 0$ is fulfilled when $x_* = x_R$, that is, when the missing values in x_* are reestimated exactly (without error) by x_R. This seems to be an interesting result to be taken into account in the general mixed regression framework in the sense that additional linear stochastic restrictions of type $r = R\beta + v_*$ should not be incorporated without using a prior weight λ (and $\lambda < 1$ in general).
Furthermore, it may be conjectured that the weighted mixed regression becomes equivalent (in a sense to be specified) to the familiar (unweighted) mixed regression, when the former is related to a strong MDE criterion and the latter is related to a weaker MDE criterion.
Now, $\lambda = 1$ may be caused by $\sigma^2 \to \infty$ also. Since σ^2 is the variance common to both y_c and $y_*, \sigma^2 \to \infty$ leads to unreliable (imprecise) estimators in the complete model $y_c = X_c'\beta + \epsilon_c$ as well as in the enlarged mixed model (8.42).

(iii) In general, an increase in δ decreases the weight λ of the additional stochastic relation $y_* = x_R'\beta + v_*$. If $\delta \to \infty$, then

$$\lambda \to 0 \quad \text{and} \quad \lim_{\lambda \to 0} b(\lambda) = b_c . \tag{8.54}$$

8.8.6 The Two-Stage WMRE

To bring the mixed estimator $b(\lambda)$ with λ from (8.109) in an operational form, σ^2 and δ have to be estimated by $\hat{\sigma}^2$ and $\hat{\delta}$, resulting in $\hat{\lambda} = 1/(1 + \hat{\sigma}^{-2}\hat{\delta}^2)$ and $b(\hat{\lambda})$. By using the consistent estimators

$$\hat{\sigma}^2 = \frac{1}{t_c - K}(y_c - X_c b_c)'(y_c - X_c b_c)$$

and

$$\hat{\delta} = y_* - x_R' b_c \,,$$

we investigate the properties of the resulting two-stage WMRE $b(\hat{\lambda})$. This will depend on the statistical properties (e.g., mean and variance) of $\hat{\lambda}$ itself. The bootstrap method is one of the nonparametric methods in estimating variance and bias of a statistic of interest. By following the presentation of Efron (1979) for the one-sample situation, the starting point is the sample of size m based on the complete model

$$y_{ci} = x_{ci}'\beta + \epsilon_{ci} \,, \quad \epsilon_{ci} \sim F \,, \quad (i = 1, \ldots, m) \,.$$

The random sample is

$$\epsilon_c = (\epsilon_{c1}, \ldots, \epsilon_{cm})' \,.$$

In the notation of Efron, the parameter of interest is

$$\theta(F) = \lambda = \frac{1}{1 + \sigma^{-2}(x_*'\beta - x_R'\beta)^2} \,,$$

and its estimator is

$$t(\epsilon_c) = \hat{\lambda} = \frac{1}{1 + \hat{\sigma}^{-2}(y_* - x_R' b_c)^2} \,.$$

Now the sample probability distribution $\hat{F}$ may be defined by putting mass $1/m$ at each residual

$$\hat{\epsilon}_{ci} = y_{ci} - x_{ci}' b_c \,, \quad (i = 1, \ldots, m)$$

(for $m = t_c$ this yields $\bar{\hat{\epsilon}} = 0$; if $m < t_c$ the estimated residuals have to be centered around mean 0).

With $(b_c, \hat{F})$ fixed, draw a random sample of size m from $\hat{F}$ and call this the bootstrap sample:

$$y_{\text{Boot}} = X_{\text{Boot}} b_c + \epsilon_i^* \,, \quad \epsilon_i^* \sim \hat{F} \quad (i = 1, \ldots, m) \,. \tag{8.55}$$

Each realization of (8.55) yields a realization of a bootstrap estimator $\hat{\beta}^*$ of β:

$$\hat{\beta}^* = \min_{\beta}(y_{\text{Boot}} - X_{\text{Boot}}\beta)'(y_{\text{Boot}} - X_{\text{Boot}}\beta) \,. \tag{8.56}$$

Repeating this procedure N times independently results in a random bootstrap sample $\hat{\beta}^{*1}, \ldots, \hat{\beta}^{*N}$, which can be used to construct a bootstrap sample $\hat{\lambda}^{*1}, \ldots, \hat{\lambda}^{*N}$ of the weight λ:

$$\hat{\lambda}^{*v} = \frac{1}{1 + \hat{\sigma}_{*v}^{-2}(y_* - x_{Rv}^{*'}\hat{\beta}^{*v})^2}, \quad v = 1, \ldots, N.$$

Here

$$\hat{\sigma}_{*v}^2 = \frac{1}{m-K}(y_{\text{Boot},v} - X_{\text{Boot},v}\hat{\beta}^{*v})'(y_{\text{Boot},v} - X_{\text{Boot},v}\hat{\beta}^{*v})$$

is the bootstrap estimator of σ^2, and x_{Rv}^*, $(v = 1, \ldots, N)$ is the vector replacement for x_* owing to dependence on the matrix $X_{\text{Boot},v}$ which comes from X_{c} by the vth bootstrap step. Now the random sample $\hat{\lambda}^{*1}, \ldots, \hat{\lambda}^{*N}$ can be used to estimate the bootstrap distribution of $t(\epsilon_{\text{c}}) = \hat{\lambda}$.

A problem of interest is to compare the bootstrap distributions of $\hat{\lambda}$ or $b(\hat{\lambda})$ for the different missing-values methods, keeping in mind that $\lambda = \lambda(x_R)$ and $\hat{\lambda} = \hat{\lambda}(x_R)$ are dependent on the chosen method for finding x_R. This investigation has to be based on a Monte Carlo experiment for specific patterns. Toutenburg, Heumann, Fieger, and Park (1995) have presented some results, which indicate that (1) using weights λ yields (MDE) better estimates and (2) the weights $\hat{\lambda}$ are biased, which means that further improvements can be achieved by using some sort of bias correction.

8.9 Assumptions about the Missing Mechanism

Complete case analysis requires that missingness is independent of the response y. Least-squares estimation using imputed values yields valid estimates if missingness depends on the fully observed covariates, and the assumption that missing covariates have a linear relationship on the observed covariates holds, that is, the auxiliary regression models are correctly specified. The maximum-likelihood methods require the MAR assumption to hold, which includes the case that missingness depends on the (fully observed) response y.

8.10 Regression Diagnostics to Identify Non-MCAR Processes

In the preceding sections we have discussed various methods to handle incomplete X-matrices. In general they are based on assumptions on the missing data mechanism. The most restrictive one is the assumption based on the requirement that missingness is independent of the data (observed and nonobserved). Less restrictive is the MAR assumption that allows missingness to be dependent of the observed data.

In the following, we discuss the MCAR assumption in more detail and especially under the aspect of how to test this assumption. The idea presented here was first discussed by Simonoff (1988), who used diagnostic measures known from the sensitivity analysis. These measures are adopted to the context of missing values. This enables us to identify some well-defined non-MCAR processes that cannot be detected by standard tests, such as the comparison of the means of the complete and the incomplete data sets.

8.10.1 Comparison of the Means

Cohen and Cohen (1983) proposed to compare the sample mean $\bar{y}_c$ of the observations y_i of the complete-case model and the sample mean $\bar{y}_*$ of the model with partially nonobserved data.

For the case in which missing values x_* in the matrix X_* are of type MCAR, then the partition of y in y_c and y_* is random, indicating that there might be no significant difference between the corresponding sample means.

If their difference would significantly differ from zero, this might be interpreted as a contradiction to a MCAR assumption. Hence, the hypothesis H_0: MCAR would be rejected.

8.10.2 Comparing the Variance-Covariance Matrices

The idea to compare the variance-covariance matrices of the parameter estimates $\hat{\beta}$ for the various methods that react on missing X-values is based on the work of Evans, Cooley, and Piserchia (1979). They propose, to compare $\mathrm{V}(b_c)$ and $\mathrm{V}(\hat{\beta})$ where $\hat{\beta}$ is the estimator of β in the repaired model. Severe differences are interpreted again as a signal against the MCAR-assumption.

8.10.3 Diagnostic Measures from Sensitivity Analysis

In the context of sensitivity analysis we have discussed measures that may detect the influence of the ith observation by comparing some scalar statistic based either on the full data set or on the data set reduced by the ith observation (called the "leave-one-out" strategy). Adapting this idea for the purpose of detecting "influential" missingness means to redefine these measures such that the complete-case model and the filled-up models are compared to each other.

Let $\hat{\beta}_R$ denote the estimator of β for the linear model $y = \begin{pmatrix} X_c \\ X_R \end{pmatrix} + \epsilon$, where X_R is the matrix X_* filled up by some method.

Cook's Distance: Adapting Cook's distance C_i (cf. (7.59)) gives

$$D = \frac{(\hat{\beta}_R - \hat{\beta}_c)'(X'X)(\hat{\beta}_R - \hat{\beta}_c)}{Ks^2} \geq 0 \qquad (8.57)$$

where the estimation s^2 is based on the completed data.

Change of the Residual Sum of Squares: Adapting a measure for the change in the residual sum of squares to our problem results in

$$DRSS = \frac{(RSS_R - RSS_c)/n_R}{RSS_c/(T - n_r - K + 1)} \in [0, \infty]\,. \qquad (8.58)$$

Large values of *DRSS* will indicate departure from the MCAR assumption.

Change of the Determinant: Adaption of the kernel of the Andrews-Pregibon statistic AP_i (cf. (7.70)) to our problem gives the change of determinant *DXX* as

$$DXX = \frac{|X_c'X_c|}{|X'X|} \in [0, 1]\,. \qquad (8.59)$$

where small values of *DXX* will indicate departure from the MCAR assumption.

8.10.4 Distribution of the Measures and Test Procedure

To construct a test procedure for testing H_0: MCAR against H_1: Non-MCAR we need the null distributions of the three measures. These distributions are dependent on the matrix X of regressors, on the variance σ^2 and on the parameter β. In this way, no closed-form solution is available and we have to estimate these null distributions by Monte Carlo simulations with the following steps:

At first, missing values in X_* are filled up by suitable MCAR substitutes. Then with the estimations $\hat{\beta}_c$ and s^2 and with the matrix X_R, updated data $y_*^s = X_R\hat{\beta}_c + \epsilon^s$ (superscript s stands for simulation) are calculated where $\epsilon \sim N(0, s^2 I)$ are pseudorandom numbers. Finally, a MCAR mechanism is selecting cells from the matrix X as missing. This way we get a data set with missing values that are due to a MCAR mechanism, independent of whether the real missing values in X_* were MCAR. Based on these data, the diagnostic measures are calculated. This process is repeated N times using an updated ϵ^s in each step, so that the null distribution f_0 of the diagnostic measure of interest may be estimated.

Test Procedure: With the estimated null distribution we get a critical value that is the $N(1-\alpha)$th-order statistic for D and $DRSS$ or the $N\alpha$th-order statistic for DXX, respectively. H_0: MCAR is rejected if D (or $DRSS$) $\geq f_{0,N(1-\alpha)}$ or if $DXX \leq f_{0,N\alpha}$, respectively.

8.11 Exercises

Exercise 1. Consider the model specified by $y_c = X_c\beta + \epsilon_c$ and $y_{\text{mis}} = X_*\beta + \epsilon_*$. If every element of y_{mis} is replaced by the mean of the elements of y_c, find the least-squares estimator of β from the repaired model and discuss its properties.

Exercise 2. Consider the model $y_{\text{c}} = X_{\text{c}}\beta + \epsilon_{\text{c}}$ and $y_{\text{mis}} = X_*\beta + \epsilon_*$. If $\delta X_* b_{\text{c}}$ with $0 < \delta < 1$ is used in the repaired model and β is estimated by the least-squares method, show that the estimator of β is biased. Also find its dispersion matrix.

Exercise 3. Given the model $y_{\text{c}} = x_{\text{c}}\beta + \epsilon_{\text{c}}$ and $y_* = x_{\text{mis}}\beta + \epsilon_*$, suppose that we regress x_{c} on y_{c} and use the estimated equation to find the imputed values for x_{mis}. Obtain the least-squares estimator of the scalar β from the repaired model.

Exercise 4. For the model $y_{\text{c}} = X_{\text{c}}\beta + Z_{\text{c}}\gamma + \epsilon_{\text{c}}$ and $y_* = X_{\text{mis}}\beta + Z_*\gamma + \epsilon_*$, examine whether the least-squares estimator of β from the complete model is equal to the least-squares estimator from the filled-in model using a first order regression method.

Exercise 5. Suppose that β is known in the preceding exercise. How will you estimate γ then?

Exercise 6. Consider the model $y_{\text{c}} = Z_{\text{c}}\beta + \alpha x_{\text{c}} + \epsilon_{\text{c}}$ and $y_* = Z_*\beta + \alpha x_{\text{mis}} + \epsilon_*$. If the regressor associated with α is assumed to be stochastic and the regression of x_{c} on Z_{c} is used to find the predicted values for missing observations, what are the properties of these imputed values?

Exercise 7. For the set-up in the preceding exercise, obtain the least-squares estimator $\hat{\alpha}$ of α from the repaired model. How does it differ from $\tilde{\alpha}$, the least-squares estimator of α using the complete observations alone?

Exercise 8. Offer your remarks on the estimation of α and β in a bivariate model $y_{\text{mis}} = \alpha\mathbf{1} + \beta x_* + \epsilon_*$ and $y_{**} = \alpha\mathbf{1} + \beta x_{\text{mis}} + \epsilon_{**}$, where $\mathbf{1}$ denotes a column vector (of appropriate length) with all elements unity.

9 Robust Regression

9.1 Overview

Consider the multivariate linear model

$$Y_i = X_i'\beta + E_i, \quad i = 1, \ldots, n, \tag{9.1}$$

where $Y_i : p \times 1$ is the observation on the ith individual, $X_i : q \times p$ is the design matrix with known elements, $\beta : q \times 1$ is a vector of unknown regression coefficients, and $E_i : p \times 1$ is the unobservable random error that is usually assumed to be suitably centered and to have a p-variate distribution. A central problem in linear models is estimating the regression vector β. Note that model (9.1) reduces to the univariate regression model when $p = 1$, which we can write as

$$y_i = x_i'\beta + \epsilon_i, \quad i = 1, \ldots, n, \tag{9.2}$$

where x_i is now a q-vector. Model (9.1) becomes the classical multivariate regression, also called MANOVA model, when $X_i : q \times p$ is of the special form

$$X_i = \begin{pmatrix} x_i & 0 & \ldots & 0 \\ 0 & x_i & \ldots & 0 \\ \vdots & \vdots & \ldots & \vdots \\ 0 & 0 & \ldots & x_i \end{pmatrix}, \tag{9.3}$$

where $x_i : m \times 1$ and $q = mp$. Our discussion of the general model will cover both classical cases considered in the literature.

When E_i has a p-variate normal distribution, the least-squares method provides the most efficient estimators of the unknown parameters. In addition, we have an elegant theory for inference on the unknown parameters. However, recent investigations have shown that the LS method is sensitive to departures from the basic normality of the distribution of errors and to the presence of outliers even if normality holds.

The general method, called the M-estimation, was introduced by Huber (1964) to achieve robustness in data analysis. This has generated considerable research in recent times. It may be pointed out that a special case of M-estimation based on the L_1-norm—estimation by minimizing the sum of absolute deviations rather than the sum of squares, called the least absolute deviation (LAD) method—was developed and was the subject of active discussion. The earliest uses of the LAD method may be found in the seventeenth- and eighteenth-century works of Galilie (1632), Boscovich (1757), and Laplace (1793). However, because of computational difficulties in obtaining LAD estimates and lack of exact sampling theory based on such estimates, the LAD method lay in the background and the LS method became popular. The two basic papers, one by Charnes, Cooper, and Ferguson (1955) reducing the LAD method of estimation to a linear programming problem, and another by Bassett and Koenker (1978) developing the asymptotic theory of LAD estimates, have cleared some of the difficulties and opened up the possibilities of replacing the LS theory by more robust techniques using the L_1-norm or more general discrepancy functions in practical applications.

In this chapter, we review some of the recent contributions to the theory of robust estimation and inference in linear models. In the following section, we consider the problem of consistency of the LAD and, in general, of M-estimators. Furthermore we review some contributions to the asymptotic normality and tests of significance for the univariate and multivariate LAD and M-estimators.

9.2 Least Absolute Deviation Estimators—Univariate Case

Consider model (9.2). Let $\hat{\beta}_n$ be any solution to the minimization problem

$$\sum_{i=1}^{n} |y_i - x_i'\hat{\beta}_n| = \min_{\beta} \sum_{i=1}^{n} |y_i - x_i'\beta| . \tag{9.4}$$

$\hat{\beta}_n$ is called the LAD estimator of the vector parameter β. Some general properties of the LAD estimators can be found in Rao (1988). An iterative algorithm for the numerical solution of problem (9.4) is given in Birkes and Dodge (1993, Chapter 4).

In almost all related papers, the weak consistency of the LAD estimators is established under the same conditions that guarantee its asymptotic normality, although intuitively it should be true under weaker conditions than those required for the latter. So in the present section we mainly discuss the strong consistency, except for some remarks in discussing the significance of the conditions for weak consistency.

Bloomfield and Steiger (1983) give a proof of the strong consistency of $\hat{\beta}_n$, where $\{x_i\}$ is a sequence of i.i.d. observations of a random vector x. Dupaková (1987) and Dupaková and Wets (1988) consider the strong consistency of LAD estimators under linear constraints, when the x_i's are random.

It is easy to see that the strong consistency under the random case is a simple consequence of that for the nonrandom case. In the following we present several recent results in the set-up of nonrandom x_i. Write

$$S_n = \sum_{i=1}^n x_i x_i', \quad \rho_n = \text{the smallest eigenvalue of } S_n\,, \tag{9.5}$$

$$d_n = \max\{1, \|x_1\|, \dots, \|x_n\|\}\,, \tag{9.6}$$

where $\|a\|$ denotes the Euclidean norm of the vector a. Wu (1988) proves the following theorem.

Theorem 9.1 *Suppose that the following conditions are satisfied:*

(i)
$$\frac{\rho_n}{(d_n^2 \log n)} \to \infty\,. \tag{9.7}$$

(ii) *There exists a constant $k > 1$ such that*

$$\frac{d_n}{n^{k-1}} \to 0\,. \tag{9.8}$$

(iii) *$\epsilon_1, \epsilon_2, \dots$ are independent random variables, and med$(\epsilon_i) = 0$, $i = 1, 2, \dots$, where med$(\cdot)$ denotes the median.*

(iv) *There exist constants $c_1 > 0, c_2 > 0$ such that*

$$P\{-h < \epsilon_i < 0\} \geq c_2 h \tag{9.9}$$

$$P\{0 < \epsilon_i < h\} \geq c_2 h \tag{9.10}$$

for all $i = 1, 2, \dots$ and $h \in (0, c_1)$. Then we have (cf. Definition A.101 (ii))

$$\lim_{n\to\infty} \hat{\beta}_n = \beta_0 \quad a.s.\,, \tag{9.11}$$

(a.s. almost surely) where β_0 is the true value of β. Further, under the additional condition that for some constant $M > 0$

$$\frac{\rho_n}{d_n^2} \geq Mn \quad \text{for large } n\,, \tag{9.12}$$

$\hat{\beta}_n$ converges to β_0 at an exponential rate in the following sense: For arbitrarily given $\epsilon > 0$, there exists a constant $c > 0$ independent of n such that

$$P\{||\hat{\beta}_n - \beta_0|| \geq \epsilon\} \leq O(e^{-cn}). \tag{9.13}$$

The above result was sharpened in Wu (1989) to apply to the case in which conditions (9.7) and (9.8) are replaced by

$$\Delta_n^2 \log n \to 0 \tag{9.14}$$

and (9.12) by

$$n\Delta_n^2 \leq M, \tag{9.15}$$

where

$$\Delta_n^2 = \max_{i \leq n}\{x_i' S_n^{-1} x_i\}. \tag{9.16}$$

Now consider the inhomogeneous linear model

$$y_i = \alpha_0 + x_i'\beta_0 + \epsilon_i, \quad i = 1, 2, \ldots, n. \tag{9.17}$$

Theoretically speaking, the inhomogeneous model is merely a special case of the homogeneous model (9.2) in which the first element of each x_i is equal to 1. So the strong consistency of the LAD estimators for an inhomogeneous model should follow from Theorem 9.1. However, although Theorem 9.2 looks like Theorem 9.1, we have not yet proved that it is a consequence of Theorem 9.1.

Theorem 9.2 *Suppose we have model (9.17), and the conditions of Theorem 9.1 are satisfied, except that here we define S_n as $\sum_{i=1}^n (x_i - \bar{x}_n)(x_i - \bar{x}_n)'$ where $\bar{x}_n = 1/n \sum_{i=1}^n x_i$. Then*

$$\lim_{n\to\infty} \hat{\alpha} = \alpha_0 \quad a.s., \qquad \lim_{n\to\infty} \hat{\beta} = \beta_0 \quad a.s. \tag{9.18}$$

Also, under the additional assumption (9.12) for arbitrarily given $\epsilon > 0$, we can find a constant $c > 0$ independent of n such that

$$P\{||\hat{\alpha}_n - \alpha_0||^2 + ||\hat{\beta}_n - \beta_0||^2 \geq \epsilon^2\} \leq O(e^{-cn}). \tag{9.19}$$

As in Theorem 9.1, Theorem 9.2 can be improved to the case in which conditions (9.7), (9.8), and (9.12) are replaced by (9.14) and (9.15) with Δ_n redefined as

$$\tilde{\Delta}_n^2 = \max_{i \leq n}\{(x_i - \bar{x}_n)'\bar{S}_n^{-1}(x_i - \bar{x}_n)\}, \tag{9.20}$$

$$\bar{S}_n = \sum_{i=1}^n (x_i - \bar{x}_n)(x_i - \bar{x}_n)', \quad \bar{x}_n = \frac{1}{n}\sum_{i=1}^n x_i. \tag{9.21}$$

Remark: Conditions (9.9) and (9.10) stipulate that the random errors should not be "too thinly" distributed around their median zero. It is likely that they are not necessary and that further improvement is conceivable, yet they cannot be totally eliminated.

Example 9.1: Take the simplest case in which we know that $\alpha_0 = 0$ and all x_i are zero. In this case the minimum L_1-norm principle gives

$$\hat{\alpha} = \text{med}(y_1, \ldots, y_n) \tag{9.22}$$

as an estimate of α_0. Suppose that $\epsilon_1, \epsilon_2, \ldots$ are mutually independent; then ϵ_i has the following density function:

$$f_i(x) = \begin{cases} \dfrac{|x|}{i^2}, & 0 \le |x| \le i, \quad i = 1, 2, \ldots, \\ 0, & \text{otherwise}. \end{cases} \tag{9.23}$$

Then

$$P\{\epsilon_i \ge 1\} = \frac{1}{2} - \frac{1}{(2i^2)}, \quad i = 1, 2, \ldots. \tag{9.24}$$

Denote by ξ_n the number of ϵ_i's for which $\sqrt{n} \le i \le n$ and $\epsilon_i \ge 1$. An application of the central limit theorem shows that for some $\delta \in (0, \frac{1}{2})$ we have

$$P\left\{\xi_n > \frac{n}{2}\right\} \ge \delta \tag{9.25}$$

for n sufficiently large. This implies that

$$P\{\hat{\alpha} \ge 1\} \ge \delta \tag{9.26}$$

for n sufficiently large, and hence $\hat{\alpha}_n$ is a consistent estimate of α_0.

Remark: In the case of LS estimation, the condition for consistency is

$$\lim_{n \to \infty} S_n^{-1} = 0 \tag{9.27}$$

whereas that for the LAD estimates is much stronger. However, (9.27) does not guarantee the strong consistency for LAD estimates, even if the error sequence consists of i.i.d. random variables.

Example 9.2: This example shows that even when $\epsilon_1, \epsilon_2, \ldots$ are i.i.d., consistency may not hold in case d_n tends to infinity too fast.

Suppose that in model (9.2) the true parameter $\beta_0 = 0$, the random errors are i.i.d. with a common distribution $P\{\epsilon_i = 10^k\} = P\{\epsilon_i = -10^k\} = 1/[k(k+1)], k = 6, 7, \ldots$, and ϵ_i is uniformly distributed over $(-\frac{1}{3}, \frac{1}{3})$ with density 1. Let $x_i = 10^i$, $i = 1, 2, \ldots$. We can prove that $\hat{\beta}_n$ is not strongly consistent.

When the errors are i.i.d., we do not know whether (9.27) implies the weak consistency of the LAD estimates. However, if we do not assume that the errors are i.i.d., then we have the following counterexample.

Example 9.3: Suppose that in model (9.2), the random errors $\epsilon_1, \epsilon_2, \ldots$ are independent, $P\{\epsilon_i = 10^i\} = P\{\epsilon_i = -10^i\} = \frac{1}{6}$, and ϵ_i is uniformly distributed over the interval $(-\frac{1}{3}, \frac{1}{3})$ with density 1. For convenience assume that the true parameter value $\beta_0 = 0$. Let $x_i = 10^i, i = 1, 2, \ldots$. Then the weak consistency does not hold.

9.3 M-Estimates: Univariate Case

Let ρ be a suitably chosen function on $\mathbb{R}$. Consider the minimization problem

$$\sum_{i=1}^{n} \rho(y_i - x_i'\hat{\beta}_n) = \min_{\beta} \sum_{i=1}^{n} \rho(y_i - x_i'\beta). \tag{9.28}$$

Following Huber (1964), $\hat{\beta}_n$ is called the M-estimate of β_0.

If ρ is continuously differentiable everywhere, then $\hat{\beta}_n$ is one of the solutions to the following equation:

$$\sum_{i=1}^{n} x_i \rho'(y_i - x_i'\beta) = 0 \tag{9.29}$$

When ρ' is not continuous or ρ' equals the derivative of ρ except at finite or countably infinitely many points, the following two cases may be met. First, (9.29) may not have any solution at all, even with a probability arbitrarily close to 1. In such a situation, the solution of (9.28) cannot be characterized by that of (9.29). Second, even if (9.29) has solutions, $\hat{\beta}_n$ may not belong to the set of solutions of (9.28). Such a situation leading to a wrong solution of (9.28) frequently happens when ρ is not convex. This may result in serious errors in practical applications. So in this chapter we always consider the M-estimates to be the solution of (9.28), instead of being that of (9.29).

Chen and Wu (1988) established the following results. First consider the case where $x_1, x_2, \ldots$ are i.i.d. random vectors.

Theorem 9.3 *Suppose that $(x_1', y_1), (x_2', y_2), \ldots$ are i.i.d. observations of a random vector (x', y), and the following conditions are satisfied:*

(i) *The function ρ is continuous everywhere on $\mathbb{R}$, nondecreasing on $[0, \infty)$, nonincreasing on $(-\infty, 0]$, and $\rho(0) = 0$.*

(ii) *Either $\rho(\infty) = \rho(-\infty) = \infty$ and*

$$P\{\alpha + x'\beta = 0\} < 1 \quad \text{where} \quad (\alpha, \beta') \neq (0, 0'), \tag{9.30}$$

or $\rho(\infty) = \rho(-\infty) \in (0, \infty)$ and

$$P\{\alpha + x'\beta = 0\} = 0 \quad \text{where} \quad (\alpha, \beta') \neq (0, 0'). \tag{9.31}$$

(iii) *For every* $(\alpha, \beta') \in \mathbb{R}^{p+1}$, *we have*

$$Q(\alpha, \beta') \equiv \mathrm{E}\,\rho(y - \alpha - x'\beta) < \infty\,, \tag{9.32}$$

and Q *attains its minimum uniquely at* (α_0, β_0'). *Then*

$$\hat{\alpha}_n \to \alpha_0\,, \quad \hat{\beta}_n \to \beta_0\,, \quad a.s.\ as\ n \to \infty\,. \tag{9.33}$$

When ρ is a convex function, condition (9.32) can be somewhat weakened.

Theorem 9.4 *If* ρ *is a convex function, then (9.33) is still true when condition (i) of Theorem 9.3 is satisfied, condition (ii) is deleted, and condition (iii) is replaced by condition (iii′):*

(iii′) *For every* $(\alpha, \beta') \in \mathbb{R}^{p+1}$,

$$Q^*(\alpha, \beta') \equiv \mathrm{E}\{\rho(y - \alpha - x'\beta) - \rho(y - \alpha_0 - x'\beta_0)\} \tag{9.34}$$

exists and is finite, and

$$Q^*(\alpha, \beta') > 0, \textit{for any } (\alpha, \beta') \neq (\alpha_0, \beta_0')\,. \tag{9.35}$$

The following theorem gives an exponential convergence rate of the estimate $(\hat{\alpha}_n, \hat{\beta}_n')$.

Theorem 9.5 *Suppose that the conditions of Theorem 9.3 are met, and in addition the moment-generating function of* $\rho(y - \alpha - x'\beta)$ *exists in some neighborhood of 0. Then for arbitrarily given* $\epsilon > 0$, *there exists a constant* $c > 0$ *independent of* n *such that*

$$P\{|\hat{\alpha}_n - \alpha_0| \geq \epsilon\} = O(e^{-cn}), \quad P\{||\hat{\beta}_n - \beta_0|| \geq \epsilon\} = O(e^{-cn})\,. \tag{9.36}$$

This conclusion remains valid if the conditions of Theorem 9.4 are met and the moment-generating function of $\rho(y - \alpha - x'\beta) - \rho(y - \alpha_0 - x'\beta_0)$ exists in some neighborhood of 0.

Next we consider the case where $x_1, x_2, \ldots$ are nonrandom q-vectors.

Theorem 9.6 *Suppose that in model (9.17)* $x_1, x_2, \ldots$ *are nonrandom* q*-vectors and the following conditions are satisfied:*

(i) *Condition (i) of Theorem 9.3 is true and* $\rho(\infty) = \rho(-\infty) = \infty$.

(ii) $\{x_i\}$ *is bounded, and if* λ_n *denotes the smallest eigenvalue of the matrix* $\sum_{i=1}^n (x_i - \bar{x}_n)(x_i - \bar{x}_n)'$, *where* $\bar{x}_n = \frac{1}{n}\sum_{i=1}^n x_i$, *then*

$$\liminf_{n \to \infty} \frac{\lambda_n}{n} > 0\,. \tag{9.37}$$

(iii) $\{\epsilon_i\}$ *is a sequence of i.i.d. random errors.*

(iv) *For any* $t \in \mathbb{R}, \mathrm{E}\,\rho(\epsilon_1 + t) < \infty, \mathrm{E}\{\rho(\epsilon_1 + t) - \rho(\epsilon_1)\} > 0$ *for any* $t \neq 0$, *and there exists a constant* $c_1 > 0$ *such that*

$$\mathrm{E}\{\rho(\epsilon_1 + t) - \rho(\epsilon_1)\} \geq c_1 t^2 \tag{9.38}$$

for $|t|$ sufficiently small.

Then (9.33) is true. This conclusion remains valid if (i) and (ii) in Theorem 9.6 are replaced by (i') and (ii'):

(i$'$) Condition (i) of Theorem 9.3 is true,

$$0 < \rho(\infty) = \rho(-\infty) < \infty. \tag{9.39}$$

(ii$'$)

$$\lim_{\epsilon \to 0} \limsup_{n \to \infty} \frac{\sharp\{i : 1 \le i \le n, |\alpha + x_i'\beta| \le \epsilon\}}{n} = 0, (\alpha, \beta') \ne (0, 0'). \tag{9.40}$$

where $\sharp B$ denotes the number of elements in a set B. Note that condition (9.40) corresponds to condition (9.31) of Theorem 9.3.

Also, when ρ is convex, the condition $\mathrm{E}\,\rho(\epsilon_1 + t) < \infty$ can be weakened to $\mathrm{E}\,|\rho(\epsilon_1 + t) - \rho(\epsilon)| < \infty$.

Now we make some comments concerning the conditions assumed in these theorems:

1. Condition (iii) of Theorem 9.3, which stipulates that Q attains its minimum uniquely at the point (α_0, β_0'), is closely related to the interpretation of regression. The essence is that the selection of ρ must be compatible with the type of regression considered. For example, when $\alpha_0 + x'\beta_0$ is the conditional median of Y given $X = x$ (median regression), we may choose $\rho(u) = |u|$. Likewise, when $\alpha_0 + x'\beta_0 = \mathrm{E}(Y|X = x)$ (the usual mean regression), we may choose $\rho(u) = |u|^2$. This explains the reason why we say at the beginning of this chapter that the errors are suitably centered. An important case is that of the conditional distribution of Y given $X = x$ being symmetric and unimodal with the center at $\alpha_0 + x'\beta_0$. In this case, ρ can be chosen as any function satisfying condition (i), and such that $\rho(t) > 0$ when $t \ne 0$. This gives us some freedom in the choice of ρ with the aim of obtaining more robust estimates.

2. Condition (9.38) of Theorem 9.6 reveals a difference between the two cases of $\{x_i\}$ mentioned earlier. In the case that $\{x_i\}$ is a sequence of nonrandom vectors, we can no longer assume that only 0 is the unique minimization point of $\mathrm{E}\,\rho(\epsilon_1 + u)$, as shown in the counterexample given in Bai, Chen, Wu, and Zhao (1987) for $\rho(u) = |u|$.

Condition (9.38) holds automatically when $\rho(u) = u^2$ and $\mathrm{E}(\epsilon_1) = 0$. When $\rho(u) = |u|$, it holds when ϵ_1 has median 0 and a density that is bounded away from 0 in some neighborhood of 0. When ρ is even and ϵ_1 is symmetric and unimodal with center 0, (9.38) holds if one of the following two conditions is satisfied:

(i) $\inf\left\{\dfrac{(\rho(u_2) - \rho(u_1))}{(u_2 - u_1)} : \epsilon \le u_1 < u_2 < \infty\right\} > 0$ for any $\epsilon > 0$.

(ii) There exist positive constants $a < b$ and c, such that

$$\frac{\rho(u_2) - \rho(u_1)}{u_2 - u_1} \geq c, \quad \frac{|f(u_2) - f(u_1)|}{u_2 - u_1} \geq c$$

for any $a \leq u_1 < u_2 \leq b$, where f is the density of ϵ_1.

9.4 Asymptotic Distributions of LAD Estimators

9.4.1 Univariate Case

The asymptotic distribution of LAD estimates was first given by Bassett and Koenker (1978) and later by Amemiya (1982) and Bloomfield and Steiger (1983, p. 62). Bloomfield and Steiger (1983) also pointed out that the limiting distribution of the LAD estimate of β (except the constant term, but the model may have a constant term) follows from a result on a class of R-estimates due to Jaeckel (1972), who proved the asymptotic equivalence between his estimates and those introduced and studied by Jureckova (1971). Heiler and Willers (1988) removed some complicated conditions on the x_i-vectors made by Jureckova (1971) and hence greatly improved Jaeckel's result. However, it should be noted that all the above results about the limiting distribution of LAD estimates are special cases of those of Ruppert and Carroll (1980), who derived the limiting distribution of quantile estimates in linear models.

Bai et al. (1987) derived the limiting distribution of the LAD estimates under mild conditions. The results are given below.

Theorem 9.7 *Suppose that in model (9.2), $\epsilon_1, \ldots, \epsilon_n$ are i.i.d. with a common distribution function F, and the following two conditions are satisfied:*

(i) *There is a constant $\Delta > 0$ such that $f(u) = F'(u)$ exists when $|u| \leq \Delta$, f is continuous and strictly positive at zero, and $F(0) = \frac{1}{2}$.*

(ii) *The matrix $S_n = x_1x_1' + \ldots + x_nx_n'$ is nonsingular for some n and*

$$\lim_{n\to\infty} \max_{1\leq i\leq n} x_i'S_n^{-1}x_i = 0 .$$

Then

$$2f(0)S_n^{\frac{1}{2}}(\hat{\beta}_n - \beta) \xrightarrow{L} N(0, I_q), \tag{9.41}$$

where $\hat{\beta}_n$ is the LAD estimator of β.

The distribution (9.41) is derived by using the Bahadur-type representation

$$2f(0)S_n^{\frac{1}{2}}(\hat{\beta}_n - \beta) - \sum_{i=1}^{n}(\text{sign}\epsilon_i)S_n^{-\frac{1}{2}}x_i = o_p(1), \tag{9.42}$$

which is valid under the conditions of Theorem 9.7.

Bai et al. (1987) established the following theorem when ϵ_i are not i.i.d.

Theorem 9.8 *Suppose that in model (9.2), $\epsilon_1, \ldots, \epsilon_n$ are independent; the distribution function F_i of ϵ_i is differentiable over the interval $(-\Delta, \Delta)$; $F_i(0) = \frac{1}{2}, i = 1, 2, \ldots$; and $\Delta > 0$ does not depend on i. Write $f_i(x) = F_i'(x)$. Suppose that $\{f_i(x)\}$ is equicontinuous at $x = 0$,*

$$0 < \inf_i f_i(0) \leq \sup_i f_i(0) < \infty ,$$

$S_n = x_1 x_1' + \ldots + x_n x_n'$ is nonsingular for some n, and

$$\lim_{n\to\infty} \max_{1\leq i\leq n} x_i' S_n^{-1} x_i = 0 .$$

Then as $n \to \infty$,

$$2S_n^{-\frac{1}{2}}[\sum_{i=1}^{n} f_i(0) x_i x_i'](\hat{\beta}_n - \beta) \xrightarrow{L} N(0, I_q) .$$

9.4.2 Multivariate Case

Consider model (9.1). Define $\hat{\beta}_n = \hat{\beta}_n(y_1, \ldots, y_n)$ as the LD (least-distances) estimate of β if it minimizes

$$\sum_{i=1}^{n} \|y_i - x_i'\beta\| , \tag{9.43}$$

where $\|\cdot\|$ denotes the Euclidean norm.

For the special case where $X_1 = \ldots = X_n = I_q$, the LD estimate of β reduces to the spatial median defined by Haldane (1948) and studied by Brown (1983), Gower (1974), and others. Recently, the limiting distribution of the general case was obtained by Bai, Chen, Miao, and Rao (1988), whose results are given below.

We make the following assumptions about model (9.1):

(i) The random errors $E_1, E_2, \ldots$ are i.i.d. with a common distribution function F having a bounded density on the set $\{u : \|u\| < \delta\}$ for some $\delta > 0$ and $P\{c'E_1 = 0\} < 1$ for every $c \neq 0$.

(ii) $\int u\|u\|^{-1} dF(u) = 0$.

(iii) There exists an integer n_0 such that the matrix $(X_1, \ldots, X_{n_0})$ has rank q.

(iv) Define the matrices

$$A = \int \|u\|^{-1}(I_p - uu'\|u\|^{-2}) dF(u) , \tag{9.44}$$

$$B = \int uu'\|u\|^{-2} dF(u) . \tag{9.45}$$

Condition (i) ensures that A and B exist and are positive definite when $p \geq 2$, so that by condition (iii), the matrices

$$S_n = \sum_{i=1}^{n} X_i B X_i' \quad \text{and} \quad T_n = \sum_{i=1}^{n} X_i A X_i' \tag{9.46}$$

are positive definite when $n \geq n_0$. We assume that

$$\lim_{n\to\infty} d_n = 0 \quad \text{where} \quad d_n = \max_{1\leq i\leq n} \|S_n^{-1/2} X_i\|. \tag{9.47}$$

Theorem 9.9 *If $p \geq 2$ and conditions (i)–(iv) above are met, then as $n \to \infty$ we have*

$$S_n^{-\frac{1}{2}} T_n(\hat{\beta}_n - \beta) \xrightarrow{L} N(0, I_q).$$

In the limiting distribution, the unknown matrices A and B are involved. Bai et al. (1988) also proposed the following estimates of A and B:

$$\hat{A} = \frac{1}{n}\sum_{i=1}^{n} \|\hat{\epsilon}_{ni}\|^{-1}(I_p - \|\hat{\epsilon}_{ni}\|^{-2}\hat{\epsilon}_{ni}\hat{\epsilon}_{ni}'), \tag{9.48}$$

$$\hat{B} = \frac{1}{n}\sum_{i=1}^{n} \|\hat{\epsilon}_{ni}\|^{-2}\hat{\epsilon}_{ni}\hat{\epsilon}_{ni}', \tag{9.49}$$

where $\hat{\epsilon}_{ni} = Y_i - X_i'\hat{\beta}_n, i = 1, 2, \ldots, n$, and proved the following theorem.

Theorem 9.10 *Under the conditions of Theorem 9.9, $\hat{A}$ and $\hat{B}$ are weakly consistent estimates of A and B, respectively.*

Remark: The asymptotic distribution in Theorem 9.9 holds when S and T are computed by substituting $\hat{A}$ and $\hat{B}$ for the unknown matrices A and B.

9.5 General M-Estimates

A number of papers in the literature have been devoted to M-estimation. Basically speaking, there are two kinds of M-estimation:

1. Simple form:

$$\min_{\beta} \sum_{i=1}^{n} \rho(Y_i - X_i'\beta). \tag{9.50}$$

2. General form:

$$\min_{\rho,\sigma} \sum_{i=1}^{n} \left[\rho\left(\frac{Y_i - X_i'\beta}{\sigma}\right) + \log\sigma\right]. \tag{9.51}$$

When ρ is differentiable, the solutions of the two forms can be obtained by solving

$$\sum_{i=1}^{n} X_i \Psi(Y_i - X_i'\beta) = 0, \tag{9.52}$$

$$\sum_{i=1}^{n} X_i \Psi\left(\frac{Y_i - X_i'\beta}{\sigma}\right) = 0, \tag{9.53}$$

$$\sum_{i=1}^{n} \chi\left(\frac{Y_i - X_i'\beta}{\sigma}\right) = 0, \tag{9.54}$$

where Ψ is the derivative (or gradient) of ρ and

$$\chi(t) = t\Psi(t) - 1\,.$$

Huber (1964) proposed the M-estimation methodology, which includes the usual likelihood estimation as a special case. Maronna and Yohai (1981) generalized Huber's equation as

$$\sum_{i=1}^{n} X_i \phi\left(X_i, \frac{Y_i - X_i'\beta}{\sigma}\right) = 0, \tag{9.55}$$

$$\sum_{i=1}^{n} \chi\left(\left|\frac{Y_i - X_i'\beta}{\sigma}\right|\right) = 0, \tag{9.56}$$

without reference to any minimization problem such as (9.51).

In view of Bickel (1975), once the M-estimate satisfies the above equations, it can be regarded as Bickel's one-step estimate and hence its asymptotic normality follows by showing that it is a square-root-n consistent estimate.

However, in many practical situations of M-estimation the derivative of ρ is not continuous. Hence, according to Bai et al. (1987), as mentioned in Section 9.2, equations (9.52) or (9.53) and (9.54) may have no solutions with a large probability. Therefore the M-estimate cannot be regarded as Bickel's one-step estimate. More important, even though the above equations have solutions, the set of their solutions may not contain the M-estimate of the original form (9.50) or (9.51).

A more general form than the classical M-esimation, called quadratic dispersion or discrepancy, is discussed by Basawa and Koul (1988). Under a series of broad assumptions, they established the asymptotic properties of estimates based on quadratic dispersion, and indicated how these assumptions can be established in some specific applications.

Here we confine ourselves to the case of (9.50). For the univariate case, Bai, Rao, and Wu (1989) prove some general basic results under the following assumptions:

(U.1) $\rho(x)$ is convex.

(U.2) The common distribution function F of ϵ_i has no atoms on D.

This last condition is imposed to provide unique values for certain functionals of Ψ, which appear in our discussion, and it automatically holds when ρ is differentiable. (For instance, if $\rho(x) = |x|^p, p > 1$, the condition does not impose any restriction on F.) We conjecture that this condition is crucial for asymptotic normality but not necessary for consistency of M-estimates.

(U.3) With Ψ as the derivative (or gradient) of ρ

$$\mathrm{E}[\Psi(\epsilon_1 + a)] = \lambda a + o(a) \text{ as } a \to 0\,, \tag{9.57}$$

where $\lambda > 0$ is a constant.

When (U.2) is true, it is easy to see that if (U.3) holds for one choice of Ψ, then it holds for all choices of Ψ with the same constant λ. Conversely, it is not difficult to give an example showing that the constant λ in (U.3) depends on the choice of Ψ when (U.2) fails. This shows the essence of assumption (U.2).

(U.4)

$$g(a) = \mathrm{E}[\Psi(\epsilon_1 + a) - \Psi(\epsilon_1)]^2 \tag{9.58}$$

exists for all small a, and g is continuous at $a = 0$.

(U.5)

$$\mathrm{E}[\Psi^2(\epsilon_1)] = \sigma^2 \in (0, \infty)\,. \tag{9.59}$$

(U.6) $S_n = x_1 x_1' + \ldots + x_n x_n'$ is nonsingular for $n \geq n_0$ (some value of n) and

$$d_n^2 = \max_{1 \leq i \leq n} x_i' S_n^{-1} x_i \to 0 \quad \text{as } n \to \infty\,. \tag{9.60}$$

Theorem 9.11 *Under assumptions (U.1)–(U.5), for any $c > 0$,*

$$\begin{aligned} \sup_{|S_n^{1/2}(\beta - \beta_0)| \leq c} & \Bigg| \sum_{i=1}^{n} \big[\rho(y_i - x_i'\beta) - \rho(y_i - x_i'\beta_0) \\ & + x_i'(\beta - \beta_0)\Psi(y_i - x_i'\beta_0)\big] \\ & - \frac{\lambda}{2}(\beta - \beta_0)' S_n (\beta - \beta_0) \Bigg| \to 0 \quad \textit{in probability}, \end{aligned} \tag{9.61}$$

where β_0 is the true value for model (9.2), $S_n = \sum_{i=1}^n x_i x_i'$ is assumed to be positive definite (for all $n \geq n_0$), and λ is as defined in (9.57).

Theorem 9.12 *Under assumptions (U.1)–(U.6),*

$$\hat{\beta}_n \to \beta_0 \quad \textit{in probability}.$$

Theorem 9.13 *Under assumptions (U.1)–(U.6), we have for any* $c > 0$

$$\sup_{|S_n^{1/2}(\beta-\beta_0)|\le c} \left| \sum_{i=1}^{n} \left[\Psi(y_i - x_i'\beta) - \Psi(y_i - x_i'\beta_0)\right] S_n^{\frac{1}{2}} x_i + \lambda S_n^{\frac{1}{2}} (\beta - \beta_0) \right| \to 0 \quad \textit{in probability}, \tag{9.62}$$

where λ *is defined in (9.57).*

Theorem 9.14 *Under assumptions* (U_1)*-*(U_6)*,*

$$S_n^{\frac{1}{2}}(\hat{\beta}_n - \beta_0) \xrightarrow{L} N(0, \lambda^{-2}\sigma^2 I_q), \tag{9.63}$$

where σ^2 *is as defined in (9.59).*

For the multivariate case, in the same paper, the following results are established.

The assumptions used in the multivariate case are summarized below, where Ψ represents any (vector) gradient function of ρ.

(M.1) ρ is convex, $\rho(0) = 0, \rho(u) > 0$ for any p-vector $u \neq 0$.

(M.2) $F(D) = 0$, where F is the distribution function of E_1 and D is the set of points at which F is not differentiable.

(M.3) $\mathrm{E}[\Psi(E_1 + a)] = Aa + o(a)$ as $a \to 0, A > 0$. (Note that if (M_3) holds for one choice of Ψ, then it holds for all choices of Ψ with the same matrix A).

(M.4) $g(a) = \mathrm{E}\,\|\Psi(E_1 + a) - \Psi(E_1)\|^2 < \infty$ for small a, and g is continuous at $a = 0$, where $\|\cdot\|$ denotes the Euclidean norm.

(M.5) $B = \mathrm{cov}[\psi(E_1)] > 0$.

(M.6) $d_n^2 = \max_{1\le i\le n} |X_i' S_n^{-1} X_i| \to 0$

where $S_n = X_1X_1' + \ldots + X_nX_n'$ is supposed to be positive definite for $n \ge n_0$ (some value).

Theorems analogous to those in the univariate case can be extended to the multivariate case as follows. We use the additional notation

$$T = \sum_{i=1}^{n} X_i B X_i', \quad K = \sum_{i=1}^{n} X_i A X_i' \tag{9.64}$$

where the matrices A and B are as defined in assumptions (M.3) and (M.5), respectively.

Theorem 9.15 *Under assumptions (M.1)–(M.5), we have for each* $c > 0$

$$\sup_{T^{1/2}(\beta-\beta_0)|\le c} \left| \sum_{i=1}^{n} \left[\rho(Y_i - X_i'\beta) - \rho(Y_i - X_i'\beta_0) + (\beta - \beta_0')X_i\Psi(Y_i - X_i'\beta_0)\right]\right.$$

$$-\frac{1}{2}(\beta-\beta_0)'K(\beta-\beta_0)\Big| \to 0 \quad \textit{in probability}.$$

Theorem 9.16 *Under assumptions (M.1)–(M.6), we have for any* $c_n \to \infty$,

$$P\{|T^{\frac{1}{2}}(\hat\beta_n-\beta_0)| \geq c_n\} \to 0 \quad \Rightarrow \quad \hat\beta_n \to \beta_0 \quad \textit{in probability}.$$

Theorem 9.17 *Under assumptions (M.1)–(M.6), we have for each* $c > 0$

$$\sup_{|T^{-\frac{1}{2}}(\beta-\beta_0)|\leq c} \Big|\sum_{i=1}^{n} [T^{-\frac{1}{2}}X_i[\Psi(Y_i-X_i'\beta)-\Psi(Y_i-X_i'\beta_0)]]$$

$$+T^{-\frac{1}{2}}K(\beta-\beta_0)\Big| \to 0 \quad \textit{in probability}.$$

Theorem 9.18 *Under assumptions (M.1)–(M.6),*

$$T^{-\frac{1}{2}}K(\hat\beta_n-\beta_0) \xrightarrow{L} N(0, I_q).$$

9.6 Tests of Significance

The test of significance of LAD estimates (univariate case) and for LD estimates (multivariate case) were considered in Bai et al. (1987) and Bai et al. (1988), respectively. Because both of the above results are special cases of those considered in Bai et al. (1989) in this section, we present results only for the latter.

For the univariate case, we consider a test of the hypothesis H_0: $H\beta = r$ where H is a $m \times q$-matrix of rank m. Let $\tilde\beta_n$ denote the solution of

$$\min_{H\beta=r} \sum_{i=1}^{n} \rho(y_i - x_i'\beta) \tag{9.65}$$

and $\hat\beta_n$ the solution for the unrestricted minimum.

Theorem 9.19 *Under assumptions (U.1)–(U.6), we have*

(i)

$$\frac{2\lambda}{\sigma^2}\sum_{i=1}^{n} [\rho(y_i-x_i'\tilde\beta_n)-\rho(y_i-x_i'\hat\beta_n)] \xrightarrow{L} \chi^2_m, \tag{9.66}$$

where χ^2_m *represents the chi-square distribution on* m *degrees of freedom.*

(ii)

$$\frac{\lambda^2}{\sigma^2}(H\hat\beta_n-r)'(HS_n^{-1}H')^{-1}(H\hat\beta_n-r) \xrightarrow{L} \chi^2_m. \tag{9.67}$$

The asymptotic distribution (9.66) involves the nuisance parameters λ and σ^2, which may be unknown. In such a case we suggest the following procedure. Consider an extended linear model

$$y_i = x_i'\beta + Z_i'\gamma + \epsilon_i, \quad i = 1, \ldots, n, \tag{9.68}$$

where the Z_i are s-vectors satisfying the conditions

$$Z'X = 0, \quad Z'Z = I_s, \quad d_n = \max_{1 \le i \le n} |Z_i| \to 0 \tag{9.69}$$

with $Z = (Z_1, \ldots, Z_n)'$ and $X = (x_1, \ldots, x_n)'$. Let (β_n^*, γ_n^*) be a solution of

$$\min_{\beta,\gamma} \sum_{i=1}^{n} \rho(y_i - x_i'\beta - Z_i'\gamma). \tag{9.70}$$

By Theorem 9.19, under model (9.2),

$$2\lambda\sigma^{-2} \sum_{i=1}^{n} [\rho(y_i - x_i'\hat{\beta}_n) - \rho(y_i - x_i'\beta_n^* - Z_i'\gamma_n^*)] \xrightarrow{L} \chi_s^2 \tag{9.71}$$

whether or not the hypothesis H is true. Then we have the following theorem.

Theorem 9.20 *For model (9.2), under assumptions (U.1)–(U.6),*

$$\frac{s \sum_{i=1}^{n} [\rho(y_i - x_i'\tilde{\beta}_n) - \rho(y_i - x_i'\hat{\beta}_n)]}{q \sum_{i=1}^{n} [\rho(y_i - x_i'\hat{\beta}_n) - \rho(y_i - x_i'\beta_n^* - Z_i'\gamma_n^*)]} \xrightarrow{L} F(m, s), \tag{9.72}$$

where $F(m, s)$ denotes the F-distribution on m and s degrees of freedom.

Now we consider the multivariate case. Let $\tilde{\beta}$ be a solution of the minimization problem

$$\min_{H\beta = r} \left[\sum_{i=1}^{n} \rho(Y_i - X_i'\beta) \right]. \tag{9.73}$$

Then we have the following theorem.

Theorem 9.21 *Under assumptions (M.1)–(M.6),*

$$\left| \sum_{i=1}^{n} \left[\rho(Y_i - X_i'\tilde{\beta}_n) - \rho(Y_i - X_i'\hat{\beta}_n) \right] - \frac{1}{2} \left| Q' \sum_{i=1}^{n} X_i \Psi(E_i) \right|^2 \right| \to 0 \quad \text{in probability},$$

where Q is a $q \times m$-matrix such that

$$\begin{aligned} Q'KQ &= I_m, \\ Q'KG &= 0, \end{aligned} \tag{9.74}$$

with G as a $q \times (q - m)$-matrix determined by

$$\begin{aligned} G'TG &= I_{q-m}, \\ G'H &= 0. \end{aligned} \tag{9.75}$$

(Note that Q and G may not be uniquely determined by (9.74) and (9.75), but the product $Q'T^{-1}Q$ is the same for all choices of Q. In fact, $QQ' = K^{-1}H(H'K^{-1}H)^{-1}H'K^{-1}$.)

Remark: If $m = q$, we take $G = 0$ and $Q = K^{-\frac{1}{2}}$, whereas if $m = 0$, we take $H = 0, Q = 0$, and $G = T^{-\frac{1}{2}}$. With such choices, Theorem 9.21 is still true.

Remark: The test statistic

$$\sum_{i=1}^{n} \rho(Y_i - X_i'\tilde{\beta}_n) - \sum_{i=1}^{n} \rho(Y_i - X_i'\hat{\beta}_n) \tag{9.76}$$

has the same asymptotic distribution as $2^{-1}|Q' \sum_{i=1}^{n} X_i \Psi(E_i)|^2$, which, in general is a mixture of chi-squares.

10

Models for Categorical Response Variables

10.1 Generalized Linear Models

10.1.1 Extension of the Regression Model

Generalized linear models are a generalization of the classical linear models of the regression analysis and analysis of variance, which model the relationship between the expectation of a response variable and unknown predictor variables according to

$$\begin{aligned} \mathrm{E}(y_i) &= x_{i1}\beta_1 + \ldots + x_{ip}\beta_p \\ &= x_i'\beta \, . \end{aligned} \qquad (10.1)$$

The parameters are estimated according to the principle of least squares and are optimal according to minimum dispersion theory, or in case of a normal distribution, are optimal according to the ML theory (cf. Chapter 3).

Assuming an additive random error ϵ_i, the density function can be written as

$$f(y_i) = f_{\epsilon_i}(\, y_i - x_i'\beta) \, , \qquad (10.2)$$

where $\eta_i = x_i'\beta$ is the linear predictor. Hence, for continuous normally distributed data, we have the following distribution and mean structure:

$$y_i \sim N(\mu_i, \sigma^2), \quad \mathrm{E}(y_i) = \mu_i \, , \quad \mu_i = \eta_i = x_i'\beta \, . \qquad (10.3)$$

In analyzing categorical response variables, three major distributions may arise: the binomial, multinomial, and Poisson distributions, which belong to the natural exponential family (along with the normal distribution).

In analogy to the normal distribution, the effect of covariates on the expectation of the response variables may be modeled by linear predictors for these distributions as well.

Binomial Distribution

Assume that I predictors $\eta_i = x_i'\beta$ ($i = 1, \ldots, I$) and N_i realizations y_{ij}, $j = 1, \ldots, N_i$, respectively, are given, and furthermore, assume that the response has a binomial distribution

$$y_i \sim B(N_i, \pi_i) \quad \text{with} \quad \mathrm{E}(y_i) = N_i \pi_i = \mu_i\,.$$

Let $g(\pi_i) = \mathrm{logit}(\pi_i)$ be the chosen link function between μ_i and η_i:

$$\begin{aligned} \mathrm{logit}(\pi_i) &= \ln\left(\frac{\pi_i}{1-\pi_i}\right) \\ &= \ln\left(\frac{N_i\pi_i}{N_i - N_i\pi_i}\right) = x_i'\beta\,. \end{aligned} \tag{10.4}$$

With the inverse function $g^{-1}(x_i'\beta)$ we then have

$$N_i\pi_i = \mu_i = N_i \frac{\exp(x_i'\beta)}{1+\exp(x_i'\beta)} = g^{-1}(\eta_i)\,. \tag{10.5}$$

Poisson Distribution

Let y_i ($i = 1, \ldots, I$) have a Poisson distribution with $\mathrm{E}(y_i) = \mu_i$

$$P(y_i) = \frac{e^{-\mu_i}\mu_i^{y_i}}{y_i!} \quad \text{for } y_i = 0, 1, 2, \ldots\,. \tag{10.6}$$

The link function can then be chosen as $\ln(\mu_i) = x_i'\beta$.

Contingency Tables

The cell frequencies y_{ij} of an $I \times J$ contingency table of two categorical variables can have a Poisson, multinomial, or binomial distribution (depending on the sampling design). By choosing appropriate design vectors x_{ij}, the expected cell frequencies can be described by a loglinear model

$$\begin{aligned} \ln(m_{ij}) &= \mu + \alpha_i^A + \beta_j^B + (\alpha\beta)_{ij}^{AB} \\ &= x_{ij}'\beta \end{aligned} \tag{10.7}$$

and hence we have

$$\mu_{ij} = m_{ij} = \exp(x_{ij}'\beta) = \exp(\eta_{ij})\,. \tag{10.8}$$

In contrast to the classical model of the regression analysis, where $\mathrm{E}(y)$ is linear in the parameter vector β, so that $\mu = \eta = x'\beta$ holds, the generalized models are of the following form:

$$\mu = g^{-1}(x'\beta)\,, \tag{10.9}$$

where g^{-1} is the inverse function of the link function. Furthermore, the additivity of the random error is no longer a necessary assumption, so that in general

$$f(y) = f(y, x'\beta) \tag{10.10}$$

is assumed, instead of (10.2).

10.1.2 Structure of the Generalized Linear Model

The *generalized linear model (GLM)* (cf. Nelder and Wedderburn, 1972) is defined as follows. A GLM consists of three components:

- the *random component*, which specifies the probability distribution of the response variable,
- the *systematic component*, which specifies a linear function of the explanatory variables,
- the *link function*, which describes a functional relationship between the systematic component and the expectation of the random component.

The three components are specified as follows:

1. The random component Y consists of N independent observations $y' = (y_1, y_2, \ldots, y_N)$ of a distribution belonging to the natural exponential family (cf. Agresti, 1990, p. 80). Hence, each observation y_i has—in the simplest case of a one-parametric exponential family—the following probability density function:

$$f\left(y_i, \theta_i\right) = a\left(\theta_i\right) b\left(y_i\right) \exp\left(y_i\, Q\left(\theta_i\right)\right)\,. \tag{10.11}$$

Remark: The parameter θ_i can vary over $i = 1, 2, \ldots, N$, depending on the value of the explanatory variable, which influences y_i through the systematic component.

Special distributions of particular importance in this family are the Poisson and the binomial distribution. $Q(\theta_i)$ is called the *natural parameter* of the distribution. Likewise, if the y_i are independent, the joint distribution is a member of the exponential family.

A more general parameterization allows inclusion of scaling or nuisance variables. For example, an alternative parameterization with an additional scaling parameter ϕ (the so-called dispersion parameter) is given by

$$f(y_i|\theta_i,\phi) = \exp\left\{\frac{y_i\theta_i - b(\theta_i)}{a(\phi)} + c(y_i,\phi)\right\}, \tag{10.12}$$

where θ_i is called the natural parameter. If ϕ is known, (10.12) represents a linear exponential family. If, on the other hand, ϕ is unknown, then (10.12) is called an *exponential dispersion model*. With ϕ and θ_i, (10.12) is a two-parametric distribution for $i = 1, \ldots, N$, which is used for normal or gamma distributions, for instance. Introducing y_i and θ_i as vector-valued parameters rather than scalars leads to multivariate generalized models, which include multinomial response models as special case (cf. Fahrmeir and Tutz, 1994, Chapter 3) .

2. The systematic component relates a vector $\eta = (\eta_1, \eta_2, \ldots, \eta_N)$ to a set of explanatory variables through a linear model

$$\eta = X\beta\,. \tag{10.13}$$

Here η is called the linear predictor, $X : N \times p$ is the matrix of observations on the explanatory variables, and β is the $(p \times 1)$-vector of parameters.

3. The link function connects the systematic component with the expectation of the random component. Let $\mu_i = \mathrm{E}(y_i)$; then μ_i is linked to η_i by $\eta_i = g(\mu_i)$. Here g is a monotonic and differentiable function:

$$g(\mu_i) = \sum_{j=1}^{p} \beta_j x_{ij} \quad i = 1, 2, \ldots, N\,. \tag{10.14}$$

Special cases:

(i) $g(\mu) = \mu$ is called the *identity link*. We get $\eta_i = \mu_i$.

(ii) $g(\mu) = Q(\theta_i)$ is called the *canonical (natural) link*. We have $Q(\theta_i) = \sum_{j=1}^{p} \beta_j x_{ij}$.

Properties of the Density Function (10.12)

Let

$$l_i = l(\theta_i, \phi; y_i) = \ln f(y_i; \theta_i, \phi) \tag{10.15}$$

be the contribution of the ith observation y_i to the loglikelihood. Then

$$l_i = [y_i\theta_i - b(\theta_i)]/a(\phi) + c(y_i; \phi) \tag{10.16}$$

holds and we get the following derivatives with respect to θ_i

$$\frac{\partial l_i}{\partial \theta_i} = \frac{[y_i - b'(\theta_i)]}{a(\phi)}, \tag{10.17}$$

$$\frac{\partial^2 l_i}{\partial \theta_i^2} = \frac{-b''(\theta_i)}{a(\phi)}, \tag{10.18}$$

where $b'(\theta_i) = \partial b(\theta_i)/\partial\theta_i$ and $b''(\theta_i) = \partial^2 b(\theta_i)/\partial\theta_i^2$ are the first and second derivatives of the function $b(\theta_i)$, assumed to be known. By equating (10.17) to zero, it becomes obvious that the solution of the likelihood equations is independent of $a(\phi)$. Since our interest belongs to the estimation of θ and β in $\eta = x'\beta$, we could assume $a(\phi) = 1$ without any loss of generality (this corresponds to assuming $\sigma^2 = 1$ in the case of a normal distribution). For the present, however, we retain $a(\phi)$.

Under certain assumptions of regularity, the order of integration und differentiation may be interchangeable, so that

$$\mathrm{E}\left(\frac{\partial l_i}{\partial\theta_i}\right) = 0 \tag{10.19}$$

$$-\mathrm{E}\left(\frac{\partial^2 l_i}{\partial\theta_i^2}\right) = \mathrm{E}\left(\frac{\partial l_i}{\partial\theta_i}\right)^2. \tag{10.20}$$

Hence we have from (10.17) and (10.19)

$$\mathrm{E}(y_i) = \mu_i = b'(\theta_i). \tag{10.21}$$

Similarly, from (10.18) and (10.20), we find

$$\begin{aligned} \frac{b''(\theta_i)}{a(\phi)} &= \mathrm{E}\{\frac{[y_i - b'(\theta_i)]^2}{a^2(\phi)}\} \\ &= \frac{\mathrm{var}(y_i)}{a^2(\phi)}, \end{aligned} \tag{10.22}$$

since $\mathrm{E}[y_i - b'(\theta_i)] = 0$, and hence

$$V(\mu_i) = \mathrm{var}(y_i) = b''(\theta_i)a(\phi). \tag{10.23}$$

Under the assumption that the y_i ($i = 1, \ldots, N$) are independent, the loglikelihood of $y' = (y_1, \ldots, y_N)$ equals the sum of $l_i(\theta_i, \phi; y_i)$. Let $\theta' = (\theta_1, \ldots, \theta_N)$, $\mu' = (\mu_1, \ldots, \mu_N)$, $X = \begin{pmatrix} x_1' \\ \vdots \\ x_N' \end{pmatrix}$, and $\eta = (\eta_1, \ldots, \eta_N)' = X\beta$. We then have, from (10.21),

$$\mu = \frac{\partial b(\theta)}{\partial\theta} = \left(\frac{\partial b(\theta_1)}{\partial\theta_1}, \ldots, \frac{\partial b(\theta_1)}{\partial\theta_N}\right)', \tag{10.24}$$

and in analogy to (10.23) for the covariance matrix of $y' = (y_1, \ldots, y_N)$,

$$\mathrm{cov}(y) = V(\mu) = \frac{\partial^2 b(\theta)}{\partial\theta\partial\theta'} = a(\phi)\mathrm{diag}(b''(\theta_1), \ldots, b''(\theta_N)). \tag{10.25}$$

These relations hold in general, as we show in the following discussion.

10.1.3 Score Function and Information Matrix

The likelihood of the random sample is the product of the density functions:

$$L(\theta, \phi; y) = \prod_{i=1}^{N} f(y_i; \theta_i, \phi)\,. \tag{10.26}$$

The loglikelihood $\ln L(\theta, \phi; y)$ for the sample y of independent y_i ($i = 1, \ldots, N$) is of the form

$$l = l(\theta, \phi; y) = \sum_{i=1}^{N} l_i = \sum_{i=1}^{N} \left\{ \frac{(y_i\theta_i - b(\theta_i))}{a(\phi)} + c(y_i; \phi) \right\}. \tag{10.27}$$

The vector of first derivatives of l with respect to θ_i is needed for determining the ML estimates. This vector is called the *score function.* For now, we neglect the parameterization with ϕ in the representation of l and L and thus get the score function as

$$s(\theta; y) = \frac{\partial}{\partial \theta} l(\theta; y) = \frac{1}{L(\theta; y)} \frac{\partial}{\partial \theta} L(\theta; y)\,. \tag{10.28}$$

Let

$$\frac{\partial^2 l}{\partial\theta\partial\theta'} = \left(\frac{\partial^2 l}{\partial\theta_i\partial\theta_j}\right)_{\substack{i=1,\ldots,N\\ j=1,\ldots,N}}$$

be the matrix of the second derivatives of the loglikelihood. Then

$$F_{(N)}(\theta) = \mathrm{E}\left(\frac{-\partial^2 l(\theta; y)}{\partial\theta\partial\theta'}\right) \tag{10.29}$$

is called the expected *Fisher-information matrix* of the sample $y' = (y_1, \ldots, y_N)$, where the expectation is to be taken with respect to the following density function

$$f(y_1, \ldots, y_N|\theta_i) = \prod f(y_i|\theta_i) = L(\theta; y)\,.$$

In case of regular likelihood functions (where regular means: exchange of integration and differentiation is possible), to which the exponential families belong, we have

$$\mathrm{E}(s(\theta; y)) = 0 \tag{10.30}$$

and

$$F_{(N)}(\theta) = \mathrm{E}(s(\theta; y)s'(\theta; y)) = \mathrm{cov}(s(\theta; y))\,, \tag{10.31}$$

Relation (10.30) follows from

$$\int f(y_1, \ldots, y_N|\theta)\mathrm{d}y_1 \cdots \mathrm{d}y_N = \int L(\theta; y)\mathrm{d}y = 1\,, \tag{10.32}$$

by differentiating with respect to θ using (10.28):

$$\begin{aligned}\int \frac{\partial L(\theta;y)}{\partial\theta}\mathrm{d}y &= \int \frac{\partial l(\theta;y)}{\partial\theta}L(\theta;y)\mathrm{d}y \\ &= \mathrm{E}(s(\theta;y)) = 0\,. \end{aligned} \tag{10.33}$$

Differentiating (10.33) with respect to θ', we get

$$\begin{aligned} 0 &= \int \frac{\partial^2 l(\theta;y)}{\partial\theta\partial\theta'}L(\theta;y)\mathrm{d}y \\ &\quad + \int \frac{\partial l(\theta;y)}{\partial\theta}\frac{\partial l(\theta;y)}{\partial\theta'}L(\theta;y)\mathrm{d}y \\ &= -F_{(N)}(\theta) + \mathrm{E}(s(\theta;y)s'(\theta;y))\,, \end{aligned}$$

and hence (10.31), because $\mathrm{E}(s(\theta;y)) = 0$.

10.1.4 Maximum-Likelihood Estimation

Let $\eta_i = x_i'\beta = \sum_{j=1}^{p} x_{ij}\beta_j$ be the predictor of the ith observation of the response variable ($i = 1,\ldots,N$) or—in matrix representation—

$$\eta = \begin{pmatrix}\eta_1\\ \vdots \\ \eta_N\end{pmatrix} = \begin{pmatrix}x_1'\beta\\ \vdots \\ x_N'\beta\end{pmatrix} = X\beta\,. \tag{10.34}$$

Assume that the predictors are linked to $\mathrm{E}(y) = \mu$ by a monotonic differentiable function $g(\cdot)$:

$$g(\mu_i) = \eta_i \quad (i = 1\ldots,N)\,, \tag{10.35}$$

or, in matrix representation,

$$g(\mu) = \begin{pmatrix}g(\mu_1)\\ \vdots \\ g(\mu_N)\end{pmatrix} = \eta\,. \tag{10.36}$$

The parameters θ_i and β are then linked by the relation (10.21), that is $\mu_i = b'(\theta_i)$, with $g(\mu_i) = x_i'\beta$. Hence we have $\theta_i = \theta_i(\beta)$. Since we are interested only in estimating β, we write the loglikelihood (10.27) as a function of β:

$$l(\beta) = \sum_{i=1}^{N} l_i(\beta)\,. \tag{10.37}$$

We can find the derivatives $\partial l_i(\beta)/\partial\beta_j$ according to the chain rule:

$$\frac{\partial l_i(\beta)}{\partial\beta_j} = \frac{\partial l_i}{\partial\theta_i}\frac{\partial\theta_i}{\partial\mu_i}\frac{\partial\mu_i}{\partial\eta_i}\frac{\partial\eta_i}{\partial\beta_j}\,. \tag{10.38}$$

The partial results are as follows:

$$\begin{aligned}
\frac{\partial l_i}{\partial \theta_i} &= \frac{[y_i - b'(\theta_i)]}{a(\phi)} \quad \text{[cf. (10.17)]} \\
&= \frac{[y_i - \mu_i]}{a(\phi)} \quad \text{[cf. (10.21)]}, \qquad (10.39) \\
\mu_i &= b'(\theta_i)\,, \\
\frac{\partial \mu_i}{\partial \theta_i} &= b''(\theta_i) = \frac{\operatorname{var}(y_i)}{a(\phi)} \quad \text{[cf. (10.23)]}, \qquad (10.40) \\
\frac{\partial \eta_i}{\partial \beta_j} &= \frac{\partial \sum_{k=1}^{p} x_{ik}\beta_k}{\partial \beta_j} = x_{ij}\,. \qquad (10.41)
\end{aligned}$$

Because $\eta_i = g(\mu_i)$, the derivative $\partial\mu_i/\partial\eta_i$ is dependent on the link function $g(\cdot)$, or rather its inverse $g^{-1}(\cdot)$. Hence, it cannot be specified until the link is defined.

Summarizing, we now have

$$\frac{\partial l_i}{\partial \beta_j} = \frac{(y_i - \mu_i)x_{ij}}{\operatorname{Var}(y_i)}\frac{\partial \mu_i}{\partial \eta_i}, \quad j = 1, \ldots, p \qquad (10.42)$$

using the rule

$$\frac{\partial \theta_i}{\partial \mu_i} = \left(\frac{\partial \mu_i}{\partial \theta_i}\right)^{-1}$$

for inverse functions ($\mu_i = b'(\theta_i), \theta_i = (b')^{-1}(\mu_i)$). The likelihood equations for finding the components β_j are now

$$\sum_{i=1}^{N} \frac{(y_i - \mu_i)x_{ij}}{\operatorname{var}(y_i)}\frac{\partial \mu_i}{\partial \eta_i} = 0\,, \quad j = 1 \ldots, p. \qquad (10.43)$$

The loglikelihood is nonlinear in β. Hence, the solution of (10.43) requires iterative methods. For the second derivative with respect to components of β, we have, in analogy to (10.20), with (10.42),

$$\begin{aligned}
\mathrm{E}\left(\frac{\partial^2 l_i}{\partial \beta_j \partial \beta_h}\right) &= -\mathrm{E}\left(\frac{\partial l_i}{\partial \beta_j}\right)\left(\frac{\partial l_i}{\partial \beta_h}\right) \\
&= -\mathrm{E}\left[\frac{(y_i - \mu_i)(y_i - \mu_i)x_{ij}x_{ih}}{(\operatorname{var}(y_i))^2}\left(\frac{\partial \mu_i}{\partial \eta_i}\right)^2\right] \\
&= -\frac{x_{ij}x_{ih}}{\operatorname{var}(y_i)}\left(\frac{\partial \mu_i}{\partial \eta_i}\right)^2, \qquad (10.44)
\end{aligned}$$

and hence

$$\mathrm{E}\left(-\frac{\partial^2 l(\beta)}{\partial \beta_j \partial \beta_h}\right) = \sum_{i=1}^{N} \frac{x_{ij}x_{ih}}{\operatorname{var}(y_i)}\left(\frac{\partial \mu_i}{\partial \eta_i}\right)^2 \qquad (10.45)$$

and in matrix representation for all (j, h)-combinations

$$F_{(N)}(\beta) = \mathrm{E}\left(-\frac{\partial^2 l(\beta)}{\partial\beta\partial\beta'}\right) = X'WX \tag{10.46}$$

with

$$W = \mathrm{diag}(w_1 \ldots, w_N) \tag{10.47}$$

and the weights

$$w_i = \frac{\left(\frac{\partial\mu_i}{\partial\eta_i}\right)^2}{\mathrm{var}(y_i)} . \tag{10.48}$$

Fisher-Scoring Algorithm

For the iterative determination of the ML estimate of β, the method of iterative reweighted least squares is used. Let $\beta^{(k)}$ be the kth approximation of the ML estimate $\hat{\beta}$. Furthermore, let $q^{(k)}(\beta) = \partial l(\beta)/\partial\beta$ be the vector of the first derivatives at $\beta^{(k)}$ (cf. (10.42)). Analogously, we define $W^{(k)}$. The formula of the Fisher-scoring algorithm is then

$$(X'W^{(k)}X)\beta^{(k+1)} = (X'W^{(k)}X)\beta^{(k)} + q^{(k)} . \tag{10.49}$$

The vector on the right side of (10.49) has the components (cf. (10.45) and (10.42))

$$\sum_h \left[\sum_i \frac{x_{ij}x_{ih}}{\mathrm{var}(y_i)}\left(\frac{\partial\mu_i}{\partial\eta_i}\right)^2 \beta_h^{(k)}\right] + \sum_i \frac{(y_i - \mu_i^{(k)})x_{ij}}{\mathrm{var}(y_i)}\left(\frac{\partial\mu_i}{\partial\eta_i}\right) . \qquad (j = 1, \ldots, p) \tag{10.50}$$

The entire vector (10.50) can now be written as

$$X'W^{(k)}z^{(k)} , \tag{10.51}$$

where the $(N \times 1)$-vector $z^{(k)}$ has the jth element as follows:

$$\begin{aligned} z_i^{(k)} &= \sum_{j=1}^{p} x_{ij}\beta_j^{(k)} + (y_i - \mu_i^{(k)})\left(\frac{\partial\eta_i^{(k)}}{\partial\mu_i^{(k)}}\right) \\ &= \eta_i^{(k)} + (y_i - \mu_i^{(k)})\left(\frac{\partial\eta_i^{(k)}}{\partial\mu_i^{(k)}}\right) . \end{aligned} \tag{10.52}$$

Hence, the equation of the Fisher-scoring algorithm (10.49) can now be written as

$$(X'W^{(k)}X)\beta^{(k+1)} = X'W^{(k)}z^{(k)} . \tag{10.53}$$

This is the likelihood equation of a generalized linear model with the response vector $z^{(k)}$ and the random error covariance matrix $(W^{(k)})^{-1}$. If

$\text{rank}(X) = p$ holds, we obtain the ML estimate $\hat{\beta}$ as the limit of

$$\hat{\beta}^{(k+1)} = (X'W^{(k)}X)^{-1}X'W^{(k)}z^{(k)} \tag{10.54}$$

for $k \to \infty$, with the asymptotic covariance matrix

$$\text{V}(\hat{\beta}) = (X'\hat{W}X)^{-1} = F^{-1}_{(N)}(\hat{\beta}), \tag{10.55}$$

where $\hat{W}$ is determined at $\hat{\beta}$. Once a solution is found, then $\hat{\beta}$ is consistent for β, asymptotically normal, and asymptotically efficient (see. Fahrmeir and Kaufmann (1985) and Wedderburn (1976) for existence and uniqueness of the solutions). Hence we have $\hat{\beta} \overset{\text{as.}}{\sim} N(\beta, \text{V}(\hat{\beta}))$.

Remark: In case of a canonical link function, that is for $g(\mu_i) = \theta_i$, the ML equations simplify and the Fisher-scoring algorithm is identical to the Newton-Raphson algorithm (cf. Agresti, 1990, p. 451). If the values $a(\phi)$ are identical for all observations, then the ML equations are

$$\sum_i x_{ij} y_i = \sum_i x_{ij} \mu_i \,. \tag{10.56}$$

If, on the other hand, $a(\phi) = a_i(\phi) = a_i \phi$ $(i = 1, \ldots, N)$ holds, then the ML equations are

$$\sum_i \frac{x_{ij} y_i}{a_i} = \sum_i \frac{x_{ij} \mu_i}{a_i} \,. \tag{10.57}$$

As starting values for the Fisher-scoring algorithm the estimates $\hat{\beta}^{(0)} = (X'X)^{-1}X'y$ or $\hat{\beta}^{(0)} = (X'X)^{-1}X'g(y)$ may be used.

10.1.5 Testing of Hypotheses and Goodness of Fit

A generalized linear model $g(\mu_i) = x_i'\beta$ is—besides the distributional assumptions—determined by the link function $g(\cdot)$ and the explanatory variables $X_1, \ldots, X_p$, as well as their number p, which determines the length of the parameter vector β to be estimated. If $g(\cdot)$ is chosen, then the model is defined by the design matrix X.

Testing of Hypotheses

Let X_1 and X_2 be two design matrices (models), and assume that the hierarchical order $X_1 \subset X_2$ holds; that is, we have $X_2 = (X_1, X_3)$ with some matrix X_3 and hence $\mathcal{R}(X_1) \subset \mathcal{R}(X_2)$. Let β_1, β_2, and β_3 be the corresponding parameter vectors to be estimated. Further let $g(\hat{\mu}_1) = \hat{\eta}_1 = X_1\hat{\beta}_1$ and $g(\hat{\mu}_2) = \hat{\eta}_2 = X_2\tilde{\beta}_2 = X_1\tilde{\beta}_1 + X_3\tilde{\beta}_3$, where $\hat{\beta}_1$ and $\tilde{\beta}_2 = (\tilde{\beta}_1', \tilde{\beta}_3')'$ are the maximum-likelihood estimates under the two models, and $\text{rank}(X_1) = r_1$, $\text{rank}(X_2) = r_2$, and $(r_2 - r_1) = r = df$. The likelihood ratio statistic, which compares a larger model X_2 with a (smaller) submodel X_1, is then defined as follows (where L is the likelihood function)

$$\Lambda = \frac{\max_{\beta_1} L(\beta_1)}{\max_{\beta_2} L(\beta_2)} . \tag{10.58}$$

Wilks (1938) showed that $-2\ln\Lambda$ has a limiting χ^2_{df}-distribution where the degrees of freedom *df* equal the difference in the dimensions of the two models. Transforming (10.58) according to $-2\ln\Lambda$, with l denoting the loglikelihood, and inserting the maximum likelihood estimates gives

$$-2\ln\Lambda = -2[l(\hat{\beta}_1) - l(\tilde{\beta}_2)] . \tag{10.59}$$

In fact one tests the hypotheses $H_0 : \beta_3 = 0$ against $H_1 : \beta_3 \neq 0$. If H_0 holds, then $-2\ln\Lambda \sim \chi^2_r$. Therefore H_0 is rejected if the loglikelihood is significantly higher under the greater model using X_2. According to Wilks, we write

$$G^2 = -2\ln\Lambda$$

Goodness of Fit

Let X be the design matrix of the saturated model that contains the same number of parameters as observations. Denote by $\tilde{\theta}$ the estimate of θ that belongs to the estimates $\tilde{\mu}_i = y_i$ $(i = 1, \ldots, N)$ in the saturated model. For every submodel X_j that is not saturated, we then have (assuming, again, that $a(\phi) = a_i(\phi) = a_i\phi$)

$$\begin{aligned} G^2(X_j|X) &= 2\sum \frac{1}{a_i} \frac{y_i(\tilde{\theta}_i - \hat{\theta}_i) - b(\tilde{\theta}_i) + b(\hat{\theta}_i)}{\phi} \\ &= \frac{D(y;\hat{\mu}_j)}{\phi} \end{aligned} \tag{10.60}$$

as a measure for the loss in goodness of fit of the model X_j compared to the perfect fit achieved by the saturated model. The statistic $D(y;\hat{\mu}_j)$ is called the *deviance* of the model X_j. We then have

$$G^2(X_1|X_2) = G^2(X_1|X) - G^2(X_2|X) = \frac{D(y;\hat{\mu}_1) - D(y;\hat{\mu}_2)}{\phi} . \tag{10.61}$$

That is, the test statistic for comparing the model X_1 with the larger model X_2 equals the difference of the goodness-of-fit statistics of the two models, weighted with $1/\phi$.

10.1.6 Overdispersion

In samples of a Poisson or multinomial distribution, it may occur that the elements show a larger variance than that given by the distribution. This may be due to a violation of the assumption of independence, as, for example, a positive correlation in the sample elements. A frequent cause for this is the cluster-structure of the sample. Examples are

- the behavior of families of insects in the case of the influence of insecticides (Agresti, 1990, p. 42), where the family (cluster, batch) shows a *collective* (correlated) survivorship (many survive or most of them die) rather than an independent survivorship, due to dependence on cluster-specific covariables such as the temperature,
- the survivorship of dental implants when two or more implants are incorporated for each patient,
- the developement of diseases or social behavior of the members of a family,
- heterogeneity not taken into account, which is, for example, caused by having not measured important covariates for the linear predictor.

The existence of a larger variation (inhomogeneity) in the sample than in the sample model is called *overdispersion*. Overdispersion is in the simplest way modeled by multiplying the variance with a constant $\phi > 1$, where ϕ is either known (e.g., $\phi = \sigma^2$ for a normal distribution), or has to be estimated from the sample (cf. Fahrmeir and Tutz, 1994, Section 10.1.7, for alternative approaches).

Example (McCullagh and Nelder, 1989, p. 125): Let N individuals be divided into N/k clusters of equal cluster size k. Assume that the individual response is binary with $P(Y_i = 1) = \pi_i$, so that the total response

$$Y = Z_1 + Z_2 + \cdots + Z_{N/k}$$

equals the sum of independent $B(k; \pi_i)$-distributed binomial variables Z_i $(i = 1, \ldots, N/k)$. The π_i's vary across the clusters and assume that $\mathrm{E}(\pi_i) = \pi$ and $\mathrm{var}(\pi_i) = \tau^2\pi(1-\pi)$ with $0 \le \tau^2 \le 1$. We then have

$$\begin{aligned} \mathrm{E}(Y) &= N\pi \\ \mathrm{var}(Y) &= N\pi(1-\pi)\{1 + (k-1)\tau^2\} \\ &= \phi N\pi(1-\pi)\,. \end{aligned} \qquad (10.62)$$

The dispersion parameter $\phi = 1 + (k-1)\tau^2$ is dependent on the cluster size k and on the variability of the π_i, but not on the sample size N. This fact is essential for interpreting the variable Y as the sum of binomial variables Z_i and for estimating the dispersion parameter ϕ from the residuals. Because of $0 \le \tau^2 \le 1$, we have

$$1 \le \phi \le k \le N\,. \qquad (10.63)$$

Relationship (10.62) means that

$$\frac{\mathrm{var}(Y)}{N\pi(1-\pi)} = 1 + (k-1)\tau^2 = \phi \qquad (10.64)$$

is constant. An alternative model—the beta-binomial distribution—has the property that the quotient in (10.64), that is ϕ, is a linear function of the sample size N. By plotting the residuals against N, it is easy to recognize which of the two models is more likely. Rosner (1984) used the the beta-binomial distribution for estimation in clusters of size $k = 2$.

10.1.7 Quasi Loglikelihood

The generalized models assume a distribution of the natural exponential family for the data as the random component (cf. (10.11)). If this assumption does not hold, an alternative approach can be used to specify the functional relationship between the mean and the variance. For exponential families, the relationship (10.23) between variance and expectation holds. Assume the general approach

$$\operatorname{var}(Y) = \phi V(\mu)\,, \tag{10.65}$$

where $V(\cdot)$ is an appropriately chosen function.

In the quasi-likelihood approach (Wedderburn, 1974), only assumptions about the first and second moments of the random variables are made. It is not necessary for the distribution itself to be specified. The starting point in estimating the influence of covariables is the score function (10.28), or rather the system of ML equations (10.43). If the general specification (10.65) is inserted into (10.43), we get the system of *estimating equations* for β

$$\sum_{i=1}^{N} \frac{(y_i - \mu_i)}{V(\mu_i)} x_{ij} \frac{\partial \mu_i}{\partial \eta_i} = 0 \quad (j = 1, \ldots, p)\,, \tag{10.66}$$

which is of the same form as as the likelihood equations (10.43) for GLMs. However, system (10.66) is an ML equation system only if the y_i's have a distribution of the natural exponential family.

In the case of independent response, the modeling of the influence of the covariables X on the mean response $\mathrm{E}(y) = \mu$ is done according to McCullagh and Nelder (1989, p. 324) as follows. Assume that for the response vector we have

$$y \sim (\mu, \phi V(\mu)) \tag{10.67}$$

where $\phi > 0$ is an unknown dispersion parameter and $V(\mu)$ is a matrix of known functions. Expression $\phi V(\mu)$ is called the *working variance.*

If the components of y are assumed to be independent, the covariance matrix $\phi V(\mu)$ has to be diagonal, that is,

$$V(\mu) = \operatorname{diag}(V_1(\mu), \ldots, V_N(\mu))\,. \tag{10.68}$$

Here it is realistic to assume that the variance of each random variable y_i is dependent only on the ith component μ_i of μ, meaning thereby

$$V(\mu) = \operatorname{diag}(V_1(\mu_1), \ldots, V_N(\mu_N)). \tag{10.69}$$

A dependency on all components of μ according to (10.68) is difficult to interpret in practice if independence of the y_i is demanded as well. (Nevertheless, situations as in (10.68) are possible.) In many applications it is reasonable to assume, in addition to functional independency (10.69), that the V_i functions are identical, so that

$$V(\mu) = \operatorname{diag}(v(\mu_1), \ldots, v(\mu_N)) \tag{10.70}$$

holds, with $V_i = v(\cdot)$.

Under the above assumptions, the following function for a component y_i of y:

$$U = u(\mu_i, y_i) = \frac{y_i - \mu_i}{\phi v(\mu_i)} \tag{10.71}$$

has the properties

$$\mathrm{E}(U) = 0, \tag{10.72}$$

$$\operatorname{var}(U) = \frac{1}{\phi v(\mu_i)}, \tag{10.73}$$

$$\frac{\partial U}{\partial \mu_i} = \frac{-\phi v(\mu_i) - (y_i - \mu_i)\phi \frac{\partial v(\mu_i)}{\partial \mu_i}}{\phi^2 v^2(\mu_i)}$$

$$-\mathrm{E}\left(\frac{\partial U}{\partial \mu_i}\right) = \frac{1}{\phi v(\mu_i)}. \tag{10.74}$$

Hence U has the same properties as the derivative of a loglikelihood, which, of course, is the score function (10.28). Property (10.47) corresponds to (10.31), whereas property (10.74) in combination with (10.73) corresponds to (10.31). Therefore,

$$Q(\mu; y) = \sum_{i=1}^{N} Q_i(\mu_i; y_i) \tag{10.75}$$

with

$$Q_i(\mu_i; y_i) = \int_{y_i}^{\mu_i} \frac{\mu_i - t}{\phi v(t)} \mathrm{d}t \tag{10.76}$$

(cf. McCullagh and Nelder, 1989, p. 325) is the analogue of the loglikelihood function. $Q(\mu; y)$ is called *quasi loglikelihood.* Hence, the *quasi score function*, which is obtained by differentiating $Q(\mu; y)$, equals

$$U(\beta) = \phi^{-1} D' V^{-1} (y - \mu), \tag{10.77}$$

with $D = (\partial \mu_i / \partial \beta_j)$ $(i = 1, \ldots, N, j = 1, \ldots, p)$ and $V = \operatorname{diag}(v_1, \ldots, v_N)$. The quasi-likelihood estimate $\hat{\beta}$ is the solution of $U(\hat{\beta}) = 0$. It has the asymptotic covariance matrix

$$\operatorname{cov}(\hat{\beta}) = \phi(D'V^{-1}D)^{-1}. \tag{10.78}$$

The dispersion parameter ϕ is estimated by

$$\hat{\phi} = \frac{1}{N-p}\frac{\sum(y_i - \hat{\mu}_i)^2}{v(\hat{\mu}_i)} = \frac{X^2}{N-p}, \tag{10.79}$$

where X^2 is the so-called Pearson statistic. In the case of overdispersion (or assumed overdispersion), the influence of covariables (i.e., of the vector β) is to be estimated by a quasi-likelihood approach (10.66) rather than by a likelihood approach.

10.2 Contingency Tables

10.2.1 Overview

This section deals with contingency tables and the appropriate models. We first consider so-called two-way contingency tables. In general, a bivariate relationship is described by the joint distribution of the two associated random variables. The two marginal distributions are obtained by integrating (summing) the joint distribution over the respective variables. Likewise, the conditional distributions can be derived from the joint distribution.

Definition 10.1 (Contingency Table) *Let X and Y denote two categorical variables, with X at I levels and Y at J levels. When we observe subjects with the variables X and Y, there are $I \times J$ possible combinations of classifications. The outcomes $(X;Y)$ of a sample with sample size n are displayed in an $I \times J$ (contingency) table. (X,Y) are realizations of the joint two-dimensional distribution:*

$$P(X = i, Y = j) = \pi_{ij}. \tag{10.80}$$

The set $\{\pi_{ij}\}$ forms the joint distribution of X and Y. The marginal distributions are obtained by summing over rows or columns:

		Y				
		1	*2*	...	*J*	*Marginal distribution of X*
X	*1*	π_{11}	π_{12}	...	π_{1J}	π_{1+}
	2	π_{21}	π_{22}	...	π_{2J}	π_{2+}
	$\vdots$	$\vdots$	$\vdots$		$\vdots$	$\vdots$
	I	π_{I1}	π_{I2}	...	π_{IJ}	π_{I+}
Marginal distribution of Y		π_{+1}	π_{+2}	...	π_{+J}	

$$\begin{aligned} \pi_{+j} &= \sum_{i=1}^{I} \pi_{ij}\,, \quad j = 1, \ldots, J\,, \\ \pi_{i+} &= \sum_{j=1}^{J} \pi_{ij}\,, \quad i = 1, \ldots, I\,, \\ \sum_{i=1}^{I} \pi_{i+} &= \sum_{j=1}^{J} \pi_{+j} = 1\,. \end{aligned}$$

In many contingency tables the explanatory variable X is fixed, and only the response Y is a random variable. In such cases, the main interest is not the joint distribution, but rather the conditional distribution. $\pi_{j|i} = P(Y = j|X = i)$ is the conditional probability, and $\{\pi_{1|i}, \pi_{2|i}, \ldots, \pi_{J|i}\}$ with $\sum_{j=1}^{J} \pi_{j|i} = 1$ is the conditional distribution of Y, given $X = i$.

A general aim of many studies is the comparison of the conditional distributions of Y at various levels i of X.

Suppose that X as well as Y are random response variables, so that the joint distribution describes the association of the two variables. Then, for the conditional distribution $Y|X$, we have

$$\pi_{j|i} = \frac{\pi_{ij}}{\pi_{i+}} \qquad \forall i, j\,. \tag{10.81}$$

Definition 10.2 *Two variables are called independent if*

$$\pi_{ij} = \pi_{i+}\pi_{+j} \qquad \forall i, j. \tag{10.82}$$

If X and Y are independent, we obtain

$$\pi_{j|i} = \frac{\pi_{ij}}{\pi_{i+}} = \frac{\pi_{i+}\pi_{+j}}{\pi_{i+}} = \pi_{+j}\,. \tag{10.83}$$

The conditional distribution is equal to the marginal distribution and thus is independent of i.

Let $\{p_{ij}\}$ denote the sample joint distribution. They have the following properties, with n_{ij} being the cell frequencies and $n = \sum_{i=1}^{I}\sum_{j=1}^{J} n_{ij}$:

$$\left.\begin{aligned} p_{ij} &= \frac{n_{ij}}{n}\,, \\ p_{j|i} &= \frac{p_{ij}}{p_{i+}} = \frac{n_{ij}}{n_{i+}}\,, & p_{i|j} &= \frac{p_{ij}}{p_{+j}} = \frac{n_{ij}}{n_{+j}}\,, \\ p_{i+} &= \frac{\sum_{j=1}^{J} n_{ij}}{n}\,, & p_{+j} &= \frac{\sum_{i=1}^{I} n_{ij}}{n}\,, \\ n_{i+} &= \sum_{j=1}^{J} n_{ij} = np_{i+}\,, & n_{+j} &= \sum_{i=1}^{I} n_{ij} = np_{+j}\,. \end{aligned}\right\} \tag{10.84}$$

10.2.2 Ways of Comparing Proportions

Suppose that Y is a binary response variable (Y can take only the values 0 or 1), and let the outcomes of X be grouped. When row i is fixed, $\pi_{1|i}$ is the probability for response ($Y = 1$), and $\pi_{2|i}$ is the probability for nonresponse ($Y = 0$). The conditional distribution of the binary response variable Y, given $X = i$, then is

$$(\pi_{1|i}; \pi_{2|i}) = (\pi_{1|i}, (1 - \pi_{1|i})). \tag{10.85}$$

We can now compare two rows, say i and h, by calculating the difference in proportions for response, or nonresponse, respectively:

$$\begin{array}{rlcl} \text{Response:} & \pi_{1|h} - \pi_{1|i} & \text{and} & \\ \text{Nonresponse:} & \pi_{2|h} - \pi_{2|i} & = & (1 - \pi_{1|h}) - (1 - \pi_{1|i}) \\ & & = & -(\pi_{1|h} - \pi_{1|i}) \,. \end{array}$$

The differences have different signs, but their absolute values are identical. Additionally, we have

$$-1.0 \leq \pi_{1|h} - \pi_{1|i} \leq 1.0 \,. \tag{10.86}$$

The difference equals zero if the conditional distributions of the two rows i and h coincide. From this, one may conjecture that the response variable Y is independent of the row classification when

$$\pi_{1|h} - \pi_{1|i} = 0 \quad \forall (h, i) \quad i, h = 1, 2, \ldots, I \,, \quad i \neq h \,. \tag{10.87}$$

In a more general setting, with the response variable Y having J categories, the variables X and Y are independent if

$$\pi_{j|h} - \pi_{j|i} = 0 \quad \forall j \,, \forall (h, i) \quad i, h = 1, 2, \ldots, I \,, \quad i \neq h \,. \tag{10.88}$$

Definition 10.3 (Relative Risk) *Let Y denote a binary response variable. The ratio $\pi_{1|h}/\pi_{1|i}$ is called the relative risk for response of category h in relation to category i.*

For 2×2 tables the relative risk (for response) is

$$0 \leq \frac{\pi_{1|1}}{\pi_{1|2}} < \infty \,. \tag{10.89}$$

The relative risk is a nonnegative real number. A relative risk of 1 corresponds to independence. For nonresponse, the relative risk is

$$\frac{\pi_{2|1}}{\pi_{2|2}} = \frac{1 - \pi_{1|1}}{1 - \pi_{1|2}} \,. \tag{10.90}$$

Definition 10.4 (Odds) *The odds are defined as the ratio of the probability of response in relation to the probability of nonresponse, within one category of X.*

For 2×2 tables, the odds in row 1 equal

$$\Omega_1 = \frac{\pi_{1|1}}{\pi_{2|1}} . \tag{10.91}$$

Within row 2, the corresponding odds equal

$$\Omega_2 = \frac{\pi_{1|2}}{\pi_{2|2}} . \tag{10.92}$$

Hint: For the joint distribution of two binary variables, the definition is

$$\Omega_i = \frac{\pi_{i1}}{\pi_{i2}}, \quad i = 1, 2 . \tag{10.93}$$

In general, Ω_i is nonnegative. When $\Omega_i > 1$, response is more likely than nonresponse. If, for instance, $\Omega_1 = 4$, then response in the first row is four times as likely as nonresponse. The *within-row conditional distributions* are independent when $\Omega_1 = \Omega_2$. This implies that the two variables are independent:

$$X, Y \ \ independent \quad \Leftrightarrow \quad \Omega_1 = \Omega_2 . \tag{10.94}$$

Definition 10.5 (Odds Ratio) *The odds ratio is defined as:*

$$\theta = \frac{\Omega_1}{\Omega_2} . \tag{10.95}$$

From the definition of the odds using joint probabilities, we have

$$\theta = \frac{\pi_{11}\pi_{22}}{\pi_{12}\pi_{21}} . \tag{10.96}$$

Another terminology for θ is the cross-product ratio. X and Y are independent when the odds ratio equals 1:

$$X, Y \ \ independent \quad \Leftrightarrow \quad \theta = 1 . \tag{10.97}$$

When all the cell probabilities are greater than 0 and $1 < \theta < \infty$, response for the subjects in the first row is more likely than for the subjects in the second row, that is, $\pi_{1|1} > \pi_{1|2}$. For $0 < \theta < 1$, we have $\pi_{1|1} < \pi_{1|2}$ (with a reverse interpretation).

The sample version of the odds ratio for the 2×2 table

		Y 1	2	
X	1	n_{11}	n_{12}	n_{1+}
	2	n_{21}	n_{22}	n_{2+}
		n_{+1}	n_{+2}	n

is

$$\hat{\theta} = \frac{n_{11}n_{22}}{n_{12}n_{21}} . \tag{10.98}$$

Odds Ratios for $I \times J$ Tables

From any given $I \times J$ table, 2×2 tables can be constructed by picking two different rows and two different columns. There are $I(I-1)/2$ pairs of rows and $J(J-1)/2$ pairs of columns; hence an $I \times J$ table contains $IJ(I-1)(J-1)/4$ tables. The set of all 2×2 tables contains much redundant information; therefore, we consider only neighboring 2×2 tables with the local odds ratios

$$\theta_{ij} = \frac{\pi_{i,j}\pi_{i+1,j+1}}{\pi_{i,j+1}\pi_{i+1,j}}, \quad i = 1, 2, \ldots, I-1, \quad j = 1, 2, \ldots, J-1. \tag{10.99}$$

These $(I-1)(J-1)$ odds ratios determine all possible odds ratios formed from all pairs of rows and all pairs of columns.

10.2.3 Sampling in Two-Way Contingency Tables

Variables having nominal or ordinal scale are denoted as categorical variables. In most cases, statistical methods assume a multinomial or a Poisson distribution for categorical variables. We now elaborate these two sample models. Suppose that we observe counts n_i $(i = 1, 2, \ldots, N)$ in the N cells of a contingency table with a single categorical variable or in $N = I \times J$ cells of a two-way contingency table.

We assume that the n_i are random variables with a distribution in $\mathbb{R}^+$ and the expected values $\mathrm{E}(n_i) = m_i$, which are called expected frequencies.

Poisson Sample

The Poisson distribution is used for counts of events (such as response to a medical treatment) that occur randomly over time when outcomes in disjoint periods are independent. The Poisson distribution may be interpreted as the limit distribution of the binomial distribution $b(n;p)$ if $\lambda = n \cdot p$ is fixed for increasing n. For each of the N cells of a contingency table $\{n_i\}$, we have

$$P(n_i) = \frac{e^{-m_i} m_i^{n_i}}{n_i!}, \quad n_i = 0, 1, 2, \ldots, \quad i = 1, \ldots, N. \tag{10.100}$$

This is the probability mass function of the Poisson distribution with the parameter m_i. It satisfies the identities $\mathrm{var}(n_i) = E(n_i) = m_i$.

The Poisson model for $\{n_i\}$ assumes that the n_i are independent. The joint distribution for $\{n_i\}$ then is the product of the distributions for n_i in the N cells. The total sample size $n = \sum_{i=1}^N n_i$ also has a Poisson distribution with $E(n) = \sum_{i=1}^N m_i$ (the rule for summing up independent random variables with Poisson distribution).

The Poisson model is used if rare events are independently distributed over disjoint classes.

Let $n = \sum_{i=1}^{N} n_i$ be fixed. The conditional probability of a contingency table $\{n_i\}$ that satisfies this condition is

$$P\big(n_i \text{ observations in cell } i, \quad i = 1,2,\ldots,N \mid \sum_{i=1}^{N} n_i = n\big) =$$

$$= \frac{P(n_i \text{ observations in cell } i, \quad i = 1,2,\ldots,N)}{P(\sum_{i=1}^{N} n_i = n)}$$

$$= \frac{\prod_{i=1}^{N} e^{-m_i} \frac{m_i^{n_i}}{n_i!}}{\exp(-\sum_{j=1}^{N} m_j) \frac{(\sum_{j=1}^{N} m_j)^n}{n!}}$$

$$= \left(\frac{n!}{\prod_{i=1}^{N} n_i!}\right) \cdot \prod_{i=1}^{N} \pi_i^{n_i}, \quad \text{with} \quad \pi_i = \frac{m_i}{\sum_{i=1}^{N} m_i}. \qquad (10.101)$$

For $N = 2$, this is the binomial distribution. For the multinomial distribution for $(n_1, n_2, \ldots, n_N)$, the marginal distribution for n_i is a binomial distribution with $E(n_i) = n\pi_i$ and $\text{var}(n_i) = n\pi_i(1 - \pi_i)$.

Independent Multinomial Sample

Suppose we observe on a categorical variable Y at various levels of an explanatory variable X. In the cell $(X = i, Y = j)$ we have n_{ij} observations. Suppose that $n_{i+} = \sum_{j=1}^{J} n_{ij}$, the number of observations of Y for fixed level i of X, is fixed in advance (and thus not random) and that the n_{i+} observations are independent and have the distribution $(\pi_{1|i}, \pi_{2|i}, \ldots, \pi_{J|i})$. Then the cell counts in row i have the multinomial distribution

$$\left(\frac{n_{i+}!}{\prod_{j=1}^{J} n_{ij}!}\right) \cdot \prod_{j=1}^{J} \pi_{j|i}^{n_{ij}}. \qquad (10.102)$$

Furthermore, if the samples are independent for different i, then the joint distribution for the n_{ij} in the $I \times J$ table is the product of the multinomial distributions (10.102). This is called *product multinomial sampling* or *independent multinomial sampling*.

10.2.4 Likelihood Function and Maximum-Likelihood Estimates

For the observed cell counts $\{n_i, i = 1,2,\ldots,N\}$, the likelihood function is defined as the probability of $\{n_i, i = 1,2,\ldots,N\}$ for a given sampling model. This function in general is dependent on an unknown parameter θ—here, for instance, $\theta = \{\pi_{j|i}\}$. The maximum-likelihood estimate for this vector of parameters is the value for which the likelihood function of the observed data takes its maximum.

To illustrate, we now look at the estimates of the category probabilities $\{\pi_i\}$ for multinomial sampling. The joint distribution $\{n_i\}$ is (cf. (10.102) and the notation $\{\pi_i\}$, $i = 1, \ldots, N$, $N = I \cdot J$, instead of $\pi_{j|i}$)

$$\frac{n!}{\prod_{i=1}^{N} n_i!} \underbrace{\prod_{i=1}^{N} \pi_i^{n_i}}_{\text{kernel}} . \tag{10.103}$$

It is proportional to the so-called kernel of the likelihood function. The kernel contains all unknown parameters of the model. Hence, maximizing the likelihood is equivalent to maximizing the kernel of the loglikelihood function:

$$\ln(\text{kernel}) = \sum_{i=1}^{N} n_i \ln(\pi_i) \rightarrow \max_{\pi_i} . \tag{10.104}$$

Under the condition $\pi_i > 0$, $i = 1, 2, \ldots, N$, $\sum_{i=1}^{N} \pi_i = 1$, we have $\pi_N = 1 - \sum_{i=1}^{N-1} \pi_i$ and hence

$$\frac{\partial \pi_N}{\partial \pi_i} = -1, \quad i = 1, 2, \ldots, N-1, \tag{10.105}$$

$$\frac{\partial \ln \pi_N}{\partial \pi_i} = \frac{1}{\pi_N} \cdot \frac{\partial \pi_N}{\partial \pi_i} = \frac{-1}{\pi_N}, \quad i = 1, 2, \ldots, N-1, \tag{10.106}$$

$$\frac{\partial L}{\partial \pi_i} = \frac{n_i}{\pi_i} - \frac{n_N}{\pi_N} = 0, \quad i = 1, 2, \ldots, N-1. \tag{10.107}$$

From (10.107) we get

$$\frac{\hat{\pi}_i}{\hat{\pi}_N} = \frac{n_i}{n_N}, \quad i = 1, 2, \ldots, N-1, \tag{10.108}$$

and thus

$$\hat{\pi}_i = \hat{\pi}_N \frac{n_i}{n_N} . \tag{10.109}$$

Using

$$\sum_{i=1}^{N} \hat{\pi}_i = 1 = \frac{\hat{\pi}_N \sum_{i=1}^{N} n_i}{n_N}, \tag{10.110}$$

we obtain the solutions

$$\hat{\pi}_N = \frac{n_N}{n} = p_N . \tag{10.111}$$

$$\hat{\pi}_i = \frac{n_i}{n} = p_i, \quad i = 1, 2, \ldots, N-1. \tag{10.112}$$

The ML estimates are the proportions (relative frequencies) p_i.

For contingency tables, we have for independent X and Y:

$$\pi_{ij} = \pi_{i+} \pi_{+j} . \tag{10.113}$$

The ML estimates under this condition are

$$\hat{\pi}_{ij} = p_{i+}p_{+j} = \frac{n_{i+}n_{+j}}{n^2} \tag{10.114}$$

with the expected cell frequencies

$$\hat{m}_{ij} = n\hat{\pi}_{ij} = \frac{n_{i+}n_{+j}}{n}. \tag{10.115}$$

Because of the similarity of the likelihood functions, the ML estimates for Poisson, multinomial, and product multinomial sampling are identical (as long as no further assumptions are made).

10.2.5 Testing the Goodness of Fit

A principal aim of the analysis of contingency tables is to test whether the observed and the expected cell frequencies (specified by a model) coincide. For instance, Pearson's χ^2 statistic compares the observed and the expected cell frequencies from (10.115) for independent X and Y.

Testing a Specified Multinomial Distribution (Theoretical Distribution)

We first want to compare a multinomial distribution, specified by $\{\pi_{i0}\}$, with the observed distribution $\{n_i\}$ for N classes.

The hypothesis for this problem is

$$H_0 : \pi_i = \pi_{i0}\,, \quad i = 1, 2, \ldots, N\,, \tag{10.116}$$

whereas for the π_i we have the restriction

$$\sum_{i=1}^{N} \pi_i = 1\,. \tag{10.117}$$

When H_0 is true, the expected cell frequencies are

$$m_i = n\pi_{i0}\,, \quad i = 1, 2, \ldots, N\,. \tag{10.118}$$

The appropriate test statistic is Pearson's χ^2, where

$$\chi^2 = \sum_{i=1}^{N} \frac{(n_i - m_i)^2}{m_i} \overset{\text{approx.}}{\sim} \chi^2_{N-1}\,. \tag{10.119}$$

This can be justified as follows: Let $p = (n_1/n, \ldots, n_{N-1}/n)$ and $\pi_0 = (\pi_{1_0}, \ldots, \pi_{N-1_0})$. By the central limit theorem we then have for $n \to \infty$,

$$\sqrt{n}\,(p - \pi_0) \to N\left(0, \Sigma_0\right)\,, \tag{10.120}$$

and so

$$n\,(p - \pi_0)'\,\Sigma_0^{-1}\,(p - \pi_0) \to \chi^2_{N-1}\,. \tag{10.121}$$

The asymptotic covariance matrix has the form

$$\Sigma_0 = \Sigma_0(\pi_0) = \text{diag}(\pi_0) - \pi_0\pi_0'\,. \tag{10.122}$$

Its inverse can be written as

$$\Sigma_0^{-1} = \frac{1}{\pi_{N0}} 11' + \operatorname{diag}\left(\frac{1}{\pi_{10}}, \ldots, \frac{1}{\pi_{N-1,0}}\right). \tag{10.123}$$

The equivalence of (10.119) and (10.121) is proved by direct calculation. To illustrate, we choose $N = 3$. Using the relationship $\pi_1 + \pi_2 + \pi_3 = 1$, we have

$$\begin{aligned}
\Sigma_0 &= \begin{pmatrix} \pi_1 & 0 \\ 0 & \pi_2 \end{pmatrix} - \begin{pmatrix} \pi_1^2 & \pi_1\pi_2 \\ \pi_1\pi_2 & \pi_2^2 \end{pmatrix}, \\
\Sigma_0^{-1} &= \begin{pmatrix} \pi_1(1-\pi_1) & -\pi_1\pi_2 \\ -\pi_1\pi_2 & \pi_2(1-\pi_2) \end{pmatrix}^{-1} \\
&= \frac{1}{\pi_1\pi_2\pi_3} \begin{pmatrix} \pi_2(1-\pi_2) & \pi_1\pi_2 \\ \pi_1\pi_2 & \pi_1(1-\pi_1) \end{pmatrix} \\
&= \begin{pmatrix} \frac{1}{\pi_1} + \frac{1}{\pi_3} & \frac{1}{\pi_3} \\ \frac{1}{\pi_3} & \frac{1}{\pi_2} + \frac{1}{\pi_3} \end{pmatrix}.
\end{aligned}$$

The left side of (10.121) now is

$$\begin{aligned}
& n\left(\frac{n_1}{n} - \frac{m_1}{n}, \frac{n_2}{n} - \frac{m_2}{n}\right) \begin{pmatrix} \frac{n}{m_1} + \frac{n}{m_3} & \frac{n}{m_3} \\ \frac{n}{m_3} & \frac{n}{m_2} + \frac{n}{m_3} \end{pmatrix} \begin{pmatrix} \frac{n_1}{n} - \frac{m_1}{n} \\ \frac{n_2}{n} - \frac{m_2}{n} \end{pmatrix} \\
&= \frac{(n_1 - m_1)^2}{m_1} + \frac{(n_2 - m_2)^2}{m_2} + \frac{1}{m_3}\left[(n_1 - m_1) + (n_2 - m_2)\right]^2 \\
&= \sum_{i=1}^{3} \frac{(n_i - m_i)^2}{m_i}.
\end{aligned}$$

Goodness of Fit for Estimated Expected Frequencies

When the unknown parameters are replaced by the ML estimates for a specified model, the test statistic is again approximately distributed as χ^2 with the number of degrees of freedom reduced by the number of estimated parameters.

The degrees of freedom are $(N - 1) - t$, if t parameters are estimated.

Testing for Independence

In two-way contingency tables with multinomial sampling, the hypothesis H_0 : *X and Y are statistically independent* is equivalent to $H_0 : \pi_{ij} = \pi_{i+}\pi_{+j} \quad \forall i, j$. The test statistic is Pearson's χ^2 in the following form:

$$\chi^2 = \sum_{\substack{i=1,2,\ldots,I \\ j=1,2,\ldots,J}} \frac{(n_{ij} - m_{ij})^2}{m_{ij}}, \tag{10.124}$$

where $m_{ij} = n\pi_{ij} = n\pi_{i+}\pi_{+j}$ (expected cell frequencies under H_0) are unknown.

Given the estimates $\hat{m}_{ij} = np_{i+}p_{+j}$, the χ^2 statistic then equals

$$\chi^2 = \sum_{\substack{i=1,2,\ldots,I \\ j=1,2,\ldots,J}} \frac{(n_{ij} - \hat{m}_{ij})^2}{\hat{m}_{ij}} \tag{10.125}$$

with $(I-1)(J-1) = (IJ-1) - (I-1) - (J-1)$ degrees of freedom. The numbers $(I-1)$ and $(J-1)$ correspond to the $(I-1)$ independent row proportions $(\pi_{i+})'$ and $(J-1)$ independent column proportions (π_{+j}) estimated from the sample.

Likelihood-Ratio Test

The likelihood-ratio test (LRT) is a general-purpose method for testing H_0 against H_1. The main idea is to compare $\max_{H_0} L$ and $\max_{H_1 \vee H_0} L$ with the corresponding parameter spaces $\omega \subseteq \Omega$. As test statistic, we have

$$\Lambda = \frac{\max_\omega L}{\max_\Omega L} \leq 1\,. \tag{10.126}$$

It follows that for $n \to \infty$ (Wilks, 1932)

$$G^2 = -2 \ln \Lambda \to \chi^2_d \tag{10.127}$$

with $d = \dim(\Omega) - \dim(\omega)$ as the degrees of freedom.

For multinomial sampling in a contingency table, the kernel of the likelihood function is

$$K = \prod_{i=1}^{I} \prod_{j=1}^{J} \pi_{ij}^{n_{ij}}\,, \tag{10.128}$$

with the constraints for the parameters:

$$\pi_{ij} \geq 0 \quad \text{and} \quad \sum_{i=1}^{I} \sum_{j=1}^{J} \pi_{ij} = 1\,. \tag{10.129}$$

Under the null hypothesis $H_0 : \pi_{ij} = \pi_{i+}\pi_{+j}$, K is maximum for $\hat{\pi}_{i+} = n_{i+}/n$, $\hat{\pi}_{+j} = n_{+j}/n$, and $\hat{\pi}_{ij} = n_{i+}n_{+j}/n^2$. Under $H_0 \vee H_1$, K is maximum for $\hat{\pi}_{ij} = n_{ij}/n$. We then have

$$\Lambda = \frac{\prod_{i=1}^{I} \prod_{j=1}^{J} (n_{i+}n_{+j})^{n_{ij}}}{n^n \prod_{i=1}^{I} \prod_{j=1}^{J} n_{ij}^{n_{ij}}}\,. \tag{10.130}$$

It follows that Wilks's G^2 is given by

$$G^2 = -2 \ln \Lambda = 2 \sum_{i=1}^{I} \sum_{j=1}^{J} n_{ij} \ln \left(\frac{n_{ij}}{\hat{m}_{ij}} \right) \sim \chi^2_{(I-1)(J-1)} \tag{10.131}$$

with $\hat{m}_{ij} = n_{i+}n_{+j}/n$ (estimate under H_0).

If H_0 holds, Λ will be large, that is near 1, and G^2 will be small. This means that H_0 is to be rejected for large G^2.

10.3 GLM for Binary Response

10.3.1 Logit Models and Logistic Regression

Let Y be a binary random variable, that is, Y has only two categories (for instance, success/failure or case/control). Hence the response variable Y can always be coded as ($Y = 0$, $Y = 1$). Y_i has a Bernoulli distribution, with $P(Y_i = 1) = \pi_i = \pi_i(x_i)$ and $P(Y_i = 0) = 1 - \pi_i$, where $x_i = (x_{i1}, x_{i2}, \ldots, x_{ip})'$ denotes a vector of *prognostic factors*, which we believe influence the success probability $\pi(x_i)$, and $i = 1, \ldots, N$ denotes individuals as usual. With these assumptions it immediately follows that

$$\begin{aligned} E(Y_i) &= 1 \cdot \pi_i + 0 \cdot (1 - \pi_i) = \pi_i \ , \\ E(Y_i^2) &= 1^2 \cdot \pi_i + 0^2 \cdot (1 - \pi_i) = \pi_i \ , \\ \text{var}(Y_i) &= E(Y_i^2) - (E(Y_i))^2 = \pi_i - \pi_i^2 = \pi_i(1 - \pi_i) \, . \end{aligned}$$

The likelihood contribution of an individual i is further given by

$$\begin{aligned} f(y_i; \pi_i) &= \pi_i^{y_i} (1 - \pi_i)^{1 - y_i} \\ &= (1 - \pi_i) \left(\frac{\pi_i}{1 - \pi_i} \right)^{y_i} \\ &= (1 - \pi_i) \exp\left(y_i \ln\left(\frac{\pi_i}{1 - \pi_i} \right) \right) . \end{aligned}$$

The natural parameter $Q(\pi_i) = \ln[\pi_i/(1 - \pi_i)]$ is the log odds of response 1 and is called the logit of π_i.

A GLM with the *logit link* is called a logit model or *logistic regression model*. The model is, on an individual basis, given by

$$\ln\left(\frac{\pi_i}{1 - \pi_i} \right) = x_i'\beta \, . \qquad (10.132)$$

This parametrization guarantees a monotonic course (S-curve) of the probability π_i, under inclusion of the linear approach $x_i'\beta$ over the range of definition [0,1]:

$$\pi_i = \frac{\exp(x_i'\beta)}{1 + \exp(x_i'\beta)} \, . \qquad (10.133)$$

Grouped Data

If possible (for example, if prognostic factors are themselves categorical), patients can be grouped along the strata defined by the number of possible factor combinations. Let n_j, $j = 1, \ldots, G$, $G \leq N$, be the number of patients falling in strata j. Then we observe y_j patients having response $Y = 1$ and $n_j - y_j$ patients with response $Y = 0$. Then a natural estimate for π_j is $\hat{\pi}_j = y_j/n_j$. This corresponds to a saturated model, that is, a model in which main effects and all interactions between the factors are

TABLE 10.1. 5×2 table of loss of abutment teeth by age groups (Example 10.1)

j	Age group	Loss yes	Loss no	n_j
1	< 40	4	70	74
2	$40 - 50$	28	147	175
3	$50 - 60$	38	207	245
4	$60 - 70$	51	202	253
5	> 70	32	92	124
		153	718	871

included. But one should note that this is reasonable only if the number of strata is low compared to N so that n_j is not too low. Whenever $n_j = 1$ these estimates degenerate, and more smoothing of the probabilities and thus a more parsimonious model is necessary.

The Simplest Case and an Example

For simplicity, we assume now that $p = 1$, that is, we consider only one explanatory variable. The model in this simplest case is given by

$$\ln\left(\frac{\pi_i}{1-\pi_i}\right) = \alpha + \beta x_i \,. \tag{10.134}$$

For this special situation we get for the odds

$$\frac{\pi_i}{1-\pi_i} = \exp(\alpha + \beta x_i) = e^{\alpha}\left(e^{\beta}\right)^{x_i} \,, \tag{10.135}$$

that is, if x_i increases by one unit, the odds increases by e^{β}.

An advantage of this link is that the effects of X can be estimated, whether the study of interest is retrospective or prospective (cf. Toutenburg, 1992, Chapter 5). The effects in the logistic model refer to the odds. For two different x-values, $\exp(\alpha + \beta x_1)/\exp(\alpha + \beta x_2)$ is an odds ratio.

To find the appropriate form for the systematic component of the logistic regression, the sample logits are plotted against x.

Remark: Let x_j be chosen (j being a group index). For n_j observations of the response variable Y, let 1 be observed y_j times at this setting. Hence $\hat{\pi}(x_j) = y_j/n_j$ and $\ln[\hat{\pi}_j/(1-\hat{\pi}_j)] = \ln[y_j/(n_j - y_j)]$ is the sample logit.

This term, however, is not defined for $y_j = 0$ or $n_j = 0$. Therefore, a correction is introduced, and we utilize the smoothed logit:

$$\ln\left[(y_j + \tfrac{1}{2})/(n_j - y_j + \tfrac{1}{2})\right].$$

Example 10.1: We examine the risk (Y) for the loss of abutment teeth by extraction in dependence on age (X) (Walther and Toutenburg, 1991). From Table 10.1, we calculate $\chi^2_4 = 15.56$, which is significant at the 5 % level ($\chi^2_{4;0.95} = 9.49$). Using the unsmoothed sample logits results in the following table:

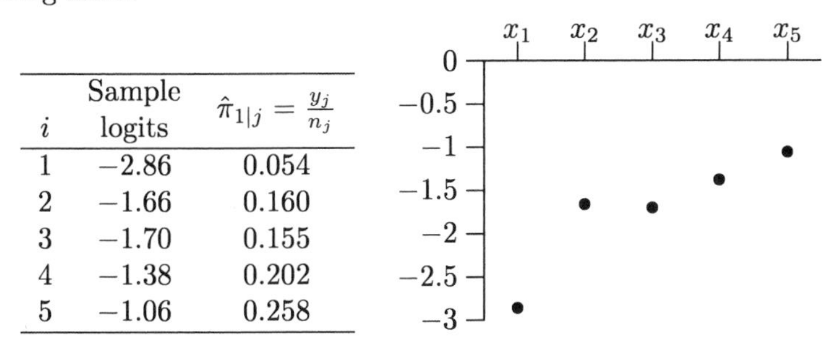

i	Sample logits	$\hat{\pi}_{1\|j} = \frac{y_j}{n_j}$
1	−2.86	0.054
2	−1.66	0.160
3	−1.70	0.155
4	−1.38	0.202
5	−1.06	0.258

$\hat{\pi}_{1|j}$ is the estimated risk for loss of abutment teeth. It increases linearly with age group. For instance, age group 5 has five times the risk of age group 1.

Modeling with the *logistic regression*

$$\ln\left(\frac{\hat{\pi}_1(x_j)}{1-\hat{\pi}_1(x_j)}\right) = \alpha + \beta x_j$$

results in

x_j	Sample logits	Fitted logits	$\hat{\pi}_1(x_j)$	Expected $n_j\hat{\pi}_1(x_j)$	Observed y_j
35	−2.86	−2.22	0.098	7.25	4
45	−1.66	−1.93	0.127	22.17	28
55	−1.70	−1.64	0.162	39.75	38
65	−1.38	−1.35	0.206	51.99	51
75	−1.06	−1.06	0.257	31.84	32

with the ML estimates

$$\hat{\alpha} = -3.233\,,$$
$$\hat{\beta} = 0.029\,.$$

10.3.2 *Testing the Model*

Under general conditions the maximum-likelihood estimates are asymptotically normal. Hence tests of significance and setting up of confidence limits can be based on the normal theory.

The significance of the effect of the variable X on π is equivalent to the significance of the parameter β. The hypothesis β *is significant* or $\beta \neq 0$ is tested by the statistical hypothesis $H_0 : \beta = 0$ against $H_1 : \beta \neq 0$. For this test, we compute the Wald statistic $Z^2 = \hat{\beta}'(\text{cov}_{\hat{\beta}})^{-1}\hat{\beta} \sim \chi^2_{df}$, where *df* is the number of components of the vector β.

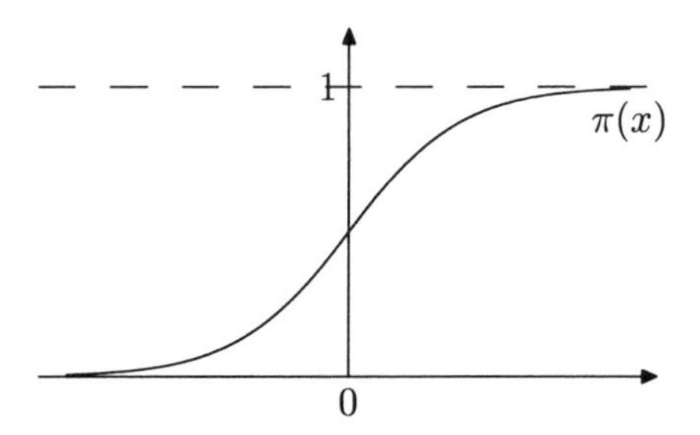

FIGURE 10.1. Logistic function $\pi(x) = \exp(x)/(1+\exp(x))$

In the above Example 10.1, we have $Z^2 = 13.06 > \chi^2_{1;0.95} = 3.84$ (the upper 5% value), which leads to a rejection of $H_0 : \beta = 0$ so that the trend is seen to be significant.

10.3.3 Distribution Function as a Link Function

The logistic function has the shape of the cumulative distribution function of a continuous random variable.

This suggests a class of models for binary responses having the form

$$\pi(x) = F(\alpha + \beta x)\,, \tag{10.136}$$

where F is a standard, continuous, cumulative distribution function. If F is strictly monotonically increasing over the entire real line, we have

$$F^{-1}(\pi(x)) = \alpha + \beta x\,. \tag{10.137}$$

This is a GLM with F^{-1} as the link function. F^{-1} maps the $[0,1]$ range of probabilities onto $(-\infty, \infty)$.

The cumulative distribution function of the logistic distribution is

$$F(x) = \frac{\exp\left(\frac{x-\mu}{\tau}\right)}{1+\exp\left(\frac{x-\mu}{\tau}\right)}\,, \quad -\infty < x < \infty\,, \tag{10.138}$$

with μ as the location parameter and $\tau > 0$ as the scale parameter.

The distribution is symmetric with mean μ and standard deviation $\tau\pi/\sqrt{3}$ (bell-shaped curve, similar to the standard normal distribution). The logistic regression $\pi(x) = F(\alpha + \beta x)$ belongs to the standardized logistic distribution F with $\mu = 0$ and $\tau = 1$. Thus, the logistic regression has mean $-\alpha/\beta$ and standard deviation $\pi/|\beta|\sqrt{3}$.

If F is the standard normal cumulative distribution function, $\pi(x) = F(\alpha + \beta x) = \Phi(\alpha + \beta x)$, $\pi(x)$ is called the *probit model.*

10.4 Logit Models for Categorical Data

The explanatory variable X can be continuous or categorical. Assume X to be categorical and choose the logit link; then the logit models are equivalent to *loglinear models (categorical regression)*, which are discussed in detail in Section 10.6. For the explanation of this equivalence we first consider the logit model.

Logit Models for $I \times 2$ Tables

Let X be an explanatory variable with I categories. If response/nonresponse is the Y factor, we then have an $I \times 2$ table. In row i the probability for response is $\pi_{1|i}$ and for nonresponse $\pi_{2|i}$, with $\pi_{1|i} + \pi_{2|i} = 1$.

This leads to the following logit model:

$$\ln\left(\frac{\pi_{1|i}}{\pi_{2|i}}\right) = \alpha + \beta_i \,. \tag{10.139}$$

Here the x-values are not included explicitly but only through the category i. β_i describes the effect of category i on the response. When $\beta_i = 0$, there is no effect. This model resembles the one-way analysis of variance and, likewise, we have the constraints for identifiability $\sum \beta_i = 0$ or $\beta_I = 0$. Then $I - 1$ of the parameters $\{\beta_i\}$ suffice for characterization of the model. For the constraint $\sum \beta_i = 0$, α is the overall mean of the logits and β_i is the deviation from this mean for row i. The higher β_i is, the higher is the logit in row i, and the higher is the value of $\pi_{1|i}$ (= chance for response in category i).

When the factor X (in I categories) has no effect on the response variable, the model simplifies to the model of statistical independence of the factor and response:

$$\ln\left(\frac{\pi_{1|i}}{\pi_{2|i}}\right) = \alpha \quad \forall i\,,$$

We now have $\beta_1 = \beta_2 = \cdots = \beta_I = 0$, and thus $\pi_{1|1} = \pi_{1|2} = \cdots = \pi_{1|I}$.

Logit Models for Higher Dimensions

As a generalization to two or more categorical factors that have an effect on the binary response, we now consider the two factors A and B with I and J levels. Let $\pi_{1|ij}$ and $\pi_{2|ij}$ denote the probabilities for response and nonresponse for the combination ij of factors so that $\pi_{1|ij} + \pi_{2|ij} = 1$. For the $I \times J \times 2$ table, the logit model

$$\ln\left(\frac{\pi_{1|ij}}{\pi_{2|ij}}\right) = \alpha + \beta_i^A + \beta_j^B \tag{10.140}$$

represents the effects of A and B without interaction. This model is equivalent to the two-way analysis of variance without interaction.

10.5 Goodness of Fit—Likelihood-Ratio Test

For a given model M, we can use the estimates of the parameters $(\widehat{\alpha+\beta_i})$ and $(\hat{\alpha}, \hat{\beta})$ to predict the logits, to estimate the probabilities of response $\hat{\pi}_{1|i}$, and hence to calculate the expected cell frequencies $\hat{m}_{ij} = n_{i+}\hat{\pi}_{j|i}$.

We can now test the goodness of fit of a model M with Wilks's G^2-statistic

$$G^2(M) = 2\sum_{i=1}^{I}\sum_{j=1}^{J} n_{ij} \ln\left(\frac{n_{ij}}{\hat{m}_{ij}}\right). \tag{10.141}$$

The $\hat{m}_{ij}$ are calculated by using the estimated model parameters. The degrees of freedom equal the number of logits minus the number of independent parameters in the model M.

We now consider three models for binary response (cf. Agresti, 1990, p. 95).

1. Independence model:

$$M = I: \quad \ln\left(\frac{\pi_{1|i}}{\pi_{2|i}}\right) = \alpha. \tag{10.142}$$

Here we have I logits and one parameter, that is, $I-1$ degrees of freedom.

2. Logistic model:

$$M = L: \quad \ln\left(\frac{\pi_{1|i}}{\pi_{2|i}}\right) = \alpha + \beta x_i. \tag{10.143}$$

The number of degrees of freedom equals $I-2$.

3. Logit model:

$$M = S: \quad \ln\left(\frac{\pi_{1|i}}{\pi_{2|i}}\right) = \alpha + \beta_i. \tag{10.144}$$

The model has I logits and I independent parameters. The number of degrees of freedom is 0, so it has perfect fit. This model, with equal numbers of parameters and observations, is called a *saturated model.*

The likelihood-ratio test compares a model M_1 with a simpler model M_2 (in which a few parameters equal zero). The test statistic then is

$$\Lambda = \frac{L(M_2)}{L(M_1)} \tag{10.145}$$

$$\text{or} \quad G^2(M_2|M_1) = -2\left(\ln L(M_2) - \ln L(M_1)\right). \tag{10.146}$$

The statistic $G^2(M)$ is a special case of this statistic, in which $M_2 = M$ and M_1 is the saturated model. If we want to test the goodness of fit with

$G^2(M)$, this is equivalent to testing whether all the parameters that are in the saturated model, but not in the model M, are equal to zero.

Let l_S denote the maximized loglikelihood function for the saturated model. Then we have

$$\begin{aligned} G^2(M_2|M_1) &= -2\,(\ln L(M_2) - \ln L(M_1)) \\ &= -2\,(\ln L(M_2) - l_S) - [-2(\ln L(M_1) - l_S)] \\ &= G^2(M_2) - G^2(M_1)\,. \end{aligned} \quad (10.147)$$

That is, the statistic $G^2(M_2|M_1)$ for comparing two models is identical to the difference of the goodness-of-fit statistics for the two models.

Example 10.2: In Example 10.1 "Loss of abutment teeth/age" we have for the logistic model:

Age	Loss		No loss	
group	observed	expected	observed	expected
1	4	7.25	70	66.75
2	28	22.17	147	152.83
3	38	39.75	207	205.25
4	51	51.99	202	201.01
5	32	31.84	92	92.16

and get $G^2(L) = 3.66$, $df = 5 - 2 = 3$.

For the independence model, we get $G^2(I) = 17.25$ with $df = 4 = (I-1)(J-1) = (5-1)(2-1)$. The test statistic for testing $H_0 : \beta = 0$ in the logistic model then is

$$G^2(I|L) = G^2(I) - G^2(L) = 17.25 - 3.66 = 13.59\,, \quad df = 4 - 3 = 1\,.$$

This value is significant, which means that the logistic model, compared to the independence model, holds.

10.6 Loglinear Models for Categorical Variables

10.6.1 Two-Way Contingency Tables

The previous models focused on bivariate response, that is, on $I \times 2$ tables. We now generalize this set-up to $I \times J$ and later to $I \times J \times K$ tables.

Suppose that we have a realization (sample) of two categorical variables with I and J categories and sample size n. This yields observations in $N = I \times J$ cells of the contingency table. The number in the (i,j)-th cell is denoted by n_{ij}.

The probabilities π_{ij} of the multinomial distribution form the joint distribution. Independence of the variables is equivalent to

$$\pi_{ij} = \pi_{i+}\pi_{+j} \quad \text{(for all } i,j). \quad (10.148)$$

If this is applied to the expected cell frequencies $m_{ij} = n\pi_{ij}$, the condition of independence is equivalent to

$$m_{ij} = n\pi_{i+}\pi_{+j} \,. \tag{10.149}$$

The modeling of the $I \times J$ table is based on this relation as an independence model on the logarithmic scale:

$$\ln(m_{ij}) = \ln n + \ln \pi_{i+} + \ln \pi_{+j} \,. \tag{10.150}$$

Hence, the effects of the rows and columns on $\ln(m_{ij})$ are additive. An alternative expression, following the models of analysis of variance of the form

$$y_{ij} = \mu + \alpha_i + \beta_j + \varepsilon_{ij} \,, \quad \left(\sum \alpha_i = \sum \beta_j = 0\right) , \tag{10.151}$$

is given by

$$\ln m_{ij} = \mu + \lambda_i^X + \lambda_j^Y \tag{10.152}$$

with

$$\lambda_i^X = \ln \pi_{i+} - \frac{1}{I}\left(\sum_{k=1}^{I} \ln \pi_{k+}\right) , \tag{10.153}$$

$$\lambda_j^Y = \ln \pi_{+j} - \frac{1}{J}\left(\sum_{k=1}^{J} \ln \pi_{+k}\right) , \tag{10.154}$$

$$\mu = \ln n + \frac{1}{I}\left(\sum_{k=1}^{I} \ln \pi_{k+}\right) + \frac{1}{J}\left(\sum_{k=1}^{J} \ln \pi_{+k}\right) . \tag{10.155}$$

The parameters satisfy the constraints

$$\sum_{i=1}^{I} \lambda_i^X = \sum_{j=1}^{J} \lambda_j^Y = 0 \,, \tag{10.156}$$

which make the parameters identifiable.

Model (10.152) is called *loglinear model of independence* in a two-way contingency table.

The related saturated model contains the additional interaction parameters λ_{ij}^{XY}:

$$\ln m_{ij} = \mu + \lambda_i^X + \lambda_j^Y + \lambda_{ij}^{XY} \,. \tag{10.157}$$

This model describes the perfect fit. The interaction parameters satisfy

$$\sum_{i=1}^{I} \lambda_{ij}^{XY} = \sum_{j=1}^{J} \lambda_{ij}^{XY} = 0 \,. \tag{10.158}$$

Given the λ_{ij} in the first $(I-1)(J-1)$ cells, these constraints determine the λ_{ij} in the last row or the last column. Thus, the saturated model contains

$$\underbrace{1}_{\mu}+\underbrace{(I-1)}_{\lambda_i^X}+\underbrace{(J-1)}_{\lambda_j^Y}+\underbrace{(I-1)(J-1)}_{\lambda_{ij}^{XY}}=I\,J \tag{10.159}$$

independent parameters.

For the independence model, the number of independent parameters equals

$$1+(I-1)+(J-1)=I+J-1\,. \tag{10.160}$$

Interpretation of the Parameters

Loglinear models estimate the effects of rows and columns on $\ln m_{ij}$. For this, no distinction is made between explanatory and response variables. The information of the rows or columns influence m_{ij} symmetrically.

Consider the simplest case—the $I \times 2$ table (independence model). According to (10.160), the logit of the binary variable equals

$$\begin{aligned}\ln\left(\frac{\pi_{1|i}}{\pi_{2|i}}\right) &= \ln\left(\frac{m_{i1}}{m_{i2}}\right)\\ &= \ln(m_{i1})-\ln(m_{i2})\\ &= (\mu+\lambda_i^X+\lambda_1^Y)-(\mu+\lambda_i^X+\lambda_2^Y)\\ &= \lambda_1^Y-\lambda_2^Y\,. \end{aligned}\tag{10.161}$$

The logit is the same in every row and hence independent of X or the categories $i=1,\ldots,I$, respectively.

For the constraints

$$\begin{aligned}\lambda_1^Y+\lambda_2^Y=0 \quad &\Rightarrow \quad \lambda_1^Y=-\lambda_2^Y\,,\\ &\Rightarrow \quad \ln\left(\frac{\pi_{1|i}}{\pi_{2|i}}\right)=2\lambda_1^Y \quad (i=1,\ldots,I)\,.\end{aligned}$$

Hence we obtain

$$\frac{\pi_{1|i}}{\pi_{2|i}}=\exp(2\lambda_1^Y) \qquad (i=1,\ldots,I)\,. \tag{10.162}$$

In each category of X, the odds that Y is in category 1 rather than in category 2 are equal to $\exp(2\lambda_1^Y)$, when the independence model holds.

The following relationship exists between the odds ratio in a 2×2 table and the saturated loglinear model:

$$\begin{aligned}\ln\theta &= \ln\left(\frac{m_{11}\,m_{22}}{m_{12}\,m_{21}}\right)\\ &= \ln(m_{11})+\ln(m_{22})-\ln(m_{12})-\ln(m_{21})\\ &= (\mu+\lambda_1^X+\lambda_1^Y+\lambda_{11}^{XY})+(\mu+\lambda_2^X+\lambda_2^Y+\lambda_{22}^{XY})\\ &\quad -(\mu+\lambda_1^X+\lambda_2^Y+\lambda_{12}^{XY})-(\mu+\lambda_2^X+\lambda_1^Y+\lambda_{21}^{XY})\end{aligned}$$

TABLE 10.2. $2 \times 2 \times 2$-table for endodontic risk

Age group	Form of construction	Endodontic treatment yes	no
< 60	H	62	1041
	B	23	463
≥ 60	H	70	755
	B	30	215
Σ		185	2474

$$= \lambda_{11}^{XY} + \lambda_{22}^{XY} - \lambda_{12}^{XY} - \lambda_{21}^{XY} .$$

Since $\sum_{i=1}^{2} \lambda_{ij}^{XY} = \sum_{j=1}^{2} \lambda_{ij}^{XY} = 0$, we have $\lambda_{11}^{XY} = \lambda_{22}^{XY} = -\lambda_{12}^{XY} = -\lambda_{21}^{XY}$ and thus $\ln \theta = 4\lambda_{11}^{XY}$. Hence the odds ratio in a 2×2 table equals

$$\theta = \exp(4\lambda_{11}^{XY}) , \tag{10.163}$$

and is dependent on the association parameter in the saturated model. When there is no association, that is $\lambda_{ij} = 0$, we have $\theta = 1$.

10.6.2 Three-Way Contingency Tables

We now consider three categorical variables X, Y, and Z. The frequencies of the combinations of categories are displayed in the $I \times J \times K$ contingency table. We are especially interested in $I \times J \times 2$ contingency tables, where the last variable is a bivariate risk or response variable. Table 10.2 shows the risk for an endodontic treatment depending on the age of patients and the type of construction of the denture (Walther and Toutenburg, 1991).

In addition to the bivariate associations, we want to model an overall association. The three variables are mutually independent if the following independence model for the cell frequencies m_{ijk} (on a logarithmic scale) holds:

$$\ln(m_{ijk}) = \mu + \lambda_i^X + \lambda_j^Y + \lambda_k^Z . \tag{10.164}$$

(In the above example we have X: age group, Y: type of construction, Z: endodontic treatment.) The variable Z is independent of the joint distribution of X and Y (jointly independent) if

$$\ln(m_{ijk}) = \mu + \lambda_i^X + \lambda_j^Y + \lambda_k^Z + \lambda_{ij}^{XY} . \tag{10.165}$$

A third type of independence (conditional independence of two variables given a fixed category of the third variable) is expressed by the following model (j fixed!):

$$\ln(m_{ijk}) = \mu + \lambda_i^X + \lambda_j^Y + \lambda_k^Z + \lambda_{ij}^{XY} + \lambda_{jk}^{YZ} . \tag{10.166}$$

This is the approach for the conditional independence of X and Z at level j of Y. If they are conditionally independent for all $j = 1, \ldots, J$, then X and Z are called conditionally independent given Y. Similarly, if X and Y are conditionally independent at level k of Z, the parameters λ_{ij}^{XY} and λ_{jk}^{YZ} in (10.166) are replaced by the parameters λ_{ik}^{XZ} and λ_{jk}^{YZ}. The parameters with two subscripts describe two-way interactions. The appropriate conditions for the cell probabilities are

(a) mutual independence of X, Y, Z

$$\pi_{ijk} = \pi_{i++}\pi_{+j+}\pi_{++k} \quad \text{(for all } i, j, k). \tag{10.167}$$

(b) joint independence
Y is jointly independent of X and Z when

$$\pi_{ijk} = \pi_{i+k}\pi_{+j+} \quad \text{(for all } i, j, k). \tag{10.168}$$

(c) conditional independence
X and Y are conditionally independent of Z when

$$\pi_{ijk} = \frac{\pi_{i+k}\pi_{+jk}}{\pi_{++k}} \quad \text{(for all } i, j, k). \tag{10.169}$$

The most general loglinear model (saturated model) for three-way tables is the following:

$$\ln(m_{ijk}) = \mu + \lambda_i^X + \lambda_j^Y + \lambda_k^Z + \lambda_{ij}^{XY} + \lambda_{ik}^{XZ} + \lambda_{jk}^{YZ} + \lambda_{ijk}^{XYZ} . \tag{10.170}$$

The last parameter describes the three-factor interaction.

All association parameters describing the deviation from the general mean μ, satisfy the constraints

$$\sum_{i=1}^{I} \lambda_{ij}^{XY} = \sum_{j=1}^{J} \lambda_{ij}^{XY} = \ldots = \sum_{k=1}^{K} \lambda_{ijk}^{XYZ} = 0 . \tag{10.171}$$

Similarly, for the main factor effects we have:

$$\sum_{i=1}^{I} \lambda_i^X = \sum_{j=1}^{J} \lambda_j^Y = \sum_{k=1}^{K} \lambda_k^Z = 0 . \tag{10.172}$$

From the general model (10.170), submodels can be constructed. For this, the hierarchical principle of construction is preferred. A model is called hierarchical when, in addition to significant higher-order effects, it contains all lower-order effects of the variables included in the higher-order effects, even if these parameter estimates are not statistically significant. For instance, if the model contains the association parameter λ_{ik}^{XZ}, it must also contain λ_i^X and λ_k^Z:

$$\ln(m_{ijk}) = \mu + \lambda_i^X + \lambda_k^Z + \lambda_{ik}^{XZ} . \tag{10.173}$$

A symbol is assigned to the various hierarchical models (Table 10.3).

TABLE 10.3. Symbols of the hierarchical models for three-way contingency tables (Agresti, 1990, p. 144).

Loglinear model			Symbol
$\ln(m_{ij+})$	$=$	$\mu + \lambda_i^X + \lambda_j^Y$	(X, Y)
$\ln(m_{i+k})$	$=$	$\mu + \lambda_i^X + \lambda_k^Z$	(X, Z)
$\ln(m_{+jk})$	$=$	$\mu + \lambda_j^Y + \lambda_k^Z$	(Y, Z)
$\ln(m_{ijk})$	$=$	$\mu + \lambda_i^X + \lambda_j^Y + \lambda_k^Z$	(X, Y, Z)
$\ln(m_{ijk})$	$=$	$\mu + \lambda_i^X + \lambda_j^Y + \lambda_k^Z + \lambda_{ij}^{XY}$	(XY, Z)
	$\vdots$		$\vdots$
$\ln(m_{ijk})$	$=$	$\mu + \lambda_i^X + \lambda_j^Y + \lambda_{ij}^{XY}$	(XY)
	$\vdots$		$\vdots$
$\ln(m_{ijk})$	$=$	$\mu + \lambda_i^X + \lambda_j^Y + \lambda_k^Z + \lambda_{ij}^{XY} + \lambda_{ik}^{XZ}$	(XY, XZ)
	$\vdots$		$\vdots$
$\ln(m_{ijk})$	$=$	$\mu + \lambda_i^X + \lambda_j^Y + \lambda_k^Z + \lambda_{ij}^{XY} + \lambda_{ik}^{XZ} + \lambda_{jk}^{YZ}$	(XY, XZ, YZ)
	$\vdots$		$\vdots$
$\ln(m_{ijk})$	$=$	$\mu + \lambda_i^X + \lambda_j^Y + \lambda_k^Z + \lambda_{ij}^{XY} + \lambda_{ik}^{XZ} + \lambda_{jk}^{YZ} + \lambda_{ijk}^{XYZ}$	(XYZ)

Similar to 2×2 tables, a close relationship exists between the parameters of the model and the odds ratios. Given a $2 \times 2 \times 2$ table, we have, under the constraints (10.171) and (10.172), for instance

$$\frac{\theta_{11(1)}}{\theta_{11(2)}} = \frac{\frac{\pi_{111}\pi_{221}}{\pi_{211}\pi_{121}}}{\frac{\pi_{112}\pi_{222}}{\pi_{212}\pi_{122}}} = \exp(8\lambda_{111}^{XYZ}) . \tag{10.174}$$

This is the conditional odds ratio of X and Y given the levels $k = 1$ (numerator) and $k = 2$ (denominator) of Z. The same holds for X and Z under Y and for Y and Z under X. In the population, we thus have for the three-way interaction λ_{111}^{XYZ},

$$\frac{\theta_{11(1)}}{\theta_{11(2)}} = \frac{\theta_{1(1)1}}{\theta_{1(2)1}} = \frac{\theta_{(1)11}}{\theta_{(2)11}} = \exp(8\lambda_{111}^{XYZ}) . \tag{10.175}$$

In the case of independence in the equivalent subtables, the odds ratios (of the population) equal 1. The sample odds ratio gives a first hint at a deviation from independence.

Consider the conditional odds ratio (10.175) for Table 10.2 assuming that X is the variable "age group," Y is the variable "form of construction," and Z is the variable "endodontic treatment."

We then have a value of 1.80. This indicates a positive tendency for an increased risk of endodontic treatment in comparing the following subtables for endodontic treatment (left) versus no endodontic treatment (right):

	H	B
< 60	62	23
≥ 60	70	30

	H	B
< 60	1041	463
≥ 60	755	215

The relationship (10.102) is also valid for the sample version. Thus a comparison of the following subtables for < 60 (left) versus ≥ 60 (right):

	treatment yes	treatment no
H	62	1041
B	23	463

	treatment yes	treatment no
H	70	755
B	30	215

or for H (left) versus B (right):

	treatment yes	treatment no
< 60	62	1041
≥ 60	70	755

	treatment yes	treatment no
< 60	23	463
≥ 60	30	215

leads to the same sample value 1.80 and hence $\hat{\lambda}_{111}^{XYZ} = 0.073$.

Calculations for Table 10.2:

$$\frac{\hat{\theta}_{11(1)}}{\hat{\theta}_{11(2)}} = \frac{\frac{n_{111}n_{221}}{n_{211}n_{121}}}{\frac{n_{112}n_{222}}{n_{212}n_{122}}} = \frac{\frac{62\cdot 30}{70\cdot 23}}{\frac{1041\cdot 215}{755\cdot 463}} = \frac{1.1553}{0.6403} = 1.80\,,$$

$$\frac{\hat{\theta}_{(1)11}}{\hat{\theta}_{(2)11}} = \frac{\frac{n_{111}n_{122}}{n_{121}n_{112}}}{\frac{n_{211}n_{222}}{n_{221}n_{212}}} = \frac{\frac{62\cdot 463}{23\cdot 1041}}{\frac{70\cdot 215}{30\cdot 755}} = \frac{1.1989}{0.6645} = 1.80\,,$$

$$\frac{\hat{\theta}_{1(1)1}}{\hat{\theta}_{1(2)1}} = \frac{\frac{n_{111}n_{212}}{n_{211}n_{112}}}{\frac{n_{121}n_{222}}{n_{221}n_{122}}} = \frac{\frac{62\cdot 755}{70\cdot 1041}}{\frac{23\cdot 215}{30\cdot 463}} = \frac{0.6424}{0.3560} = 1.80\,.$$

10.7 The Special Case of Binary Response

If one of the variables is a binary response variable (in our example Z: endodontic treatment) and the others are explanatory categorical variables (in our example X: age group and Y: type of construction), these models lead to the already known logit model.

Given the independence model

$$\ln(m_{ijk}) = \mu + \lambda_i^X + \lambda_j^Y + \lambda_k^Z\,, \tag{10.176}$$

we then have for the logit of the response variable Z

$$\ln\left(\frac{m_{ij1}}{m_{ij2}}\right) = \lambda_1^Z - \lambda_2^Z \,. \tag{10.177}$$

With the constraint $\sum_{k=1}^{2} \lambda_k^Z = 0$ we thus have

$$\ln\left(\frac{m_{ij1}}{m_{ij2}}\right) = 2\lambda_1^Z \qquad (\text{for all } i, j)\,. \tag{10.178}$$

The higher the value of λ_1^Z is, the higher is the risk for category $Z = 1$ (endodontic treatment), independent of the values of X and Y.

In case the other two variables are also binary, implying a $2 \times 2 \times 2$ table, and if the constraints

$$\lambda_2^X = -\lambda_1^X\,, \quad \lambda_2^Y = -\lambda_1^Y\,, \quad \lambda_2^Z = -\lambda_1^Z$$

hold, then the model (10.176) can be expressed as follows:

$$\begin{pmatrix} \ln(m_{111}) \\ \ln(m_{112}) \\ \ln(m_{121}) \\ \ln(m_{122}) \\ \ln(m_{211}) \\ \ln(m_{212}) \\ \ln(m_{221}) \\ \ln(m_{222}) \end{pmatrix} = \begin{pmatrix} 1 & 1 & 1 & 1 \\ 1 & 1 & 1 & -1 \\ 1 & 1 & -1 & 1 \\ 1 & 1 & -1 & -1 \\ 1 & -1 & 1 & 1 \\ 1 & -1 & 1 & -1 \\ 1 & -1 & -1 & 1 \\ 1 & -1 & -1 & -1 \end{pmatrix} \begin{pmatrix} \mu \\ \lambda_1^X \\ \lambda_1^Y \\ \lambda_1^Z \end{pmatrix}, \tag{10.179}$$

which is equivalent to $\ln(m) = X\beta$.

This corresponds to the effect coding of categorical variables (Section 10.8). The ML equation is

$$X'n = X'\hat{m}\,. \tag{10.180}$$

The estimated asymptotic covariance matrix for Poisson sampling reads as follows:

$$\widehat{\text{cov}}(\hat{\beta}) = \left[X'(\text{diag}(\hat{m}))X\right]^{-1}\,. \tag{10.181}$$

where $\text{diag}(\hat{m})$ has the elements $\hat{m}$ on the main diagonal. The solution of the ML equation (10.180) is obtained by the Newton-Raphson or any other iterative algorithm—for instance, the iterative proportional fitting (IPF).

The IPF method (Deming and Stephan, 1940; cf. Agresti, 1990, p. 185) adjusts initial estimates $\{\hat{m}_{ijk}^{(0)}\}$ successively to the respective expected marginal table of the model until a prespecified accuracy is achieved. For the independence model the steps of iteration are

$$\hat{m}_{ijk}^{(1)} = \hat{m}_{ijk}^{(0)}\left(\frac{n_{i++}}{\hat{m}_{i++}^{(0)}}\right),$$

$$\hat{m}^{(2)}_{ijk} = \hat{m}^{(1)}_{ijk}\left(\frac{n_{+j+}}{\hat{m}^{(1)}_{+j+}}\right),$$

$$\hat{m}^{(3)}_{ijk} = \hat{m}^{(2)}_{ijk}\left(\frac{n_{++k}}{\hat{m}^{(2)}_{++k}}\right).$$

Example 10.3 (Tartar-Smoking Analysis): A study cited in Toutenburg (1992, p. 42) investigates to what extent smoking influences the development of tartar. The 3×3 contingency table (Table 10.4) is modeled by the loglinear model

$$\ln(m_{ij}) = \mu + \lambda_i^{\text{Smoking}} + \lambda_j^{\text{Tartar}} + \lambda_{ij}^{\text{Smoking/Tartar}},$$

with $i, j = 1, 2$. Here we have

$$\begin{aligned} \lambda_1^{\text{Smoking}} &= \text{Effect nonsmoker} \\ \lambda_2^{\text{Smoking}} &= \text{Effect light smoker} \\ \lambda_3^{\text{Smoking}} = -(\lambda_1^{\text{Smoking}} + \lambda_2^{\text{Smoking}}) &= \text{Effect heavy smoker}. \end{aligned}$$

For the development of tartar, analogous expressions are valid.

(i) Model of independence. For the null hypothesis

$$H_0 : \ln(m_{ij}) = \mu + \lambda_i^{\text{Smoking}} + \lambda_j^{\text{Tartar}},$$

we receive $G^2 = 76.23 > 9.49 = \chi^2_{4;0.95}$. This leads to a clear rejection of this model.

(ii) Saturated model. Here we have $G^2 = 0$. The estimates of the parameters are (values in parantheses are standardized values)

$$\begin{aligned} \lambda_1^{\text{Smoking}} &= -1.02 \quad (-25.93) \\ \lambda_2^{\text{Smoking}} &= 0.20 \quad (7.10) \\ \lambda_3^{\text{Smoking}} &= 0.82 \quad (\text{—}) \end{aligned}$$

$$\begin{aligned} \lambda_1^{\text{Tartar}} &= 0.31 \quad (11.71) \\ \lambda_2^{\text{Tartar}} &= 0.61 \quad (23.07) \\ \lambda_3^{\text{Tartar}} &= -0.92 \quad (\text{—}) \end{aligned}$$

All single effects are highly significant. The interaction effects are

	Tartar 1	2	3	$\sum$
Smoking 1	0.34	−0.14	−0.20	0
2	−0.12	0.06	0.06	0
3	−0.22	0.08	0.14	0
$\sum$	0	0	0	

TABLE 10.4. Smoking and development of tartar

		Tartar		
		none	middle	heavy
	no	284	236	48
Smoking	middle	606	983	209
	heavy	1028	1871	425

The main diagonal is very well marked, which is an indication for a trend. The standardized interaction effects are significant as well:

	1	2	3
1	7.30	−3.05	—
2	−3.51	1.93	—
3	—	—	—

10.8 Coding of Categorical Explanatory Variables

10.8.1 Dummy and Effect Coding

If a bivariate response variable Y is connected to a linear model $x'\beta$, with x being categorical, by an appropriate link, the parameters β are always to be interpreted in terms of their dependence on the x-scores. To eliminate this arbitrariness, an appropriate coding of x is chosen. Here two ways of coding are suggested (partly in analogy to the analysis of variance).

Dummy Coding

Let A be a variable in I categories. Then the $I-1$ dummy variables are defined as follows:

$$x_i^A = \begin{cases} 1 & \text{for category } i \text{ of variable A} \\ 0 & \text{for others} \end{cases} \tag{10.182}$$

with $i = 1, \ldots, I-1$.

The category I is implicitly taken into account by $x_1^A = \ldots = x_{I-1}^A = 0$. Thus, the vector of explanatory variables belonging to variable A is of the following form:

$$x^A = (x_1^A, x_2^A, \ldots x_{I-1}^A)' \,. \tag{10.183}$$

The parameters β_i, which go into the final regression model proportional to $x'^A\beta$, are called main effects of A.

Example:

(i) Sex male/female, with male: category 1, female: category 2

$$\begin{aligned} x_1^{\text{Sex}} &= (1) \quad \Rightarrow \quad \text{Person is male} \\ x_2^{\text{Sex}} &= (0) \quad \Rightarrow \quad \text{Person is female}. \end{aligned}$$

(ii) Age groups $i = 1, \ldots 5$

$$\begin{aligned} x^{\text{Age}} &= (1,0,0,0)' \quad \Rightarrow \quad \text{Age group is 1} \\ x^{\text{Age}} &= (0,0,0,0)' \quad \Rightarrow \quad \text{Age group is 5}. \end{aligned}$$

Let y be a bivariate response variable. The probability of response $(y = 1)$ dependent on a categorical variable A in I categories can be modeled as follows:

$$P(y = 1 \mid x^A) = \beta_0 + \beta_1 x_1^A + \cdots + \beta_{I-1} x_{I-1}^A . \tag{10.184}$$

Given category i (age group i), we have

$$P(y = 1 \mid x^A \textit{ represents the } i\textit{-th age group}) = \beta_0 + \beta_i \,,$$

as long as $i = 1, 2, \ldots, I - 1$ and, for the implicitly coded category I, we get

$$P(y = 1 \mid x^A \textit{ represents the } I\textit{-th age group}) = \beta_0 . \tag{10.185}$$

Hence for each category i another probability of response $P(y = 1 \mid x^A)$ is possible.

Effect Coding

For an explanatory variable A in I categories, effect coding is defined as follows:

$$x_i^A = \begin{cases} 1 & \text{for category } i,\ i = 1, \ldots I - 1, \\ -1 & \text{for category } I, \\ 0 & \text{for others.} \end{cases} \tag{10.186}$$

Consequently, we have

$$\beta_I = -\sum_{i=1}^{I-1} \beta_i \,, \tag{10.187}$$

which is equivalent to

$$\sum_{i=1}^{I} \beta_i = 0 . \tag{10.188}$$

In analogy to the analysis of variance, the model for the probability of response has the following form:

$$P(y = 1 | x^A \textit{ represents the } i\textit{-th age group}) = \beta_0 + \beta_i \tag{10.189}$$

for $i = 1, \ldots, I$ and with the constraint (10.188).

Example: $I = 3$ age groups A1, A2, A3. A person in A1 is coded $(1, 0)$, a person in A2 is coded $(0, 1)$ for both dummy and effect coding. A person in A3 is coded $(0, 0)$ using dummy coding or $(-1, -1)$ using effect coding. The two ways of coding categorical variables generally differ only for category I.

Inclusion of More than One Variable

If more than one explanatory variable is included in the model, the categories of A, B, C (with I, J, and K categories, respectively), for example, are combined in a common vector

$$x' = (x_1^A, \ldots, x_{I-1}^A, x_1^B, \ldots, x_{J-1}^B, x_1^C, \ldots, x_{K-1}^C) \,. \tag{10.190}$$

In addition to these main effects, the interaction effects $x_{ij}^{AB}, \ldots, x_{ijk}^{ABC}$ can be included. The codings of the $x_{ij}^{AB}, \ldots, x_{ijk}^{ABC}$ are chosen in consideration of constraints (10.171).

Example: In case of effect coding, we obtain for the saturated model (10.157) with binary variables A and B,

$$\begin{pmatrix} \ln(m_{11}) \\ \ln(m_{12}) \\ \ln(m_{21}) \\ \ln(m_{22}) \end{pmatrix} = \begin{pmatrix} 1 & 1 & 1 & 1 \\ 1 & 1 & -1 & -1 \\ 1 & -1 & 1 & -1 \\ 1 & -1 & -1 & 1 \end{pmatrix} \begin{pmatrix} \mu \\ \lambda_1^A \\ \lambda_1^B \\ \lambda_{11}^{AB} \end{pmatrix},$$

from which we receive the following values for x_{ij}^{AB}, recoded for parameter λ_{11}^{AB}:

(i,j)		Parameter	Constraints	Recoding for λ_{11}^{AB}
(1,1)	$x_{11}^{AB} = 1$	λ_{11}^{AB}		
(1,2)	$x_{12}^{AB} = 1$	λ_{12}^{AB}	$\lambda_{12}^{AB} = -\lambda_{11}^{AB}$	$x_{12}^{AB} = -1$
(2,1)	$x_{21}^{AB} = 1$	λ_{21}^{AB}	$\lambda_{21}^{AB} = \lambda_{12}^{AB} = -\lambda_{11}^{AB}$	$x_{21}^{AB} = -1$
(2,2)	$x_{22}^{AB} = 1$	λ_{22}^{AB}	$\lambda_{22}^{AB} = -\lambda_{21}^{AB} = \lambda_{11}^{AB}$	

Thus the interaction effects develop from multiplying the main effects.

Let L be the number of possible (different) combinations of variables. If, for example, we have three variables A, B, C in I, J, K categories, L equals IJK.

Consider a complete factorial experimental design (as in an $I \times J \times K$ contingency table). Now L is known, and the design matrix X (in effect or dummy coding) for the main effects can be specified (independence model).

Example (Fahrmeir and Hamerle, 1984, p. 507): Reading habits of women (preference for a specific magazine: yes/no) are to be analyzed in terms of dependence on employment (A: yes/no), age group (B: 3 categories), and education (C: 4 categories). The complete design matrix X (Figure 10.2) is of dimension $IJK \times \{1 + (I-1) + (J-1) + (K-1)\}$, therefore $(2 \cdot 3 \cdot$

$$X = \begin{pmatrix} \beta_0 & x_1^A & x_1^B & x_2^B & x_1^C & x_2^C & x_3^C \\ 1 & 1 & 1 & 0 & 1 & 0 & 0 \\ 1 & 1 & 1 & 0 & 0 & 1 & 0 \\ 1 & 1 & 1 & 0 & 0 & 0 & 1 \\ 1 & 1 & 1 & 0 & -1 & -1 & -1 \\ 1 & 1 & 0 & 1 & 1 & 0 & 0 \\ 1 & 1 & 0 & 1 & 0 & 1 & 0 \\ 1 & 1 & 0 & 1 & 0 & 0 & 1 \\ 1 & 1 & 0 & 1 & -1 & -1 & -1 \\ 1 & 1 & -1 & -1 & 1 & 0 & 0 \\ 1 & 1 & -1 & -1 & 0 & 1 & 0 \\ 1 & 1 & -1 & -1 & 0 & 0 & 1 \\ 1 & 1 & -1 & -1 & -1 & -1 & -1 \\ 1 & -1 & 1 & 0 & 1 & 0 & 0 \\ 1 & -1 & 1 & 0 & 0 & 1 & 0 \\ 1 & -1 & 1 & 0 & 0 & 0 & 1 \\ 1 & -1 & 1 & 0 & -1 & -1 & -1 \\ 1 & -1 & 0 & 1 & 1 & 0 & 0 \\ 1 & -1 & 0 & 1 & 0 & 1 & 0 \\ 1 & -1 & 0 & 1 & 0 & 0 & 1 \\ 1 & -1 & 0 & 1 & -1 & -1 & -1 \\ 1 & -1 & -1 & -1 & 1 & 0 & 0 \\ 1 & -1 & -1 & -1 & 0 & 1 & 0 \\ 1 & -1 & -1 & -1 & 0 & 0 & 1 \\ 1 & -1 & -1 & -1 & -1 & -1 & -1 \end{pmatrix}$$

FIGURE 10.2. Design matrix for the main effects of a $2 \times 3 \times 4$ contingency table

$4) \times (1+1+2+3) = 24 \times 7$. In this case, the number of columns m is equal to the number of parameters in the independence model (cf. Figure 10.2).

10.8.2 Coding of Response Models

Let

$$\pi_i = P(y = 1 \mid x_i)\,, \quad i = 1, \ldots, L$$

be the probability of response dependent on the level x_i of the vector of covariates x. Summarized in matrix representation we then have

$$\underset{L,1}{\pi} = \underset{L,m}{X}\ \underset{m,1}{\beta}\ . \tag{10.191}$$

N_i observations are made for the realization of covariates coded by x_i. Thus, the vector $\{y_i^{(j)}\}$, $j = 1, \dots N_i$ is observed, and we get the ML estimate

$$\hat{\pi}_i = \hat{P}(y = 1 \mid x_i) = \frac{1}{N_i} \sum_{j=1}^{N_i} y_i^{(j)} \qquad (10.192)$$

for π_i $(i = 1, \dots, L)$. For contingency tables the cell counts with binary response $N_i^{(1)}$ and $N_i^{(0)}$ are given from which $\hat{\pi}_i = N_i^{(1)}/(N_i^{(1)} + N_i^{(0)})$ is calculated.

The problem of finding an appropriate link function $h(\hat{\pi})$ for estimating

$$h(\hat{\pi}) = X\beta + \varepsilon \qquad (10.193)$$

has already been discussed in several previous sections. If model (10.191) is chosen, that is, the identity link, the parameters β_i are to be interpreted as the percentages with which the categories contribute to the conditional probabilities.

The logit link

$$h(\hat{\pi}_i) = \ln\left(\frac{\hat{\pi}_i}{1 - \hat{\pi}_i}\right) = x_i'\beta \qquad (10.194)$$

is again equivalent to the logistic model for $\hat{\pi}_i$:

$$\hat{\pi}_i = \frac{\exp(x_i'\beta)}{1 + \exp(x_i'\beta)} \,. \qquad (10.195)$$

The design matrices under inclusion of various interactions (up to the saturated model) are obtained as an extension of the designs for effect-coded main effects.

10.8.3 Coding of Models for the Hazard Rate

The analysis of lifetime data, given the variables $Y = 1$ (event) and $Y = 0$ (censored), is an important special case of the application of binary response in long-term studies.

The Cox model is often used as a semiparametric model for the modeling of failure time. Under inclusion of the vector of covariates x, this model can be written as follows:

$$\lambda(t \mid x) = \lambda_0(t) \exp(x'\beta) \,. \qquad (10.196)$$

If the hazard rates of two vectors of covariates x_1, x_2 are to be compared with each other (for example, stratification according to therapy x_1, x_2), the following relation is valid

$$\frac{\lambda(t \mid x_1)}{\lambda(t \mid x_2)} = \exp((x_1 - x_2)'\beta) \,. \qquad (10.197)$$

In order to be able to realize tests for quantitative or qualitative interactions between types of therapy and groups of patients, J subgroups of patients are defined (for example, stratification according to prognostic factors). Let therapy Z be bivariate, that is $Z = 1$ (therapy A) and $Z = 0$ (therapy B). For a fixed group of patients the hazard rate $\lambda_j(t \mid Z)$ $j = 1, \ldots, J$, for instance, is determined according to the Cox approach:

$$\lambda_j(t \mid Z) = \lambda_{0j}(t) \exp(\beta_j Z) . \tag{10.198}$$

In the case of $\hat{\beta}_j > 0$, the risk is higher for $Z = 1$ than for $Z = 0$ (jth stratum).

Test for Quantitative Interaction

We test H_0: effects of therapy is identical across the J strata, that is, $H_0 : \beta_1 = \ldots = \beta_J = \beta$, against the alternative $H_1 : \beta_i \lessgtr \beta_j$ for at least one pair (i, j). Under H_0, the test statistic

$$\chi^2_{J-1} = \sum_{j=1}^{J} \frac{\left(\hat{\beta}_j - \bar{\hat{\beta}}\right)^2}{\operatorname{var}(\hat{\beta}_j)} \tag{10.199}$$

with

$$\bar{\hat{\beta}} = \frac{\sum_{j=1}^{J} \left[\frac{\hat{\beta}_j}{\operatorname{var}(\hat{\beta}_j)}\right]}{\sum_{j=1}^{J} \left[\frac{1}{\operatorname{var}(\hat{\beta}_j)}\right]} \tag{10.200}$$

is distributed according to χ^2_{J-1}.

Test for Qualitative Differences

The null hypothesis H_0: therapy B ($Z = 0$) is better than therapy A ($Z = 1$) means $H_0 : \beta_j \leq 0 \ \forall j$. We define the sum of squares of the standardized estimates

$$Q^- = \sum_{j:\beta_j<0} \left[\frac{\hat{\beta}_j}{\operatorname{var}(\hat{\beta}_j)}\right]^2 \tag{10.201}$$

and

$$Q^+ = \sum_{j:\beta_j>0} \left[\frac{\hat{\beta}_j}{\operatorname{var}(\hat{\beta}_j)}\right]^2 , \tag{10.202}$$

as well as the test statistic

$$Q = \min(Q^-, Q^+) . \tag{10.203}$$

H_0 is rejected if $Q > c$ (Table 10.5).

TABLE 10.5. Critical values for the Q-test for $\alpha = 0.05$ (Gail and Simon, 1985).

J	2	3	4	5
c	2.71	4.23	5.43	6.50

Starting with the logistic model for the probability of response

$$P(Y = 1 \mid x) = \frac{\exp(\theta + x'\beta)}{1 + \exp(\theta + x'\beta)}, \tag{10.204}$$

and

$$P(Y = 0 \mid x) = 1 - P(Y = 1 \mid x) = \frac{1}{1 + \exp(\theta + x'\beta)} \tag{10.205}$$

with the binary variable

$$\begin{array}{llcl} Y = 1: & \{T = t \mid T \geq t, x\} & \Rightarrow & \text{failure at time } t \\ Y = 0: & \{T > t \mid T \geq t, x\} & \Rightarrow & \text{no failure} \end{array}$$

we obtain the model for the hazard function

$$\lambda(t \mid x) = \frac{\exp(\theta + x'\beta)}{1 + \exp(\theta + x'\beta)} \quad \text{for } t = t_1, \ldots, t_T \tag{10.206}$$

(Cox, 1972; cf. Doksum and Gasko, 1990; Lawless, 1982; Hamerle and Tutz, 1989). Thus the contribution of a patient to the likelihood (x fixed) with failure time t is

$$P(T = t \mid x) = \frac{\exp(\theta_t + x'\beta)}{\prod_{i=1}^{t} (1 + \exp(\theta_i + x'\beta))}. \tag{10.207}$$

Example 10.4: Assume that a patient has an event in the 4 failure times (for example, loss of abutment teeth by extraction). Let the patient have the following categories of the covariates: sex = 1 and age group=5 (60–70 years). The model is then $l = \theta + x'\beta$:

$$\begin{pmatrix} 0 \\ 0 \\ 0 \\ 1 \end{pmatrix} = \begin{pmatrix} 1 & 0 & 0 & 0 & 1 & 5 \\ 0 & 1 & 0 & 0 & 1 & 5 \\ 0 & 0 & 1 & 0 & 1 & 5 \\ 0 & 0 & 0 & 1 & \underbrace{1}_{\text{Sex}} & \underbrace{5}_{\text{Age}} \end{pmatrix} \begin{pmatrix} \theta_1 \\ \theta_2 \\ \theta_3 \\ \theta_4 \\ \beta_{11} \\ \beta_{12} \end{pmatrix} \begin{matrix} \left.\right\} \theta_t \\ \left.\right\} \beta \end{matrix} \tag{10.208}$$

(The last two columns, Sex and Age, form x.)

For N patients we have the model

$$\begin{pmatrix} l_1 \\ l_2 \\ \vdots \\ l_N \end{pmatrix} = \begin{pmatrix} I_1 & x_1 \\ I_2 & x_2 \\ \vdots & \\ I_N & x_N \end{pmatrix} \begin{pmatrix} \theta \\ \beta \end{pmatrix},$$

The dimension of the identity matrices I_j (patient j) is the number of survived failure times plus 1 (failure time of the jth patient). The vectors l_j for the jth patient contain as many zeros as the number of survived failure times of the other patients and the value 1 at the failure time of the jth patient.

The numerical solutions (for instance, according to Newton-Raphson) for the ML estimates $\hat{\theta}$ and $\hat{\beta}$ are obtained from the product of the likelihood functions (10.207) of all patients.

10.9 Extensions to Dependent Binary Variables

Although loglinear models are sufficiently rich to model any dependence structure between categorical variables, if one is interested in a regression of multivariate binary responses on a set of possibly continuous covariates, alternative models, which are better suited and have easier parameter interpretation, exist. Two often used-models in applications are marginal models and random effects models. In the following, we emphasize the idea of marginal models, because these seem to be a natural extension of the logistic regression model to more than one response variable. The first approach we describe in detail is called the quasi-likelihood approach (cf. Section 10.1.7), because the distribution of the binary response variables is not fully specified. We start describing these models in detail in Section 10.9.3. Then the generalized estimating equations (GEE) approach (Liang and Zeger, 1986) is introduced and two examples are given. The third approach is a full likelihood approach (Section 10.9.12). That section mainly gives an overview of the recent literature.

10.9.1 Overview

We now extend the problems of categorical response to the situations of correlation within the response values. These correlations are due to classification of the individuals into clusters of "related" elements. As already mentioned in Section 10.1.6, a positive correlation among related elements in a cluster leads to overdispersion if independence among these elements is falsely assumed.

Examples:

- Two or more implants or abutment teeth in dental reconstructions (Walther and Toutenburg, 1991).
- Response of a patient in cross-over in case of significant carry-over effect.

- Repeated categorical measurement of a response such as function of the lungs, blood pressure, or performance in training (repeated measures design or panel data).
- Measurement of paired organs (eyes, kidneys, etc.)
- Response of members of a family.

Let y_{ij} be the categorical response of the jth individual in the ith cluster:

$$y_{ij}\,, \quad i = 1, \ldots, N, \quad j = 1, \ldots, n_i\,. \tag{10.209}$$

We assume that the expectation of the response y_{ij} is dependent on prognostic variables (covariables) x_{ij} by a regression, that is,

$$\mathrm{E}(y_{ij}) = \beta_0 + \beta_1 x_{ij}\,. \tag{10.210}$$

Assume $\mathrm{var}(y_{ij}) = \sigma^2$ and

$$\mathrm{cov}(y_{ij}, y_{ij'}) = \sigma^2\rho \quad (j \neq j'). \tag{10.211}$$

The response of individuals from different clusters is assumed to be uncorrelated. Let us assume that the covariance matrix for the response of every cluster equals

$$\mathrm{V}\begin{pmatrix} y_{i1} \\ \vdots \\ y_{in_i} \end{pmatrix} = \mathrm{V}(y_i) = \sigma^2(1-\rho)I_{n_i} + \sigma^2\rho J_{n_i} \tag{10.212}$$

and thus has a compound symmetric structure. Hence, the covariance matrix of the entire sample vector is block-diagonal

$$W = \mathrm{V}\begin{pmatrix} y_1 \\ \vdots \\ y_N \end{pmatrix} = \mathrm{diag}(\mathrm{V}(y_1), \ldots, \mathrm{V}(y_N))\,. \tag{10.213}$$

Notice that the matrix W itself does not have a compound symmetric structure. Hence, we have a generalized regression model. The best linear unbiased estimate of $\beta = (\beta_0, \beta_1)'$ is given by the Gauss-Markov-Aitken estimator (4.64):

$$b = (X'W^{-1}X)^{-1}X'W^{-1}y\,, \tag{10.214}$$

and does not coincide with the OLS estimator, because the preconditions of Theorem 4.6 are not fulfilled. The choice of an incorrect covariance structure leads, according to our remarks in Section (4.3), to a bias in the estimate of the variance. On the other hand, the unbiasedness or consistency of the estimator of β stays untouched even in case of incorrect choice of the covariance matrix. Liang and Zeger (1993) examined the bias of $\mathrm{var}(\hat{\beta}_1)$ for the wrong choice of $\rho = 0$. In the case of positive correlation within the cluster, the variance is underestimated. This corresponds to the results of Goldberger (1964) for positive autocorrelation in econometric models.

The following problems arise in practice:

(i) identification of the covariance structure,

(ii) estimation of the correlation,

(iii) application of an Aitken-type estimate.

However, it is no longer possible to assume the usual GLM approach, because this does not take the correlation structure into consideration. Various approaches were developed as extensions of the GLM approach, in order to be able to include the correlation structure in the response:

- marginal model,
- random-effects model,
- observation-driven model,
- conditional model.

For binary response, simplifications arise (Section 10.9.8). Liang and Zeger (1989) proved that the joint distribution of the y_{ij} can be descibed by n_i logistic models for y_{ij} given y_{ik} $(k \neq j)$. Rosner (1984) used this approach and developed beta-binomial models.

10.9.2 Modeling Approaches for Correlated Response

The modeling approaches can be ordered according to diverse criteria.

Population-Averaged versus Subject-Specific Models

The essential difference between population-averaged (PA) and subject-specific (SS) models lies in the answer to the question of whether the regression coefficients vary for the individuals. In PA models, the β's are independent of the specific individual i. Examples are the marginal and conditional models. In SS models, the β's are dependent on the specific i and are therefore written as β_i. An example for a SS model is the random-effects model.

Marginal, Conditional, and Random-Effects Models

In the marginal model, the regression is modeled seperately from the dependence within the measurement in contrast to the two other approaches. The marginal expectation $\mathrm{E}(y_{ij})$ is modeled as a function of the explanatory variables and is interpreted as the mean response over the population of individuals with the same x. Hence, marginal models are mainly suitable for the analysis of covariable effects in a population.

The random-effects model, often also titled the mixed model, assumes that there are fixed effects, as in the marginal model, as well as individual specific effects. The dependent observations on each individual are assumed to be *conditionally independent given the subject-specific effects.*

Hence random-effects models are useful if one is interested in subject-specific behavior. But, concerning interpretation, only the *linear mixed model* allows an easy interpretation of fixed effect parameters as population-averaged effects and the others as subject-specific effects. *Generalized linear mixed models* are more complex, and even if a parameter is estimated as a fixed effect it may not be easily interpreted as a population-averaged effect.

For the conditional model (observation-driven model), a time-dependent response y_{it} is modeled as a function of the covariables and of the past response values $y_{it-1}, \ldots, y_{i1}$. This is done by assuming a specific correlation structure among the response values. Conditional models are useful if the main point of interest is the conditional probability of a state or the transition of states.

10.9.3 Quasi-Likelihood Approach for Correlated Binary Response

The following sections are dedicated to binary response variables and especially the bivariate case (that is, cluster size $n_i = 2$ for all $i = 1, \ldots, N$).

In case of a violation of independence or in case of a missing distribution assumption of the natural exponential family, the core of the ML method, namely the score function, may be used, nevertheless, for parameter estimation. We now want to specify the so-called quasi-score function (10.77) for the binary response (cf. Section 10.1.7).

Let $y_i' = (y_{i1}, \ldots, y_{in_i})$ be the response vector of the ith cluster ($i = 1, \ldots, N$) with the true covariance matrix $\operatorname{cov}(y_i)$ and let x_{ij} be the $p \times 1$-vector of the covariable corresponding to y_{ij}. Assume the variables y_{ij} are binary with values 1 and 0, and assume $P(y_{ij} = 1) = \pi_{ij}$. We then have $\mu_{ij} = \pi_{ij}$. Let $\pi_i' = (\pi_{i1}, \ldots, \pi_{in_i})$. Suppose that the link function is $g(\cdot)$, that is,

$$g(\pi_{ij}) = \eta_{ij} = x_{ij}'\beta\,.$$

Let $h(\cdot)$ be the inverse function, that is,

$$\mu_{ij} = \pi_{ij} = h(\eta_{ij}) = h(x_{ij}'\beta)\,.$$

For the canonical link

$$\operatorname{logit}(\pi_{ij}) = \ln\left(\frac{\pi_{ij}}{1 - \pi_{ij}}\right) = g(\pi_{ij}) = x_{ij}'\beta$$

we have

$$\pi_{ij} = h(\eta_{ij}) = \frac{\exp(\eta_{ij})}{1 + \exp(\eta_{ij})} = \frac{\exp(x_{ij}'\beta)}{1 + \exp(x_{ij}'\beta)}\,.$$

Hence

$$D = \left(\frac{\partial \mu_{ij}}{\partial \beta}\right) = \left(\frac{\partial \pi_{ij}}{\partial \beta}\right).$$

We have

$$\frac{\partial \pi_{ij}}{\partial \beta} = \frac{\partial \pi_{ij}}{\partial \eta_{ij}} \frac{\partial \eta_{ij}}{\partial \beta} = \frac{\partial h(\eta_{ij})}{\partial \eta_{ij}} x_{ij},$$

and hence, for $i = 1, \ldots, N$ and the $p \times n_i$-matrix $X_i' = (x_{i1}, \ldots, x_{in_i})$

$$D_i = \tilde{D}_i X_i \qquad \text{with} \quad \tilde{D}_i = \left(\frac{\partial h(\eta_{ij})}{\partial \eta_{ij}}\right).$$

For the quasi-score function for all N clusters, we now get

$$U(\beta) = \sum_{i=1}^{N} X_i' \tilde{D}_i' \, \mathrm{V}_i^{-1} (y_i - \pi_i), \tag{10.215}$$

where V_i is the matrix of the working variances and covariances of the y_{ij} of the ith cluster. The solution of $U(\hat{\beta}) = 0$ is found iteratively under further specifications, which we describe in the next section.

10.9.4 The GEE Method by Liang and Zeger

The variances are modeled as a function of the mean, that is,

$$v_{ij} = \mathrm{var}(y_{ij}) = v(\pi_{ij})\phi. \tag{10.216}$$

(In the binary case, the form of the variance of the binomial distribution is often chosen: $v(\pi_{ij}) = \pi_{ij}(1 - \pi_{ij})$.) With these, the following matrix is formed

$$A_i = \mathrm{diag}(v_{i1}, \ldots, v_{in_i}). \tag{10.217}$$

Since the structure of dependence is not known, an $n_i \times n_i$ *quasi-correlation matrix* $R_i(\alpha)$ is chosen for the vector of the ith cluster $y_i' = (y_{i1}, \ldots, y_{in_i})$ according to

$$R_i(\alpha) = \begin{pmatrix} 1 & \rho_{i12}(\alpha) & \cdots & \rho_{i1n_i}(\alpha) \\ \rho_{i21}(\alpha) & 1 & \cdots & \rho_{i2n_i}(\alpha) \\ \vdots & & & \vdots \\ \rho_{in_i1}(\alpha) & \rho_{in_i2}(\alpha) & \cdots & 1 \end{pmatrix}, \tag{10.218}$$

where the $\rho_{ikl}(\alpha)$ are the correlations as function of α (α may be a scalar or a vector). $R_i(\alpha)$ may vary for the clusters.

By multiplying the quasi-correlation matrix $R_i(\alpha)$ with the root diagonal matrix of the variances A_i, we obtain a working covariance matrix

$$\mathrm{V}_i(\beta, \alpha, \phi) = A_i^{\frac{1}{2}} R_i(\alpha) A_i^{\frac{1}{2}}, \tag{10.219}$$

which is no longer completely specified by the expectations, as in the case of independent response. We have $\mathrm{V}_i(\beta, \alpha, \phi) = \mathrm{cov}(y_i)$ if and only if $R_i(\alpha)$ is the true correlation matrix of y_i.

If the matrices V_i in (10.215) are replaced by the matrices $\mathrm{V}_i(\beta, \alpha, \phi)$ from (10.219), we get the *generalized estimating equations* by Liang and Zeger (1986), that is,

$$U(\beta, \alpha, \phi) = \sum_{i=1}^{N} \left(\frac{\partial \pi_i}{\partial \beta} \right)' \mathrm{V}_i^{-1}(\beta, \alpha, \phi)(y_i - \pi_i) = 0\,. \tag{10.220}$$

The solutions are denoted by $\hat{\beta}_G$. For the quasi-Fisher matrix, we have

$$F_G(\beta, \alpha) = \sum_{i=1}^{N} \left(\frac{\partial \pi_i}{\partial \beta} \right)' \mathrm{V}_i^{-1}(\beta, \alpha, \phi) \left(\frac{\partial \pi_i}{\partial \beta} \right)\,. \tag{10.221}$$

To avoid the dependence of α in determining $\hat{\beta}_G$, Liang and Zeger (1986) propose to replace α by a $N^{\frac{1}{2}}$-consistent estimate $\hat{\alpha}(y_1, \ldots, y_N, \beta, \phi)$ and ϕ by $\hat{\phi}$ (10.79) and to determine $\hat{\beta}_G$ from $U(\beta, \hat{\alpha}, \hat{\phi}) = 0$.

Remark: The iterative estimating procedure for GEE is described in detail in Liang and Zeger (1986). For the computational translation, a SAS macro by Karim and Zeger (1988) and a program by Kastner, Fieger, and Heumann (1997) exist.

If $R_i(\alpha) = I_{n_i}$ for $i = 1, \ldots, N$, is chosen, then the GEE are reduced to the *independence estimating equations (IEE)* . The IEE are

$$U(\beta, \phi) = \sum_{i=1}^{N} \left(\frac{\partial \pi_i}{\partial \beta} \right)' A_i^{-1}(y_i - \pi_i) = 0 \tag{10.222}$$

with $A_i = \mathrm{diag}(v(\pi_{ij})\phi)$. The solution is denoted by $\hat{\beta}_I$. Under some weak conditions, we have (Theorem 1 in Liang and Zeger, 1986) that $\hat{\beta}_I$ *is asymptotically consistent if the expectation* $\pi_{ij} = h(x'_{ij}\beta)$ *is correctly specified and the dispersion parameter* ϕ *is consistantly estimated.*

$\hat{\beta}_I$ is asymptotically normal

$$\hat{\beta}_I \overset{\text{a.s.}}{\sim} N(\beta; F_Q^{-1}(\beta, \phi) F_2(\beta, \phi) F_Q^{-1}(\beta, \phi)), \tag{10.223}$$

where

$$F_Q^{-1}(\beta, \phi) = \left[\sum_{i=1}^{N} \left(\frac{\partial \pi_i}{\partial \beta} \right)' A_i^{-1} \left(\frac{\partial \pi_i}{\partial \beta} \right) \right]^{-1},$$

$$F_2(\beta, \phi) = \sum_{i=1}^{N} \left(\frac{\partial \pi_i}{\partial \beta} \right)' A_i^{-1} \,\mathrm{cov}(y_i) A_i^{-1} \left(\frac{\partial \pi_i}{\partial \beta} \right)$$

and $\mathrm{cov}(y_i)$ is the true covariance matrix of y_i.

A consistent estimate for the variance of $\hat{\beta}_I$ is found by replacing β_I by $\hat{\beta}_I$, $\text{cov}(y_i)$ by its estimate $(y_i - \hat{\pi}_i)(y_i - \hat{\pi}_i)'$, and ϕ by $\hat{\phi}$ from (10.79), if ϕ is an unknown nuisance parameter. The consistency is independent of the correct specification of the covariance.

The advantages of $\hat{\beta}_I$ are that $\hat{\beta}_I$ is easy to calculate with available software for generalized linear models (see Appendix C) and that in case of correct specification of the regression model, $\hat{\beta}_I$ and $\text{cov}(\hat{\beta}_I)$ are consistent estimates. However, $\hat{\beta}_I$ loses in efficiency if the correlation between the clusters is large.

10.9.5 Properties of the GEE Estimate $\hat{\beta}_G$

Liang and Zeger (1986, Theorem 2) state that under some weak assumptions and under the conditions

(i) $\hat{\alpha}$ is $N^{\frac{1}{2}}$-consistent for α, given β and ϕ

(ii) $\hat{\phi}$ is a $N^{\frac{1}{2}}$-consistent estimate for ϕ, given β

(iii) the derivation $\partial\hat{\alpha}(\beta,\phi)/\partial\phi$ is independent of ϕ and α and is of stochastic order $O_p(1)$

the estimate $\hat{\beta}_G$ is consistent and asymptotic normal:

$$\hat{\beta}_G \overset{\text{as.}}{\sim} N(\beta,\ \text{V}_G) \tag{10.224}$$

with the asymptotic covariance matrix

$$\text{V}_G = F_Q^{-1}(\beta,\alpha)F_2(\beta,\alpha)F_Q^{-1}(\beta,\alpha), \tag{10.225}$$

where

$$F_Q^{-1}(\beta,\alpha) = \left(\sum_{i=1}^{N}\left(\frac{\partial\pi_i}{\partial\beta}\right)' \ \text{V}_i^{-1}\left(\frac{\partial\pi_i}{\partial\beta}\right)\right)^{-1},$$

$$F_2(\beta,\alpha) = \sum_{i=1}^{N}\left(\frac{\partial\pi_i}{\partial\beta}\right)' \ \text{V}_i^{-1}\,\text{cov}(y_i)\ \text{V}_i^{-1}\left(\frac{\partial\pi_i}{\partial\beta}\right)$$

and $\text{cov}(y_i) = \text{E}[(y_i - \pi_i)(y_i - \pi_i)']$ is the true covariance matrix of y_i. A short outline of the proof may be found in the appendix of Liang and Zeger (1986).

The asymptotic properties hold only for $N \to \infty$. Hence, it should be remembered that the estimation procedure should be used only for a large number of clusters.

An estimate $\hat{\text{V}}_G$ for the covariance matrix V_G may be found by replacing β, ϕ, α by their consistent estimates in (10.225), or by replacing $\text{cov}(y_i)$ by $(y_i - \hat{\pi}_i)(y_i - \hat{\pi}_i)'$.

If the covariance structure is specified correctly so that $V_i = \text{cov}(y_i)$, then the covariance of $\hat{\beta}_G$ is the inverse of the expected Fisher information matrix

$$V_G = \left(\sum_{i=1}^{N} \left(\frac{\partial \pi_i}{\partial \beta} \right)' V_i^{-1} \left(\frac{\partial \pi_i}{\partial \beta} \right) \right)^{-1} = F^{-1}(\beta, \alpha).$$

The estimate of this matrix is more stable than that of (10.225), but it has a loss in efficiency if the correlation structure is specified incorrectly (cf. Prentice, 1988, p. 1040).

The method of Liang and Zeger leads to an asymptotic variance of $\hat{\beta}_G$ that is independent of the choice of the estimates $\hat{\alpha}$ and $\hat{\phi}$ within the class of the $N^{\frac{1}{2}}$-consistent estimates. This is true for the asymptotic distribution of $\hat{\beta}_G$ as well.

In case of correct specification of the regression model, the estimates $\hat{\beta}_G$ and $\hat{V}_G$ are consistent, independent of the choice of the quasi-correlation matrix $R_i(\alpha)$. This means that even if $R_i(\alpha)$ is specified incorrectly, $\hat{\beta}_G$ and $\hat{V}_G$ stay consistent as long as $\hat{\alpha}$ and $\hat{\phi}$ are consistent. This robustness of the estimates is important, because the admissibility of the working covariance matrix V_i is difficult to check for small n_i. An incorrect specification of $R_i(\alpha)$ can reduce the efficiency of $\hat{\beta}_G$.

If the identity matrix is assumed for $R_i(\alpha)$, that is, $R_i(\alpha) = I$, $i = 1, \cdots, N$, then the estimating equations for β are reduced to the IEE. If the variances of the binomial distribution are chosen, as is usually done in the binary case, then the IEE and the ML score function (with binomially distributed variables) lead to the same estimates for β. However, the IEE method should be preferred in general, because the ML estimation procedure leads to incorrect variances for $\hat{\beta}_G$ and, hence, for example, incorrect test statistics and *p-values*. This leads to incorrect conclusions, for instance, related to significance or nonsignificance of the covariables (cf. Liang and Zeger, 1993).

Diggle, Liang, and Zeger (1994, Chapter 7.5) have proposed checking the consistency of $\hat{\beta}_G$ by fitting an appropriate model with various covariance structures. The estimates $\hat{\beta}_G$ and their consistent variances are then compared. If these differ too much, the modeling of the covariance structure calls for more attention.

10.9.6 Efficiency of the GEE and IEE Methods

Liang and Zeger (1986) stated the following about the comparison of $\hat{\beta}_I$ and $\hat{\beta}_G$. $\hat{\beta}_I$ is almost as efficient as $\hat{\beta}_G$ if the true correlation α is small.

$\hat{\beta}_I$ is very efficient if α is small and the data are binary.

If α is large, then $\hat{\beta}_G$ is more efficient than $\hat{\beta}_I$, and the efficiency of $\hat{\beta}_G$ can be increased if the correlation matrix is specified correctly.

In case of a high correlation within the blocks, the loss of efficiency of $\hat{\beta}_I$ compared to $\hat{\beta}_G$ is larger if the number of subunits n_i, $i = 1, \cdots, N$, varies between the clusters than if the clusters are all of the same size.

10.9.7 Choice of the Quasi-Correlation Matrix $R_i(\alpha)$

The working correlation matrix $R_i(\alpha)$ is chosen according to considerations such as simplicity, efficiency, and amount of existing data. Furthermore, assumptions about the structure of the dependence among the data should be considered by the choice. As mentioned before, the importance of the correlation matrix is due to the fact that it influences the variance of the estimated parameters.

The simplest specification is the assumption that the repeated observations of a cluster are uncorrelated, that is,

$$R_i(\alpha) = I, \qquad i = 1, \cdots, N.$$

This assumption leads to the IEE equations for uncorrelated response variables.

Another special case, which is the most efficient according to Liang and Zeger (1986, §4) but may be used only if the number of observations per cluster is small and the same for all clusters (e.g., equals n), is given by the choice

$$R_i(\alpha) = R(\alpha)$$

where $R(\alpha)$ is left totally unspecified and may be estimated by the empirical correlation matrix. The $n(n-1)/2$ parameters have to be estimated.

If it is assumed that the same pairwise dependencies exist among all response variables of one cluster, then the *exchangeable correlation structure* may be chosen:

$$\operatorname{Corr}(y_{ik}, y_{il}) = \alpha, \qquad k \neq l, \quad i = 1, \ldots, N.$$

This corresponds to the correlation assumption in random-effects models.

If $\operatorname{Corr}(y_{ik}, y_{il}) = \alpha(|k-l|)$ is chosen, then the correlations are stationary. The specific form $\alpha(|k-l|) = \alpha^{|l-k|}$ corresponds to the autocorrelation function of an AR(1)-process.

Further methods for parameter estimation in quasi-likelihood approaches are the GEE1 method by Prentice (1988) that estimates the α and β simultaneously from the GEE for α and β; the modified GEE1 method by Fitzmaurice and Laird (1993) based on conditional odds ratios; those by Lipsitz, Laird, and Harrington (1991) and Liang, Zeger, and Qaqish (1992) based on marginal odds ratios for modeling the cluster correlation; the GEE2 method by Liang et al. (1992) that estimates $\delta' = (\beta', \alpha)$ simultaneously as a joint parameter; and the pseudo-ML method by Zhao and Prentice (1990) and Prentice and Zhao (1991).

10.9.8 Bivariate Binary Correlated Response Variables

The previous sections introduced various methods developed for regression analysis of correlated binary data. They were described in a general form for N blocks (clusters) of size n_i. These methods may, of course, be used for bivariate binary data as well. This has the advantage that it simplifies the matter.

In this section, the GEE and IEE methods are developed for the bivariate binary case. Afterwards, an example demonstrates for the case of bivariate binary data the difference between a naive ML estimate and the GEE method by Liang and Zeger (1986).

We have: $y_i = (y_{i1}, y_{i2})'$, $i = 1, \cdots, N$. Each response variable y_{ij}, $j = 1, 2$, has its own vector of covariables $x'_{ij} = (x_{ij1}, \cdots, x_{ijp})$. The chosen link function for modeling the relationship between $\pi_{ij} = P(y_{ij} = 1)$ and x_{ij} is the logit link

$$\operatorname{logit}(\pi_{ij}) = \ln\left(\frac{\pi_{ij}}{1-\pi_{ij}}\right) = x'_{ij}\beta . \qquad (10.226)$$

Let

$$\pi'_i = (\pi_{i1}, \pi_{i2}) , \quad \eta_{ij} = x'_{ij}\beta , \quad \eta' = (\eta_{i1}, \eta_{i2}) . \qquad (10.227)$$

The logistic regression model has become the standard method for regression analysis of binary data.

10.9.9 The GEE Method

From Section 10.9.4 it can be seen that the form of the estimating equations for β is as follows:

$$U(\beta, \alpha, \phi) = S(\beta, \alpha) = \sum_{i=1}^{N} \left(\frac{\partial \pi_i}{\partial \beta}\right)' \mathrm{V}_i^{-1}(y_i - \pi_i) = 0 , \qquad (10.228)$$

where $\mathrm{V}_i = A_i^{\frac{1}{2}} R_i(\alpha) A_i^{\frac{1}{2}}$, $A_i = \operatorname{diag}(v(\pi_{ij})\phi)$, $j = 1, 2$, and $R_i(\alpha)$ is the working correlation matrix. Since only one correlation coefficient $\rho_i = \operatorname{Corr}(y_{i1}, y_{i2})$, $i = 1, \cdots, N$, has to be specified for bivariate binary data, and this is assumed to be constant, we have for the correlation matrix:

$$R_i(\alpha) = \begin{pmatrix} 1 & \rho \\ \rho & 1 \end{pmatrix}, \qquad i = 1, \cdots, N . \qquad (10.229)$$

For the matrix of derivatives we have:

$$\begin{aligned} \left(\frac{\partial \pi_i}{\partial \beta}\right)' &= \left(\frac{\partial h(\eta_i)}{\partial \beta}\right)' = \left(\frac{\partial \eta_i}{\partial \beta}\right)' \left(\frac{\partial h(\eta_i)}{\partial \eta_i}\right)' \\ &= \begin{pmatrix} x'_{i1} \\ x'_{i2} \end{pmatrix}' \begin{pmatrix} \frac{\partial h(\eta_{i1})}{\partial \eta_{i1}} & 0 \\ 0 & \frac{\partial h(\eta_{i2})}{\partial \eta_{i2}} \end{pmatrix} . \end{aligned}$$

Since $h(\eta_{i1}) = \pi_{i1} = \frac{\exp(x'_{i1}\beta)}{1+\exp(x'_{i1}\beta)}$ and $\exp(x'_{i1}\beta) = \frac{\pi_{i1}}{1-\pi_{i1}}$, we have $1 + \exp(x'_{i1}\beta) = 1 + \frac{\pi_{i1}}{1-\pi_{i1}} = \frac{1}{1-\pi_{i1}}$, and

$$\frac{\partial h(\eta_{i1})}{\partial \eta_{i1}} = \frac{\pi_{i1}}{1+\exp(x'_{i1}\beta)} = \pi_{i1}(1-\pi_{i1}). \tag{10.230}$$

holds. Analogously we have:

$$\frac{\partial h(\eta_{i2})}{\partial \eta_{i2}} = \pi_{i2}(1-\pi_{i2}). \tag{10.231}$$

If the variance is specified as $\text{var}(y_{ij}) = \pi_{ij}(1-\pi_{ij})$, $\phi = 1$, then we get

$$\left(\frac{\partial \pi_i}{\partial \beta}\right)' = x'_i \begin{pmatrix} \text{var}(y_{i1}) & 0 \\ 0 & \text{var}(y_{i2}) \end{pmatrix} = x'_i \Delta_i$$

with $x'_i = (x_{i1}, x_{i2})$ and $\Delta_i = \begin{pmatrix} \text{var}(y_{i1}) & 0 \\ 0 & \text{var}(y_{i2}) \end{pmatrix}$. For the covariance matrix V_i we have:

$$\begin{aligned} \mathrm{V}_i &= \begin{pmatrix} \text{var}(y_{i1}) & 0 \\ 0 & \text{var}(y_{i2}) \end{pmatrix}^{\frac{1}{2}} \begin{pmatrix} 1 & \rho \\ \rho & 1 \end{pmatrix} \begin{pmatrix} \text{var}(y_{i1}) & 0 \\ 0 & \text{var}(y_{i2}) \end{pmatrix}^{\frac{1}{2}} \\ &= \begin{pmatrix} \text{var}(y_{i1}) & \rho(\text{var}(y_{i1})\,\text{var}(y_{i2}))^{\frac{1}{2}} \\ \rho(\text{var}(y_{i1})\,\text{var}(y_{i2}))^{\frac{1}{2}} & \text{var}(y_{i2}) \end{pmatrix} \end{aligned} \tag{10.232}$$

and for the inverse of V_i:

$$\begin{aligned} \mathrm{V}_i^{-1} &= \frac{1}{(1-\rho^2)\,\text{var}(y_{i1})\,\text{var}(y_{i2})} \\ &\quad \begin{pmatrix} \text{var}(y_{i2}) & -\rho(\text{var}(y_{i1})\,\text{var}(y_{i2}))^{\frac{1}{2}} \\ -\rho(\text{var}(y_{i1})\,\text{var}(y_{i2}))^{\frac{1}{2}} & \text{var}(y_{i1}) \end{pmatrix} \\ &= \frac{1}{1-\rho^2} \begin{pmatrix} [\text{var}(y_{i1})]^{-1} & -\rho(\text{var}(y_{i1})\,\text{var}(y_{i2}))^{-\frac{1}{2}} \\ -\rho(\text{var}(y_{i1})\,\text{var}(y_{i2}))^{-\frac{1}{2}} & [\text{var}(y_{i2})]^{-1} \end{pmatrix}. \end{aligned} \tag{10.233}$$

If Δ_i is multiplied by V_i^{-1}, we obtain

$$W_i = \Delta_i\, \mathrm{V}_i^{-1} = \frac{1}{1-\rho^2} \begin{pmatrix} 1 & -\rho\left(\frac{\text{var}(y_{i1})}{\text{var}(y_{i2})}\right)^{\frac{1}{2}} \\ -\rho\left(\frac{\text{var}(y_{i2})}{\text{var}(y_{i1})}\right)^{\frac{1}{2}} & 1 \end{pmatrix} \tag{10.234}$$

and for the GEE method for β in the bivariate binary case:

$$S(\beta, \alpha) = \sum_{i=1}^{N} x'_i W_i (y_i - \pi_i) = 0. \tag{10.235}$$

According to by Liang and Zeger (1986, Theorem 2), under some weak conditions and under the assumption that the correlation parameter was consistently estimated, the solution $\hat{\beta}_G$ is consistent and asymptotic normal with expectation β and covariance matrix (10.225).

10.9.10 The IEE Method

If it is assumed that the response variables of each of the blocks are independent, that is, $R_i(\alpha) = I$ and $V_i = A_i$, then GEE method is reduced to IEE method.

$$U(\beta, \phi) = S(\beta) = \sum_{i=1}^{N} \left(\frac{\partial \pi_i}{\partial \beta} \right)' A_i^{-1} (y_i - \pi_i) = 0. \tag{10.236}$$

As we just showed, we have for the bivariate binary case:

$$\left(\frac{\partial \pi_i}{\partial \beta} \right)' = x_i' \Delta_i = x_i' \begin{pmatrix} \text{var}(y_{i1}) & 0 \\ 0 & \text{var}(y_{i2}) \end{pmatrix} \tag{10.237}$$

with $\text{var}(y_{ij}) = \pi_{ij}(1 - \pi_{ij})$, $\phi = 1$, and

$$A_i^{-1} = \begin{pmatrix} [\text{var}(y_{i1})]^{-1} & 0 \\ 0 & [\text{var}(y_{i2})]^{-1} \end{pmatrix} .$$

The IEE method then simplifies to

$$S(\beta) = \sum_{i=1}^{N} x_i' (y_i - \pi_i) = 0. \tag{10.238}$$

The solution $\hat{\beta}_I$ is consistent and asymptotic normal, according to by Liang and Zeger (1986, Theorem 1).

10.9.11 An Example from the Field of Dentistry

In this section, we demonstrate the procedure of the GEE method by means of a "twin" data set, that was documented by the Dental Clinic in Karlsruhe, Germany (Walther, 1992). The focal point is to show the difference between a robust estimate (GEE method) that takes the correlation of the response variables into account and the naive ML estimate. For the parameter estimation with the GEE method, a SAS macro is available (Karim and Zeger, 1988), as well as a procedure by Kastner et al. (1997).

Description of the "Twin" Data Set

During the examined interval, 331 patients were provided with two conical crowns each in the Dental Clinic in Karlsruhe. Since 50 conical crowns showed missing values and since the SAS macro for the GEE method needs complete data sets, these patients were excluded. Hence, for estimation of the regression parameters, the remaining 612 completely observed twin data sets were used. In this example, the twin pairs make up the clusters and the twins themselves (1.twin, 2.twin) are the subunits of the clusters.

The Response Variable

For all twin pairs in this study, the lifetime of the conical crowns was recorded in days. This lifetime is chosen as the response and is transformed into a binary response variable y_{ij} of the jth twin ($j = 1, 2$) in the ith cluster with

$$y_{ij} = \begin{cases} 1\,, & \text{if the conical crown is in function longer than } x \text{ days} \\ 0\,, & \text{if the conical crown is not in function longer than } x \text{ days.} \end{cases}$$

Different values may be defined for x. In the example, the values, in days, of 360 (1 year), 1100 (3 years), and 2000 (5 years) were chosen. Because the response variable is binary, the response probability of y_{ij} is modeled by the logit link (logistic regression). The model for the log-odds (i.e., the logarithm of the odds $\pi_{ij}/(1 - \pi_{ij})$ of the response $y_{ij} = 1$) is linear in the covariables, and in the model for the odds itself, the covariables have a multiplicative effect on the odds. Aim of the analysis is to find out whether the prognostic factors have a significant influence on the response probability.

Prognostic Factors

The covariables that were included in the analysis with the SAS macro, are

- age (in years)
- sex (1: male, 2: female)
- jaw (1: upper jaw, 2: lower jaw)
- type (1: dentoalveolar design, 2: transversal design)

All covariables, except for the covariable age, are dichotomous. The two types of conical crown constructions, dentoalveolar and transversal design, are distinguished as follows (cf. Walther, 1992):

- The dentoalveolar design connects all abutments exclusively by a rigid connection that runs on the alveolar ridge.
- The transversal design is used if the parts of reconstruction have to be connected by a transversal bar. This is the case if teeth in the front area are not included in the construction.

A total of 292 conical crowns were included in a dentoalveolar designs and 320 in a transversal design. Of these, 258 conical crowns were placed in the upper jaw, and 354 in the lower jaw.

The GEE Method

A problem that arises for the twin data is that the twins of a block are correlated. If this correlation is not taken into account, then the estimates

$\hat{\beta}$ stay unchanged but the variance of the $\hat{\beta}$ is underestimated. In case of positive correlation in a cluster, we have:

$$\text{var}(\hat{\beta})_{\text{naive}} < \text{var}(\hat{\beta})_{\text{robust}}.$$

Therefore,

$$\frac{\hat{\beta}}{\sqrt{\text{var}(\hat{\beta})_{\text{naive}}}} > \frac{\hat{\beta}}{\sqrt{\text{var}(\hat{\beta})_{\text{robust}}}},$$

which leads to incorrect tests and possibly to significant effects that might not be significant in a correct analysis (e.g., GEE). For this reason, appropriate methods that estimate the variance correctly should be chosen if the response variables are correlated.

The following regression model without interaction is assumed:

$$\ln \frac{P(\text{Lifetime} \geq \text{x})}{P(\text{Lifetime} < \text{x})} = \beta_0 + \beta_1 \cdot \text{Age} + \beta_2 \cdot \text{Sex} + \beta_3 \cdot \text{Jaw} + \beta_4 \cdot \text{Type}. \quad (10.239)$$

Additionally, we assume that the dependencies between the twins are identical and hence the exchangeable correlation structure is suitable for describing the dependencies.

To demonstrate the effects of various correlation assumptions on the estimation of the parameters, the following logistic regression models, which differ only in the assumed association parameter, are compared:

Model 1: naive (incorrect) ML estimation

Model 2: robust (correct) estimation, where independence is assumed, that is, $R_i(\alpha) = I$

Model 3: robust estimation with exchangeable correlation structure ($\rho_{ikl} = \text{Corr}(y_{ik}, y_{il}) = \alpha$, $k \neq l$)

Model 4: robust estimation with unspecified correlation structure ($R_i(\alpha) = R(\alpha)$).

As a test statistic (z-naive and z-robust) the ratio of estimate and standard error is calculated.

Results

Table 10.6 summarizes the estimated regression parameters, the standard errors, the z-statistics, and the p-values of models 2, 3, and 4 of the response variables

$$y_{ij} = \begin{cases} 1, & \text{if the conical crown is in function longer than 360 days} \\ 0, & \text{if the conical crown is in function not longer than 360 days.} \end{cases}$$

It turns out that the $\hat{\beta}$-values and the z-statistics are identical, independent of the choice of R_i, even though a high correlation between the twins

TABLE 10.6. Results of the robust estimates for models 2, 3, and 4 for $x = 360$.

	Model 2 (independence assump.)		Model 3 (exchangeable)		Model 4 (unspecified)	
Age	0.017[1)]	(0.012)[2)]	0.017	(0.012)	0.017	(0.012)
	1.33[3)]	(0.185)[4)]	1.33	(0.185)	1.33	(0.185)
Sex	−0.117	(0.265)	−0.117	(0.265)	−0.117	(0.265)
	−0.44	(0.659)	−0.44	(0.659)	−0.44	(0.659)
Jaw	0.029	(0.269)	0.029	(0.269)	0.029	(0.269)
	0.11	(0.916)	0.11	(0.916)	0.11	(0.916)
Type	−0.027	(0.272)	−0.027	(0.272)	−0.027	(0.272)
	−0.10	(0.920)	−0.10	(0.920)	−0.10	(0.920)

1) estimated regression values $\hat{\beta}$ 2) standard errors of $\hat{\beta}$
3) z-statistic 4) p-value

TABLE 10.7. Comparison of the standard errors, the z-statistics, and the p-values of models 1 and 2 for $x = 360$. (* indicates significace at the 10% level)

	Model 1 (naive)			Model 2 (robust)		
	σ	z	p-value	σ	z	p-value
Age	0.008	1.95	0.051*	0.012	1.33	0.185
Sex	0.190	−0.62	0.538	0.265	−0.44	0.659
Jaw	0.192	0.15	0.882	0.269	0.11	0.916
Type	0.193	−0.14	0.887	0.272	−0.10	0.920

exists. The exchangeable correlation model yields the value 0.9498 for the estimated correlation parameter $\hat{\alpha}$. In the model with the unspecified correlation structure, ρ_{i12} and ρ_{i21} were estimated as 0.9498 as well. The fact that the estimates of models 2, 3, and 4 coincide was observed in the analyses of the response variables with $x = 1100$ and $x = 2000$ as well. This means that the choice of R_i has no influence on the estimation procedure in the case of bivariate binary response. The GEE method is robust with respect to various correlation assumptions.

Table 10.7 compares the results of models 1 and 2. A striking difference between the two methods is that the covariate age in case of a naive ML estimation (model 1) is significant at the 10% level, even though this significance does not turn up if the robust method with the assumption of independence (model 2) is used. In the case of coinciding estimated regression parameters, the robust variances of $\hat{\beta}$ are larger and, accordingly, the robust z-statistics are smaller than the naive z-statistics. This result shows clearly that the ML method, which is incorrect in this case, underestimates the variances of $\hat{\beta}$ and hence leads to an incorrect age effect.

Tables 10.8 and 10.9 summarize the results with x-values 1100 and 2000. Table 10.8 shows that if the response variable is modeled with $x = 1100$, then none of the observed covariables is significant. As before, the estimated

TABLE 10.8. Comparison of the standard errors, the z-statistics, and the p-values of models 1 and 2 for $x = 1100$.

		Model 1 (naive)			Model 2 (robust)		
	$\hat{\beta}$	σ	z	p-value	σ	z	p-value
Age	0.0006	0.008	0.08	0.939	0.010	0.06	0.955
Sex	−0.0004	0.170	−0.00	0.998	0.240	−0.00	0.999
Jaw	0.1591	0.171	0.93	0.352	0.240	0.66	0.507
Type	0.0369	0.172	0.21	0.830	0.242	0.15	0.878

TABLE 10.9. Comparison of the standard errors, the z-statistics, and the p-values of models 1 and 2 for $x = 2000$. (* indicates significace at the 10% level)

		Model 1 (naive)			Model 2 (robust)		
	$\hat{\beta}$	σ	z	p-value	σ	z	p-value
Age	−0.0051	0.013	−0.40	0.691	0.015	−0.34	0.735
Sex	−0.2177	0.289	−0.75	0.452	0.399	−0.55	0.586
Jaw	0.0709	0.287	0.25	0.805	0.412	0.17	0.863
Type	0.6531	0.298	2.19	0.028*	0.402	1.62	0.104

correlation parameter $\hat{\alpha} = 0.9578$ indicates a strong dependency between the twins. In Table 10.9, the covariable "type" has a significant influence in the case of naive estimation. In the case of the GEE method ($R = I$), it might be significant with a p-value $= 0.104$ (10% level). The result $\hat{\beta}_{\text{type}} = 0.6531$ indicates that a dentoalveolar design significantly increases the log-odds of the response variable

$$y_{ij} = \begin{cases} 1, & \text{if the conical crown is in function longer than 2000 days} \\ 0, & \text{if the conical crown is in function not longer than 2000 days.} \end{cases}$$

Assuming the model

$$\frac{P(\text{Lifetime} \geq 2000)}{P(\text{Lifetime} < 2000)} = \exp(\beta_0 + \beta_1 \cdot \text{Age} + \beta_2 \cdot \text{Sex} + \beta_3 \cdot \text{Jaw} + \beta_4 \cdot \text{Type})$$

the odds $\frac{P(\text{Lifetime} \geq 2000)}{P(\text{Lifetime} < 2000)}$ for a dentoalveolar design is higher than the odds for a transversal design by the factor $\exp(\beta_4) = \exp(0.6531) = 1.92$, or alternatively, the odds ratio equals 1.92. The correlation parameter yields the value 0.9035.

In summary, it can be said that age and type are significant but not time-dependent covariables. The robust estimation yields no significant interaction, and a high correlation α exists between the twins of a pair.

Problems

The GEE estimations, which were carried out stepwise, have to be compared with caution, because they are not independent due to the time effect in the response variables. In this context, time-adjusted GEE methods that could be applied in this example are still missing. Therefore, further efforts are necessary in the field of survivorship analysis, in order to be able to complement the standard procedures, such as the Kaplan-Meier estimate and log-rank test, which are based on the independence of the response variables.

10.9.12 Full Likelihood Approach for Marginal Models

A useful full likelihood approach for marginal models in the case of multivariate binary data was proposed by Fitzmaurice and Laird (1993). Their starting point is the joint density

$$f(y; \Psi, \Omega) = P(Y_1 = y_1, \dots, Y_T = y_T; \Psi, \Omega) = \exp\{y'\Psi + w'\Omega - A(\Psi, \Omega)\} \tag{10.240}$$

with $y = (y_1, \dots, y_T)'$, $w = (y_1y_2, y_1y_3, \dots, y_{T-1}y_T, \dots, y_1y_2\cdots y_T)'$, $\Psi = (\Psi_1, \dots, \Psi_T)'$ and $\Omega = (\omega_{12}, \omega_{13}, \dots, \omega_{T-1T}, \dots, \omega_{12\cdots T})'$. Further

$$\exp\{A(\Psi, \Omega)\} = \sum_{y=(0,0,\dots,0)}^{y=(1,1,\dots,1)} \exp\{y'\Psi + w'\Omega\}$$

is a normalizing constant. Note that this is essentially the saturated parameterization in a loglinear model for T binary responses, since interactions of order 2 to T are included. A model that considers only all pairwise interactions, that is $w = (y_1y_2), \dots, (y_{T-1}y_T)$ and $\Omega = (\omega_{12}, \omega_{13}, \dots, \omega_{T-1,T})$, was already proposed by Cox (1972) and Zhao and Prentice (1990). The models are special cases of the so-called partial exponential families that were introduced by Zhao, Prentice, and Self (1992). The idea of Fitzmaurice and Laird (1993) was then to make a one-to-one transformation of the canonical parameter vector Ψ to the mean vector μ, which then can be linked to covariates via link functions such as in logistic regression. This idea of transforming canonical parameters one-to-one into (eventually centralized) moment parameters can be generalized to higher moments and to dependent categorical variables with more than two categories. Because the details, theoretically and computationally, are somewhat complex, we refer the reader to Lang and Agresti (1994), Molenberghs and Lesaffre (1994), Glonek (1996), Heagerty and Zeger (1996), and Heumann (1998). Each of these sources gives different possibilities on how to model the pairwise and higher interactions.

10.10 Exercises

Exercise 1. Let two models be defined by their design matrices X_1 and $X_2 = (X_1, X_3)$. Name the test statistic for testing H_0: Model X_1 holds and its distribution.

Exercise 2. What is meant by overdispersion? How is it parameterized in case of a binomial distribution?

Exercise 3. Why would a quasi-loglikelihood approach be chosen? How is the correlation in cluster data parameterized?

Exercise 4. Compare the models of two-way classification for continuous, normal data (ANOVA) and for categorical data. What are the reparametrization conditions in each case?

Exercise 5. The following table gives G^2-analysis of a two-way model with all submodels:

Model	G^2	p-value
A	200	0.00
B	100	0.00
A+B	20	0.10
A*B	0	1.00

Which model is valid?

Exercise 6. The following $I \times 2$-table gives frequencies for the age group X and the binary response Y:

	1	0
< 40	10	8
40–50	15	12
50–60	20	12
60–70	30	20
> 70	30	25

Analyze the trend of the sample logits.

Exercise 7. Consider the likelihood model of Section 10.9.12 for the case $T = 2$. Derive the Jacobian matrix J of the one-to-one transformation $(\psi_1, \psi_2, \omega_{12})$ to $(\mu_1, \mu_2, \gamma_{12})$ where $\gamma_{12} = \omega_{12}$, and

$$J = \begin{pmatrix} \frac{\partial \mu_1}{\partial \Psi_1} & \frac{\partial \mu_1}{\partial \Psi_2} & \frac{\partial \mu_1}{\partial \omega_{12}} \\ \frac{\partial \mu_2}{\partial \Psi_1} & \frac{\partial \mu_2}{\partial \Psi_2} & \frac{\partial \mu_2}{\partial \omega_{12}} \\ \frac{\partial \gamma_{12}}{\partial \Psi_1} & \frac{\partial \gamma_{12}}{\partial \Psi_2} & \frac{\partial \gamma_{12}}{\partial \omega_{12}} \end{pmatrix}.$$

Also derive its inverse J^{-1}.

Appendix A
Matrix Algebra

There are numerous books on matrix algebra that contain results useful for the discussion of linear models. See for instance books by Graybill (1961); Mardia, Kent, and Bibby (1979); Searle (1982); Rao (1973a); Rao and Mitra (1971); and Rao and Rao (1998), to mention a few. We collect in this Appendix some of the important results for ready reference. Proofs are generally omitted. References to original sources are given wherever necessary.

A.1 Overview

Definition A.1 *An $m \times n$-matrix A is a rectangular array of elements in m rows and n columns.*

In the context of the material treated in the book and in this Appendix, the elements of a matrix are taken as real numbers. We indicate an $m \times n$-matrix by writing $A : m \times n$ or $\underset{m,n}{A}$.

Let a_{ij} be the element in the ith row and the jth column of A. Then A may be represented as

$$A = \begin{pmatrix} a_{11} & a_{12} & \cdots & a_{1n} \\ a_{21} & a_{22} & \cdots & a_{2n} \\ \vdots & \vdots & & \cdots \\ a_{m1} & a_{m2} & \cdots & a_{mn} \end{pmatrix} = (a_{ij}) .$$

A matrix with $n = m$ rows and columns is called a square matrix. A square matrix having zeros as elements below (above) the diagonal is called an upper (lower) triangular matrix.

Definition A.2 *The transpose $A' : n \times m$ of a matrix $A : m \times n$ is given by interchanging the rows and columns of A. Thus*

$$A' = (a_{ji}) .$$

Then we have the following rules:

$$(A')' = A , \quad (A+B)' = A' + B' , \quad (AB)' = B'A' .$$

Definition A.3 *A square matrix is called symmetric if $A' = A$.*

Definition A.4 *An $m \times 1$ matrix a is said to be an m-vector and written as a column*

$$a = \begin{pmatrix} a_1 \\ \vdots \\ a_m \end{pmatrix} .$$

Definition A.5 *A $1 \times n$-matrix a' is said to be a row vector*

$$a' = (a_1, \cdots, a_n).$$

$A : m \times n$ may be written alternatively in a partitioned form as

$$A = (a_{(1)}, \ldots, a_{(n)}) = \begin{pmatrix} a'_1 \\ \vdots \\ a'_m \end{pmatrix}$$

with

$$a_{(j)} = \begin{pmatrix} a_{1j} \\ \vdots \\ a_{mj} \end{pmatrix} , \quad a_i = \begin{pmatrix} a_{i1} \\ \vdots \\ a_{in} \end{pmatrix} .$$

Definition A.6 *The $n \times 1$ row vector $(1, \cdots, 1)'$ is denoted by $1'_n$ or $1'$.*

Definition A.7 *The matrix $A : m \times m$ with $a_{ij} = 1$ (for all i,j) is given the symbol J_m, that is,*

$$J_m = \begin{pmatrix} 1 & \cdots & 1 \\ \vdots & & \vdots \\ 1 & \vdots & 1 \end{pmatrix} = 1_m 1'_m .$$

Definition A.8 *The n-vector*

$$e_i = (0, \cdots, 0, 1, 0, \cdots, 0)'$$

with the ith component as 1 and all the others as 0, is called the ith unit vector.

Definition A.9 *A $n \times n$ (square) matrix with elements 1 on the main diagonal and zeros off the diagonal is called the identity matrix I_n.*

Definition A.10 *A square matrix $A : n \times n$ with zeros in the off diagonal is called a diagonal matrix. We write*

$$A = \operatorname{diag}(a_{11}, \cdots, a_{nn}) = \operatorname{diag}(a_{ii}) = \begin{pmatrix} a_{11} & & 0 \\ & \ddots & \\ 0 & & a_{nn} \end{pmatrix}.$$

Definition A.11 *A matrix A is said to be partitioned if its elements are arranged in submatrices.*

Examples are

$$\underset{m,n}{A} = (\underset{m,r}{A_1}, \underset{m,s}{A_2}) \quad \text{with} \quad r + s = n$$

or

$$\underset{m,n}{A} = \begin{pmatrix} \underset{r,n-s}{A_{11}} & \underset{r,s}{A_{12}} \\ \underset{m-r,n-s}{A_{21}} & \underset{m-r,s}{A_{22}} \end{pmatrix}.$$

For partitioned matrices we get the transposess as

$$A' = \begin{pmatrix} A_1' \\ A_2' \end{pmatrix}, \quad A' = \begin{pmatrix} A_{11}' & A_{21}' \\ A_{12}' & A_{22}' \end{pmatrix},$$

respectively.

A.2 Trace of a Matrix

Definition A.12 *Let $a_{11}, \ldots, a_{nn}$ be the elements on the main diagonal of a square matrix $A : n \times n$. Then the trace of A is defined as the sum*

$$\operatorname{tr}(A) = \sum_{i=1}^{n} a_{ii} .$$

Theorem A.13 *Let A and B be square $n \times n$ matrices, and let c be a scalar factor. Then we have the following rules:*

(i) $\operatorname{tr}(A \pm B) = \operatorname{tr}(A) \pm \operatorname{tr}(B)$;

(ii) $\operatorname{tr}(A') = \operatorname{tr}(A)$;

(iii) $\operatorname{tr}(cA) = c \operatorname{tr}(A)$;

(iv) $\operatorname{tr}(AB) = \operatorname{tr}(BA)$ *(here A and B can be rectangular matrices of the form $A : m \times n$ and $B : n \times m$);*

(v) $\operatorname{tr}(AA') = \operatorname{tr}(A'A) = \sum_{i,j} a_{ij}^2$;

(vi) *If* $a = (a_1, \ldots, a_n)'$ *is an* n*-vector, then its squared norm may be written as*

$$\|a\|^2 = a'a = \sum_{i=1}^{n} a_i^2 = \operatorname{tr}(aa').$$

Note, that rules (iv) and (v) also hold for the case $A : n \times m$ and $B : m \times n$.

A.3 Determinant of a Matrix

Definition A.14 *Let* $n > 1$ *be a positive integer. The determinant of a square matrix* $A : n \times n$ *is defined by*

$$|A| = \sum_{i=1}^{n} (-1)^{i+j} a_{ij} |M_{ij}| \quad \textit{(for any } j\textit{, } j \textit{ fixed)},$$

with $|M_{ij}|$ *being the minor of the element* a_{ij}. $|M_{ij}|$ *is the determinant of the remaining* $(n-1) \times (n-1)$ *matrix when the* i*th row and the* j*th column of* A *are deleted.* $A_{ij} = (-1)^{i+j}|M_{ij}|$ *is called the cofactor of* a_{ij}.

Examples:

For $n = 2$: $|A| = a_{11}a_{22} - a_{12}a_{21}$.

For $n = 3$ (first column ($j = 1$) fixed):

$$\begin{aligned}
A_{11} &= (-1)^2 \begin{vmatrix} a_{22} & a_{23} \\ a_{32} & a_{33} \end{vmatrix} = (-1)^2 M_{11} \\
A_{21} &= (-1)^3 \begin{vmatrix} a_{12} & a_{13} \\ a_{32} & a_{33} \end{vmatrix} = (-1)^3 M_{21} \\
A_{31} &= (-1)^4 \begin{vmatrix} a_{12} & a_{13} \\ a_{22} & a_{23} \end{vmatrix} = (-1)^4 M_{31} \\
\Rightarrow |A| &= a_{11}A_{11} + a_{21}A_{21} + a_{31}A_{31}.
\end{aligned}$$

Note: As an alternative one may fix a row and develop the determinant of A according to

$$|A| = \sum_{j=1}^{n} (-1)^{i+j} a_{ij} |M_{ij}| \quad \text{(for any } i\text{, } i \text{ fixed)}.$$

Definition A.15 *A square matrix* A *is said to be regular or nonsingular if* $|A| \neq 0$. *Otherwise* A *is said to be singular.*

Theorem A.16 *Let* A *and* B *be* $n \times n$ *square matrices, and* c *be a scalar. Then we have*

(i) $|A'| = |A|$,

(ii) $|cA| = c^n |A|$,

(iii) $|AB| = |A||B|$,

(iv) $|A^2| = |A|^2$,

(v) *If A is diagonal or triangular, then*

$$|A| = \prod_{i=1}^{n} a_{ii} \, .$$

(vi) *For* $D = \begin{pmatrix} \underset{n,n}{A} & \underset{n,m}{C} \\ \underset{m,n}{0} & \underset{m,m}{B} \end{pmatrix}$ *we have*

$$\begin{vmatrix} A & C \\ 0 & B \end{vmatrix} = |A||B|,$$

and analogously

$$\begin{vmatrix} A' & 0' \\ C' & B' \end{vmatrix} = |A||B|.$$

(vii) *If A is partitioned with $A_{11} : p \times p$ and $A_{22} : q \times q$ square and non-singular, then*

$$\begin{aligned} \begin{vmatrix} A_{11} & A_{12} \\ A_{21} & A_{22} \end{vmatrix} &= |A_{11}||A_{22} - A_{21}A_{11}^{-1}A_{12}| \\ &= |A_{22}||A_{11} - A_{12}A_{22}^{-1}A_{21}|. \end{aligned}$$

Proof: Define the following matrices

$$Z_1 = \begin{pmatrix} I & -A_{12}A_{22}^{-1} \\ 0 & I \end{pmatrix} \quad \text{and} \quad Z_2 = \begin{pmatrix} I & 0 \\ -A_{22}^{-1}A_{21} & I \end{pmatrix},$$

where $|Z_1| = |Z_2| = 1$ by (vi). Then we have

$$Z_1 A Z_2 = \begin{pmatrix} A_{11} - A_{12}A_{22}^{-1}A_{21} & 0 \\ 0 & A_{22} \end{pmatrix}$$

and [using (iii) and (iv)]

$$|Z_1 A Z_2| = |A| = |A_{22}||A_{11} - A_{12}A_{22}^{-1}A_{21}|.$$

(viii) $\begin{vmatrix} A & x \\ x' & c \end{vmatrix} = |A|(c - x'A^{-1}x)$ *where x is an n-vector.*

Proof: Use (vii) with A instead of A_{11} and c instead of A_{22}.

(ix) *Let $B : p \times n$ and $C : n \times p$ be any matrices and $A : p \times p$ a nonsingular matrix. Then*

$$\begin{aligned} |A + BC| &= |A||I_p + A^{-1}BC| \\ &= |A||I_n + CA^{-1}B|. \end{aligned}$$

Proof: The first relationship follows from (iii) and

$$(A + BC) = A(I_p + A^{-1}BC)$$

immediately. The second relationship is a consequence of (vii) applied to the matrix

$$\begin{vmatrix} I_p & -A^{-1}B \\ C & I_n \end{vmatrix} = |I_p||I_n + CA^{-1}B| = |I_n||I_p + A^{-1}BC|.$$

(x) $|A + aa'| = |A|(1 + a'A^{-1}a)$, *if* A *is nonsingular.*

(xi) $|I_p + BC| = |I_n + CB|$, *if* $B : p \times n$ *and* $C : n \times p$.

A.4 Inverse of a Matrix

Definition A.17 *A matrix* $B : n \times n$ *is said to be an inverse of* $A : n \times n$ *if* $AB = I$. *If such a* B *exists, it is denoted by* A^{-1}. *It is easily seen that* A^{-1} *exists if and only if* A *is nonsingular. It is easy to establish that if* A^{-1} *exists; then* $AA^{-1} = A^{-1}A = I$.

Theorem A.18 *If all the inverses exist, we have*

(i) $(cA)^{-1} = c^{-1}A^{-1}$.

(ii) $(AB)^{-1} = B^{-1}A^{-1}$.

(iii) *If* $A : p \times p$, $B : p \times n$, $C : n \times n$ *and* $D : n \times p$ *then*

$$(A + BCD)^{-1} = A^{-1} - A^{-1}B(C^{-1} + DA^{-1}B)^{-1}DA^{-1}.$$

(iv) *If* $1 + b'A^{-1}a \neq 0$, *then we get from (iii)*

$$(A + ab')^{-1} = A^{-1} - \frac{A^{-1}ab'A^{-1}}{1 + b'A^{-1}a}.$$

(v) $|A^{-1}| = |A|^{-1}$.

Theorem A.19 (Inverse of a partitioned matrix) *For partitioned regular* A

$$A = \begin{pmatrix} E & F \\ G & H \end{pmatrix},$$

where $E : (n_1 \times n_1)$, $F : (n_1 \times n_2)$, $G : (n_2 \times n_1)$ *and* $H : (n_2 \times n_2)$ $(n_1 + n_2 = n)$ *are such that* E *and* $D = H - GE^{-1}F$ *are regular, the partitioned inverse is given by*

$$A^{-1} = \begin{pmatrix} E^{-1}(I + FD^{-1}GE^{-1}) & -E^{-1}FD^{-1} \\ -D^{-1}GE^{-1} & D^{-1} \end{pmatrix} = \begin{pmatrix} A^{11} & A^{12} \\ A^{21} & A^{22} \end{pmatrix}.$$

Proof: Check that the product of A and A^{-1} reduces to the identity matrix, that is,

$$AA^{-1} = A^{-1}A = I.$$

A.5 Orthogonal Matrices

Definition A.20 *A square matrix $A : n \times n$ is said to be orthogonal if $AA' = I = A'A$. For orthogonal matrices, we have*

(i) $A' = A^{-1}$.

(ii) $|A| = \pm 1$.

(iii) *Let $\delta_{ij} = 1$ for $i = j$ and 0 for $i \neq j$ denote the Kronecker symbol. Then the row vectors a_i and the column vectors $a_{(i)}$ of A satisfy the conditions*

$$a_i a_j' = \delta_{ij}\,, \quad a_{(i)}' a_{(j)} = \delta_{ij}\,.$$

(iv) *AB is orthogonal if A and B are orthogonal.*

Theorem A.21 *For $A : n \times n$ and $B : n \times n$ symmetric matrices, there exists an orthogonal matrix H such that $H'AH$ and $H'BH$ become diagonal if and only if A and B commute, that is,*

$$AB = BA.$$

A.6 Rank of a Matrix

Definition A.22 *The rank of $A : m \times n$ is the maximum number of linearly independent rows (or columns) of A. We write* $\text{rank}(A) = p$.

Theorem A.23 (Rules for ranks)

(i) $0 \leq \text{rank}(A) \leq \min(m, n)$.

(ii) $\text{rank}(A) = \text{rank}(A')$.

(iii) $\text{rank}(A + B) \leq \text{rank}(A) + \text{rank}(B)$.

(iv) $\text{rank}(AB) \leq \min\{\text{rank}(A), \text{rank}(B)\}$.

(v) $\text{rank}(AA') = \text{rank}(A'A) = \text{rank}(A) = \text{rank}(A')$.

(vi) *For nonsingular $B : m \times m$ and $C : n \times n$, we have* $\text{rank}(BAC) = \text{rank}(A)$.

(vii) *For $A : n \times n$,* $\text{rank}(A) = n$ *if and only if A is nonsingular.*

(viii) *If* $A = \text{diag}(a_i)$, *then* $\text{rank}(A)$ *equals the number of the* $a_i \neq 0$.

A.7 Range and Null Space

Definition A.24

(i) *The range* $\mathcal{R}(A)$ *of a matrix* $A : m \times n$ *is the vector space spanned by the column vectors of* A*, that is,*

$$\mathcal{R}(A) = \left\{ z : z = Ax = \sum_{i=1}^{n} a_{(i)} x_i, \quad x \in \mathbb{R}^n \right\} \subset \mathbb{R}^m ,$$

where $a_{(1)}, \ldots, a_{(n)}$ *are the column vectors of* A.

(ii) *The null space* $\mathcal{N}(A)$ *is the vector space defined by*

$$\mathcal{N}(A) = \{x \in \mathbb{R}^n \quad \text{and} \quad Ax = 0\} \subset \mathbb{R}^n .$$

Theorem A.25

(i) $\operatorname{rank}(A) = \dim \mathcal{R}(A)$*, where* $\dim \mathcal{V}$ *denotes the numb·r of basis vectors of a vector space* $\mathcal{V}$.

(ii) $\dim \mathcal{R}(A) + \dim \mathcal{N}(A) = n$.

(iii) $\mathcal{N}(A) = \{\mathcal{R}(A')\}^{\perp}$. *(*$\mathcal{V}^{\perp}$ *is the orthogonal complement of a vector space* $\mathcal{V}$ *defined by* $\mathcal{V}^{\perp} = \{x : x'y = 0 \quad \forall y \in \mathcal{V}\}$.*)*

(iv) $\mathcal{R}(AA') = \mathcal{R}(A)$.

(v) $\mathcal{R}(AB) \subseteq \mathcal{R}(A)$ *for any* A *and* B.

(vi) *For* $A \geq 0$ *and any* B, $\mathcal{R}(BAB') = \mathcal{R}(BA)$.

A.8 Eigenvalues and Eigenvectors

Definition A.26 *If* $A : p \times p$ *is a square matrix, then*

$$q(\lambda) = |A - \lambda I|$$

is a pth order polynomial in λ*. The* p *roots* $\lambda_1, \ldots, \lambda_p$ *of the characteristic equation* $q(\lambda) = |A - \lambda I| = 0$ *are called eigenvalues or characteristic roots of* A.

The eigenvalues possibly may be complex numbers. Since $|A - \lambda_i I| = 0$, $A - \lambda_i I$ is a singular matrix. Hence, there exists a nonzero vector $\gamma_i \neq 0$ satisfying $(A - \lambda_i I)\gamma_i = 0$, that is,

$$A\gamma_i = \lambda_i \gamma_i .$$

γ_i is called the (right) eigenvector of A for the eigenvalue λ_i. If λ_i is complex, then γ_i may have complex components. An eigenvector γ with real components is called standardized if $\gamma'\gamma = 1$.

Theorem A.27

(i) *If x and y are nonzero eigenvectors of A for λ_i, and α and β are any real numbers, then $\alpha x + \beta y$ also is an eigenvector for λ_i, that is,*

$$A(\alpha x + \beta y) = \lambda_i(\alpha x + \beta y).$$

Thus the eigenvectors for any λ_i span a vector space, which is called the eigenspace of A for λ_i.

(ii) *The polynomial $q(\lambda) = |A - \lambda I|$ has the normal form in terms of the roots*

$$q(\lambda) = \prod_{i=1}^{p} (\lambda_i - \lambda).$$

Hence, $q(0) = \prod_{i=1}^{p} \lambda_i$ and

$$|A| = \prod_{i=1}^{p} \lambda_i.$$

(iii) *Matching the coefficients of λ^{n-1} in $q(\lambda) = \prod_{i=1}^{p}(\lambda_i - \lambda)$ and $|A - \lambda I|$ gives*

$$tr(A) = \sum_{i=1}^{p} \lambda_i.$$

(iv) *Let $C : p \times p$ be a regular matrix. Then A and CAC^{-1} have the same eigenvalues λ_i. If γ_i is an eigenvector for λ_i, then $C\gamma_i$ is an eigenvector of CAC^{-1} for λ_i.*

Proof: As C is nonsingular, it has an inverse C^{-1} with $CC^{-1} = I$. We have $|C^{-1}| = |C|^{-1}$ and

$$\begin{aligned} |A - \lambda I| &= |C||A - \lambda C^{-1} C||C^{-1}| \\ &= |CAC^{-1} - \lambda I|. \end{aligned}$$

Thus, A and CAC^{-1} have the same eigenvalues. Let $A\gamma_i = \lambda_i \gamma_i$, and multiply from the left by C:

$$CAC^{-1}C\gamma_i = (CAC^{-1})(C\gamma_i) = \lambda_i(C\gamma_i).$$

(v) *The matrix $A + \alpha I$ with α a real number has the eigenvalues $\tilde{\lambda}_i = \lambda_i + \alpha$, and the eigenvectors of A and $A + \alpha I$ coincide.*

(vi) *Let λ_1 denote any eigenvalue of $A : p \times p$ with eigenspace H of dimension r. If k denotes the multiplicity of λ_1 in $q(\lambda)$, then*

$$1 \leq r \leq k.$$

Remarks:

(a) For symmetric matrices A, we have $r = k$.

(b) If A is not symmetric, then it is possible that $r < k$. Example:

$$A = \begin{pmatrix} 0 & 1 \\ 0 & 0 \end{pmatrix}, \quad A \neq A'$$

$$|A - \lambda I| = \begin{vmatrix} -\lambda & 1 \\ 0 & -\lambda \end{vmatrix} = \lambda^2 = 0.$$

The multiplicity of the eigenvalue $\lambda = 0$ is $k = 2$.

The eigenvectors for $\lambda = 0$ are $\gamma = \alpha \begin{pmatrix} 1 \\ 0 \end{pmatrix}$ and generate an eigenspace of dimension 1.

(c) If for any particular eigenvalue λ, $\dim(H) = r = 1$, then the standardized eigenvector for λ is unique (up to the sign).

Theorem A.28 *Let $A : n \times p$ and $B : p \times n$ with $n \geq p$ be any two matrices. Then from Theorem A.16 (vii),*

$$\begin{vmatrix} -\lambda I_n & -A \\ B & I_p \end{vmatrix} = (-\lambda)^{n-p}|BA - \lambda I_p| = |AB - \lambda I_n|.$$

Hence the n eigenvalues of AB are equal to the p eigenvalues of BA plus the eigenvalue 0 with multiplicity $n-p$. Suppose that $x \neq 0$ is an eigenvector of AB for any particular $\lambda \neq 0$. Then $y = Bx$ is an eigenvector of BA for this λ and we have $y \neq 0$, too.

Corollary: A matrix $A = aa'$ with a as a nonnull vector has all eigenvalues 0 except one, with $\lambda = a'a$ and the corresponding eigenvector a.

Corollary: The nonzero eigenvalues of AA' are equal to the nonzero eigenvalues of $A'A$.

Theorem A.29 *If A is symmetric, then all the eigenvalues are real.*

A.9 Decomposition of Matrices

Theorem A.30 (Spectral decomposition theorem) *Any symmetric matrix $A : (p \times p)$ can be written as*

$$A = \Gamma\Lambda\Gamma' = \sum \lambda_i \gamma_{(i)}\gamma'_{(i)},$$

where $\Lambda = \mathrm{diag}(\lambda_1, \ldots, \lambda_p)$ is the diagonal matrix of the eigenvalues of A, and $\Gamma = (\gamma_{(1)}, \ldots, \gamma_{(p)})$ is the orthogonal matrix of the standardized eigenvectors $\gamma_{(i)}$.

Theorem A.31 *Suppose A is symmetric and $A = \Gamma\Lambda\Gamma'$. Then*

(i) *A and Λ have the same eigenvalues (with the same multiplicity).*

(ii) *From $A = \Gamma\Lambda\Gamma'$ we get $\Lambda = \Gamma' A \Gamma$.*

(iii) *If $A : p \times p$ is a symmetric matrix, then for any integer n, $A^n = \Gamma\Lambda^n\Gamma'$ and $\Lambda^n = \operatorname{diag}(\lambda_i^n)$. If the eigenvalues of A are positive, then we can define the rational powers*

$$A^{\frac{r}{s}} = \Gamma\Lambda^{\frac{r}{s}}\Gamma' \quad \textit{with} \quad \Lambda^{\frac{r}{s}} = \operatorname{diag}(\lambda_i^{\frac{r}{s}})$$

for integers $s > 0$ and r. Important special cases are (when $\lambda_i > 0$)

$$A^{-1} = \Gamma\Lambda^{-1}\Gamma' \quad \textit{with} \quad \Lambda^{-1} = \operatorname{diag}(\lambda_i^{-1});$$

the symmetric square root decomposition of A (when $\lambda_i \geq 0$)

$$A^{\frac{1}{2}} = \Gamma\Lambda^{\frac{1}{2}}\Gamma' \quad \textit{with} \quad \Lambda^{\frac{1}{2}} = \operatorname{diag}(\lambda_i^{\frac{1}{2}})$$

and if $\lambda_i > 0$

$$A^{-\frac{1}{2}} = \Gamma\Lambda^{-\frac{1}{2}}\Gamma' \quad \textit{with} \quad \Lambda^{-\frac{1}{2}} = \operatorname{diag}(\lambda_i^{-\frac{1}{2}}).$$

(iv) *For any square matrix A, the rank of A equals the number of nonzero eigenvalues.*

Proof: According to Theorem A.23 (vi) we have $\operatorname{rank}(A) = \operatorname{rank}(\Gamma\Lambda\Gamma')$ $= \operatorname{rank}(\Lambda)$. But $\operatorname{rank}(\Lambda)$ equals the number of nonzero λ_i's.

(v) *A symmetric matrix A is uniquely determined by its distinct eigenvalues and the corresponding eigenspaces. If the distinct eigenvalues λ_i are ordered as $\lambda_1 \geq \cdots \geq \lambda_p$, then the matrix Γ is unique (up to sign).*

(vi) *$A^{\frac{1}{2}}$ and A have the same eigenvectors. Hence, $A^{\frac{1}{2}}$ is unique.*

(vii) *Let $\lambda_1 \geq \lambda_2 \geq \cdots \geq \lambda_k > 0$ be the nonzero eigenvalues and $\lambda_{k+1} = \cdots = \lambda_p = 0$. Then we have*

$$A = (\Gamma_1 \Gamma_2) \begin{pmatrix} \Lambda_1 & 0 \\ 0 & 0 \end{pmatrix} \begin{pmatrix} \Gamma_1' \\ \Gamma_2' \end{pmatrix} = \Gamma_1 \Lambda_1 \Gamma_1'$$

with $\Lambda_1 = \operatorname{diag}(\lambda_1, \cdots, \lambda_k)$ and $\Gamma_1 = (\gamma_{(1)}, \cdots, \gamma_{(k)})$, whereas $\Gamma_1'\Gamma_1 = I_k$ holds so that Γ_1 is column-orthogonal.

(viii) *A symmetric matrix A is of rank 1 if and only if $A = aa'$ where $a \neq 0$.*

Proof: If $\operatorname{rank}(A) = \operatorname{rank}(\Lambda) = 1$, then $\Lambda = \begin{pmatrix} \lambda & 0 \\ 0 & 0 \end{pmatrix}$, $A = \lambda\gamma\gamma' = aa'$ with $a = \sqrt{\lambda}\gamma$. If $A = aa'$, then by Theorem A.23 (v) we have $\operatorname{rank}(A) = \operatorname{rank}(a) = 1$.

Theorem A.32 (Singular-value decomposition of a rectangular matrix) *Let A : $n \times p$ be a rectangular matrix of rank r. Then we have*

$$\underset{n,p}{A} = \underset{n,r}{U}\,\underset{r,r}{L}\,\underset{r,p}{V'}$$

with $U'U = I_r$, $V'V = I_r$, and $L = \mathrm{diag}(l_1, \cdots, l_r)$, $l_i > 0$.

For a proof, see Rao (1973, p. 42).

Theorem A.33 *If $A : p \times q$ has $\mathrm{rank}(A) = r$, then A contains at least one nonsingular (r, r)-submatrix X, such that A has the so-called normal presentation*

$$\underset{p,q}{A} = \begin{pmatrix} \underset{r,r}{X} & \underset{r,q-r}{Y} \\ \underset{p-r,r}{Z} & \underset{p-r,q-r}{W} \end{pmatrix}.$$

All square submatrices of type $(r + s, r + s)$ with $(s \geq 1)$ are singular.

Proof: As $\mathrm{rank}(A) = \mathrm{rank}(X)$ holds, the first r rows of (X, Y) are linearly independent. Then the $p-r$ rows (Z, W) are linear combinations of (X, Y); that is, there exists a matrix F such that

$$(Z, W) = F(X, Y).$$

Analogously, there exists a matrix H satisfying

$$\begin{pmatrix} Y \\ W \end{pmatrix} = \begin{pmatrix} X \\ Z \end{pmatrix} H.$$

Hence we get $W = FY = FXH$, and

$$\begin{aligned} A = \begin{pmatrix} X & Y \\ Z & W \end{pmatrix} &= \begin{pmatrix} X & XH \\ FX & FXH \end{pmatrix} \\ &= \begin{pmatrix} I \\ F \end{pmatrix} X(I, H) \\ &= \begin{pmatrix} X \\ FX \end{pmatrix} (I, H) = \begin{pmatrix} I \\ F \end{pmatrix} (X, XH). \end{aligned}$$

As X is nonsingular, the inverse X^{-1} exists. Then we obtain $F = ZX^{-1}$, $H = X^{-1}Y$, $W = ZX^{-1}Y$, and

$$\begin{aligned} A = \begin{pmatrix} X & Y \\ Z & W \end{pmatrix} &= \begin{pmatrix} I \\ ZX^{-1} \end{pmatrix} X(I, X^{-1}Y) \\ &= \begin{pmatrix} X \\ Z \end{pmatrix} (I, X^{-1}Y) \\ &= \begin{pmatrix} I \\ ZX^{-1} \end{pmatrix} (X\,Y). \end{aligned}$$

Theorem A.34 (Full rank factorization)

(i) *If* $A : p \times q$ *has* $\text{rank}(A) = r$, *then* A *may be written as*

$$\underset{p,q}{A} = \underset{p,r}{K} \underset{r,q}{L}$$

with K *of full column rank* r *and* L *of full row rank* r.

Proof: Theorem A.33.

(ii) *If* $A : p \times q$ *has* $\text{rank}(A) = p$, *then* A *may be written as*

$$A = M(I, H), \quad \text{where } M : p \times p \text{ is regular.}$$

Proof: Theorem A.34 (i).

A.10 Definite Matrices and Quadratic Forms

Definition A.35 *Suppose* $A : n \times n$ *is symmetric and* $x : n \times 1$ *is any vector. Then the quadratic form in* x *is defined as the function*

$$Q(x) = x'Ax = \sum_{i,j} a_{ij} x_i x_j .$$

Clearly, $Q(0) = 0$.

Definition A.36 *The matrix* A *is called positive definite (p.d.) if* $Q(x) > 0$ *for all* $x \neq 0$. *We write* $A > 0$.

Note: If $A > 0$, then $(-A)$ is called negative definite.

Definition A.37 *The quadratic form* $x'Ax$ *(and the matrix* A*, also) is called positive semidefinite (p.s.d.) if* $Q(x) \geq 0$ *for all* x *and* $Q(x) = 0$ *for at least one* $x \neq 0$.

Definition A.38 *The quadratic form* $x'Ax$ *(and* A*) is called nonnegative definite (n.n.d.) if it is either p.d. or p.s.d., that is, if* $x'Ax \geq 0$ *for all* x. *If* A *is n.n.d., we write* $A \geq 0$.

Theorem A.39 *Let the* $n \times n$ *matrix* $A > 0$. *Then*

(i) A *has all eigenvalues* $\lambda_i > 0$.

(ii) $x'Ax > 0$ *for any* $x \neq 0$.

(iii) A *is nonsingular and* $|A| > 0$.

(iv) $A^{-1} > 0$.

(v) $\text{tr}(A) > 0$.

(vi) *Let* $P : n \times m$ *be of* $\text{rank}(P) = m \leq n$. *Then* $P'AP > 0$ *and in particular* $P'P > 0$, *choosing* $A = I$.

(vii) *Let* $P : n \times m$ *be of* $\text{rank}(P) < m \leq n$. *Then* $P'AP \geq 0$ *and* $P'P \geq 0$.

Theorem A.40 *Let* $A : n \times n$ *and* $B : n \times n$ *such that* $A > 0$ *and* $B : n \times n \geq 0$. *Then*

(i) $C = A + B > 0$.

(ii) $A^{-1} - (A + B)^{-1} \geq 0$.

(iii) $|A| \leq |A + B|$.

Theorem A.41 *Let* $A \geq 0$. *Then*

(i) $\lambda_i \geq 0$.

(ii) $\text{tr}(A) \geq 0$.

(iii) $A = A^{\frac{1}{2}} A^{\frac{1}{2}}$ *with* $A^{\frac{1}{2}} = \Gamma \Lambda^{\frac{1}{2}} \Gamma'$.

(iv) *For any matrix* $C : n \times m$ *we have* $C'AC \geq 0$.

(v) *For any matrix* C, *we have* $C'C \geq 0$ *and* $CC' \geq 0$.

Theorem A.42 *For any matrix* $A \geq 0$, *we have* $0 \leq \lambda_i \leq 1$ *if and only if* $(I - A) \geq 0$.

Proof: Write the symmetric matrix A in its spectral form as $A = \Gamma \Lambda \Gamma'$. Then we have

$$(I - A) = \Gamma(I - \Lambda)\Gamma' \geq 0$$

if and only if

$$\Gamma'\Gamma(I - \Lambda)\Gamma'\Gamma = I - \Lambda \geq 0.$$

(a) If $I - \Lambda \geq 0$, then for the eigenvalues of $I - A$ we have $1 - \lambda_i \geq 0$ (i.e., $0 \leq \lambda_i \leq 1$).

(b) If $0 \leq \lambda_i \leq 1$, then for any $x \neq 0$,

$$x'(I - \Lambda)x = \sum x_i^2 (1 - \lambda_i) \geq 0,$$

that is, $I - \Lambda \geq 0$.

Theorem A.43 (Theobald, 1974) *Let* $D : n \times n$ *be symmetric. Then* $D \geq 0$ *if and only if* $\text{tr}\{CD\} \geq 0$ *for all* $C \geq 0$.

Proof: D is symmetric, so that

$$D = \Gamma \Lambda \Gamma' = \sum \lambda_i \gamma_i \gamma_i',$$

and hence

$$\begin{aligned} \operatorname{tr}\{CD\} &= \operatorname{tr}\left\{\sum \lambda_i C \gamma_i \gamma_i'\right\} \\ &= \sum \lambda_i \gamma_i' C \gamma_i . \end{aligned}$$

(a) Let $D \geq 0$, and, hence, $\lambda_i \geq 0$ for all i. Then $\operatorname{tr}(CD) \geq 0$ if $C \geq 0$.

(b) Let $\operatorname{tr}\{CD\} \geq 0$ for all $C \geq 0$. Choose $C = \gamma_i \gamma_i'$ $(i = 1, \ldots, n,\ i \text{ fixed})$ so that

$$\begin{aligned} 0 \leq \operatorname{tr}\{CD\} &= \operatorname{tr}\{\gamma_i \gamma_i' (\sum_j \lambda_j \gamma_j \gamma_j')\} \\ &= \lambda_i \qquad (i = 1, \cdots, n) \end{aligned}$$

and $D = \Gamma \Lambda \Gamma' \geq 0$.

Theorem A.44 *Let $A : n \times n$ be symmetric with eigenvalues $\lambda_1 \geq \cdots \geq \lambda_n$. Then*

$$\sup_x \frac{x'Ax}{x'x} = \lambda_1, \quad \inf_x \frac{x'Ax}{x'x} = \lambda_n .$$

Proof: See Rao (1973, p. 62).

Theorem A.45 *Let $A : n \times r = (A_1, A_2)$, with A_1 of order $n \times r_1$, A_2 of order $n \times r_2$, and $\operatorname{rank}(A) = r = r_1 + r_2$. Define the orthogonal projectors $M_1 = A_1(A_1'A_1)^{-1}A_1'$ and $M = A(A'A)^{-1}A'$. Then*

$$M = M_1 + (I - M_1)A_2(A_2'(I - M_1)A_2)^{-1}A_2'(I - M_1).$$

Proof: M_1 and M are symmetric idempotent matrices fulfilling the conditions $M_1 A_1 = 0$ and $MA = 0$. Using Theorem A.19 for partial inversion of $A'A$, that is,

$$(A'A)^{-1} = \begin{pmatrix} A_1'A_1 & A_1'A_2 \\ A_2'A_1 & A_2'A_2 \end{pmatrix}^{-1}$$

and using the special form of the matrix D defined in A.19, that is,

$$D = A_2'(I - M_1)A_2,$$

straightforward calculation concludes the proof.

Theorem A.46 *Let $A : n \times m$ with $\operatorname{rank}(A) = m \leq n$ and $B : m \times m$ be any symmetric matrix. Then*

$$ABA' \geq 0 \quad \textit{if and only if } B \geq 0.$$

Proof:

(a) $B \geq 0 \Rightarrow ABA' \geq 0$ for all A.

(b) Let $\text{rank}(A) = m \leq n$ and assume $ABA' \geq 0$, so that $x'ABA'x \geq 0$ for all $x \in \mathbb{R}^n$.
We have to prove that $y'By \geq 0$ for all $y \in \mathbb{R}^m$. As $\text{rank}(A) = m$, the inverse $(A'A)^{-1}$ exists. Setting $z = A(A'A)^{-1}y$, we have $A'z = y$ and $y'By = z'ABA'z \geq 0$ so that $B \geq 0$.

Definition A.47 *Let $A : n \times n$ and $B : n \times n$ be any matrices. Then the roots $\lambda_i = \lambda_i^B(A)$ of the equation*

$$|A - \lambda B| = 0$$

are called the eigenvalues of A in the metric of B. For $B = I$ we obtain the usual eigenvalues defined in Definition A.26 (cf. Dhrymes, 1974, p. 581).

Theorem A.48 *Let $B > 0$ and $A \geq 0$. Then $\lambda_i^B(A) \geq 0$.*

Proof: $B > 0$ is equivalent to $B = B^{\frac{1}{2}}B^{\frac{1}{2}}$ with $B^{\frac{1}{2}}$ nonsingular and unique (A.31 (iii)). Then we may write

$$0 = |A - \lambda B| = |B^{\frac{1}{2}}|^2 |B^{-\frac{1}{2}}AB^{-\frac{1}{2}} - \lambda I|$$

and $\lambda_i^B(A) = \lambda_i^I(B^{-\frac{1}{2}}AB^{-\frac{1}{2}}) \geq 0$, as $B^{-\frac{1}{2}}AB^{-\frac{1}{2}} \geq 0$.

Theorem A.49 (Simultaneous diagonalization) *Let $B > 0$ and $A \geq 0$, and denote by $\Lambda = \text{diag}\left(\lambda_i^B(A)\right)$ the diagonal matrix of the eigenvalues of A in the metric of B. Then there exists a nonsingular matrix W such that*

$$B = W'W \quad \text{and} \quad A = W'\Lambda W.$$

Proof: From the proof of Theorem A.48 we know that the roots $\lambda_i^B(A)$ are the usual eigenvalues of the matrix $B^{-\frac{1}{2}}AB^{-\frac{1}{2}}$. Let X be the matrix of the corresponding eigenvectors:

$$B^{-\frac{1}{2}}AB^{-\frac{1}{2}}X = X\Lambda,$$

that is,

$$A = B^{\frac{1}{2}}X\Lambda X'B^{\frac{1}{2}} = W'\Lambda W$$

with $W' = B^{\frac{1}{2}}X$ regular and

$$B = W'W = B^{\frac{1}{2}}XX'B^{\frac{1}{2}} = B^{\frac{1}{2}}B^{\frac{1}{2}}.$$

Theorem A.50 *Let $A > 0$ (or $A \geq 0$) and $B > 0$. Then*

$$B - A > 0 \quad \text{if and only if} \quad \lambda_i^B(A) < 1.$$

Proof: Using Theorem A.49, we may write

$$B - A = W'(I - \Lambda)W,$$

namely,

$$x'(B - A)x = x'W'(I - \Lambda)Wx$$

$$\begin{aligned} &= y'(I-\Lambda)y \\ &= \sum \left(1-\lambda_i^B(A)\right) y_i^2 \end{aligned}$$

with $y = Wx$, W regular, and hence $y \neq 0$ for $x \neq 0$. Then $x'(B-A)x > 0$ holds if and only if

$$\lambda_i^B(A) < 1.$$

Theorem A.51 *Let $A > 0$ (or $A \geq 0$) and $B > 0$. Then*

$$A - B \geq 0$$

if and only if

$$\lambda_i^B(A) \leq 1\,.$$

Proof: Similar to Theorem A.50.

Theorem A.52 *Let $A > 0$ and $B > 0$. Then*

$$B - A > 0 \quad \textit{if and only if} \quad A^{-1} - B^{-1} > 0.$$

Proof: From Theorem A.49 we have

$$B = W'W, \quad A = W'\Lambda W.$$

Since W is regular, we have

$$B^{-1} = W^{-1}W'^{-1}, \quad A^{-1} = W^{-1}\Lambda^{-1}W'^{-1},$$

that is,

$$A^{-1} - B^{-1} = W^{-1}(\Lambda^{-1} - I)W'^{-1} > 0,$$

as $\lambda_i^B(A) < 1$ and, hence, $\Lambda^{-1} - I > 0$.

Theorem A.53 *Let $B - A > 0$. Then $|B| > |A|$ and $\operatorname{tr}(B) > \operatorname{tr}(A)$. If $B - A \geq 0$, then $|B| \geq |A|$ and $\operatorname{tr}(B) \geq \operatorname{tr}(A)$.*

Proof: From Theorems A.49 and A.16 (iii), (v), we get

$$\begin{aligned} |B| &= |W'W| = |W|^2, \\ |A| &= |W'\Lambda W| = |W|^2|\Lambda| = |W|^2 \prod \lambda_i^B(A), \end{aligned}$$

that is,

$$|A| = |B| \prod \lambda_i^B(A).$$

For $B - A > 0$, we have $\lambda_i^B(A) < 1$ (i.e., $|A| < |B|$). For $B - A \geq 0$, we have $\lambda_i^B(A) \leq 1$ (i.e., $|A| \leq |B|$). $B - A > 0$ implies $\operatorname{tr}(B - A) > 0$, and $\operatorname{tr}(B) > \operatorname{tr}(A)$. Analogously, $B - A \geq 0$ implies $\operatorname{tr}(B) \geq \operatorname{tr}(A)$.

Theorem A.54 (Cauchy-Schwarz inequality) *Let x, y be real vectors of the same dimension. Then*

$$(x'y)^2 \leq (x'x)(y'y),$$

with equality if and only if x and y are linearly dependent.

Theorem A.55 *Let x, y be real vectors and $A > 0$. Then we have the following results:*

(i) $(x'Ay)^2 \leq (x'Ax)(y'Ay)$.

(ii) $(x'y)^2 \leq (x'Ax)(y'A^{-1}y)$.

Proof:

(a) $A \geq 0$ is equivalent to $A = BB$ with $B = A^{\frac{1}{2}}$ (Theorem A.41 (iii)). Let $Bx = \tilde{x}$ and $By = \tilde{y}$. Then (i) is a consequence of Theorem A.54.

(b) $A > 0$ is equivalent to $A = A^{\frac{1}{2}}A^{\frac{1}{2}}$ and $A^{-1} = A^{-\frac{1}{2}}A^{-\frac{1}{2}}$. Let $A^{\frac{1}{2}}x = \tilde{x}$ and $A^{-\frac{1}{2}}y = \tilde{y}$; then (ii) is a consequence of Theorem A.54.

Theorem A.56 *Let $A > 0$ and T be any square matrix. Then*

(i) $\sup_{x \neq 0} \frac{(x'y)^2}{x'Ax} = y'A^{-1}y$.

(ii) $\sup_{x \neq 0} \frac{(y'Tx)^2}{x'Ax} = y'TA^{-1}T'y$.

Proof: Use Theorem A.55 (ii).

Theorem A.57 *Let $I : n \times n$ be the identity matrix and let a be an n-vector. Then*

$$I - aa' \geq 0 \quad \textit{if and only if} \quad a'a \leq 1.$$

Proof: The matrix aa' is of rank 1 and $aa' \geq 0$. The spectral decomposition is $aa' = C\Lambda C'$ with $\Lambda = \operatorname{diag}(\lambda, 0, \cdots, 0)$ and $\lambda = a'a$. Hence, $I - aa' = C(I - \Lambda)C' \geq 0$ if and only if $\lambda = a'a \leq 1$ (see Theorem A.42).

Theorem A.58 *Assume $MM' - NN' \geq 0$. Then there exists a matrix H such that $N = MH$.*

Proof (Milliken and Akdeniz, 1977) : Let M (n, r) of $\operatorname{rank}(M) = s$, and let x be any vector $\in \mathcal{R}(I - MM^-)$, implying $x'M = 0$ and $x'MM'x = 0$. As NN' and $MM' - NN'$ (by assumption) are n.n.d., we may conclude that $x'NN'x \geq 0$ and

$$x'(MM' - NN')x = -x'NN'x \geq 0,$$

so that $x'NN'x = 0$ and $x'N = 0$. Hence, $N \subset \mathcal{R}(M)$ or, equivalently, $N = MH$ for some matrix H (r, k).

Theorem A.59 *Let A be an $n \times n$-matrix and assume $(-A) > 0$. Let a be an n-vector. In case $n \geq 2$, the matrix $A + aa'$ is never n.n.d.*

Proof (Guilkey and Price, 1981) : The matrix aa' is of rank ≤ 1. In case $n \geq 2$, there exists a nonzero vector w such that $w'aa'w = 0$, implying $w'(A + aa')w = w'Aw < 0$.

A.11 Idempotent Matrices

Definition A.60 *A square matrix A is called idempotent if it satisfies*

$$A^2 = AA = A\,.$$

An idempotent matrix A is called an orthogonal projector if $A = A'$. Otherwise, A is called an oblique projector.

Theorem A.61 *Let $A : n \times n$ be idempotent with* $\text{rank}(A) = r \leq n$. *Then we have:*

(i) *The eigenvalues of A are 1 or 0.*

(ii) $\text{tr}(A) = \text{rank}(A) = r$.

(iii) *If A is of full rank n, then $A = I_n$.*

(iv) *If A and B are idempotent and if $AB = BA$, then AB is also idempotent.*

(v) *If A is idempotent and P is orthogonal, then PAP' is also idempotent.*

(vi) *If A is idempotent, then $I - A$ is idempotent and*

$$A(I - A) = (I - A)A = 0.$$

Proof:

(a) The characteristic equation

$$Ax = \lambda x$$

multiplied by A gives

$$AAx = Ax = \lambda Ax = \lambda^2 x.$$

Multiplication of both equations by x' then yields

$$x'Ax = \lambda x'x = \lambda^2 x'x,$$

that is,

$$\lambda(\lambda - 1) = 0.$$

(b) From the spectral decomposition

$$A = \Gamma\Lambda\Gamma'\,,$$

we obtain

$$\text{rank}(A) = \text{rank}(\Lambda) = \text{tr}(\Lambda) = r,$$

where r is the number of characteristic roots with value 1.

(c) Let $\text{rank}(A) = \text{rank}(\Lambda) = n$, then $\Lambda = I_n$ and

$$A = \Gamma\Lambda\Gamma' = I_n\,.$$

(a)–(c) follow from the definition of an idempotent matrix.

A.12 Generalized Inverse

Definition A.62 *Let A be an $m \times n$-matrix. Then a matrix $A^- : n \times m$ is said to be a generalized inverse of A if*

$$AA^-A = A$$

holds (see Rao (1973), p. 24).

Theorem A.63 *A generalized inverse always exists although it is not unique in general.*

Proof: Assume $\text{rank}(A) = r$. According to the singular-value decomposition (Theorem A.32), we have

$$\underset{m,n}{A} = \underset{m,r}{U}\,\underset{r,r}{L}\,\underset{r,n}{V'}$$

with $U'U = I_r$ and $V'V = I_r$ and

$$L = \text{diag}(l_1, \cdots, l_r), \quad l_i > 0.$$

Then

$$A^- = V \begin{pmatrix} L^{-1} & X \\ Y & Z \end{pmatrix} U'$$

(X, Y and Z are arbitrary matrices of suitable dimensions) is a g-inverse of A. Using Theorem A.33, namely,

$$A = \begin{pmatrix} X & Y \\ Z & W \end{pmatrix}$$

with X nonsingular, we have

$$A^- = \begin{pmatrix} X^{-1} & 0 \\ 0 & 0 \end{pmatrix}$$

as a special g-inverse.

Definition A.64 (Moore-Penrose inverse) *A matrix A^+ satisfying the following conditions is called the Moore-Penrose inverse of A:*

(i) $AA^+A = A\,,$

(ii) $A^+AA^+ = A^+\,,$

(iii) $(A^+A)' = A^+A$,

(iv) $(AA^+)' = AA^+$.

A^+ is unique.

Theorem A.65 *For any matrix $A : m \times n$ and any g-inverse $A^- : m \times n$, we have*

(i) *A^-A and AA^- are idempotent.*

(ii) $\text{rank}(A) = \text{rank}(AA^-) = \text{rank}(A^-A)$.

(iii) $\text{rank}(A) \leq \text{rank}(A^-)$.

Proof:

(a) Using the definition of g-inverse,

$$(A^-A)(A^-A) = A^-(AA^-A) = A^-A.$$

(b) According to Theorem A.23 (iv), we get

$$\text{rank}(A) = \text{rank}(AA^-A) \leq \text{rank}(A^-A) \leq \text{rank}(A),$$

that is, $\text{rank}(A^-A) = \text{rank}(A)$. Analogously, we see that $\text{rank}(A) = \text{rank}(AA^-)$.

(c) $\text{rank}(A) = \text{rank}(AA^-A) \leq \text{rank}(AA^-) \leq \text{rank}(A^-)$.

Theorem A.66 *Let A be an $m \times n$-matrix. Then*

(i) *A regular $\Rightarrow A^+ = A^{-1}$.*

(ii) $(A^+)^+ = A$.

(iii) $(A^+)' = (A')^+$.

(iv) $\text{rank}(A) = \text{rank}(A^+) = \text{rank}(A^+A) = \text{rank}(AA^+)$.

(v) *A an orthogonal projector $\Rightarrow A^+ = A$.*

(vi) $\text{rank}(A) : m \times n = m \Rightarrow A^+ = A'(AA')^{-1}$ *and* $AA^+ = I_m$.

(vii) $\text{rank}(A) : m \times n = n \Rightarrow A^+ = (A'A)^{-1}A'$ *and* $A^+A = I_n$.

(viii) *If $P : m \times m$ and $Q : n \times n$ are orthogonal $\Rightarrow$ $(PAQ)^+ = Q^{-1}A^+P^{-1}$.*

(ix) $(A'A)^+ = A^+(A')^+$ *and* $(AA')^+ = (A')^+A^+$.

(x) $A^+ = (A'A)^+A' = A'(AA')^+$.

For further details see Rao and Mitra (1971).

Theorem A.67 (Baksalary, Kala, and Klaczynski (1983)) *Let $M : n \times n \geq 0$ and $N : m \times n$ be any matrices. Then*

$$M - N'(NM^+N')^+N \geq 0$$

if and only if

$$\mathcal{R}(N'NM) \subset \mathcal{R}(M).$$

Theorem A.68 *Let A be any square $n \times n$-matrix and a be an n-vector with $a \notin \mathcal{R}(A)$. Then a g-inverse of $A + aa'$ is given by*

$$\begin{aligned}(A + aa')^- &= A^- - \frac{A^-aa'U'U}{a'U'Ua} \\ &\quad - \frac{VV'aa'A^-}{a'VV'a} + \phi \frac{VV'aa'U'U}{(a'U'Ua)(a'VV'a)},\end{aligned}$$

with A^- any g-inverse of A and

$$\phi = 1 + a'A^-a, \quad U = I - AA^-, \quad V = I - A^-A.$$

Proof: Straightforward by checking $AA^-A = A$.

Theorem A.69 *Let A be a square $n \times n$-matrix. Then we have the following results:*

(i) *Assume a, b are vectors with $a, b \in \mathcal{R}(A)$, and let A be symmetric. Then the bilinear form $a'A^-b$ is invariant to the choice of A^-.*

(ii) *$A(A'A)^-A'$ is invariant to the choice of $(A'A)^-$.*

Proof:

(a) $a, b \in \mathcal{R}(A) \Rightarrow a = Ac$ and $b = Ad$. Using the symmetry of A gives

$$\begin{aligned}a'A^-b &= c'A'A^-Ad \\ &= c'Ad.\end{aligned}$$

(b) Using the rowwise representation of A as $A = \begin{pmatrix} a_1' \\ \vdots \\ a_n' \end{pmatrix}$ gives

$$A(A'A)^-A' = (a_i'(A'A)^-a_j).$$

Since $A'A$ is symmetric, we may conclude then (i) that all bilinear forms $a_i'(A'A)a_j$ are invariant to the choice of $(A'A)^-$, and hence (ii) is proved.

Theorem A.70 *Let $A : n \times n$ be symmetric, $a \in \mathcal{R}(A)$, $b \in \mathcal{R}(A)$, and assume $1 + b'A^+a \neq 0$. Then*

$$(A + ab')^+ = A^+ - \frac{A^+ab'A^+}{1 + b'A^+a}.$$

Proof: Straightforward, using Theorems A.68 and A.69.

Theorem A.71 *Let $A : n \times n$ be symmetric, a be an n-vector, and $\alpha > 0$ be any scalar. Then the following statements are equivalent:*

(i) $\alpha A - aa' \geq 0$.

(ii) $A \geq 0$, $a \in \mathcal{R}(A)$, *and* $a'A^-a \leq \alpha$, *with* A^- *being any g-inverse of* A.

Proof:

(i) $\Rightarrow$ (ii): $\alpha A - aa' \geq 0 \Rightarrow \alpha A = (\alpha A - aa') + aa' \geq 0 \Rightarrow A \geq 0$. Using Theorem A.31 for $\alpha A - aa' \geq 0$, we have $\alpha A - aa' = BB$, and, hence,

$$
\begin{aligned}
& \alpha A = BB + aa' = (B, a)(B, a)'. \\
\Rightarrow \quad & \mathcal{R}(\alpha A) = \mathcal{R}(A) = \mathcal{R}(B, a) \\
\Rightarrow \quad & a \in \mathcal{R}(A) \\
\Rightarrow \quad & a = Ac \quad \text{with} \quad c \in \mathbb{R}^n \\
\Rightarrow \quad & a'A^-a = c'Ac.
\end{aligned}
$$

As $\alpha A - aa' \geq 0 \quad \Rightarrow$

$$x'(\alpha A - aa')x \geq 0$$

for any vector x, choosing $x = c$, we have

$$
\begin{aligned}
\alpha c'Ac - c'aa'c &= \alpha c'Ac - (c'Ac)^2 \geq 0, \\
&\Rightarrow c'Ac \leq \alpha.
\end{aligned}
$$

(ii) $\Rightarrow$ (i): Let $x \in \mathbb{R}^n$ be any vector. Then, using Theorem A.54,

$$
\begin{aligned}
x'(\alpha A - aa')x &= \alpha x'Ax - (x'a)^2 \\
&= \alpha x'Ax - (x'Ac)^2 \\
&\geq \alpha x'Ax - (x'Ax)(c'Ac)
\end{aligned}
$$

$$\Rightarrow \quad x'(\alpha A - aa')x \geq (x'Ax)(\alpha - c'Ac).$$

In (ii) we have assumed $A \geq 0$ and $c'Ac = a'A^-a \leq \alpha$. Hence, $\alpha A - aa' \geq 0$.

Note: This theorem is due to Baksalary and Kala (1983). The version given here and the proof are formulated by G. Trenkler.

Theorem A.72 *For any matrix A we have*

$$A'A = 0 \quad \textit{if and only if } A = 0.$$

Proof:

(a) $A = 0 \Rightarrow A'A = 0$.

(b) Let $A'A = 0$, and let $A = (a_{(1)}, \cdots, a_{(n)})$ be the columnwise presentation. Then

$$A'A = (a'_{(i)} a_{(j)}) = 0,$$

so that all the elements on the diagonal are zero: $a'_{(i)} a_{(i)} = 0 \Rightarrow a_{(i)} = 0$ and $A = 0$.

Theorem A.73 *Let $X \neq 0$ be an $m \times n$-matrix and A an $n \times n$-matrix. Then*

$$X'XAX'X = X'X \quad \Rightarrow \quad XAX'X = X \quad \text{and} \quad X'XAX' = X'.$$

Proof: As $X \neq 0$ and $X'X \neq 0$, we have

$$\begin{aligned} X'XAX'X - X'X = (X'XA - I)X'X &= 0 \\ \Rightarrow (X'XA - I) &= 0 \end{aligned}$$

$$\begin{aligned} \Rightarrow 0 &= (X'XA - I)(X'XAX'X - X'X) \\ &= (X'XAX' - X')(XAX'X - X) = Y'Y, \end{aligned}$$

so that (by Theorem A.72) $Y = 0$, and, hence $XAX'X = X$.

Corollary: Let $X \neq 0$ be an $m \times n$-matrix and A and b $n \times n$-matrices. Then

$$AX'X = BX'X \quad \Leftrightarrow \quad AX' = BX'.$$

Theorem A.74 (Albert's theorem)
Let $A = \begin{pmatrix} A_{11} & A_{12} \\ A_{21} & A_{22} \end{pmatrix}$ *be symmetric. Then*

(i) $A \geq 0$ *if and only if*

(a) $A_{22} \geq 0$,
(b) $A_{21} = A_{22}A_{22}^{-}A_{21}$,
(c) $A_{11} \geq A_{12}A_{22}^{-}A_{21}$,

((b) and (c) are invariant of the choice of A_{22}^{-}).

(ii) $A > 0$ *if and only if*

(a) $A_{22} > 0$,
(b) $A_{11} > A_{12}A_{22}^{-1}A_{21}$.

Proof (Bekker and Neudecker, 1989) :

(i) Assume $A \geq 0$.

(a) $A \geq 0 \quad \Rightarrow \quad x'Ax \geq 0$ for any x. Choosing $x' = (0', x_2')$ $\Rightarrow x'Ax = x_2'A_{22}x_2 \geq 0$ for any $x_2 \Rightarrow A_{22} \geq 0$.
(b) Let $B' = (0, I - A_{22}A_{22}^{-}) \Rightarrow$

$$\begin{aligned} B'A &= \left((I - A_{22}A_{22}^{-})A_{21}, A_{22} - A_{22}A_{22}^{-}A_{22}\right) \\ &= \left((I - A_{22}A_{22}^{-})A_{21}, 0\right) \end{aligned}$$

and $B'AB = B'A^{\frac{1}{2}}A^{\frac{1}{2}}B = 0$. Hence, by Theorem A.72 we get $B'A^{\frac{1}{2}} = 0$.

$$\begin{aligned} \Rightarrow B'A^{\frac{1}{2}}A^{\frac{1}{2}} = B'A &= 0. \\ \Rightarrow (I - A_{22}A_{22}^-)A_{21} &= 0. \end{aligned}$$

This proves (b).

(c) Let $C' = (I, -(A_{22}^- A_{21})')$. $A \geq 0 \Rightarrow$

$$\begin{aligned} 0 \leq C'AC &= A_{11} - A_{12}(A_{22}^-)'A_{21} - A_{12}A_{22}^-A_{21} \\ &\quad + A_{12}(A_{22}^-)'A_{22}A_{22}^-A_{21} \\ &= A_{11} - A_{12}A_{22}^-A_{21}\,. \end{aligned}$$

(Since A_{22} is symmetric, we have $(A_{22}^-)' = A_{22}$.)

Now assume (a), (b), and (c). Then

$$D = \begin{pmatrix} A_{11} - A_{12}A_{22}^-A_{21} & 0 \\ 0 & A_{22} \end{pmatrix} \geq 0,$$

as the submatrices are n.n.d. by (a) and (b). Hence,

$$A = \begin{pmatrix} I & A_{12}(A_{22}^-) \\ 0 & I \end{pmatrix} D \begin{pmatrix} I & 0 \\ A_{22}^-A_{21} & I \end{pmatrix} \geq 0.$$

(ii) Proof as in (i) if A_{22}^- is replaced by A_{22}^{-1}.

Theorem A.75 *If $A : n \times n$ and $B : n \times n$ are symmetric, then*

(i) $0 \leq B \leq A$ *if and only if*

(a) $A \geq 0$,
(b) $B = AA^-B$,
(c) $B \geq BA^-B$.

(ii) $0 < B < A$ *if and only if* $0 < A^{-1} < B^{-1}$.

Proof: Apply Theorem A.74 to the matrix $\begin{pmatrix} B & B \\ B & A \end{pmatrix}$.

Theorem A.76 *Let A be symmetric and $c \in \mathcal{R}(A)$. Then the following statements are equivalent:*

(i) $\text{rank}(A + cc') = \text{rank}(A)$.

(ii) $\mathcal{R}(A + cc') = \mathcal{R}(A)$.

(iii) $1 + c'A^-c \neq 0$.

Corollary 1: Assume (i) or (ii) or (iii) holds; then

$$(A + cc')^- = A^- - \frac{A^-cc'A^-}{1 + c'A^-c}$$

for any choice of A^-.

Corollary 2: Assume (i) or (ii) or (iii) holds; then

$$\begin{aligned} c'(A+cc')^{-}c &= c'A^{-}c - \frac{(c'A^{-}c)^2}{1+c'A^{-}c} \\ &= 1 - \frac{1}{1+c'A^{-}c}. \end{aligned}$$

Moreover, as $c \in \mathcal{R}(A+cc')$, the results are invariant for any special choices of the g-inverses involved.

Proof: $c \in \mathcal{R}(A) \Leftrightarrow AA^{-}c = c \Rightarrow$

$$\mathcal{R}(A+cc') = \mathcal{R}(AA^{-}(A+cc')) \subset \mathcal{R}(A).$$

Hence, (i) and (ii) become equivalent. Proof of (iii): Consider the following product of matrices:

$$\begin{pmatrix} 1 & 0 \\ c & A+cc' \end{pmatrix} \begin{pmatrix} 1 & -c \\ 0 & I \end{pmatrix} \begin{pmatrix} 1 & 0 \\ -A^{-}c & I \end{pmatrix} = \begin{pmatrix} 1+c'A^{-}c & -c \\ 0 & A \end{pmatrix}.$$

The left-hand side has the rank

$$1 + \operatorname{rank}(A+cc') = 1 + \operatorname{rank}(A)$$

(see (i) or (ii)). The right-hand side has the rank $1+\operatorname{rank}(A)$ if and only if $1+c'A^{-}c \neq 0$.

Theorem A.77 *Let $A : n \times n$ be a symmetric and nonsingular matrix and $c \notin \mathcal{R}(A)$. Then we have*

(i) $c \in \mathcal{R}(A+cc')$.

(ii) $\mathcal{R}(A) \subset \mathcal{R}(A+cc')$.

(iii) $c'(A+cc')^{-}c = 1$.

(iv) $A(A+cc')^{-}A = A$.

(v) $A(A+cc')^{-}c = 0$.

Proof: As A is assumed to be nonsingular, the equation $Al = 0$ has a nontrivial solution $l \neq 0$, which may be standardized as $(c'l)^{-1}l$ such that $c'l = 1$. Then we have $c = (A+cc')l \in \mathcal{R}(A+cc')$, and hence (i) is proved. Relation (ii) holds as $c \notin \mathcal{R}(A)$. Relation (i) is seen to be equivalent to

$$(A+cc')(A+cc')^{-}c = c.$$

Then (iii) follows:

$$\begin{aligned} c'(A+cc')^{-}c &= l'(A+cc')(A+cc')^{-}c \\ &= l'c = 1, \end{aligned}$$

which proves (iii). From

$$
\begin{aligned}
c &= (A+cc')(A+cc')^{-}c \\
&= A(A+cc')^{-}c + cc'(A+cc')^{-}c \\
&= A(A+cc')^{-}c + c\,,
\end{aligned}
$$

we have (v).

(iv) is a consequence of the general definition of a g-inverse and of (iii) and (iv):

$$
\begin{aligned}
A+cc' &= (A+cc')(A+cc')^{-}(A+cc') \\
&= A(A+cc')^{-}A \\
&\quad + cc'(A+cc')^{-}cc' \quad [= cc' \text{ using (iii)}] \\
&\quad + A(A+cc')^{-}cc' \quad [= 0 \text{ using (v)}] \\
&\quad + cc'(A+cc')^{-}A \quad [= 0 \text{ using (v)}].
\end{aligned}
$$

Theorem A.78 *We have* $A \geq 0$ *if and only if*

(i) $A + cc' \geq 0$.

(ii) $(A+cc')(A+cc')^{-}c = c$.

(iii) $c'(A+cc')^{-}c \leq 1$.

Assume $A \geq 0$*; then*

(a) $c = 0 \Leftrightarrow c'(A+cc')^{-}c = 0$.

(b) $c \in \mathcal{R}(A) \Leftrightarrow c'(A+cc')^{-}c < 1$.

(c) $c \notin \mathcal{R}(A) \Leftrightarrow c'(A+cc')^{-}c = 1$.

Proof: $A \geq 0$ is equivalent to

$$0 \leq cc' \leq A + cc'.$$

Straightforward application of Theorem A.75 gives (i)–(iii).

Proof of (a): $A \geq 0 \Rightarrow A + cc' \geq 0$. Assume

$$c'(A+cc')^{-}c = 0\,,$$

and replace c by (ii) $\Rightarrow$

$$
\begin{aligned}
&c'(A+cc')^{-}(A+cc')(A+cc')^{-}c = 0 \Rightarrow \\
&(A+cc')(A+cc')^{-}c = 0
\end{aligned}
$$

as $(A+cc') \geq 0$. Assuming $c = 0 \Rightarrow c'(A+cc')c = 0$.

Proof of (b): Assume $A \geq 0$ and $c \in \mathcal{R}(A)$, and use Theorem A.76 (Corollary 2) $\Rightarrow$

$$c'(A+cc')^{-}c = 1 - \frac{1}{1 + c'A^{-}c} < 1.$$

The opposite direction of (b) is a consequence of (c).

Proof of (c): Assume $A \geq 0$ and $c \notin \mathcal{R}(A)$, and use Theorem A.77 (iii) $\Rightarrow$

$$c'(A + cc')^{-}c = 1.$$

The opposite direction of (c) is a consequence of (b).

Note: The proofs of Theorems A.74–A.78 are given in Bekker and Neudecker (1989).

Theorem A.79 *The linear equation $Ax = a$ has a solution if and only if*

$$a \in \mathcal{R}(A) \quad or \quad AA^{-}a = a$$

for any g-inverse A.

If this condition holds, then all solutions are given by

$$x = A^{-}a + (I - A^{-}A)w\,,$$

where w is an arbitrary m-vector. Further, $q'x$ has a unique value for all solutions of $Ax = a$ if and only if $q'A^{-}A = q'$, or $q \in \mathcal{R}(A')$.

For a proof, see Rao (1973, p. 25).

A.13 Projectors

Consider the range space $\mathcal{R}(A)$ of the matrix $A : m \times n$ with rank r. Then there exists $\mathcal{R}(A)^{\perp}$, which is the orthogonal complement of $\mathcal{R}(A)$ with dimension $m - r$. Any vector $x \in \mathbb{R}^m$ has the unique decomposition

$$x = x_1 + x_2\,, \quad x_1 \in \mathcal{R}(A)\,, \quad \text{and} \quad x_2 \in \mathcal{R}(A)^{\perp}\,,$$

of which the component x_1 is called the orthogonal projection of x on $\mathcal{R}(A)$. The component x_1 can be computed as Px, where

$$P = A(A'A)^{-}A'\,,$$

which is called the projection operator on $\mathcal{R}(A)$. Note that P is unique for any choice of the g-inverse $(A'A)^{-}$.

Theorem A.80 *For any $P : n \times n$, the following statements are equivalent:*

(i) *P is an orthogonal projection operator.*

(ii) *P is symmetric and idempotent.*

For proofs and other details, the reader is referred to Rao (1973) and Rao and Mitra (1971).

Theorem A.81 *Let X be a matrix of order $T \times K$ with rank $r < K$, and $U : (K - r) \times K$ be such that $\mathcal{R}(X') \cap \mathcal{R}(U') = \{0\}$. Then*

(i) $X(X'X + U'U)^{-1}U' = 0.$

(ii) $X'X(X'X+U'U)^{-1}X'X = X'X$; *that is,* $(X'X+U'U)^{-1}$ *is a g-inverse of* $X'X$.

(iii) $U'U(X'X+U'U)^{-1}U'U = U'U$; *that is,* $(X'X+U'U)^{-1}$ *is also a g-inverse of* $U'U$.

(iv) $U(X'X+U'U)^{-1}U'u = u$ *if* $u \in \mathcal{R}(U)$.

Proof: Since $X'X+U'U$ is of full rank, there exists a matrix A such that

$$\begin{aligned} & (X'X+U'U)A = U' \\ \Rightarrow \quad & X'XA = U' - U'UA \quad \Rightarrow \quad XA = 0 \text{ and } U' = U'UA \end{aligned}$$

since $\mathcal{R}(X')$ and $\mathcal{R}(U')$ are disjoint.

Proof of (i):

$$X(X'X+U'U)^{-1}U' = X(X'X+U'U)^{-1}(X'X+U'U)A = XA = 0\,.$$

Proof of (ii):

$$\begin{aligned} & X'X(X'X+U'U)^{-1}(X'X+U'U-U'U) \\ = \quad & X'X - X'X(X'X+U'U)^{-1}U'U = X'X\,. \end{aligned}$$

Result (iii) follows on the same lines as result (ii).

Proof of (iv):

$$U(X'X+U'U)^{-1}U'u = U(X'X+U'U)^{-1}U'Ua = Ua = u$$

since $u \in \mathcal{R}(U)$.

A.14 Functions of Normally Distributed Variables

Let $x' = (x_1, \cdots, x_p)$ be a p-dimensional random vector. Then x is said to have a p-dimensional normal distribution with expectation vector μ and covariance matrix $\Sigma > 0$ if the joint density is

$$f(x;\mu,\Sigma) = \{(2\pi)^p|\Sigma|\}^{-\frac{1}{2}} \exp\left\{-\frac{1}{2}(x-\mu)'\Sigma^{-1}(x-\mu)\right\}.$$

In such a case we write $x \sim N_p(\mu,\Sigma)$.

Theorem A.82 *Assume* $x \sim N_p(\mu,\Sigma)$, *and* $A : p \times p$ *and* $b : p \times 1$ *nonstochastic. Then*

$$y = Ax + b \sim N_q(A\mu + b, A\Sigma A') \quad \textit{with } q = \text{rank}(A).$$

Theorem A.83 *If* $x \sim N_p(0, I)$, *then*

$$x'x \sim \chi^2_p$$

(central χ^2*-distribution with* p *degrees of freedom).*

Theorem A.84 *If $x \sim N_p(\mu, I)$, then*

$$x'x \sim \chi^2_p(\lambda)$$

has a noncentral χ^2-distribution with noncentrality parameter

$$\lambda = \mu'\mu = \sum_{i=1}^{p} \mu_i^2.$$

Theorem A.85 *If $x \sim N_p(\mu, \Sigma)$, then*

(i) $x'\Sigma^{-1}x \sim \chi^2_p(\mu'\Sigma^{-1}\mu)$.

(ii) $(x-\mu)'\Sigma^{-1}(x-\mu) \sim \chi^2_p$.

Proof: $\Sigma > 0 \;\Rightarrow\; \Sigma = \Sigma^{\frac{1}{2}}\Sigma^{\frac{1}{2}}$ with $\Sigma^{\frac{1}{2}}$ regular and symmetric. Hence, $\Sigma^{-\frac{1}{2}}x = y \sim N_p(\Sigma^{-\frac{1}{2}}\mu, I) \Rightarrow$

$$x'\Sigma^{-1}x = y'y \sim \chi^2_p(\mu'\Sigma^{-1}\mu)$$

and

$$(x-\mu)'\Sigma^{-1}(x-\mu) = (y - \Sigma^{-\frac{1}{2}}\mu)'(y - \Sigma^{-\frac{1}{2}}\mu) \sim \chi^2_p.$$

Theorem A.86 *If $Q_1 \sim \chi^2_m(\lambda)$ and $Q_2 \sim \chi^2_n$, and Q_1 and Q_2 are independent, then*

(i) *The ratio*

$$F = \frac{Q_1/m}{Q_2/n}$$

has a noncentral $F_{m,n}(\lambda)$-distribution.

(ii) *If $\lambda = 0$, then $F \sim F_{m,n}$ (the central F-distribution).*

(iii) *If $m = 1$, then $\sqrt{F}$ has a noncentral $t_n(\sqrt{\lambda})$-distribution or a central t_n-distribution if $\lambda = 0$.*

Theorem A.87 *If $x \sim N_p(\mu, I)$ and $A : p \times p$ is a symmetric, idempotent matrix with* $\text{rank}(A) = r$, *then*

$$x'Ax \sim \chi^2_r(\mu'A\mu).$$

Proof: We have $A = P\Lambda P'$ (Theorem A.30) and without loss of generality (Theorem A.61 (i)) we may write $\Lambda = \begin{pmatrix} I_r & 0 \\ 0 & 0 \end{pmatrix}$, that is, $P'AP = \Lambda$ with P orthogonal. Let $P = (\underset{p,r}{P_1} \;\; \underset{p,(p-r)}{P_2})$ and

$$P'x = y = \begin{pmatrix} y_1 \\ y_2 \end{pmatrix} = \begin{pmatrix} P_1'x \\ P_2'x \end{pmatrix}.$$

Therefore

$$\begin{aligned} y &\sim N_p(P'\mu, I_p) \quad \text{(Theorem A.82)} \\ y_1 &\sim N_r(P_1'\mu, I_r) \\ \text{and } y_1'y_1 &\sim \chi_r^2(\mu' P_1 P_1' \mu) \quad \text{(Theorem A.84).} \end{aligned}$$

As P is orthogonal, we have

$$\begin{aligned} A &= (PP')A(PP') = P(P'AP)P \\ &= (P_1\ P_2) \begin{pmatrix} I_r & 0 \\ 0 & 0 \end{pmatrix} \begin{pmatrix} P_1' \\ P_2' \end{pmatrix} = P_1 P_1', \end{aligned}$$

and therefore

$$x'Ax = x'P_1P_1'x = y_1'y_1 \sim \chi_r^2(\mu' A \mu).$$

Theorem A.88 *Let $x \sim N_p(\mu, I)$, $A : p \times p$ be idempotent of rank r, and $B : n \times p$ be any matrix. Then the linear form Bx is independent of the quadratic form $x'Ax$ if and only if $BA = 0$.*

Proof: Let P be the matrix as in Theorem A.87. Then $BPP'AP = BAP = 0$, as $BA = 0$ was assumed. Let $BP = D = (D_1, D_2) = (BP_1, BP_2)$, then

$$BPP'AP = (D_1, D_2) \begin{pmatrix} I_r & 0 \\ 0 & 0 \end{pmatrix} = (D_1, 0) = (0, 0),$$

so that $D_1 = 0$. This gives

$$Bx = BPP'x = Dy = (0, D_2) \begin{pmatrix} y_1 \\ y_2 \end{pmatrix} = D_2 y_2,$$

where $y_2 = P_2'x$. Since P is orthogonal and hence regular, we may conclude that all the components of $y = P'x$ are independent $\Rightarrow$ $Bx = D_2 y_2$ and $x'Ax = y_1'y_1$ are independent.

Theorem A.89 *Let $x \sim N_p(0, I)$ and A and B be idempotent $p \times p$-matrices with $\operatorname{rank}(A) = r$ and $\operatorname{rank}(B) = s$. Then the quadratic forms $x'Ax$ and $x'Bx$ are independently distributed if and only if $BA = 0$.*

Proof: If we use P from Theorem A.87 and set $C = P'BP$ (C symmetric), we get with the assumption $BA = 0$,

$$\begin{aligned} CP'AP &= P'BPP'AP \\ &= P'BAP = 0. \end{aligned}$$

Using

$$\begin{aligned} C &= \begin{pmatrix} P_1 \\ P_2 \end{pmatrix} B(P_1'\ P_2') \\ &= \begin{pmatrix} C_1 & C_2 \\ C_2' & C_3 \end{pmatrix} = \begin{pmatrix} P_1BP_1' & P_1BP_2' \\ P_2BP_1' & P_2BP_2' \end{pmatrix}, \end{aligned}$$

this relation may be written as

$$CP'AP = \begin{pmatrix} C_1 & C_2 \\ C_2' & C_3 \end{pmatrix} \begin{pmatrix} I_r & 0 \\ 0 & 0 \end{pmatrix} = \begin{pmatrix} C_1 & 0 \\ C_2' & 0 \end{pmatrix} = 0 .$$

Therefore, $C_1 = 0$ and $C_2 = 0$,

$$\begin{aligned} x'Bx &= x'(PP')B(PP')x \\ &= x'P(P'BP)P'x \\ &= x'PCP'x \\ &= (y_1', y_2') \begin{pmatrix} 0 & 0 \\ 0 & C_3 \end{pmatrix} \begin{pmatrix} y_1 \\ y_2 \end{pmatrix} = y_2' C_3 y_2 . \end{aligned}$$

As shown in Theorem A.87, we have $x'Ax = y_1'y_1$, and therefore the quadratic forms $x'Ax$ and $x'Bx$ are independent.

A.15 Differentiation of Scalar Functions of Matrices

Definition A.90 *If $f(X)$ is a real function of an $m \times n$-matrix $X = (x_{ij})$, then the partial differential of f with respect to X is defined as the $m \times n$-matrix of partial differentials $\partial f / \partial x_{ij}$:*

$$\frac{\partial f(X)}{\partial X} = \begin{pmatrix} \frac{\partial f}{\partial x_{11}} & \cdots & \frac{\partial f}{\partial x_{1n}} \\ \vdots & & \vdots \\ \frac{\partial f}{\partial x_{m1}} & \cdots & \frac{\partial f}{\partial x_{mn}} \end{pmatrix} .$$

Theorem A.91 *Let x be an n-vector and A be a symmetric $n \times n$-matrix. Then*

$$\frac{\partial}{\partial x} x'Ax = 2Ax.$$

Proof:

$$\begin{aligned} x'Ax &= \sum_{r,s=1}^{n} a_{rs} x_r x_s , \\ \frac{\partial f}{\partial x_i} x'Ax &= \sum_{\substack{s=1 \\ (s \neq i)}}^{n} a_{is} x_s + \sum_{\substack{r=1 \\ (r \neq i)}}^{n} a_{ri} x_r + 2a_{ii} x_i \\ &= 2 \sum_{s=1}^{n} a_{is} x_s \quad (\text{as } a_{ij} = a_{ji}) \\ &= 2a_i' x \quad (a_i'\text{: } i\text{th row vector of } A). \end{aligned}$$

According to Definition A.90, we get

$$\frac{\partial x'Ax}{\partial x} = \begin{pmatrix} \frac{\partial}{\partial x_1} \\ \vdots \\ \frac{\partial}{\partial x_n} \end{pmatrix} (x'Ax) = 2 \begin{pmatrix} a_1' \\ \vdots \\ a_n' \end{pmatrix} x = 2Ax.$$

Theorem A.92 *If x is an n-vector, y is an m-vector, and C an $n \times m$-matrix, then*

$$\frac{\partial}{\partial C} x'Cy = xy'.$$

Proof:

$$\begin{aligned} x'Cy &= \sum_{r=1}^{m} \sum_{s=1}^{n} x_s c_{sr} y_r, \\ \frac{\partial}{\partial c_{k\lambda}} x'Cy &= x_k y_\lambda \quad \text{(the } (k,\lambda)\text{th element of } xy'\text{)}, \\ \frac{\partial}{\partial C} x'Cy &= (x_k y_\lambda) = xy'. \end{aligned}$$

Theorem A.93 *Let x be a K-vector, A a symmetric $T \times T$-matrix, and C a $T \times K$-matrix. Then*

$$\frac{\partial x'C'}{\partial C} x'C'ACx = 2ACxx'.$$

Proof: We have

$$\begin{aligned} x'C' &= \left(\sum_{i=1}^{K} x_i c_{1i}, \cdots, \sum_{i=1}^{K} x_i c_{Ti} \right), \\ \frac{\partial}{\partial c_{k\lambda}} &= (0, \cdots, 0, x_\lambda, 0, \cdots, 0) \quad (x_\lambda \text{ is an element of the } k\text{th column}). \end{aligned}$$

Using the product rule yields

$$\frac{\partial}{\partial c_{k\lambda}} x'C'ACx = \left(\frac{\partial}{\partial c_{k\lambda}} x'C' \right) ACx + x'C'A \left(\frac{\partial}{\partial c_{k\lambda}} Cx \right).$$

Since

$$x'C'A = \left(\sum_{t=1}^{T} \sum_{i=1}^{K} x_i c_{ti} a_{t1}, \cdots, \sum_{t=1}^{T} \sum_{i=1}^{K} x_i c_{ti} a_{Tt} \right),$$

we get

$$\begin{aligned} x'C'A \left(\frac{\partial}{\partial c_{k\lambda}} Cx \right) &= \sum_{t,i} x_i x_\lambda c_{ti} a_{kt} \\ &= \sum_{t,i} x_i x_\lambda c_{ti} a_{tk} \qquad \text{(as } A \text{ is symmetric)} \end{aligned}$$

$$= \left(\frac{\partial}{\partial c_{k\lambda}} x'C'\right) ACx.$$

But $\sum_{t,i} x_i x_\lambda c_{ti} a_{tk}$ is just the (k, λ)-th element of the matrix $ACxx'$.

Theorem A.94 *Assume $A = A(x)$ to be an $n \times n$-matrix, where its elements $a_{ij}(x)$ are real functions of a scalar x. Let B be an $n \times n$-matrix, such that its elements are independent of x. Then*

$$\frac{\partial}{\partial x} \operatorname{tr}(AB) = \operatorname{tr}\left(\frac{\partial A}{\partial x} B\right).$$

Proof:

$$\begin{aligned} \operatorname{tr}(AB) &= \sum_{i=1}^{n} \sum_{j=1}^{n} a_{ij} b_{ji}, \\ \frac{\partial}{\partial x} \operatorname{tr}(AB) &= \sum_{i} \sum_{j} \frac{\partial a_{ij}}{\partial x} b_{ji} \\ &= \operatorname{tr}\left(\frac{\partial A}{\partial x} B\right), \end{aligned}$$

where $\partial A / \partial x = (\partial a_{ij} / \partial x)$.

Theorem A.95 *For the differentials of the trace we have the following rules:*

	y	$\partial y / \partial X$
(i)	$\operatorname{tr}(AX)$	A'
(ii)	$\operatorname{tr}(X'AX)$	$(A + A')X$
(iii)	$\operatorname{tr}(XAX)$	$X'A + A'X'$
(iv)	$\operatorname{tr}(XAX')$	$X(A + A')$
(v)	$\operatorname{tr}(X'AX')$	$AX' + X'A$
(vi)	$\operatorname{tr}(X'AXB)$	$AXB + A'XB'$

Differentiation of Inverse Matrices

Theorem A.96 *Let $T = T(x)$ be a regular matrix, such that its elements depend on a scalar x. Then*

$$\frac{\partial T^{-1}}{\partial x} = -T^{-1} \frac{\partial T}{\partial x} T^{-1}.$$

Proof: We have $T^{-1}T = I$, $\partial I / \partial x = 0$, and

$$\frac{\partial (T^{-1}T)}{\partial x} = \frac{\partial T^{-1}}{\partial x} T + T^{-1} \frac{\partial T}{\partial x} = 0.$$

Theorem A.97 *For nonsingular X, we have*

$$\frac{\partial \operatorname{tr}(AX^{-1})}{\partial X} = -(X^{-1}AX^{-1})',$$

$$\frac{\partial \operatorname{tr}(X^{-1}AX^{-1}B)}{\partial X} = -(X^{-1}AX^{-1}BX^{-1} + X^{-1}BX^{-1}AX^{-1})'.$$

Proof: Use Theorems A95 and A96 and the product rule.

Differentiation of a Determinant

Theorem A.98 *For a nonsingular matrix Z, we have*

(i) $\frac{\partial}{\partial Z}|Z| = |Z|(Z')^{-1}$.

(ii) $\frac{\partial}{\partial Z}log|Z| = (Z')^{-1}$.

A.16 Miscellaneous Results, Stochastic Convergence

Theorem A.99 (Kronecker product) *Let $A : m \times n = (a_{ij})$ and $B : p \times q = (b_{rs})$ be any matrices. Then the Kronecker product of A and B is defined as*

$$\underset{mp,nq}{C} = \underset{m,n}{A} \otimes \underset{p,q}{B} = \begin{pmatrix} a_{11}B & a_{12}B & \cdots & a_{1n}B \\ \vdots & \vdots & & \cdots \\ a_{m1}B & a_{m2}B & \cdots & a_{mn}B \end{pmatrix},$$

and the following rules hold:

(i) $c(A \otimes B) = (cA) \otimes B = A \otimes (cB)$ *(c a scalar),*

(ii) $A \otimes (B \otimes C) = (A \otimes B) \otimes C$,

(iii) $A \otimes (B + C) = (A \otimes B) + (A \otimes C)$,

(iv) $(A \otimes B)' = A' \otimes B'$.

Theorem A.100 (Chebyschev's inequality) *For any n-dimensional random vector X and a given scalar $\epsilon > 0$, we have*

$$P\{|X| \geq \epsilon\} \leq \frac{\mathrm{E}\,|X|^2}{\epsilon^2}.$$

Proof: Let $F(x)$ be the joint distribution function of $X = (x_1, \ldots, x_n)$. Then

$$\begin{aligned} \mathrm{E}\,|x|^2 &= \int |x|^2 dF(x) \\ &= \int_{\{x:|x|\geq\epsilon\}} |x|^2 dF(x) + \int_{\{x:|x|<\epsilon\}} |x|^2 dF(x) \\ &\geq \epsilon^2 \int_{\{x:|x|\geq\epsilon\}} dF(x) = \epsilon^2 P\{|x| \geq \epsilon\}. \end{aligned}$$

Definition A.101 *Let* $\{x(t)\}$, $t = 1, 2, \ldots$ *be a multivariate stochastic process.*

(i) *Weak convergence: If*

$$\lim_{t \to \infty} P\{|x(t) - \tilde{x}| \geq \delta\} = 0 ,$$

where $\delta > 0$ *is any given scalar and* $\tilde{x}$ *is a finite vector, then* $\tilde{x}$ *is called the probability limit of* $\{x(t)\}$*, and we write*

$$p \lim x = \tilde{x} .$$

(ii) *Strong convergence: Assume that* $\{x(t)\}$ *is defined on a probability space* (Ω, Σ, P)*. Then* $\{x(t)\}$ *is said to be strongly convergent to* $\tilde{x}$*, that is,*

$$\{x(t)\} \to \tilde{x} \quad \textit{almost surely (a.s.)}$$

if there exists a set $T \in \Sigma$, $P(T) = 0$, *and* $x_\omega(t) \to \tilde{x}_\omega$, *as* $T \to \infty$, *for each* $\omega \in \Omega - T$

Theorem A.102 (Slutsky's theorem) *Using Definition A.101, we have*

(i) *if* $p \lim x = \tilde{x}$, *then* $\lim_{t \to \infty} \mathrm{E}\{x(t)\} = \bar{\mathrm{E}}(x) = \tilde{x}$,

(ii) *if* c *is a vector of constants, then* $p \lim c = c$,

(iii) *(Slutsky's theorem) if* $p \lim x = \tilde{x}$ *and* $y = f(x)$ *is any continuous vector function of* x, *then* $p \lim y = f(\tilde{x})$,

(iv) *if* A *and* B *are random matrices, then the following limits exist:*

$$p \lim(AB) = (p \lim A)(p \lim B)$$

and

$$p \lim(A^{-1}) = (p \lim A)^{-1} ,$$

(v) *if* $p \lim \left[\sqrt{T}(x(t) - Ex(t))\right]' \left[\sqrt{T}(x(t) - Ex(t))\right] = V$, *then the asymptotic covariance matrix is*

$$\bar{\mathrm{V}}(x, x) = \bar{\mathrm{E}}\left[x - \bar{\mathrm{E}}(x)\right]' \left[x - \bar{\mathrm{E}}(x)\right] = T^{-1} V .$$

Definition A.103 *If* $\{x(t)\}, t = 1, 2, \ldots$ *is a multivariate stochastic process satisfying*

$$\lim_{t \to \infty} \mathrm{E}\,|x(t) - \tilde{x}|^2 = 0 ,$$

then $\{x(t)\}$ *is called convergent in the quadratic mean, and we write*

$$\text{l.i.m.}\, x = \tilde{x} .$$

Theorem A.104 *If* $\text{l.i.m.}\, x = \tilde{x}$, *then* $p \lim x = \tilde{x}$.

Proof: Using Theorem A.100 we get

$$0 \leq \lim_{t\to\infty} P(|x(t) - \tilde{x}| \geq \epsilon) \leq \lim_{t\to\infty} \frac{\mathrm{E}\,|x(t) - \tilde{x}|^2}{\epsilon^2} = 0\,.$$

Theorem A.105 *If* l.i.m. $(x(t) - \mathrm{E}\,x(t)) = 0$ *and* $\lim_{t\to\infty} \mathrm{E}\,x(t) = c$, *then* $p\lim x(t) = c$.

Proof:

$$\begin{aligned}
&\lim_{t\to\infty} P(|x(t) - c| \geq \epsilon) \\
&\qquad \leq \epsilon^{-2} \lim_{t\to\infty} \mathrm{E}\,|x(t) - c|^2 \\
&\qquad = \epsilon^{-2} \lim_{t\to\infty} \mathrm{E}\,\left|x(t) - \mathrm{E}\,x(t) + \mathrm{E}\,x(t) - c\right|^2 \\
&\qquad = \epsilon^{-2} \lim_{t\to\infty} \mathrm{E}\,\left|x(t) - \mathrm{E}\,x(t)\right|^2 + \epsilon^{-2} \lim_{t\to\infty} \left|\mathrm{E}\,x(t) - c\right|^2 \\
&\qquad\quad + 2\epsilon^{-2} \lim_{t\to\infty} \left\{(\mathrm{E}\,x(t) - c)'(x(t) - \mathrm{E}\,x(t))\right\} \\
&\qquad = 0\,.
\end{aligned}$$

Theorem A.106 l.i.m. $x = c$ *if and only if*

$$\text{l.i.m.}\,\bigl(x(t) - \mathrm{E}\,x(t)\bigr) = 0 \text{ and } \lim_{t\to\infty} \mathrm{E}\,x(t) = c\,.$$

Proof: As in Theorem A.105, we may write

$$\begin{aligned}
\lim_{t\to\infty} \mathrm{E}\,\left|x(t) - c\right|^2 &= \lim_{t\to\infty} \mathrm{E}\,\left|x(t) - \mathrm{E}\,x(t)\right|^2 + \lim_{t\to\infty} \left|\mathrm{E}\,x(t) - c\right|^2 \\
&\quad + 2 \lim_{t\to\infty} \mathrm{E}\,\bigl(\mathrm{E}\,x(t) - c\bigr)'\bigl(x(t) - \mathrm{E}\,x(t)\bigr) = 0\,.
\end{aligned}$$

Theorem A.107 *Let $x(t)$ be an estimator of a parameter vector θ. Then we have the result*

$$\lim_{t\to\infty} \mathrm{E}\,x(t) = \theta \quad \text{if} \quad \text{l.i.m.}\,(x(t) - \theta) = 0\,.$$

That is, $x(t)$ is an asymptotically unbiased estimator for θ if $x(t)$ converges to θ in the quadratic mean.

Proof: Use Theorem A.106.

Theorem A.108 *Let $V : p \times p$ and n.n.d. and $X : p \times m$ matrices. Then one choice of the g-inverse of*

$$\begin{pmatrix} V & X \\ X' & 0 \end{pmatrix}$$

is

$$\begin{pmatrix} C_1 & C_2 \\ C_2' & -C_4 \end{pmatrix}$$

where, with $T = V + XX'$,

$$\begin{aligned} C_1 &= T - T^- X(X'T^- X)^- X'T^- \\ C_2' &= (X'T^- X)^- X'T^- \\ -C_4 &= (X'T^- X)^- (X'T^- X - I) \end{aligned}$$

For details, see Rao (1989).

Appendix B

Tables

TABLE B.1. Quantiles of the χ^2-distribution

	Level of significance α					
df	0.99	0.975	0.95	0.05	0.025	0.01
1	0.0001	0.001	0.004	3.84	5.02	6.62
2	0.020	0.051	0.103	5.99	7.38	9.21
3	0.115	0.216	0.352	7.81	9.35	11.3
4	0.297	0.484	0.711	9.49	11.1	13.3
5	0.554	0.831	1.15	11.1	12.8	15.1
6	0.872	1.24	1.64	12.6	14.4	16.8
7	1.24	1.69	2.17	14.1	16.0	18.5
8	1.65	2.18	2.73	15.5	17.5	20.1
9	2.09	2.70	3.33	16.9	19.0	21.7
10	2.56	3.25	3.94	18.3	20.5	23.2
11	3.05	3.82	4.57	19.7	21.9	24.7
12	3.57	4.40	5.23	21.0	23.3	26.2
13	4.11	5.01	5.89	22.4	24.7	27.7
14	4.66	5.63	6.57	23.7	26.1	29.1
15	5.23	6.26	7.26	25.0	27.5	30.6
16	5.81	6.91	7.96	26.3	28.8	32.0
17	6.41	7.56	8.67	27.6	30.2	33.4
18	7.01	8.23	9.39	28.9	31.5	34.8
19	7.63	8.91	10.1	30.1	32.9	36.2
20	8.26	9.59	10.9	31.4	34.2	37.6
25	11.5	13.1	14.6	37.7	40.6	44.3
30	15.0	16.8	18.5	43.8	47.0	50.9
40	22.2	24.4	26.5	55.8	59.3	63.7
50	29.7	32.4	34.8	67.5	71.4	76.2
60	37.5	40.5	43.2	79.1	83.3	88.4
70	45.4	48.8	51.7	90.5	95.0	100.4
80	53.5	57.2	60.4	101.9	106.6	112.3
90	61.8	65.6	69.1	113.1	118.1	124.1
100	70.1	74.2	77.9	124.3	129.6	135.8

TABLE B.2. Quantiles of the F_{df_1,df_2}-distribution with df_1 and df_2 degrees of freedom ($\alpha = 0.05$)

	df_1								
df_2	1	2	3	4	5	6	7	8	9
1	161	200	216	225	230	234	237	239	241
2	18.51	19.00	19.16	19.25	19.30	19.33	19.36	19.37	19.38
3	10.13	9.55	9.28	9.12	9.01	8.94	8.88	8.84	8.81
4	7.71	6.94	6.59	6.39	6.26	6.16	6.09	6.04	6.00
5	6.61	5.79	5.41	5.19	5.05	4.95	4.88	4.82	4.78
6	5.99	5.14	4.76	4.53	4.39	4.28	4.21	4.15	4.10
7	5.59	4.74	4.35	4.12	3.97	3.87	3.79	3.73	3.68
8	5.32	4.46	4.07	3.84	3.69	3.58	3.50	3.44	3.39
9	5.12	4.26	3.86	3.63	3.48	3.37	3.29	3.23	3.18
10	4.96	4.10	3.71	3.48	3.33	3.22	3.14	3.07	3.02
11	4.84	3.98	3.59	3.36	3.20	3.09	3.01	2.95	2.90
12	4.75	3.88	3.49	3.26	3.11	3.00	2.92	2.85	2.80
13	4.67	3.80	3.41	3.18	3.02	2.92	2.84	2.77	2.72
14	4.60	3.74	3.34	3.11	2.96	2.85	2.77	2.70	2.65
15	4.54	3.68	3.29	3.06	2.90	2.79	2.70	2.64	2.59
20	4.35	3.49	3.10	2.87	2.71	2.60	2.52	2.45	2.40
30	4.17	3.32	2.92	2.69	2.53	2.42	2.34	2.27	2.21

TABLE B.3. Quantiles of the F_{df_1,df_2}-distribution with df_1 and df_2 degrees of freedom ($\alpha = 0.05$)

	df_1							
df_2	10	11	12	14	16	20	24	30
1	242	243	244	245	246	248	249	250
2	19.39	19.40	19.41	19.42	19.43	19.44	19.45	19.46
3	8.78	8.76	8.74	8.71	8.69	8.66	8.64	8.62
4	5.96	5.93	5.91	5.87	5.84	5.80	5.77	5.74
5	4.74	4.70	4.68	4.64	4.60	4.56	4.53	4.50
6	4.06	4.03	4.00	3.96	3.92	3.87	3.84	3.81
7	3.63	3.60	3.57	3.52	3.49	3.44	3.41	3.38
8	3.34	3.31	3.28	3.23	3.20	3.15	3.12	3.08
9	3.13	3.10	3.07	3.02	2.98	2.93	2.90	2.86
10	2.97	2.94	2.91	2.86	2.82	2.77	2.74	2.70
11	2.86	2.82	2.79	2.74	2.70	2.65	2.61	2.57
12	2.76	2.72	2.69	2.64	2.60	2.54	2.50	2.46
13	2.67	2.63	2.60	2.55	2.51	2.46	2.42	2.38
14	2.60	2.56	2.53	2.48	2.44	2.39	2.35	2.31
15	2.55	2.51	2.48	2.43	2.39	2.33	2.29	2.25
20	2.35	2.31	2.28	2.23	2.18	2.12	2.08	2.04
30	2.16	2.12	2.00	2.04	1.99	1.93	1.89	1.84

Appendix C
Software for Linear Regression Models

This chapter describes computer programs that support estimation of regression models and model diagnostics (the description skips aspects that don't relate to regression models). Sections C.1 and C.2 describe available software. Section C.3 lists some sources that might be of interest to the user who has access to the Internet.

C.1 Software

Available statistical software can be divided into roughly three categories, but the categorization is not strict; several programs may fall into more than one group, and current development shows that the software that falls into one of the first two categories is often extended in the other direction as well.

- Statistical programming languages. These are programming languages that have special support for statistical problems, such as built-in datatypes for matrices or special statistical functions. These packages are generally more extensible than the members of the following group in that the user can supply code for new procedures that are not available in the base system. Prominent examples are Gauss, S-plus, Matlab, Xlisp-Stat, Minitab, and SAS.
- Statistical software with a graphical user interface. These programs allow the user to analyze models interactively. Dialogues allow specification of models and selection of different model-selection approaches.

These tools are not extensible unless some kind of programming language is also provided. This loss in flexibility is opposed by the user interface that makes the tool easier to use. Examples are SPSS, Systat, SAS, S-plus (the Windows versions), JMP, Statistica, and STATA.

- Special-purpose software. These are smaller packages that fall in one of the above categories with the difference that they provide methods only for a certain class of models. They often originate from research projects and cover the work done there (MAREG, R-Code extensions/macro packages for Xlisp-Stat, SAS, etc.). The programs shown here are meant only as examples; it is difficult to give complete coverage, which in addition would have to be updated frequently.

The following lists of features are taken from the documentation of the respective programs and cover only the basic systems (i.e., third party extensions available are not covered).

Gauss

Available for DOS, OS/2, Windows, Unix.
Information under `http://www.aptech.com/`. Gauss is a programming language especially designed to handle matrices.

Linear Regression: The linear regression module is a set of procedures for estimating single equations or a simultaneous system of equations. Constraints on coefficients can be incorporated. Two-stage least squares, three-stage least squares, and seemingly unrelated regression are available.

Gauss calculates heteroscedastic-consistent standard errors, and performs both influence and collinearity diagnostics inside the ordinary least squares routine. Performs multiple linear hypothesis testing with any form.

Loglinear Analysis: The estimation is based on the assumption that the cells of the K-way table are independent Poisson random variables. The parameters are found by applying the Newton-Raphson method using an algorithm found in Agresti (1990). User-defined design matrices can be incorporated.

S-plus

Available for DOS, Windows, Unix, Linux.
Information under `http://www.mathsoft.com/splus.html/`. S-plus is based on the S language (Becker, Chambers, and Wilks, 1988)

Linear Regression: Linear regression includes basic linear regression, polynomial regression, least-trimmed-squares regression, constrained regression, logistic regression, generalized linear models, linear mixed-effect models, minimum absolute-residual regression, and robust MM regression.

Nonlinear Regression and Maximum Likelihood: Nonlinear regression, nonlinear maximum likelihood, constrained nonlinear regression, nonlinear mixed effects.

Nonparametric Regression: Generalized additive models, local regression (loess), projection pursuit, ACE, AVAS, and tree-based models.

ANOVA: Fixed effects, random effects, rank tests, repeated measures, variance components, split-plot models, MANOVA, and multiple comparisons.

(X)Lisp-Stat

Available for Macintosh, UNIX systems running X11, Windows.
Information under `http://www.stat.umn.edu/~luke/xls/xlsinfo/`.
This package is by Luke Tierney (free).

(From the documentation): Lisp-Stat is an extensible statistical computing environment for data analysis, statistical instruction, and research, with an emphasis on providing a framework for exploring the use of dynamic graphical methods. Extensibility is achieved by basing Lisp-Stat on the Lisp language, in particular on a subset of Common Lisp.

A portable window system interface forms the basis of a dynamic graphics system that is designed to work identically in a number of different graphical user interface environments, such as the Macintosh operating system, the X window system, and Microsoft Windows.

The object-oriented programming system is also used as the basis for statistical model representations, such as linear and nonlinear regression models and generalized linear models. Many aspects of the system design were motivated by the S language.

Minitab

Available for Windows, Macintosh and for Mainframes and Workstations.
Information under `http://www.minitab.com/`.

Regression Analysis: Regression analysis includes simple and multiple linear regression, model selection using stepwise or best-subsets regression, residual plots, identification of unusual observations, model diagnostics, and prediction/confidence intervals for new observations.

Logistic Regression: Binary, ordinal, or normal data; diagnostic plots, polynomial regression, with or without log transforms.

ANOVA: General linear model for balanced, unbalanced and nested designs; fixed and random effects; and unbalanced nested designs. Multiple factor ANOVA for balanced models; fixed and random effects; multiple comparisons; multivariate analysis of variance; analysis of fully nested designs; sequential sum of squares; identification of unusual observations; model diagnostics; residual, main effects, and interaction plots; and tests of homogeneity of variances.

SAS

Available for Windows, Unix.
Information under `http://www.sas.com/`.

Regression Analysis: Regression analysis includes ridge regression; linear regression; model-selection techniques (backwards, forwards, stepwise, based on R-squared); diagnostics; hypothesis tests; partial regression leverage plots; outputs predicted values and residuals; graphics device plots; response surface regression; nonlinear regression; derivative-free; steepest-descent; Newton, modified Gauss-Newton, Marquardt and DUD methods; linear models with optimal nonlinear transformation; and partial least squares.

Analysis of Variance: ANOVA for balanced data; general linear models; unbalanced data; analysis of covariance; response-surface models; weighted regression; polynomial regression; MANOVA; repeated measurements analysis; least squares means; random effects; estimate linear functions of the parameters; test linear functions of the parameters; multiple comparison of means; homogeneity of variance testing; mixed linear models; fixed and random effects; REML; maximum likelihood, and MIVQUE0 estimation methods; least-squares means and differences; sampling-based Bayesian analysis; different covariance structures (compound symmetry, unstructured, AR(1), Toeplitz, heterogeneous AR(1), Huynh-Feldt); multiple comparison of least-squares means; repeated measurements analysis; variance components; nested models; and lattice designs.

SPSS

Available for DOS, Windows, Unix.
Information under `http://www.spss.com/`.

Regression: Multiple linear regression, curve estimation, weighted least squares regression, two-stage least squares, logistic regression, probit models, optimal scaling, nonlinear regression, model-selection techniques (backward, forward, stepwise), hypothesis tests, predicted values and residuals, residual plots, and collinearity diagnostics.

ANOVA: General linear model: general factorial, multivariate, repeated measures and variance components covers the the ANOVA and ANOVA models.

Missing Values: SPSS also provides a missing-values module. Patterns of missing data can be displayed, and t-tests and cross-tabulation of categorical and indicator variables can be used to investigate the missing mechanism. Esitmation of missing values is available via the EM algorithm, regression estimation, and listwise or pairwise estimation.

Systat

Available for Windows.
Information under `http://www.spss.com/software/science/systat/`.

Regression: Classification and regression trees, design of experiments, general linear model, linear regression, logistic regression, loglinear models, nonlinear regression, probit, and two-stage least squares.

ANOVA: One and two way ANOVA, post hoc tests, mixed models, repeated measures ANOVA and MANOVA.

JMP

Available for Windows and Macintosh.
Information under `http://www.jmpdiscovery.com/`

JMP (by SAS) is an environment for statistical visualization and exploratory data analysis. Analysis of variance and multiple regression and nonlinear fitting are offered.

BMDP

For DOS BMDP/Classic for DOS, see `http://www.spss.com/software/science/Bmdp/`. For BMDP Professional, with its Windows interface, see `http://statsol.ie/bmdp.html`.

Regression: Simple linear, multiple linear, stepwise, regression on principal components, ridge regression, and all possible subsets regression.

Nonlinear regression: Derivative-free non linear regression, polynomial regression, stepwise logistic regression, and polychotomous logistic regression.

Mathematical Software

Mathematical software such as Maple (see `http://www.maplesoft.com/`), Mathematica (see `http://mathematica.com/`), or Matlab (see `http://`

www.mathworks.com/) often comes with libraries for statistical problems. For example, Matlab's statistics library contains functions for linear models including regression diagnostics and ridge regression.

C.2 Special-Purpose Software

MAREG/WinMAREG

Available for DOS/Windows and Unix, Linux (MAREG only) (free software).
Information under http://www.stat.uni-muenchen.de/~andreas/mareg/winmareg.html.

MAREG is a tool for estimating marginal regression models. Marginal regression models are an extension of the well-known regression models to the case of correlated observations. MAREG currently handles binary, categorical, and continuous data with several link functions. Although intended for the analysis of correlated data, uncorrelated data can be analyzed. Two different approaches for these problems—generalized estimating equations and maximum-likelihood methods—are supplied. Handling of missing data is also provided.

WinMAREG is a Windows user interface for MAREG, allowing method specification, selection and coding of variables, treatment of missing values, and selection of general settings.

R-Code

The R-Code is based on Xlsip-Stat.
Information under http://stat.umn.edu/~rcode/.

The software comes with the book by Cook and Weisberg (1994), which describes the concepts of structure in regression, diagnostics, and visualization, and gives a tutorial to the software.

GLIM

Available for DOS and Unix.
Information under http://www.nag.co.uk/stats/GDGE.html.

GLIM is a specialized, interactive statistical modeling package that allows the user to fit a variety of statistical models developed by the GLIM Working Party of the Royal Statistical Society. It has a concise command language that allows the user to fit and refit simple or complex models iteratively. GLIM is better run interactively because model fitting is largely an iterative process, but GLIM may also be run noninteractively. Linear regression models, models for the analysis of designed experiments, log-linear models for contingency tables, probit analysis, and logistic regression are available.

C.3 Resources

StatLib Server

`http://lib.stat.cmu.edu/`. StatLib is a system for distributing statistical software, datasets, and information by electronic mail, FTP, and WWW, hosted by the Department of Statistics at Carnegie Mellon University. Several sites around the world serve as full or partial mirrors to StatLib.

Data and Story Library (DASL)

`http://lib.stat.cmu.edu/DASL/`. The DASL is a repository of stories and datafiles that illustrate various concepts in statistics and data analysis. The stories are organized both by content area and by methodology employed. The data can be downloaded as a space- or tab-delimited table of text, easily read by most statistics programs.

Statistics on the Web

`http://execpc.com/~helberg/statistics.html`. A very nice page and a good starting point covering several aspects of statistics that can be found on the Web. Organizations, institutes, educational resources, publications, and software are listed.

References

Afifi, A. A., and Elashoff, R. M. (1967). Missing observations in multivariate statistics: Part II: Point estimation in simple linear regression, *Journal of the American Statistical Association* **62**: 10–29.

Agresti, A. (1990). *Categorical Data Analysis*, Wiley, New York.

Aitchison, J. (1966). Expected-cover and linear-utility tolerance intervals, *Journal of the Royal Statistical Society, Series B* **28**: 57–62.

Aitchison, J., and Dunsmore, I. R. (1968). Linear-loss interval estimation for location and scale parameters, *Biometrika* **55**: 141–148.

Amemiya, T. (1982). Two stages least absolute deviations estimates, *Econometrica* **50**: 689–711.

Amemiya, T. (1983). Partially generalized least squares and two-stage least squares estimators, *Journal of Econometrics* **23**: 275–283.

Amemiya, T. (1985). *Advanced Econometrics*, Basil Blackwell, Oxford.

Andrews, D. F., and Pregibon, D. (1978). Finding outliers that matter, *Journal of the Royal Statistical Society, Series B* **40**: 85–93.

Bai, Z. D., Chen, X. R., Miao, B. Q., and Rao, C. R. (1988). Asymptotic theory of least distances estimates in multivariate analysis, *Technical Report 9*, Center for Multivariate Analysis, The Pennsylvania State University, State College.

Bai, Z. D., Chen, X. R., Wu, Y., and Zhao, L. C. (1987). Asymptotic normality of minimum L_1-norm estimates in linear model, *Technical Report*, Center for Multivariate Analysis, The Pennsylvania State University, State College.

Bai, Z. D., Rao, C. R., and Wu, Y. (1989). Unified theory of m-estimation of linear regression parameters, *Technical Report 4*, Center for Multivariate Analysis, The Pennsylvania State University, State College.

Baksalary, J. K. (1988). Criteria for the equality between ordinary least squares and best linear unbiased estimators under certain linear models, *Canadian Journal of Statistics* **16**: 97–102.

Baksalary, J. K., and Kala, R. (1983). Partial orderings between matrices one of which is of rank one, *Bulletin of the Polish Academy of Science, Mathematics* **31**: 5–7.

Baksalary, J. K., Kala, R., and Klaczynski, K. (1983). The matrix inequality $M \geq B^*MB$, *Linear Algebra and Its Applications* **54**: 77–86.

Baksalary, J. K., Liski, E. P., and Trenkler, G. (1989). Mean square error matrix improvements and admissibility of linear estimators, *Journal of Statistical Planning and Inference* **23**: 312–325.

Baksalary, J. K., Schipp, B., and Trenkler, G. (1992). Some further results on hermitian matrix inequalities, *Linear Algebra and Its Applications* **160**: 119–129.

Balestra, P. (1983). A note on Amemiya's partially generalized least squares, *Journal of Econometrics* **23**: 285–290.

Bartels, R., and Fiebig, D. G. (1991). A simple characterization of seemingly unrelated regression models in which OLSE is BLUE, *The American Statistician* **45**: 137–140.

Bartlett, M. S. (1937). Some examples of statistical methods of research in agriculture and applied botany, *Journal of the Royal Statistical Society, Series B* **4**: 137–170.

Basawa, L. V., and Koul, H. L. (1988). Large-sample statistics based on quadratic dispersion, *International Statistical Review* **56**: 199–219.

Bassett, G., and Koenker, R. (1978). Asymptotic theory of least absolute error regression, *Journal of the American Statistical Association* **73**: 618–622.

Becker, R. A., Chambers, J. M., and Wilks, A. R. (1988). *The new S language*, Wadsworth & Brooks, Pacific Grove, CA.

Beckman, R. J., and Trussel, H. J. (1974). The distribution of an arbitrary Studentized residual and the effects of updating in multiple regression, *Journal of the American Statistical Association* **69**: 199–201.

Bekker, P. A., and Neudecker, H. (1989). Albert's theorem applied to problems of efficiency and MSE superiority, *Statistica Neerlandica* **43**: 157–167.

Belsley, D. A., Kuh, E., and Welsch, R. E. (1980). *Regression Diagnostics*, Wiley, New York.

Ben-Israel, A., and Greville, T. N. E. (1974). *Generalized Inverses: Theory and Applications*, Wiley, New York.

Bibby, J. M., and Toutenburg, H. (1977). *Prediction and Improved Estimation in Linear Models*, Wiley, New York.

Bickel, P. J. (1975). One-step Huber estimates in the linear model, *Journal of the American Statistical Association* **70**: 428–434.

Birkes, D., and Dodge, Y. (1993). *Alternative Methods of Regression*, Wiley, New York.

Bloomfield, P., and Steiger, W. L. (1983). *Least Absolute Deviations: Theory, Applications, and Algorithms*, Birkhäuser, Boston.

Boscovich, R. J. (1757). De litteraria expeditione per pontificiam ditionem, et synopsis amplioris operis, *Bononiensi Scientiarum et Artum Instituto atque Academia Commentarii* **4**: 353–396.

Brown, B. M. (1983). Statistical uses of spatial median, *Journal of the Royal Statistical Society, Series B* **45**: 25–30.

Buck, S. F. (1960). A method of estimation of missing values in multivariate data suitable for use with an electronic computer, *Journal of the Royal Statistical Society, Series B* **22**: 302–307.

Buckley, J., and James, I. (1979). Linear regression with censored data, *Biometrika* **66**: 429–436.

Charnes, A., and Cooper, W. W. (1961). *Management Models and Industrial Applications of Linear Programming*, Wiley, New York.

Charnes, A., Cooper, W. W., and Ferguson, R. O. (1955). Optimal estimation of executive compensation by linear programming, *Management Science* **1**: 138–151.

Chatterjee, S., and Hadi, A. S. (1988). *Sensitivity Analysis in Linear Regression*, Wiley, New York.

Chen, X. R., and Wu, Y. (1988). Strong consistency of M-estimates in linear models, *Journal of Multivariate Analysis* **27**: 116–130.

Cheng, B., and Titterington, D. M. (1994). Neural networks: A review from a statistical perspective, *Statistical Science* **9**: 2–54.

Chipman, J. S., and Rao, M. M. (1964). The treatment of linear restrictions in regression analysis, *Econometrica* **32**: 198–209.

Chow, Y. S. (1966). Some convergence theorems for independent random variables, *Annals of Mathematical Statistics* **37**: 1482–1493.

Cochrane, D., and Orcutt, G. H. (1949). Application of least squares regression to relationships containing autocorrelated error terms, *Journal of the American Statistical Association* **44**: 32–61.

Cohen, J., and Cohen, P. (1983). *Applied Multiple Regression/Correlation Analysis for the Behavioral Sciences*, Lawrence Erlbaum, Hillsdale, NJ.

Cook, R. D. (1977). Detection of influential observations in linear regression, *Technometrics* **19**: 15–18.

Cook, R. D., and Weisberg, S. (1982). *Residuals and Influence in Regression*, Chapman and Hall, New York.

Cook, R. D., and Weisberg, S. (1989). Regression diagnostics with dynamic graphics, *Technometrics* **31**: 277–291.

Cook, R. D., and Weisberg, S. (1994). *An Introduction to Regression Graphics*, Wiley, New York.

Cox, D. R. (1972). Regression models and life-tables (with discussion), *Journal of the Royal Statistical Society, Series B* **34**: 187–202.

Cox, D. R., and Snell, E. J. (1968). A general definition of residuals, *Journal of the Royal Statistical Society, Series B* **30**: 248–275.

Dagenais, M. G. (1973). The use of incomplete observations in multiple regression analysis: A generalized least squares approach, *Journal of Econometrics* **1**: 317–328.

Deming, W. E., and Stephan, F. F. (1940). On a least squares adjustment of sampled frequency table when the expected marginal totals are known, *Annals of Mathematical Statistics* **11**: 427–444.

Dempster, A. P., Laird, N. M., and Rubin, D. B. (1977). Maximum likelihood from incomplete data via the EM algorithm, *Journal of the Royal Statistical Society, Series B* **43**: 1–22.

Dhrymes, P. J. (1974). *Econometrics*, Springer, New York.

Dhrymes, P. J. (1978). *Indroductory Econometrics*, Springer, New York.

Diggle, P. J., Liang, K.-Y., and Zeger, S. L. (1994). *Analysis of Longitudinal Data*, Chapman and Hall, London.

Doksum, K. A., and Gasko, M. (1990). On a correspondence between models in binary regression analysis and in survival analysis, *International Statistical Review* **58**: 243–252.

Draper, N. R., and Smith, H. (1966). *Applied Regression Analysis*, Wiley, New York.

Dube, M., Srivastava, V. K., Toutenburg, H., and Wijekoon, P. (1991). Stein-rule estimators under inclusion of superfluous variables in linear regression models, *Communications in Statistics, Part A—Theory and Methods* **20**: 2009–2022.

Dufour, J. M. (1989). Nonlinear hypotheses, inequality restrictions and non-nested hypotheses: Exact simultaneous tests in linear regression, *Econometrica* **57**: 335–355.

Dupaková, J. (1987). Asymptotic properties of restricted L_1-estimates of regression, *in* Y. Dodge (ed.), *Statistical Data Analysis Based on the L_1-Norm and Related Methods*, North Holland, Amsterdam, pp. 263–274.

Dupaková, J., and Wets, R. (1988). Asymptotic behavior of statistical estimators and optimal solutions of stochastic optimization problems, *Annals of Statistics* **16**: 1517–1549.

Durbin, J. (1953). A note on regression when there is extraneous information about one of the coefficients, *Journal of the American Statistical Association* **48**: 799–808.

Durbin, J., and Watson, G. S. (1950). Testing for serial correlation in least squares regression (I), *Biometrika* **37**: 409–428.

Durbin, J., and Watson, G. S. (1951). Testing for serial correlation in least squares regression (II), *Biometrika* **38**: 159–178.

Dwivedi, T. D., and Srivastava, V. K. (1978). Optimality of least squares in the seemingly unrelated regression equation model, *Journal of Econometrics* **7**: 391–395.

Eckart, G., and Young, G. (1936). The approximation of one matrix by another of lower rank, *Psychometrica* **1**: 211–218.

Efron, B. (1979). Bootstrap methods: Another look at the jackknife, *Annals of Statistics* **7**: 1–26.

Evans, R. W., Cooley, P. C., and Piserchia, P. V. (1979). A test for evaluating missing data imputation procedures, *Proceedings of the Social Statistics Section*, pp. 469–474.

Fahrmeir, L., and Hamerle, A. (eds.) (1984). *Multivariate statistische Verfahren*, de Gruyter, Berlin.

Fahrmeir, L., and Kaufmann, H. (1985). Consistency and asymptotic normality of the maximum likelihood estimator in generalized linear models, *Annals of Statistics* **13**: 342–368.

Fahrmeir, L., and Tutz, G. (1994). *Multivariate Statistical Modelling Based on Generalized Linear Models*, Springer, New York.

Fairfield Smith, H. (1936). A discriminant function for plant selection, *Annals of Eugenics* **7**: 240–250.

Farebrother, R. W. (1976). Further results on the mean square error of ridge regression, *Journal of the Royal Statistical Society, Series B* **38**: 248–250.

Farebrother, R. W. (1978). A class of shrinkage estimators, *Journal of the Royal Statistical Society, Series B* **40**: 47–49.

Fitzmaurice, G. M., and Laird, N. M. (1993). A likelihood-based method for analysing longitudinal binary responses, *Biometrika* **80**: 141–151.

Fomby, T. B., Hill, R. C., and Johnson, S. R. (1984). *Advanced Econometric Methods*, Springer, New York.

Freund, E., and Trenkler, G. (1986). Mean square error matrix comparisons between mixed estimators, *Statistica* **46**: 493–501.

Friedman, J. H., and Stuetzle, W. (1981). Projection pursuit regression, *Journal of the American Statistical Association* **76**: 817–823.

Friedman, J. H., and Tukey, J. W. (1974). A projection pursuit algorithm for exploratory data analysis, *IEEE Transactions on Computers C* **23**: 881–890.

Gail, M. H., and Simon, R. (1985). Testing for qualitative interactions between treatment effects and patient subsets, *Biometrics* **41**: 361–372.

Galilie, G. (1632). Dialogo dei misimi sistemi, *Technical Report.*

Garthwaite, P. H. (1994). An interpretation of partial least squares, *Journal of the American Statistical Association* **89**: 122–127.

Geisser, S. (1974). A predictive approach to the random effects model, *Biometrika* **61**: 101–107.

Geweke, J. (1986). Exact inference in the inequality constrained normal linear regression model, *Journal of Applied Econometrics* **1**: 127–141.

Gilchrist, W. (1976). *Statistical Forecasting*, Wiley, London.

Giles, D. E. A., and Srivastava, V. K. (1991). An unbiased estimator of the covariance matrix of the mixed regression estimator, *Journal of the American Statistical Association* **86**: 441–444.

Glonek, G. V. F. (1996). A class of regression models for multivariate categorical responses, *Biometrika* **83**: 15–28.

Goldberger, A. S. (1962). Best linear unbiased prediction in the generalized regression model, *Journal of the American Statistical Association* **57**: 369–375.

Goldberger, A. S. (1964). *Econometric Theory*, Wiley, New York.

Goldberger, A. S., Nagar, A. L., and Odeh, H. S. (1961). The covariance matrices of reduced-form coefficients and of forecasts for a structural econometric model, *Econometrica* **29**: 556–573.

Goldstein, M., and Smith, A. F. M. (1974). Ridge-type estimators for regression analysis, *Journal of the Royal Statistical Society, Series B* **36**: 284–291.

Gower, J. C. (1974). The median center, *Applied Statistics* **2**: 466–470.

Graybill, F. A. (1961). *An introduction to linear statistical models, Volume I*, McGraw-Hill, New York.

Groß, J., and Trenkler, G. (1997). On the equality of usual and Amemiya's partially generalized least squares estimators, *Communications in Statistics, Part A—Theory and Methods* **26**: 2075–2086.

Guilkey, D. K., and Price, J. M. (1981). On comparing restricted least squares estimators, *Journal of Econometrics* **15**: 397–404.

Guttmann, I. (1970). *Statistical Tolerance Regions*, Griffin, London.

Haitovsky, Y. (1968). Missing data in regression analysis, *Journal of the Royal Statistical Society, Series B* **34**: 67–82.

Haldane, J. B. S. (1948). Note on the median of a multivariate distribution, *Biometrika* **25**: 414–415.

Hamerle, A., and Tutz, G. (1989). *Diskrete Modelle zur Analyse von Verweildauern und Lebenszeiten*, Campus, Frankfurt/M.

Hartung, J. (1978). Zur Verwendung von Vorinformation in der Regressionsanalyse, *Technical Report*, Institut für Angewandte Statistik, Universität Bonn, Germany.

Hastie, T., and Tibshirani, R. J. (1990). *Generalized Additive Models*, Chapman and Hall, London.

Heagerty, P. J., and Zeger, S. L. (1996). Marginal regression models for clustered ordinal measurements, *Journal of the American Statistical Association* **91**: 1024–1036.

Heckman, J. J. (1976). The common structure of statistical models of truncation, sample selection and limited dependent variables and a simple estimator for such models, *Annals of Economic and Social Measurement* **5**: 475–492.

Heiler, S., and Willers, R. (1988). Asymptotic normality of R-estimates in the linear model, *Mathematische Operationsforschung und Statistik, Series Statistics* **19**: 173–184.

Helland, I. S. (1988). On the structure of partial least squares regression, *Journal of the Royal Statistical Society, Series B* **50**: 581–607.

Henderson, C. R. (1984). Application of linear models in animal breeding, *Technical Report*, University of Guelph.

Heumann, C. (1998). *Likelihoodbasierte marginale Regressionsmodelle für korrelierte kategoriale Daten*, Peter Lang Europäischer Verlag der Wissenschaften, Frankfurt am Main.

Hill, R. C., and Ziemer, R. F. (1983). Missing regressor values under conditions of multicollinearity, *Communications in Statistics, Part A—Theory and Methods* **12**: 2557–2573.

Hoerl, A. E., and Kennard, R. W. (1970). Ridge regression: Biased estimation for nonorthogonal problems, *Technometrics* **12**: 55–67.

Horowitz, J. L. (1986). A distribution-free least squares estimator for censored regression models, *Journal of Econometrics* **32**: 59–84.

Horowitz, J. L. (1988). The asymptotic efficiency of semiparametric estimators for censored linear regression models, *Semiparametric and Nonparametric Econometrics* **13**: 123–140.

Huang, D. S. (1970). *Regression and Econometric Methods*, Wiley, New York.

Huber, P. J. (1964). Robust estimation of a location parameter, *Annals of Mathematical Statistics* **35**: 73–101.

Huber, P. J. (1985). Projection pursuit, *Annals of Statistics* **13**: 435–475.

Jaeckel, L. (1972). Estimating regression coefficients by minimizing the dispersion of the residuals, *Annals of Mathematical Statistics* **43**: 1449–1458.

Jensen, D. (1979). Linear models without moments, *Biometrika* **66**: 611–617.

Jones, M. C., and Sibson, R. (1987). What is projection pursuit? *Journal of the Royal Statistical Society, Series A* **150**: 1–36.

Judge, G. G., and Bock, M. E. (1978). *The Statistical Implications of Pre-test and Stein-Rule Estimators in Econometrics*, North Holland, Amsterdam.

Judge, G. G., Griffiths, W. E., Hill, R. C., and Lee, T.-C. (1980). *The Theory and Practice of Econometrics*, Wiley, New York.

Judge, G. G., and Takayama, T. (1966). Inequality restrictions in regression analysis, *Journal of the American Statistical Association* **66**: 166–181.

Jureckova, J. (1971). Nonparametric estimate of regression coefficients, *Annals of Mathematical Statistics* **42**: 1328–1338.

Kakwani, N. C. (1967). The unbiasedness of Zellner's seemingly unrelated regression equations estimators, *Journal of the American Statistical Association* **62**: 141–142.

Kakwani, N. C. (1968). Note on the unbiasedness of mixed regression estimation, *Econometrica* **36**: 610–611.

Kakwani, N. C. (1974). A note on Nagar-Kakwani approximation to variance of mixed regression estimator, *The Indian Economic Journal* **22**: 105–107.

Kalman, R. E. (1960). A new approach to linear filtering and prediction problems, *Journal of Basic Engineering* **82**: 34–45.

Kalman, R. E., and Bucy, R. S. (1961). New results in linear filtering and prediction theory., *Journal of Basic Engineering* **83**: 95–108.

Karim, M., and Zeger, S. L. (1988). GEE: A SAS macro for longitudinal analysis, *Technical Report*, Department of Biostatistics, Johns Hopkins School of Hygiene and Public Health, Baltimore, MD.

Kariya, T. (1985). A nonlinear version of the Gauss-Markov theorem, *Journal of the American Statistical Association* **80**: 476–477.

Kastner, C., Fieger, A., and Heumann, C. (1997). MAREG and WinMAREG—a tool for marginal regression models, *Computational Statistics and Data Analysis* **24**: 235–241.

Kempthorne, O., and Nordskog, A. G. (1959). Restricted selection indices, *Biometrics* **15**: 10–19.

Kmenta, J. (1971). *Elements of Econometrics*, Macmillan, New York.

Koopmann, R. (1982). *Parameterschätzung bei a priori Information*, Vandenhoeck & Rupprecht, Göttingen.

Krämer, W. (1980). A note on the equality of ordinary least squares and Gauss-Markov estimates in the general linear model, *Sankhya, Series A* **42**: 130–131.

Krämer, W., and Donninger, C. (1987). Spatial autocorrelation among errors and the relative efficiency of OLS in the linear regression model, *Journal of the American Statistical Association* **82**: 577–579.

Kuks, J. (1972). A minimax estimator of regression coefficients (in Russian), *Iswestija Akademija Nauk Estonskoj SSR* **21**: 73–78.

Kuks, J., and Olman, W. (1971). Minimax linear estimation of regression coefficients (I) (in Russian), *Iswestija Akademija Nauk Estonskoj SSR* **20**: 480–482.

Kuks, J., and Olman, W. (1972). Minimax linear estimation of regression coefficients (II) (in Russian), *Iswestija Akademija Nauk Estonskoj SSR* **21**: 66–72.

Lang, J. B., and Agresti, A. (1994). Simultaneously modeling joint and marginal distributions of multivariate categorical responses, *Journal of the American Statistical Association* **89**: 625–632.

Laplace, P. S. d. (1793). Sur quelques points du système du monde, *Oeuvres* **11**: 477–558.

Larsen, W. A., and McCleary, S. J. (1972). The use of partial residual plots in regression analysis, *Technometrics* **14**: 781–790.

Lawless, J. F. (1982). *Statistical Models and Methods for Lifetime Data*, Wiley, New York.

Lehmann, E. L. (1986). *Testing Statistical Hypotheses*, 2nd ed., Wiley, New York.

Liang, K.-Y., and Zeger, S. L. (1986). Longitudinal data analysis using generalized linear models, *Biometrika* **73**: 13–22.

Liang, K.-Y., and Zeger, S. L. (1989). A class of logistic regression models for multivariate binary time series, *Journal of the American Statistical Association* **84**: 447–451.

Liang, K.-Y., and Zeger, S. L. (1993). Regression analysis for correlated data, *Annual Review of Public Health* **14**: 43–68.

Liang, K.-Y., Zeger, S. L., and Qaqish, B. (1992). Multivariate regression analysis for categorical data, *Journal of the Royal Statistical Society, Series B* **54**: 3–40.

Lipsitz, S. R., Laird, N. M., and Harrington, D. P. (1991). Generalized estimating equations for correlated binary data: Using the odds ratio as a measure of association, *Biometrika* **78**: 153–160.

Little, R. J. A. (1992). Regression with missing X's: A review, *Journal of the American Statistical Association* **87**: 1227–1237.

Little, R. J. A., and Rubin, D. B. (1987). *Statistical Analysis with Missing Data*, Wiley, New York.

Maddala, G. S. (1983). *Limited-Dependent and Qualitative Variables in Econometrics*, Cambridge University Press, Cambridge.

Mardia, K. V., Kent, J. T., and Bibby, J. M. (1979). *Multivariate Analysis*, Academic Press, London.

Maronna, R. A., and Yohai, V. J. (1981). Asymptotic behavior of general M-estimates for regression and scale with random carriers, *Zeitschrift für Wahrscheinlichkeitstheorie und verwandte Gebiete* **58**: 7–20.

Mayer, L. S., and Wilke, T. A. (1973). On biased estimation in linear models, *Technometrics* **15**: 497–508.

McCullagh, P., and Nelder, J. A. (1989). *Generalized Linear Models*, Chapman and Hall, London.

McElroy, F. W. (1967). A necessary and sufficient condition that ordinary least-squares estimators be best linear unbiased, *Journal of the American Statistical Association* **62**: 1302–1304.

Mehta, J. S., and Swamy, P. A. V. B. (1970). The finite sample distribution of Theil's mixed regression estimator and a related problem, *International Statistical Institute* **38**: 202–209.

Milliken, G. A., and Akdeniz, F. (1977). A theorem on the difference of the generalized inverse of two nonnegative matrices, *Communications in Statistics, Part A—Theory and Methods* **6**: 73–79.

Mills, T. C. (1991). *Time Series Techniques for Economists*, Cambridge University Press, Cambridge.

Mirsky, C. (1960). Symmetric gauge functions and unitarily invariant norms, *Quarterly Journal of Mathematics* **11**: 50–59.

Molenberghs, G., and Lesaffre, E. (1994). Marginal modeling of correlated ordinal data using a multivariate Plackett distribution, *Journal of the American Statistical Association* **89**: 633–644.

Moon, C.-G. (1989). A Monte-Carlo comparison of semiparametric Tobit estimators, *Journal of Applied Econometrics* **4**: 361–382.

Moors, J. J. A., and van Houwelingen, J. C. (1987). Estimation of linear models with inequality restrictions, *Technical Report 291*, Tilburg University, The Netherlands.

Nagar, A. L., and Kakwani, N. C. (1964). The bias and moment matrix of a mixed regression estimator, *Econometrica* **32**: 174–182.

Nagar, A. L., and Kakwani, N. C. (1969). Note on the use of prior information in statistical estimation of econometric relations, *Sankhya, Series A* **27**: 105–112.

Nelder, J. A., and Wedderburn, R. W. M. (1972). Generalized linear models, *Journal of the Royal Statistical Society, Series A* **135**: 370–384.

Nelson, C. R. (1973). *Applied Time Series Analysis for Managerial Forecasting*, Holden-Day, San Francisco.

Neter, J., Wassermann, W., and Kutner, M. H. (1990). *Applied Linear Statistical Models*, 3rd ed., Irwin, Boston.

Oberhofer, W., and Kmenta, J. (1974). A general procedure for obtaining maximum likelihood estimates in generalized regression models, *Econometrica* **42**: 579–590.

Park, S. H., Kim, Y. H., and Toutenburg, H. (1992). Regression diagnostics for removing an observation with animating graphics, *Statistical Papers* **33**: 227–240.

Perlman, M. D. (1972). Reduced mean square error estimation for several parameters, *Sankhya, Series B* **34**: 89–92.

Polasek, W., and Krause, A. (1994). The hierarchical Tobit model: A case study in Bayesian computing, *OR Spektrum* **16**: 145–154.

Pollard, D. (1990). Empirical processes: Theory and applications., *NSF–CBMS Regional Conference series in Probability and Statistics*, Vol. 2, NSF–CBMS Regional Conference Series in Probability and Statistics.

Pollock, D. S. G. (1979). *The Algebra of Econometrics*, Wiley, Chichester.

Powell, J. L. (1984). Least absolute deviations estimation for the censored regression model, *Journal of Econometrics* **25**: 303–325.

Prentice, R. L. (1988). Correlated binary regression with covariates specific to each binary observation, *Biometrics* **44**: 1033–1048.

Prentice, R. L., and Zhao, L. P. (1991). Estimating equations for parameters in means and covariances of multivariate discrete and continuous responses, *Biometrics* **47**: 825–839.

Puntanen, S. (1986). Comments on "on neccesary and sufficient condition for ordinary least estimators to be best linear unbiased estimators," *Journal of the American Statistical Association* **40**: 178.

Rao, C. R. (1953). Discriminant function for genetic differentation and selection, *Sankhya, Series A* **12**: 229–246.

Rao, C. R. (1962). Problems of selection with restriction, *Journal of the Royal Statistical Society, Series B* **24**: 401–405.

Rao, C. R. (1964). Problems of selection involving programming techniques, *Proceedings of the IBM Scientific Computing Symposium on Statistics*, IBM, New York, pp. 29–51.

Rao, C. R. (1967). Least squares theory using an estimated dispersion matrix and its application to measurement of signals, *Proceedings of the 5th Berkeley Symposium on Mathematical Statistics and Probability*, University of California Press, Berkeley pp. 355–372.

Rao, C. R. (1968). A note on a previous lemma in the theory of least squares and some further results, *Sankhya, Series A* **30**: 259–266.

Rao, C. R. (1973a). *Linear Statistical Inference and Its Applications*, 2 edn, Wiley, New York.

Rao, C. R. (1973b). Unified theory of least squares, *Communications in Statistics, Part A—Theory and Methods* **1**: 1–8.

Rao, C. R. (1974). Characterization of prior distributions and solution to a compound decision problem, *Discussion Paper*, Indian Statistical Institute, New Delhi.

Rao, C. R. (1977). Prediction of future observations with special reference to linear models, *Multivariate Analysis V* **4**: 193–208.

Rao, C. R. (1979). Separation theorems for singular values of matrices and their applications in multivariate analysis, *Journal of Multivariate Analysis* **9**: 362–377.

Rao, C. R. (1980). Matrix approximations and reduction of dimensionality in multivariate statistical analysis, *Multivariate Analysis V* **5**: 3–22.

Rao, C. R. (1984). Prediction of future observations in polynomial growth curve models, *Proceedings of the India Statistical Institute Golden Jubilee International Conference on Statistics: Application and Future Directions*, Indian Statistical Institute, Calcutta, pp. 512–520.

Rao, C. R. (1987). Prediction of future observations in growth curve type models, *Journal of Statistical Science* **2**: 434–471.

Rao, C. R. (1988). Methodology based on the L_1-norm in statistical inference, *Sankhya, Series A* **50**: 289–313.

Rao, C. R. (1989). A lemma on the optimization of a matrix function and a review of the unified theory of linear estimation, *in* Y. Dodge (ed.), *Statistical Data Analysis and Inference*, Elsevier, Amsterdam, pp. 397–418.

Rao, C. R. (1994). Some statistical Problems in multitarget tracking, *in* S. S. Gupta and J. O. Berger (eds.), *Statistical Decision Theory and Related Topics*, Springer, New York, pp. 513–522.

Rao, C. R., and Boudreau, R. (1985). Prediction of future observations in factor analytic type growth model, *Multivariate Analysis*, VI, pp. 449–466.

Rao, C. R., and Kleffe, J. (1988). *Estimation of Variance Components and Applications*, North Holland, Amsterdam.

Rao, C. R., and Mitra, S. K. (1971). *Generalized Inverse of Matrices and Its Applications*, Wiley, New York.

Rao, C. R., and Rao, M. B. (1998). *Matrix Algebra and Its Applications to Statistics and Econometrics*, World Scientific, Singapore.

Rao, C. R., and Zhao, L. C. (1993). Asymptotic normality of LAD estimator in censored regression models, *Mathematical Methods of Statistics* **2**: 228–239.

Rao, P. S. S. N. V. P., and Precht, M. (1985). On a conjecture of Hoerl and Kennard on a property of least squares estimates of regression coefficients, *Linear Algebra and Its Applications* **67**: 99–101.

Rosner, B. (1984). Multivariate methods in ophtalmology with application to paired-data situations, *Biometrics* **40**: 961–971.

Rubin, D. B. (1976). Inference and missing data, *Biometrika* **63**: 581–592.

Rubin, D. B. (1987). *Multiple Imputation for Nonresponse in Sample Surveys*, Wiley, New York.

Rumelhart, D. E., Hinton, G. E., and Williams, R. J. (1986). Learning internal representation by back-propagating errors, *Nature* **323**: 533–536.

Ruppert, D., and Carroll, R. J. (1980). Trimmed least squares estimation in the linear model, *Journal of the American Statistical Association* **75**: 828–838.

Schafer, J. L. (1997). *Analysis of Incomplete Multivariate Data*, Chapman and Hall, London.

Schaffrin, B. (1985). A note on linear prediction within a Gauss–Markov model linearized with respect to a random approximation, *in* T. Pukkila and S. Puntanen (eds.), *Proceedings of the First International Tampere Seminar on Linear Statistical Models and Their Applications*, pp. 285–300, Tampere, Finland.

Schaffrin, B. (1986). New estimation/prediction techniques for the determination of crustal deformations in the presence of geophysical prior information, *Technometrics* **130**: 361–367.

Schaffrin, B. (1987). Less sensitive tests by introducing stochastic linear hypotheses, *in* T. Pukkila and S. Puntanen (eds.), *Proceedings of the Second International Tampere Conference in Statistics*, pp. 647–664, Tampere, Finland.

Scheffé, H. (1959). *The Analysis of Variance*, Wiley, New York.

Schipp, B. (1990). *Minimax Schätzer im simultanen Gleichungsmodell bei vollständiger und partieller Vorinformation*, Hain, Frankfurt/M.

Schönfeld, P. (1969). *Methoden der Ökonometrie Bd. I*, Vahlen, Berlin.

Schumacher, M., Roßner, R., and Vach, W. (1996). Neural networks and logistic regression: Part 1, *Computational Statistics and Data Analysis* **21**: 661–682.

Searle, S. R. (1982). *Matrix Algebra Useful for Statistics*, Wiley, New York.

Shalabh (1995). Performance of Stein-rule procedure for simultaneous prediction of actual and average values of study variable in linear regression model, *Bulletin of the International Statistical Institute* **56**: 1375–1390.

Silvey, S. D. (1969). Multicollinearity and imprecise estimation, *Journal of the Royal Statistical Society, Series B* **35**: 67–75.

Simonoff, J. S. (1988). Regression diagnostics to detect nonrandom missingness in linear regression, *Technometrics* **30**: 205–214.

Siotani, M., Hayakawa, T., and Fujikoshi, Y. (1985). *Modern Multivariate Statistical Analysis: A Graduate Course and Handbook*, American Sciences Press, Columbus, OH.

Snedecor, G. W., and Cochran, W. G. (1967). *Statistical Methods*, 6th ed., Ames: Iowa State University Press.

Srivastava, M. S. (1971). On fixed-width confidence bounds for regression parameters, *Annals of Mathematical Statistics* **42**: 1403–1411.

Srivastava, M. S. (1972). Asymptotically most powerful rank tests for regression parameters in MANOVA, *Annals of the Institute of Statistical Mathematics* **24**: 285–297.

Srivastava, V. K. (1970). The efficiency of estimating seemingly unrelated regression equations, *Annals of the Institute of Statistical Mathematics* **22**: 483–493.

Srivastava, V. K., Chandra, R., and Chandra, R. (1985). Properties of the mixed regression estimator when disturbances are not necessarily normal, *Journal of Statistical Planning and Inference* **11**: 11–15.

Srivastava, V. K., and Giles, D. E. A. (1987). *Seemingly Unrelated Regression Equations Models, Estimation and Inference*, Marcel Dekker, New York.

Srivastava, V. K., and Maekawa, K. (1995). Efficiency properties of feasible generalized least squares estimators in SURE models under non-normal disturbances, *Journal of Econometrics* **66**: 99–121.

Srivastava, V. K., and Raj, B. (1979). The existence of the mean of the estimator in seemingly unrelated regressions, *Communications in Statistics, Part A—Theory and Methods* **48**: 713–717.

Srivastava, V. K., and Upadhyaha, S. (1975). Small-disturbance and large sample approximations in mixed regression estimation, *Eastern Economic Journal* **2**: 261–265.

Srivastava, V. K., and Upadhyaha, S. (1978). Large sample approximations in seemingly unrelated regression estimators, *Annals of the Institute of Statistical Mathematics* **30**: 89–96.

Stahlecker, P. (1987). *A priori Information und Minimax-Schätzung im linearen Regressionsmodell*, Athenäum, Frankfurt/M.

Stone, M. (1974). Cross-validatory choice and assessment of statistical predictions, *Journal of the Royal Statistical Society, Series B* **36**: 111–133.

Subramanyam, M. (1972). A property of simple least squares estimates, *Sankhya, Series B* **34**: 355–356.

Swamy, P. A. V. B., and Mehta, J. S. (1969). On Theil's mixed regression estimator, *Journal of the American Statistical Association* **64**: 273–276.

Swamy, P. A. V. B., and Mehta, J. S. (1977). A note on minimum average risk estimators for coefficients in linear models, *Communications in Statistics, Part A—Theory and Methods* **6**: 1181–1186.

Swamy, P. A. V. B., Mehta, J. S., and Rappoport, P. N. (1978). Two methods of evaluating Hoerl and Kennard's ridge regression, *Communications in Statistics, Part A—Theory and Methods* **12**: 1133–1155.

Tallis, G. M. (1962). A selection index for optimum genotype, *Biometrics* **18**: 120–122.

Teräsvirta, T. (1979a). The polynomial distributed lag revisited, *Discussion Paper 7919*, Louvain, CORE.

Teräsvirta, T. (1979b). Some results on improving the least squares estimation of linear models by a mixed estimation, *Discussion Paper 7914*, Louvain, CORE.

Teräsvirta, T. (1981). Some results on improving the least squares estimation of linear models by mixed estimation, *Scandinavian Journal of Statistics* **8**: 33–38.

Teräsvirta, T. (1982). Superiority comparisons of homogeneous linear estimators, *Communications in Statistics, Part A—Theory and Methods* **11**: 1595–1601.

Teräsvirta, T. (1986). Superiority comparisons of heterogeneous linear estimators, *Communications in Statistics, Part A—Theory and Methods* **15**: 1319–1336.

Teräsvirta, T., and Toutenburg, H. (1980). A note on the limits of a modified Theil estimator, *Biometrical Journal* **22**: 561–562.

Theil, H. (1963). On the use of incomplete prior information in regression analysis, *Journal of the American Statistical Association* **58**: 401–414.

Theil, H. (1971). *Principles of Econometrics*, Wiley, New York.

Theil, H., and Goldberger, A. S. (1961). On pure and mixed estimation in econometrics, *International Economic Review* **2**: 65–78.

Theobald, C. M. (1974). Generalizations of mean square error applied to ridge regression, *Journal of the Royal Statistical Society, Series B* **36**: 103–106.

Thisted, R. A. (1988). *Elements of Statistical Computing: Numerical Computation*, Chapman and Hall, New York.

Tibshirani, R. J. (1992). Slide functions for projection pursuit regression and neural networks, *Technical Report*, University of Toronto, ON.

Toro-Vizcarrondo, C., and Wallace, T. D. (1968). A test of the mean square error criterion for restrictions in linear regression, *Journal of the American Statistical Association* **63**: 558–572.

Toro-Vizcarrondo, C., and Wallace, T. D. (1969). Tables for the mean square error test for exact linear restrictions in regression, *Discussion paper*, Department of Economics, North Carolina State University, Raleigh.

Toutenburg, H. (1968). Vorhersage im allgemeinen linearen Regressionsmodell mit Zusatzinformation über die Koeffizienten, *Operationsforschung Mathematische Statistik*, Vol. 1, Akademie-Verlag, Berlin, pp. 107–120.

Toutenburg, H. (1970a). Probleme linearer Vorhersagen im allgemeinen linearen Regressionsmodell, *Biometrische Zeitschrift* **12**: 242–252.

Toutenburg, H. (1970b). Über die Wahl zwischen erwartungstreuen und nichterwartungstreuen Vorhersagen, *Operationsforschung Mathematische Statistik*, Vol. 2, Akademie-Verlag, Berlin, pp. 107–118.

Toutenburg, H. (1970c). Vorhersage im allgemeinen Regressionsmodell mit stochastischen Regressoren, *Mathematische Operationsforschung und Statistik* **1**: 105–116.

Toutenburg, H. (1970d). Vorhersagebereiche im allgemeinen linearen Regressionsmodell, *Biometrische Zeitschrift* **12**: 1–13.

Toutenburg, H. (1971). Probleme der Intervallvorhersage von normalverteilten Variablen, *Biometrische Zeitschrift* **13**: 261–273.

Toutenburg, H. (1973). Lineare Restriktionen und Modellwahl im allgemeinen linearen Regressionsmodell, *Biometrische Zeitschrift* **15**: 325–342.

Toutenburg, H. (1975a). The use of mixed prior information in regression analysis, *Biometrische Zeitschrift* **17**: 365–372.

Toutenburg, H. (1975b). *Vorhersage in linearen Modellen*, Akademie-Verlag, Berlin.

Toutenburg, H. (1976). Minimax-linear and MSE-estimators in generalized regression, *Biometrische Zeitschrift* **18**: 91–100.

Toutenburg, H. (1982). *Prior Information in Linear Models*, Wiley, New York.

Toutenburg, H. (1984). Modellwahl durch Vortestschätzung, *9. Sitzungsbericht*, Interessengemeinschaft Math. Statistik, Berlin.

Toutenburg, H. (1989a). Investigations on the MSE-superiority of several estimators of filter type in the dynamic linear model (i.e. Kalman model), *Technical Report 89-26*, Center for Multivariate Analysis, The Pennsylvania State University, State College.

Toutenburg, H. (1989b). Mean-square-error-comparisons between restricted least squares, mixed and weighted mixed estimators, *Forschungsbericht 89/12*, Fachbereich Statistik, Universität Dortmund, Germany.

Toutenburg, H. (1990). MSE- and minimax-risk-comparisons of minimax and least squares estimators in case of misspecified prior regions, *Technical Report*, Universität Regensburg, Germany.

Toutenburg, H. (1992). *Moderne nichtparametrische Verfahren der Risikoanalyse*, Physica, Heidelberg.

Toutenburg, H., Fieger, A., and Heumann, C. (1999). Regression modelling with fixed effects—missing values and other problems, *in* C. R. Rao and G. Szekely (eds.), *Statistics of the 21st Century*, Dekker, New York.

Toutenburg, H., Heumann, C., Fieger, A., and Park, S. H. (1995). Missing values in regression: Mixed and weighted mixed estimation, *in* V. Mammitzsch and H. Schneeweiß (eds.), *Gauss Symposium*, de Gruyter, Berlin, pp. 289–301.

Toutenburg, H., and Schaffrin, B. (1989). Biased mixed estimation and related problems, *Technical Report*, Universität Stuttgart, Germany.

Toutenburg, H., and Shalabh (1996). Predictive performance of the methods of restricted and mixed regression estimators, *Biometrical Journal* **38**: 951–959.

Toutenburg, H., Srivastava, V. K., and Fieger, A. (1996). Estimation of parameters in multiple regression with missing X-observations using first order regression procedure, *SFB386—Discussion Paper 38*, Ludwig-Maximilians-Universität München, Germany.

Toutenburg, H., Srivastava, V. K., and Fieger, A. (1997). Shrinkage estimation of incomplete regression models by Yates procedure, *SFB386—Discussion Paper 69*, Ludwig-Maximilians-Universität München, Germany.

Toutenburg, H., and Trenkler, G. (1990). Mean square error matrix comparisons of optimal and classical predictors and estimators in linear regression, *Computational Statistics and Data Analysis* **10**: 297–305.

Toutenburg, H., Trenkler, G., and Liski, E. P. (1992). Optimal estimation methods under weakened linear restrictions, *Computational Statistics and Data Analysis* **14**: 527–536.

Toutenburg, H., and Wargowske, B. (1978). On restricted 2-stage-least-squares (2-SLSE) in a system of structural equations, *Statistics* **9**: 167–177.

Trenkler, G. (1981). *Biased Estimators in the Linear Regression Model*, Hain, Königstein/Ts.

Trenkler, G. (1985). Mean square error matrix comparisons of estimators in linear regression, *Communications in Statistics, Part A—Theory and Methods* **14**: 2495–2509.

Trenkler, G. (1987). Mean square error matrix comparisons among restricted least squares estimators, *Sankhya, Series A* **49**: 96–104.

Trenkler, G., and Pordzik, P. (1988). Pre-test estimation in the linear regression model based on competing restrictions, *Technical Report*, Universität Dortmund, Germany.

Trenkler, G., and Stahlecker, P. (1987). Quasi minimax estimation in the linear regression model, *Statistics* **18**: 219–226.

Trenkler, G., and Toutenburg, H. (1990). Mean-square error matrix comparisons between biased estimators—an overview of recent results, *Statistical Papers* **31**: 165–179.

Trenkler, G., and Toutenburg, H. (1992). Pre-test procedures and forecasting in the regression model under restrictions, *Journal of Statistical Planning and Inference* **30**: 249–256.

Trenkler, G., and Trenkler, D. (1983). A note on superiority comparisons of linear estimators, *Communications in Statistics, Part A—Theory and Methods* **17**: 799–808.

Vach, W., Schumacher, M., and Roßner, R. (1996). Neural networks and logistic regression: Part 2, *Computational Statistics and Data Analysis* **21**: 683–701.

van Huffel, S., and Zha, H. (1993). The total least squares problem, *in* C. R. Rao (ed.), *Handbook of Statistics*, Elsevier, Amsterdam, pp. 377–408.

Vinod, H. D., and Ullah, A. (1981). *Recent Advances in Regression Methods*, Dekker, New York.

Wallace, T. D. (1972). Weaker criteria and tests for linear restrictions in regression, *Econometrica* **40**: 689–698.

Walther, W. (1992). Ein Modell zur Erfassung und statistischen Bewertung klinischer Therapieverfahren—entwickelt durch Evaluation des Pfeilerverlustes bei Konuskronenersatz, *Habilitationsschrift*, Universität Homburg, Germany.

Walther, W., and Toutenburg, H. (1991). Datenverlust bei klinischen Studien, *Deutsche Zahnärztliche Zeitschrift* **46**: 219–222.

Wedderburn, R. W. M. (1974). Quasi-likelihood functions, generalized linear models, and the Gauss-Newton method, *Biometrika* **61**: 439–447.

Wedderburn, R. W. M. (1976). On the existence and uniqueness of the maximum likelihood estimates for certain generalized linear models, *Biometrika* **63**: 27–32.

Weisberg, S. (1980). *Applied Linear Regression*, Wiley, New York.

Weisberg, S. (1985). *Applied Linear Regression*, 2nd ed., Wiley, New York.

Welsch, R. E., and Kuh, E. (1977). Linear regression diagnostics, *Technical Report 923*, Sloan School of Managment, Massachusetts Institute of Technology, Cambridge, MA.

Wilks, S. S. (1932). Moments and distributions of estimates of population parameters from fragmentary samples, *Annals of Mathematical Statistics* **3**: 163–195.

Wilks, S. S. (1938). The large-sample distribution of the likelihood ratio for testing composite hypotheses, *Annals of Mathematical Statistics* **9**: 60–62.

Wold, S., Wold, H. O., Dunn, W. J., and Ruhe, A. (1984). The collinearity problem in linear regression. The partial least squares (PLS) approach to generalized inverses, *SIAM Journal on Scientific and Statistical Computing* **5**: 735–743.

Wu, C. F. J. (1986). Jackknife, bootstrap and other resampling methods in regression analysis, *Annals of Statistics* **14**: 1261–1350.

Wu, Y. (1988). Strong consistency and exponential rate of the "minimum L_1-norm" estimates in linear regression models, *Computational Statistics and Data Analysis* **6**: 285–295.

Wu, Y. (1989). *Asymptotic Theory of Minimum L_1-Norm and M-Estimates in Linear Models*, PhD thesis, University of Pittsburgh, PA.

Yancey, T. A., Judge, G. G., and Bock, M. E. (1973). Wallace's weak mean square error criterion for testing linear restrictions in regression: A tighter bound, *Econometrica* **41**: 1203–1206.

Yancey, T. A., Judge, G. G., and Bock, M. E. (1974). A mean square error test when stochastic restrictions are used in regression, *Communications in Statistics, Part A—Theory and Methods* **3**: 755–768.

Yates, F. (1933). The analysis of replicated experiments when the field results are incomplete, *Empire Journal of Experimental Agriculture* **1**: 129–142.

Zellner, A. (1962). An efficient method of estimating seemingly unrelated regressions and tests for aggregation bias, *Journal of the American Statistical Association* **57**: 348–368.

Zellner, A. (1963). Estimates for seemingly unrelated regression equations: Some exact finite sample results, *Journal of the American Statistical Association* **58**: 977–992.

Zhao, L. P., and Prentice, R. L. (1990). Correlated binary regression using a generalized quadratic model, *Biometrika* **77**: 642–648.

Zhao, L. P., Prentice, R. L., and Self, S. G. (1992). Multivariate mean parameter estimation by using a partly exponential model, *Journal of the Royal Statistical Society, Series B* **54**: 805–811.

Index

Springer Series in Statistics

(continued from p. ii)

Kotz/Johnson (Eds.): Breakthroughs in Statistics Volume II.
Kotz/Johnson (Eds.): Breakthroughs in Statistics Volume III.
Kres: Statistical Tables for Multivariate Analysis.
Küchler/Sørensen: Exponential Families of Stochastic Processes.
Le Cam: Asymptotic Methods in Statistical Decision Theory.
Le Cam/Yang: Asymptotics in Statistics: Some Basic Concepts.
Longford: Models for Uncertainty in Educational Testing.
Manoukian: Modern Concepts and Theorems of Mathematical Statistics.
Miller, Jr.: Simultaneous Statistical Inference, 2nd edition.
Mosteller/Wallace: Applied Bayesian and Classical Inference: The Case of the Federalist Papers.
Parzen/Tanabe/Kitagawa: Selected Papers of Hirotugu Akaike.
Politis/Romano/Wolf: Subsampling.
Pollard: Convergence of Stochastic Processes.
Pratt/Gibbons: Concepts of Nonparametric Theory.
Ramsay/Silverman: Functional Data Analysis.
Rao/Toutenburg: Linear Models: Least Squares and Alternatives.
Read/Cressie: Goodness-of-Fit Statistics for Discrete Multivariate Data.
Reinsel: Elements of Multivariate Time Series Analysis, 2nd edition.
Reiss: A Course on Point Processes.
Reiss: Approximate Distributions of Order Statistics: With Applications to Non-parametric Statistics.
Rieder: Robust Asymptotic Statistics.
Rosenbaum: Observational Studies.
Ross: Nonlinear Estimation.
Sachs: Applied Statistics: A Handbook of Techniques, 2nd edition.
Särndal/Swensson/Wretman: Model Assisted Survey Sampling.
Schervish: Theory of Statistics.
Seneta: Non-Negative Matrices and Markov Chains, 2nd edition.
Shao/Tu: The Jackknife and Bootstrap.
Siegmund: Sequential Analysis: Tests and Confidence Intervals.
Simonoff: Smoothing Methods in Statistics.
Singpurwalla and Wilson: Statistical Methods in Software Engineering: Reliability and Risk.
Small: The Statistical Theory of Shape.
Stein: Interpolation of Spatial Data: Some Theory for Kriging
Tanner: Tools for Statistical Inference: Methods for the Exploration of Posterior Distributions and Likelihood Functions, 3rd edition.
Tong: The Multivariate Normal Distribution.
van der Vaart/Wellner: Weak Convergence and Empirical Processes: With Applications to Statistics.
Vapnik: Estimation of Dependences Based on Empirical Data.
Weerahandi: Exact Statistical Methods for Data Analysis.
West/Harrison: Bayesian Forecasting and Dynamic Models, 2nd edition.
Wolter: Introduction to Variance Estimation.
Yaglom: Correlation Theory of Stationary and Related Random Functions I: Basic Results.
Yaglom: Correlation Theory of Stationary and Related Random Functions II: Supplementary Notes and References.